ABHANDLUNGEN DER DEUTSCHEN AKADEMIE
DER WISSENSCHAFTEN ZU BERLIN

Klasse für Medizin

Jahrgang 1964 Nr. 6

Physikalische Chemie biogener Makromoleküle

II. Jenaer Symposium

18. bis 21. September 1963

Vorträge und Diskussionen

Herausgegeben von

HERMANN BERG

Mit 268 Abbildungen und 70 Tabellen

AKADEMIE-VERLAG · BERLIN
1964

Vorgelegt von Hrn. KNÖLL in der Klassensitzung vom 2. 12. 1963

Zum Druck genehmigt am gleichen Tage, ausgegeben am 15. 9. 1964

Erschienen im Akademie-Verlag GmbH, Berlin W 8, Leipziger Straße 3—4

Lizenznummer: 202 · 100/421/64

Gesamtherstellung: VEB Druckerei „Thomas Müntzer" Bad Langensalza

Bestellnummer: 2001/64/IV/6 · ES 18 G 1/18 C 5 · Preis: DM 47,—

Inhalt

A Allgemeine Grundlagen und Methoden

Generaldiskussion I

B Polysaccharide

C Proteine

Generaldiskussion II

D Nucleinsäuren

Vorwort

Das Hauptinteresse der modernen Naturwissenschaft konzentriert sich in immer stärkerem Maße auf die Erforschung meßbarer Erscheinungen der Lebenstätigkeit einfachster Systeme. So arbeiten im Rahmen der "Molekular-Biologie" mehrere Disziplinen am gemeinsamen Ziel, wobei der "Physikalischen Chemie" die Aufgabe zukommt, ihre methodischen und theoretischen Möglichkeiten zur Analyse von Elementarreaktionen und zur Ermittlung der räumlichen Anordnung insbesondere bei biogenen Makromolekülen einzusetzen. Der gegenwärtige Stand dieser Problematik und zukünftige Aussichten wurden auf Schloß Dornburg und im Institut für Mikrobiologie und experimentelle Therapie in 52 Vorträgen der folgenden Sektionen behandelt :

A) Allgemeine Grundlagen und Methoden
B) Polysaccharide
C) Proteine
D) Nucleinsäuren.

In den drei abschließenden Generaldiskussionen :

I Möglichkeiten und Grenzen der Methoden
II Zur Struktur der Proteine und Polysaccharide
III Probleme physikochemischer Untersuchungen der Nucleinsäuren

stand die Nucleinsäureforschung, wie schon bei den Vorträgen, wiederum im Mittelpunkt des Interesses.

Die Ergebnisse und Diskussionen sind nunmehr in vorliegendem Band unter Mitarbeit der Herren Dr. K.-E.REINERT, H. TRIEBEL und H. SCHÜTZ zusammengestellt mit Ausnahme von wenigen anderwärts bereits erschienenen Vorträgen.

Wir glauben, daß sich damit der Stand um die Mitte 1963 auf diesem Gebiet in groben Zügen kennzeichnen läßt. Andererseits sind wir uns bewußt, daß es kein abgerundetes Bild sein kann, da alles zu sehr im Fluß ist.

Hermann BERG

Freudig trete herein
und froh entferne dich wieder;
ziehst du als Wanderer vorbei,
segne die Pfade dir Gott.

Johann Wolfgang v. Goethe

Portal des Goethe-Schlosses

(Photo: M. BEYER)

GAVDEAT INGREDIENS LÆTETVR ET
HIS QVI PP ÆTEREVNT DET BONA CVNCTA

A Allgemeine Grundlagen und Methoden

Measurements of Liquid Thermal Diffusion using a Moving-Boundary Method

P. H. NORBERG

Institute of Physical Chemistry of the University, Uppsala

S U M M A R Y

Studies of liquid thermal diffusion with the boundary method have been performed with an optical registration using the Rayleigh interference system which permits a detailed investigation of the entire concentration variation throughout the cell. The experiments have shown that the construction of the thermal diffusion cell is very critical and the greatest possible consideration must be given to the prevention of convections. The latter may be the reason for the many discrepancies in the literature. The nature of the convections is discussed and their contributions to errors in the thermal diffusion data evaluated.

I N T R O D U C T I O N

Thermal diffusion is the relative movement of the components of a mixture due to an applied temperature gradient.

The effect of thermal diffusion can be measured in various ways, the most common of which are : the column method, the SORET method and the boundary method.

The column method or the thermogravitational thermal diffusion is a separation of components in a horizontal temperature gradient and a convective vertical flow due to the gravitational field.

In the case of the SORET method the composition gradient in an initially homogeneous mixture, placed between two plates kept at different temperatures, is studied.

Finally, in the boundary method a boundary is formed between solvent and solution at the same time keeping a vertical temperature gradient. The velocity of migration of the boundary is then a measure of the thermal diffusion coefficient.

In recent years a number of authors have described liquid phase thermal diffusion. On comparing the results of these investigations one is, however, left with a diverging picture as to the magnitude of the thermal diffusion coefficient and its concentration and molecular weight dependence. For instance MEYERHOFF and NACHTIGALL [1] who used the SORET method did not find any dependence or at least a very slight dependence of the thermal diffusion coefficient on concentration and molecular weight, while HOFFMAN and ZIMM [2] (boundary method) found a strong concentration dependence and a smaller molecular weight dependence. For concentrations above about one per cent the work in a separation column by DEBYE and BUECHE [3] is in good agreement with the results of HOFFMAN and

ZIMM. The thermal diffusion coefficient increases with decreasing concentration and molecular weight. An opposite molecular weight dependence was found by HERREN and HAM [4] and WHITMORE [5] who has also reported a strange concentration dependence. These studies have all been carried out on polystyrene in toluene. The results of the various authors are presented in Figs.1, 2 and 3. MEYERHOFF et al. have also recently published some articles[6][7] dealing with studies on thermal diffusion using the boundary method. The results are about the same as in the earlier article by MEYERHOFF and NACHTIGALL.

Both in the SORET and boundary methods optical systems have been most commonly used as a means of registration; however, WHITMORE has used another technique, sucking out small volumes at different heights in the cell with a pipette and determining their concentrations.

For precision studies of thermal diffusion where variations in both temperature and concentration are present in the diffusion cell the effect of convection must be very critical. Among others, BIERLEIN [8] has discussed this effect and its importance in thermal diffusion measurements. Two questions arise when dealing with these problems. First, what kind of convections can be present in a thermal diffusion cell and second, how will they affect the experimental results.

Some preliminary experiments with the boundary method showed a decrease in the difference in concentration between solvent and solution and confirming this a homogeneous increase in the concentration of the whole volume on the solvent side of the boundary could be detected. In addition the concentration dependence of the thermal diffusion coefficient of both low molecular weight substances and some polymer systems showed a drastic change towards lower concentrations as in some of the articles referred to. These observed effects indicated convection which can be explained by assuming a somewhat higher temperature existing within the cell contents than at the walls and windows of the upper part of the cell at the same level. Thus we can imagine that a thin layer is creeping down at the walls and that a balancing upward flow is occuring in the central parts of the cell. A convection of this nature will slowly plane off thin layers of solution and transport solute up to the solvent side of the boundary, and consequently decreases the difference in concentration between the two sides of the boundary (Fig.4).

When studying the speed of migration of the boundary in order to evaluate the thermal diffusion coefficient one gets an apparent value which is too high since the planing off of the boundary is added to the effect of pure thermal diffusion. It has turned out that the error in the thermal diffusion coefficient is increasing towards lower concentrations. this is a consequence of an increasing instability in the concentration gradient due to smaller differences in the densities and viscosities between solvent and solution.

The great errors in the thermal diffusion data caused by convection were detected thanks to the RAYLEIGH interference optical system which in comparison with many other optical systems is superior. As an illustration one of the preliminary not convection-free experiments will be presented.

The experiment was carried out on a solution of phenanthrene in 2-pentanone with a concentration of 0.259 g/dl. the temperature gradient was 12.5 deg cm^{-1}. The total number of fringes J (see section Treatment of data) at the beginning of the experiment (t = 0) was 4.12 and at t = 7200 sec 3.90. Consequently there had been a loss of 0,22 fringes. In other words according to the mechanism of the convection assumed, the concentration on the solvent side of the boundary should have shown a corresponding increase of 0.22 fringes. In very good agreement with this a parallel drift of the fringe pattern above the boundary of 0.18 fringes corresponded to the increase in concentration of the same magnitude as indicated by the loss in the total number of fringes. The total movement of the boundary during the period studied was 0.046 cm. Since the geometry of the cell was completely known the fraction of the distance which was responsible for the increase in concentration on the solvent side of the boundary could be calculated and found to be 0.024 cm. In other words the migration of the boundary due to thermal diffusion was 0.022 cm and the rest of it was due to convection. The thermal diffusion coefficient calculated without correction for the convection was $4.33.10^{-7} cm^2 sec^{-1} deg^{-1}$ and when corrected the value became $2.08.10^{-7} cm^2 sec^{-1} deg^{-1}$ which is in the right order of magnitude for phenanthrene in 2-pentanone.

The same type of convection can also be present in a SORÉT cell and in this case it will cause the concentration gradient to be less than if the cell was convection-free (Fig.5). Accordingly, the value of the SORÉT coefficient becomes too low. In other words the same type of convection in the SORÉT and moving boundary cells leads to errors in opposite directions.

E X P E R I M E N T A L

T h e c e l l

In the construction of a convection-free cell for measurements of thermal diffusion the greatest possible consideration must be payed in order to avoid variations in the temperature in horizontal layers in the cell. In our case the moving-boundary type thermal diffusion cell has been constructed for an optical registration with the RAYLEIGH interference optical system, which has been in use at the Institute for a number of years in the study of free diffusion at extremely low concentrations.

The essential principle of the cell is shown in Fig.6. An all glass two-channel interferometer cuvette open in top and bottom (Fig.7) and 10 mm long is placed between two silver-blocks with finely ground surfaces. The RAYLEIGH interference optical system requires a reference cell placed beside the diffusion cell along the optical axis. There are two different refractive index gradients in a cell for thermal diffusion, one coming from the pure concentration gradient and the other coming from the temperature gradient. By providing the diffusion channel with a reference channel of the same shape and containing pure solvent it has been possible to a considerable extent to compensate for the temperatur gradient and by this means register only the concentration gradient.

The choice of silver for the two blocks was dependent upon two reasons; first, its very high thermal conductivity and second, its resistance to most solvents used, especially water.

The windows and walls of the interferometer cuvette are 8 mm thick and the air-space outside the glass cell is closed by plexiglass plates in order to prevent disturbing air turbulences on the cell surfaces. A photograph of the cell is shown in Fig.8. It was hoped that by this construction the temperature profiles in the cell contents and the cell walls and windows would be very close and consequently the probability of a convection of the type previously described low. The silver-blocks are also insulated with thick (20 mm) plexiglass covers.

Water from two thermostats kept at different temperatures is circulated through the silver blocks in order to establish a constant temperature gradient over the diffusion cell. The magnitude of the gradient which is assumed to be linear is measured by means of thermocouples placed in holes bored in the silver blocks.

The boundary is formed by means of the sucking technique using a capillary. Immediately after the boundary formation the cell is completely closed by two stop-cocks on the outer surfaces of the silver-blocks (see Fig.6).

T R E A T M E N T O F D A T A

The photographic plates with one reference exposure, taken prior to the boundary formation, and the successive diffusion exposures were measured in a microcomparator with an accuracy in the reading of ± 1 μ. Small fixed interference fringes are produced at each end of the interference pattern in order to facilitate the orientation of the exposures parallel to each other in the comparator. In order to correct for irregularities in the cell windows and possible differences in the temperature gradients in the reference channel and the diffusion channel the reference exposure was subtracted from the diffusion exposures.

A picture of the RAYLEIGH interference pattern from one of the diffusions is presented in Fig.9. The xy-coordinates of the interference fringes from the different exposures are then translated into a fringe function, $j(x,t)$ otherwise expressed as the number of fringes and fraction of a fringe crossed when measuring the plate from one end to the other. The total number of fringes is called J. The derivative of the fringe function for two exposures from one of the experiments is plotted and presented in Fig.10. It should also be mentioned here that by providing each exposure of the interference pattern with fixed marks it is possible to detect a parallel transfer of the interference fringes on each side of the boundary. This would of course as pointed out earlier be the consequence of a convection which increases the concentration on the solvent side, and which was observed in the preliminary experiments. With the cell described however no transfer has been detected for the substances investigated.

In Fig.11 the positions of the straight parts of the interference pattern relative to the fixed marks are plotted versus time from the same experiment as in Fig.10. The total number of fringes which is proportional to difference in concentration between solvent and solution is very stable throughout the experiment (see Table 1).

Table 1 Experimental data from one of the glucose thermal diffusion

Exposure	Time after start (sec)	J	$\left[\frac{J}{dj(x,t)/dx}\right]^2_{max}$ (cm^2)	$x_{J/2}$ (cm)
1	2	3	4	5
3	150	8.46	0.0209	2.7900
5	657	8.48	0.0759	2.7913
7	1257	8.50	0.1368	2.7949
9	1857	8.49	0.2006	2.7964
11	2457	8.51	0.2589	-

These facts indicate a very good stability of the concentration gradient and consequently the absence of convection. The temperature gradients have been about 20 deg./cm.

For the evaluation of the thermal diffusion coefficient from the speed of the migrating boundary its position in the cell for $j(x,t) = J/2$ is plotted versus the exposure time (Fig.12). The position for half the difference in concentration between solvent and solution is plotted since great errors must be involved in a plot of the positions of the maxima of the "peaks", which are somewhat skewed due to the temperature dependence of the free diffusion coefficient. This skewness also affects the values of the free diffusion coefficients which are calculated from the spreading of the boundary according to the relation

$$\left[\frac{d\,[J(x,t)]}{d\,x}\right]_{max} = \frac{J}{2\sqrt{\pi D\,t}}$$

R E S U L T S

In this preliminary report some results from thermal diffusion measurements on water solutions of glucose, sucrose and raffinose are presented. As appears from Table 2, no

Table 2 Free diffusion and thermal diffusion coefficients for different sugars in water

Substance	Concentration (g/dl)	$D' \cdot 10^7$ ($cm^2 sec^{-1} deg^{-1}$)	$D_{25°C} \cdot 10^6$ ($cm^2\ sec^{-1}$)
1	2	3	4
Glucose	0.601	2.00	6.45
	0.353	1.80	6.35
	0.206	1.98	6.49
	0.118	1.95	5.81
Sucrose	0.976	1.67	4.76
	0.574	1.85	5.06
Raffinose	0.457	-	4.09
	0.352	1.39	4.07

concentration dependence of the thermal diffusion coefficient can be detected but there are strong indications of a molecular weight dependence. This is also supported by the fact that a separation between sucrose and glucose in a mixture of the two has been observed in the preliminary experiments (see Fig.13). With regard to the free diffusion coefficients calculated from the thermal diffusion experiments these are in reasonable good agreement with those in the literature from pure diffusion investigations. No corrections to the shape of the boundary regarding the effect of the temperature dependence of the free diffusion coefficient have been applied.

This paper is a preliminary report of some results obtained in a convection-free moving-boundary thermal diffusion cell. Work is in progress at this Institute dealing with a more detailed study of the behavior of the parameters evaluated from thermal diffusion measurements on low molecular weight substances as well as on macromolecular systems.

L I T E R A T U R E

1) NACHTIGALL,K., G.MEYERHOFF, Z.physik.Chem. (Frankfurt/M.) 30, 17 (1961)
2) HOFFMAN,J.D., B.H.ZIMM, J.Polymer Sci. 15, 405 (1955)
3) DEBYE,P., A.M.BUECHE, High Polymer Physics, edited by H.A.ROBINSON (Chemical Publishing Company, Brooklyn, New York, 1948)
4) HERREN,C.L., J.S.HAM, J.chem.Physics 35, 1479 (1961)
5) WHITMORE,F.C., J.appl.Physics 31, 1858 (1960)
6) MEYERHOFF,G., H.LÜTJE, B.RAUCH, Makromolekulare Chem. 44-46, 489 (1961)
7) RAUCH,B., G.MEYERHOFF, J.physic.Chem. 67, 946 (1963)
8) BIERLEIN,J.A., J.chem.Physics 36, 2793 (1962)

D I S C U S S I O N - Q U E S T I O N S

BERG : In what concentration range are you working with this apparatus in the determinations of the thermal diffusion and free diffusion coefficients ?

NORBERG : With the 1 cm cell, which we generally use, we work at concentrations between 0.2 - 1.0 g/dl, while with the 3 cm cell we are able to go down to about 0.04 g/dl.

LANGHAMMER : Für die von Prof.CLAESSON in der Diskussion geäußerte Vermutung, daß die Durchspülbarkeit des Knäuels das Thermodiffusionsverhalten des Makromoleküls bestimmt, spricht folgender Versuch, den wir vor Jahren einmal durchführten: Polystyrol zeigt in der Thermodiffusionstrennsäule starke Thermodiffusionseffekte, wenn man das gute Lösungsmittel Toluol verwendet. Verschlechtert man das Lösungsmittel durch Zusatz von Methanol, so geht der Thermodiffusionseffekt stark zurück. Wir haben die Thermodiffusion in der Form der Trennrohrmethode für die Fraktionierung Hochpolymerer angewandt. Z.B. läßt sich das wasserlösliche Polyvinylpyrrolidon gut fraktionieren, auch Polystyrol. Dagegen konnten wir bei Proteinen nur sehr geringe Effekte nachweisen. Da die Methode mit sehr engen Spalten durchgeführt werden muß, sind wegen der starken Viskositätsänderungen die Strömungsverhältnisse schwierig zu regulieren. Sowohl aus eigenen unveröffentlichten Messungen wie auch besonders aus den Messungen Prof. MEYERHOFFs und seiner Mitarbeiter geht hervor, daß der Thermodiffusionskoeffizient nur wenig vom Molekulargewicht abhängt und daß somit die Trennung im Trennrohr hauptsächlich durch die Molekulargewichtsabhängigkeit des Diffusionskoeffizienten bewirkt wird, der ja die Geschwindigkeit der Rückdiffusion in den aufsteigenden Lösungsmittelstrom bestimmt. In geeigneten Fällen und mit geeigneter Methodik (Spülstrom nach KORSCHING) gelingt es aber, mit dieser Methode außerordentlich hochmolekulare Fraktionen abzutrennen, wie sie mit dieser Deutlichkeit mittels keiner anderen Methode gewonnen werden können. Die genaue Kenntnis solcher hochmolekularen Fraktionen kann unter Umständen wichtig sein, z.B. bei der klinischen Verwendung bestimmter wasserlöslicher Polymerer wie eben z.B. das "Kollidon".

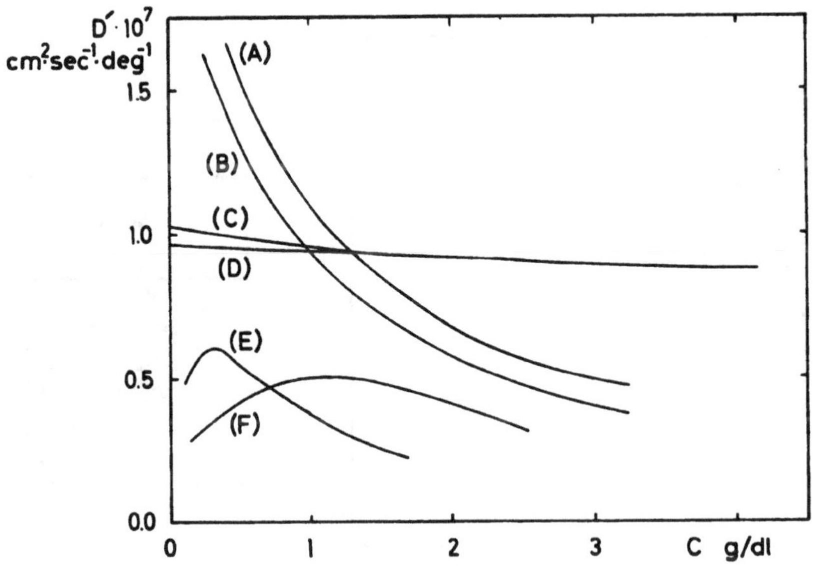

Fig.1 The concentration dependence of the thermal diffusion coefficient for polystyrene in toluene at 25°C. A and B (HOFFMAN and ZIMM) represent molecular weights 6.10^4 and $1,1.10^6$; C and D (NACHTIGALL and MEYERHOFF), molecular weights $4,4.10^4$ - $2,9.10^6$; E and F (DEBYE and BUECHE), molecular weights $1,6.10^5$ and $5,1.10^5$.

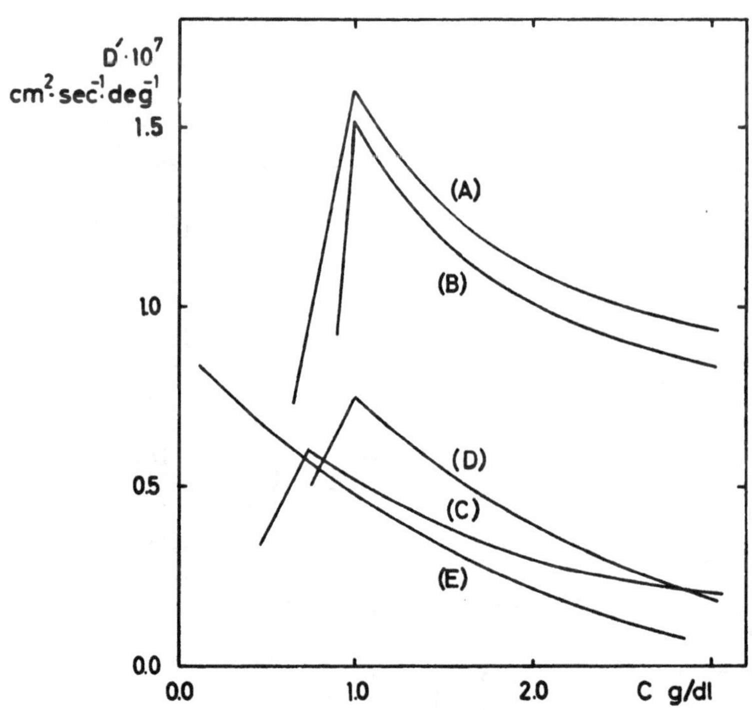

Fig.2 The concentration dependence of the thermal diffusion coefficient for some polystyrene fractions in toluene at 25°C according to WHITMORE. A: molecular weight $6,3.10^5$, B: molecular weight $5,5.10^5$, C: molecular weight $3,7.10^5$, D: molecular weight $1,3.10^5$ and E: molecular weight $5,4.10^4$

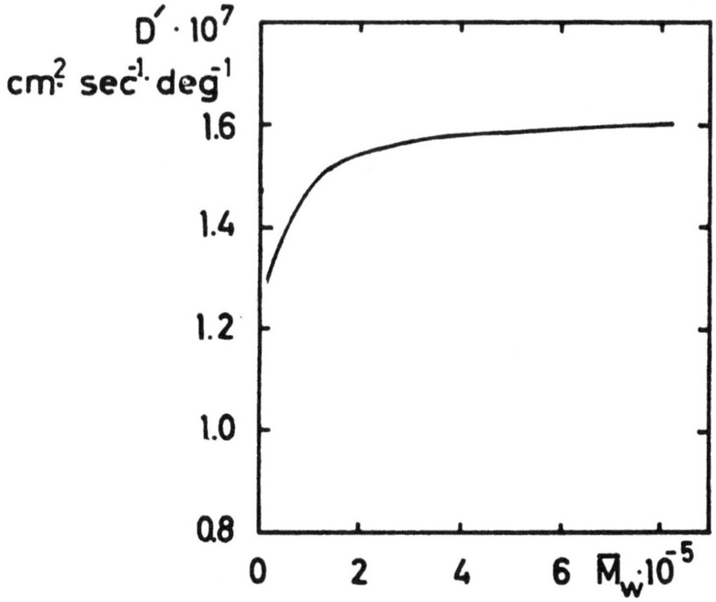

Fig.3 The molecular weight dependence of the thermal diffusion coefficient according to HERREN and HAM.

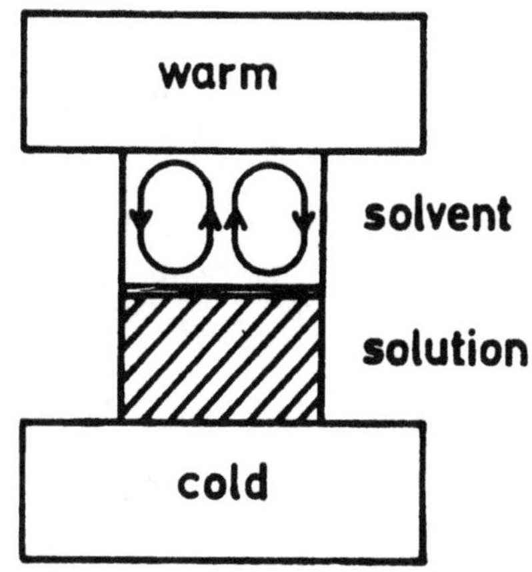

Fig.4 The form of convection assumed in a moving boundary thermal diffusion cell.

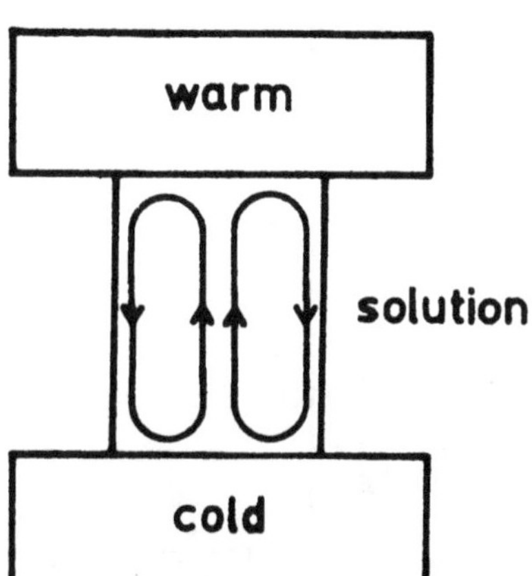

Fig.5 The form of convection assumed in a SORÉT cell.

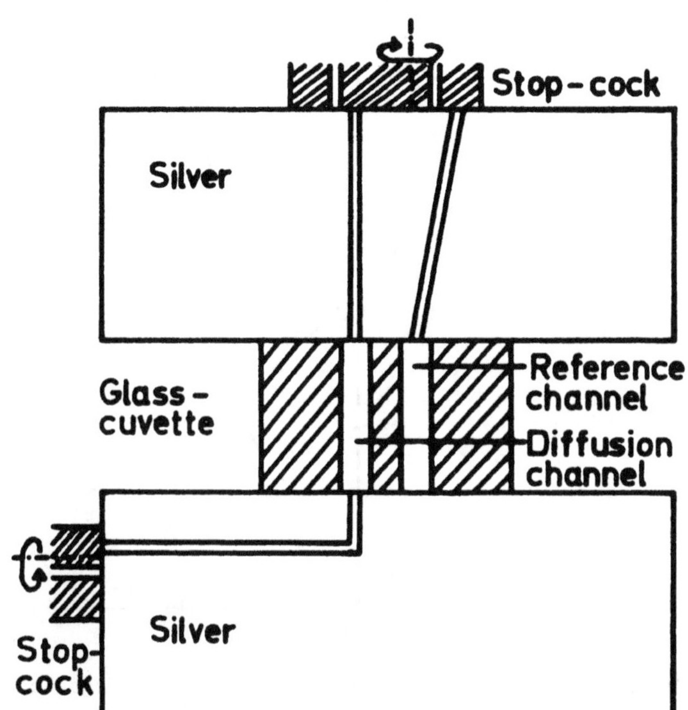

Fig.6 Schematic drawing of the thermal diffusion cell.

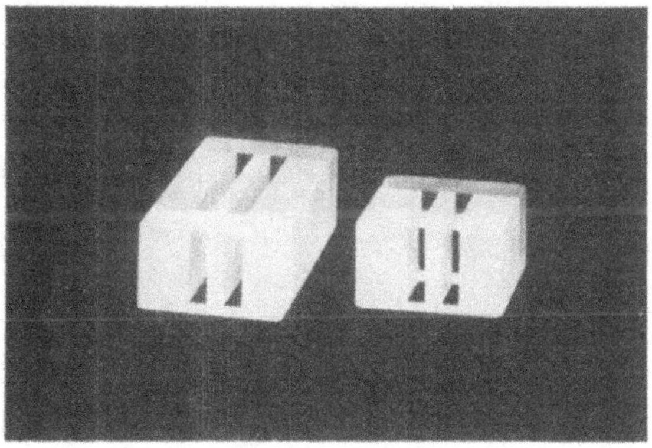

Fig. 7 Double channel interferometer glass cuvettes, one 10 mm and the other 30 mm long.

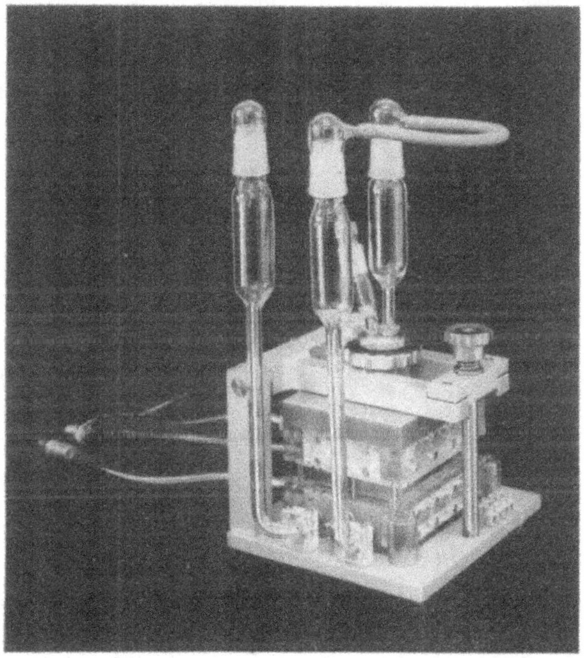

Fig. 8 Photograph of the thermal diffusion cell.

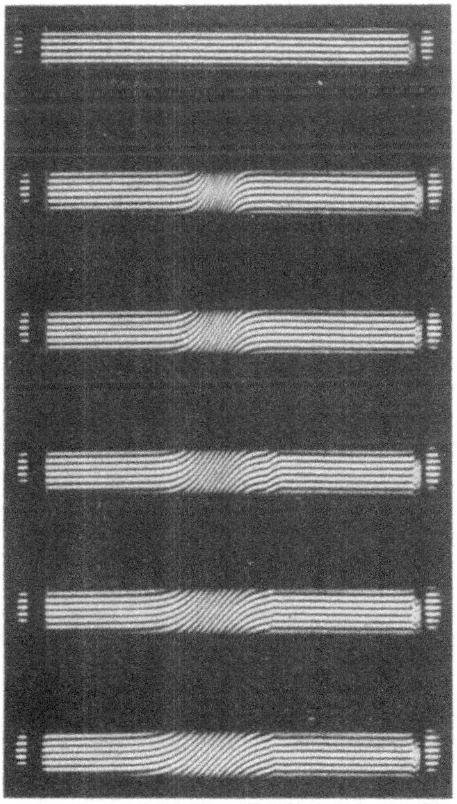

Fig. 9 A recording of a thermal diffusion experiment using the RAYLEIGH interference optical system. The fixed reference marks on each end of the interference patterns can be seen.

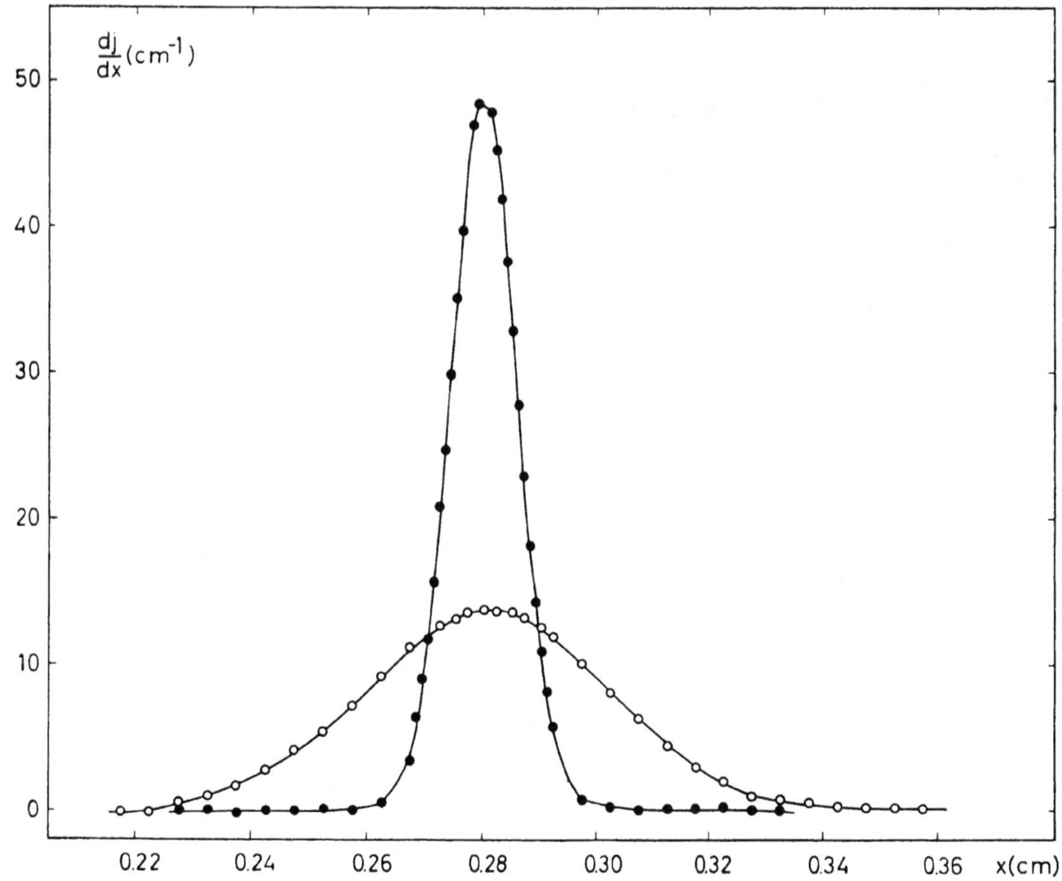

Fig. 10 The fringe gradient of the migrating boundary at 150 and 2500 sec. after start. Glucose in water; $c = 0.601$ g/dl.

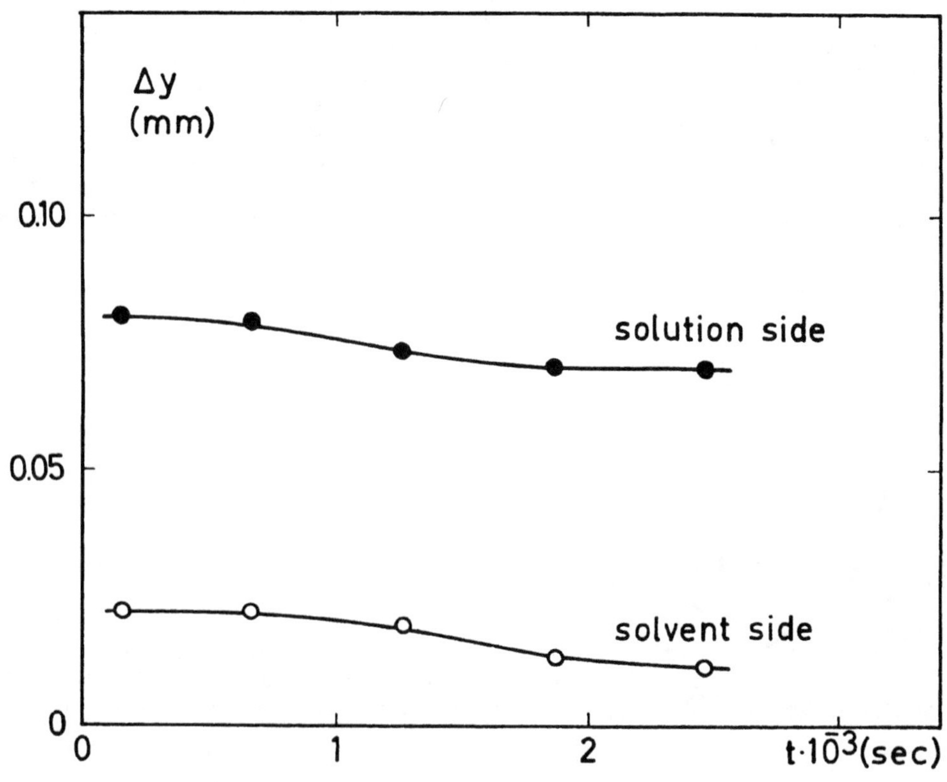

Fig. 11 The position of the fringe patterns from the homogeneous media on each side of the boundary as a function of time.

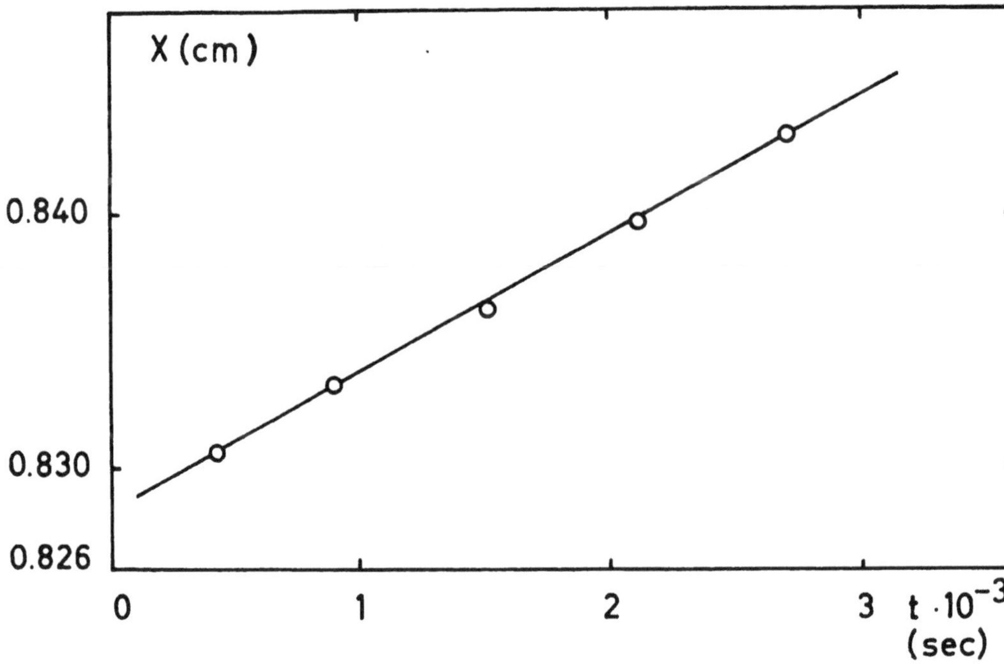

Fig. 12 The migration of the boundary due to a temperature gradient of 17.5 deg/cm. The x-position for $\frac{J}{2}$ in the fringe gradient is plotted versus time.

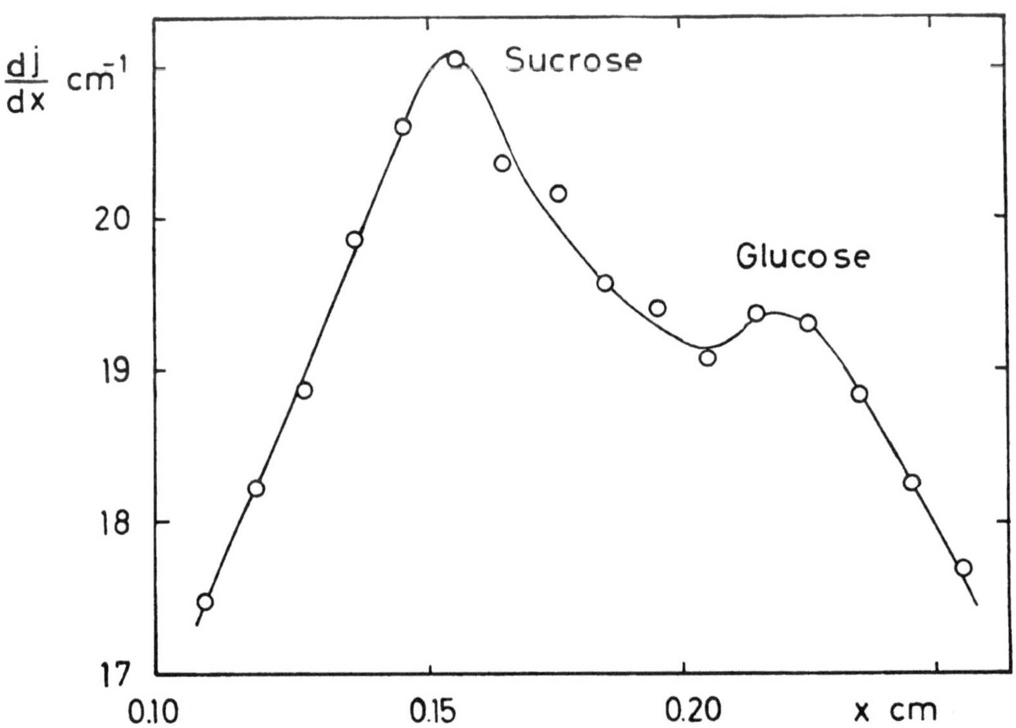

Fig. 13 The fringe gradient from an experiment with a mixture of glucose and sucrose.

A Allgemeine Grundlagen und Methoden

Conformational Changes of Biological Macromolecules and Mechanochemistry

V. I. VOROBYEV

Institute of Cytology of the Academy, Leningrad

S U M M A R Y

Results of experiments performed on different protein models including such as glycerinated muscle and collagene fibers are discussed. The experiments confirmed the theoretical assumption that small external stress stabilizes helical conformation of protein macromolecules while great forces stabilize elongated coils.

It should be suggested that specific cooperative reactions in side groups of the biopolymer chain are at the basis of mechanochemical phenomena. These reactions result in reversible alterations of conformation of the chain and its deformation (phenomena of contractility), and can also affect enzymic properties of the protein. Thus conformational changes of biomolecules not only underline contractility but can also provide for a possibility of cell activity regulation via mechanochemical processes.

Contractility in all its numerous manifestations in living nature is based on one very essential fact : mechanical work is produced in biological systems through the direct use of chemical energy. Such systems, able to perform a direct conversion of chemical energy (or generally energy of molecular interaction) into mechanical work, are called mechanochemical systems.

Mechanochemical processes can be accomplished provided the chemical reactions,taking place on the surface of a polymer chain, are followed by a change in the spatial arrangement of macromolecules or their separate parts. Mechanochemical phenomena connected with the movement of the separate parts of a molecule, i.e. with conformational changes, are caused by the reaction in side groups of a polymer chain. As the rate reaction constant for each group is governed by its position to neighbouring groups, these reactions are cooperative.

It is from this standpoint that we considered the results of our previous experiments on protein model systems in which we studied the conversion of chemical energy into mechanical work as well as a reverse conversion of mechanical energy into chemical one[1][2][3].

In fact the free energy of side groups of a polymer chain, being dependent on their position to the adjacent groups, the sizes of macromolecules (and hence of their macroscopical system), is determined by the number of the groups having taken part in the reaction. On the other hand as the rate reaction constant of a special group is dependent upon its position towards neighbouring groups, a change in the position of the latter due to deformation must change the equilibrium of the chemical reaction and hence the chemical composition of the medium (a reverse mechanochemical effect).

It should be born in mind, however, that biological systems generally are consisting of macromolecules with a regular inner structure. The mechanochemical processes connected with conformational changes affecting this ordered structure are supposed to play an essential role in biological systems. It is well known that the conformational change of the order-disorder type (e.g. helix-coil) thoroughly studied on the solutions of macromolecules accomplished according to the principle "all or none", i.e. in the narrow limits of environmental changes. Unfortunately our knowledge of such conversions studied on macroscopical systems (such as fibres or films) are not sufficient. Meanwhile from a mechanochemical point of view such data would be of special interest.

We shall discuss here three questions the consideration of which will permit to establish a connection between mechanochemistry and conformational changes in biopolymers ; first, the effect of external forces on conformational transitions in biopolymers; secondly, the effect of deformation on biological activity of proteins; and, third, very briefly, a relation between mechanochemical processes and sorption phenomena in biopolymers.

Thus let us consider the connection between deformation and conformational changes in biopolymers.

FLORY was the first who analysed the influence of an external force on the conformational changes [4)]. It was shown that the external force can change the point of conformational transition. The mechanochemical phenomena occuring at the transition helix-coil were investigated in collaboration with Dr. BIRSTEIN and Dr. PTYTZYN and reported in the work on the theory of mechanochemical phenomena connected with conformational changes of macromolecules [5)].

Theoretical considerations given in the above work lead to the conclusion that external force must stabilize the helical conformation of macromolecules of polypeptides and probably ordered conformations of fibrillar proteins and other biopolymers also. This conclusion however proves right, provided that external forces are relatively small. On the contrary great forces must promote the transition to a stretched coil conformation which in the three-dimensional system converts into ß-structure.

These theoretical considerations were analysed experimentally on the following objects: collagene fibres (in collaboration with Dr. LUTZENKO and Dr. FRENKEL), films and fibres of DNA and on glycerinated muscle.

Collagen fibres form the rat's tail tendons were treated with formalin to form interchain bonds. The fibres under a certain load were placed into a vessel containing a solvent. The experiment was started only after the diffusion equilibrium had been achieved.

It was found, according to FLORY's data [6)], that heating resulted in a sharp shortening of fibres due to transition of polypeptide chains from an ordered conformation to a conformation of random coil. Temperature ranges of the transition proved to depend on the load applied to the fiber. It is shown that the load stabilizes the ordered conformations. For example, the median point of a temperature transition range (T_m) for a fiber under the load of 0,5 g is 82,5^o, for the fiber under 7 g T_m=92,5^o, i.e. it appears to be 10° higher (fig.1, 2). However, in the case of greater loads we can observe a reverse shift of T_m towards lower temperatures: T_m for the fiber under 12 g is 86,5^o. These results are in accord with the above theoretical suggestions.

The region of transition order-disorder depends on the chemical composition of the media surrounding the fibre. The substances compeering for hydrogen bonds (e.g.urea) decrease the value of T_m, beeing in contrast to the substances strengthening the structure stabilized by hydrogene bonds (e.g. dimethylformamide and ethylenchlorhydrine), which increase the value of T_m.

The same results were obtained from the experiments on DNA. They showed that the external forces stabilized the helical conformation of DNA.

A study of the behaviour of glycerinated muscles under a load reveales relations.

In our experiments we made use of the fact (well known in physiological literature) that the resistance of muscles to injurious influences increases upon stretching. We thought that this phenomenon could be accounted for by the stabilization of the ordered configuration of muscle proteins by the stretching of muscles. Yet, if the suggested explanation is correct, the increasing muscle resistance to denaturating factors must be revealed only under small loads, whereas large loads must not exert stabilizing action.

These considerations underlie the experiments about the influence of large and small loads on the resistance of the muscles extracted in glycerol to the denaturating effect of high temperature [7].

The frog's muscles were washed out of glycerol, placed in physiological solution (0,65m NaCl) under an adequate load and then heated at different temperatures (30°, 40°, 50°, 60°) for a certain time. The ATP-ase muscle activity decreasing served as indication of the degree of muscle proteins denaturation.

The experiments showed that ATP-ase activity after 15 min heating at 40° and 50° proved to be 10,6 % and 9,9 % lesser for the muscles under large loads (35 g) compared to the muscles stretched under small forces (5 g). These differences are statistically reliable (fig.3).

Thus the results of the experiments showed that, while the stretching of glycerinated muscles by a small load increases their resistance to thermal action, a further increasing of the load is followed by a decreased resistance. Hence, the increase in muscle resistance to damage upon stretching is due to the stabilization of ordered conformations of muscle proteins by stretching.

In most experiments we used thermal factors as an agent inducing conformational changes. However in some cases, and in experiments with collagene fibers and DNA fibers and films in particular, conformational changes appeared as a result of changes in chemical composition of the media (pH, ionic strength, addition of substances which act more or less selectively on the secondary or tertiary structure of polymers).

Mechanical actions, as we have seen, influence the conformation of macromolecules of biopolymers. As biological activity of proteins is known to vary to a great extent with conformational changes [8][9], the question arises whether mechanical action may result in the change of biological activity (for instance enzymic activity). This question is of primary importance in studying of biological systems.

It is difficult to choose an object to check this suggestion experimentally, for the enzyme under investigation evidently should be an integral component of structure which might undergo mechanical deformation. It appeared that "muscle models", i.e. the muscles extracted in glycerol can serve a suitable object at the first stage of investigation [10][11].

Frog's sartorius muscles were extracted by repeated portions of 50 % glycerol solution in the cold for 3-5 days, then washed out of glycerol, measured in length and placed in solution to determine the ATP-ase activity.

First we compared ATP splitting by paired glycerinated muscles (right or left). Then the muscles were thoroughly washed out of ATP, and one of them stretched. Next we determined the rate of ATP splitting by stretched and paired unstretched muscles. Such mode of experimenting gave us the possibility to provide for a double control: we compared rates of ATP splitting on one muscle model before and after stretching, and also on paired (stretched and unstretched) models.

The results obtained show a considerable increase in the rate of enzymatic splitting of ATP upon stretching of muscle models. In one of the experiments, for instance, it was observed that paired muscles liberated 56 μg and 57 μg of phosphorus of the ATP per 100 mg of wet weight of muscle for 30 min. After stretching the first muscle split 95 μg of phosphorus per 100 mg of wet weight for the next 30 min while unstretched one gave only 62 μg. In fig.4 one can see the results of such a typical experiment. It is evident that ATP splitting by stretched glycerinated muscle (curve 1) takes place more rapidly than the same process in the presence of an unstretched muscle (curve 2).

An experimentally observed increase in enzymatic splitting of ATP by muscle models upon stretching is, probably, conditioned by two factors : 1) conformational changes of enzyme macromolecules upon stretching, and 2) increase in the rate of ATP diffusion into a muscle if it occurs in the process of stretching.

If the increase of diffusion of substrate (ATP) does occur during stretching, the difference between the amount of the ATP splitted by stretched and unstretched muscles must be eliminated with the increase of its concentration. In our experiments we failed to watch it while increasing the concentration of ATP ten times as compared with the usual one.

It is clear, however, that such experiments could not exclude completely the possibility of the effect of stretching being somehow due to the diffusion factor. Therefore it was desirable to perform a number of experiments with muscle fibres of small diameter, when the differences due to the permeability of substrate into a stretched and unstretched fibre could be reduced to minimum. Preparations for such experiments had been completed when Japanese authors [12] published the results of their experimental work performed on glycerinated muscle fibers 0,05–0,1 mm in diameter.

Their work fully confirmed the data we obtained. It was found that the rate of ATP dephosphorilation upon stretching increases.

We chose another way and decided to carry out a next stage of investigation on the level of macromolecules : experiments were made in solutions of myosin and actomyosin. Deformation of macromolecules was produced by application of a hydrodynamic field.

The experiments (made in collaboration with Dr. KUCHAREVA) were carried out in the device for the study of double refraction in flow. Experimental conditions in all cases provided the laminarity of the flow. ATP-ase activity of myosin and actomyosin solution was compared without hydrodynamic field and upon its application.

It is known that even moderate forces of the flow causes disturbances in the structure of macromolecules, nucleic acids for example [13][14]. Really, high rate gradients in our case also resulted in irreversible inactivation of myosin and actomyosin. In the range of average rate gradients (near to 2000 sec^{-1}), however, the increase of enzymic activity of both myosin and actomyosin could be observed (fig.5). This increase reached 25–30 % and even more. Small rate gradients did not produce any effect.

Thus direct experiments performed on protein systems with different levels of organisation have shown that mechanical actions, which evidently are accompanied by conformational changes of macromolecules, may influence the enzymic activity of proteins. The significance of this phenomena for biological systems, contractility including, will be revealed only in the course of future investigations.

In conclusion let us consider one more aspect of mechanochemical phenomena. Being due to conformational changes of macromolecules mechanochemical processes should be connected with the processes of sorption of molecules and ions on the surface of polymer chains as sorption phenomena depend upon the spatial arrangement of side groups of macromolecules and therefore are determined by chain conformations.

Our experiments (made in collaboration with Dr.LEV) showed the sorption of monovalent cations on the surface of DNA to depend on the conformation of macromolecules. Thus it was found that the activity of counterion in the medium increases with the helix-coil transition in DNA, i.e. the native DNA binds up more ions than the denatured one.

The helix-coil transition in DNA was followed by the contraction of fibres and films from DNA. Consequently, the contraction in this system must be connected with a decrease in counterions sorption.

As to the question of specificity of ion sorption is more interesting from the biological point of view, it is necessary to mention the theoretical and experimental studies of EISENMAN and his group on the sorption alkaline metal ions [15)16)]. It was revealed that intensity of alkaline metal ions sorption and its specificity was determined by the strength of field made by anions. Depending on that strength, cations can be sorbed in a various sequences irrespective of their hydration radii in solution.

In polymer electrolytes, which include all the biopolymers, the strength of electrostatic field per charged monomer unit varies with a change in spatial arrangement of side groups of a polymer chain, i.e. with the conformational changes of macromolecules. Therefore specificity of ion sorption on the surface of biomacromolecules should be a function of chain conformation. As conformational changes underlie mechanochemical phenomena it can be suggested that the order of ion sorption specificity must change in the process of deformation. At the present time we are making attempts to solve this problem experimentally.

The data presented in this paper show that mechanical forces influence conformational transitions of biomacromolecules. On the other hand, the conformational changes resulting from chemical reactions of side groups of macromolecules proved to be connected with the dimensional changes, i.e. contraction and lengthening. Thus, here we can clearly see interrelation of mechanical and chemical processes. Perhaps it can be concluded that conformational changes of biomacromolecules in a macroscopic systems can be regarded as mechanochemical process.

L I T E R A T U R E

1) VOROBYEV, V.I., Biochimia , 22, 597 (1957)
2) VOROBYEV, V.I., Struktur und Biologische Funktion der Eiweiße. III.Humboldt-Symposium, Berlin 21.-26.Sept.1960. 118-125 VEB Gustav Fischer Verlg.,Jena,1962
3) VOROBYEV, V.I., Doklady Akad.Nauk SSSR 137, 972 (1961)
4) FLORY, P.J., Science 124, 53 (1956)
5) BIRSTEIN,T.M., V.I.VOROBYEV, O.B.PTYTZIN, Biophysica 6, 524 (1961)
6) FLORY,P.J., J.cellular comparat.Physiol. 49, Suppl. 1, 175 (1957)
7) VOROBYEV,V.I., L.Sch.GANELINA, Biochimia, 27, 565 (1962)
8) ELODI,P., Biochim. biophysica Acta 44, 610 (1960)
9) ELODI,P., Biochim. biophysica Acta 53, 218 (1961)
10) VOROBYEV,V.I., L.Sch.GANELINA, 2nd Conference of Institute of Cytology, Acad.Sci. USSR, Reports of Communications, 17 (1962)
11) VOROBYEV,V.I., L.Sch.GANELINA, Tsitologia 5, N6 (1963)
12) OHNISHI,T., T.OHNISHI, Nature (London), 197, 184 (1963)
13) LEVINTHAL,C., P.F.DAVISON, J.Chim.physique Physica-Chim.biol. 58, 887 (1961)
14) FRISMAN,E.V., V.I.VOROBYEV, N.K.YANOVSKAYA, L.V.SHCHAGINA, Biochimia 28, 137 (1963)
15) EISENMAN,G., D.RUDIN, J.CASBY, Science 126, 831 (1957)
16) EISENMAN,G., Membrane Transport and Metabolism. Proc. of Symposium. p. 163. Praha,1960

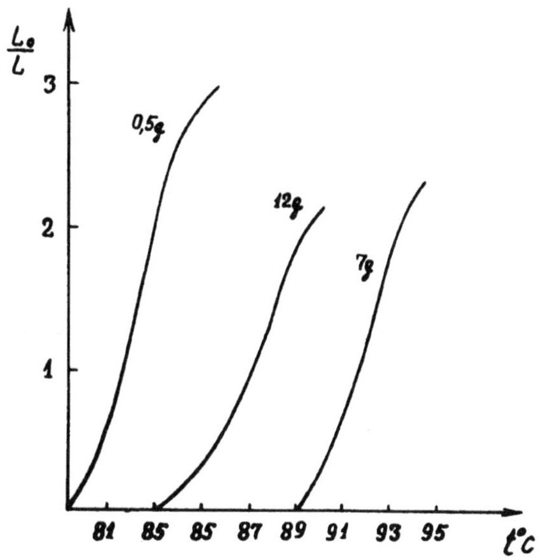

Fig.1 Temperature transition in rat tail collagen fibers.
L_0: initial length of the fiber.
L : length of contracted fiber.

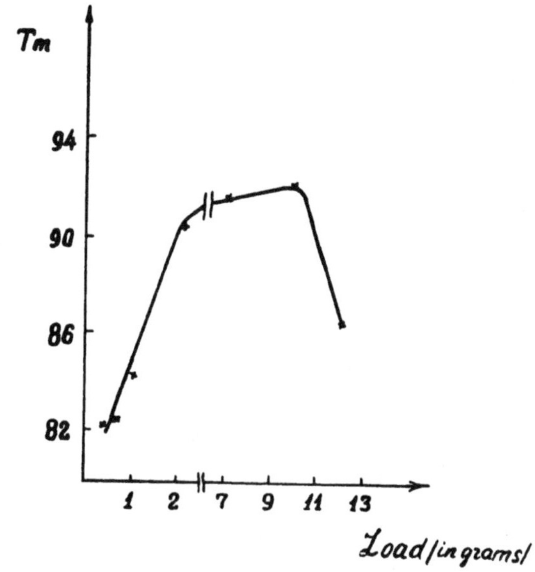

Fig.2 The relation of T_m for rat tail collagen fibers and the external force.

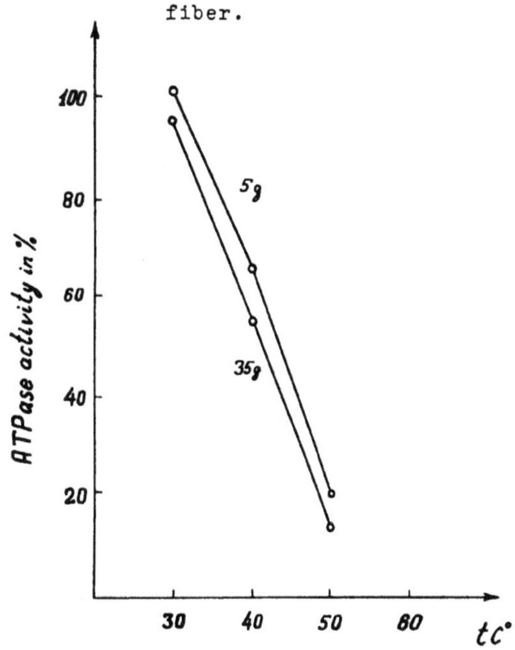

Fig.3 ATP-ase activity of homogenates of glycerinated frog sartorius muscles depending on temperature of muscle heating (according to [7]). The figures under the curves designate the load under which the "muscle model" was heated.

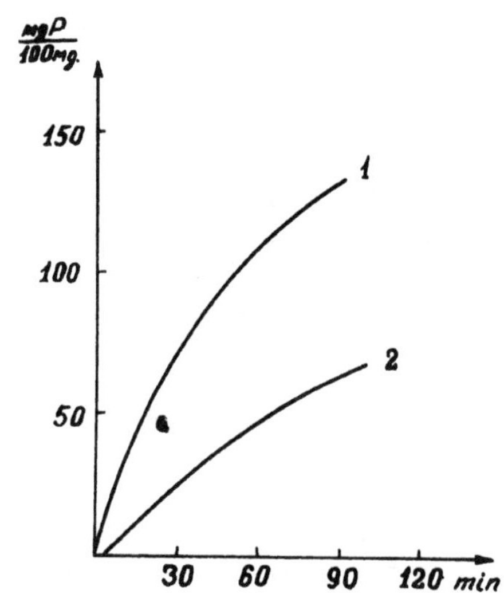

Fig.4 Kinetics of ATP-splitting by stretched (1) and unstretched (2) glycerinated frog sartorius muscle. Reaction mixture: ATP-5.10^{-3}m; Mg^{++} - 2.10^{-2}m; Glycine buffer 3.10^{-2}m, pH 8,2.

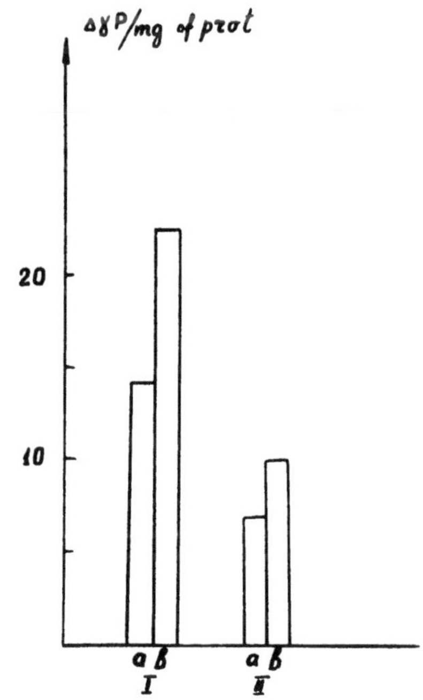

Fig.5 Change in enzymic activity of actomyosin (I) and myosin (II) at the application of hydrodynamic fields. Ordinate - μg P liberated from ATP per mg of protein (average value for three separate experiments).
a - control, b - experiment.
Flow gradient is 2280 sec^{-1}.

A Allgemeine Grundlagen und Methoden

Ein Verfahren zur Steigerung der Empfindlichkeit bei Diffusions-, Sedimentations- und Elektrophoreseuntersuchungen von Makromolekülen

K. E. REINERT

Institut für Mikrobiologie und experimentelle Therapie der Akademie, Jena

S U M M A R Y

A procedure is described for increasing sensitivity of above methods in steps to one order of magnitude. This is realized by making the parallel windows reflecting in a special manner. So it is possible to inrease the effective pathlength of light in the cuvette. The number of reflections is automatically recorded together with the main informations.

Bei der Messung der Eigenschaften der Makromoleküle in Lösung ist stets die Extrapolation der Meßwerte auf die Konzentration c = o zwecks Ausschaltung der Wechselwirkungseinflüsse notwendig ; so auch bei Diffusions- und Sedimentationsmessungen. Dabei treten u.a. 2 Hauptschwierigkeiten auf :

1. zu geringe Empfindlichkeit der refraktometrischen Meßmethoden,
2. Konvektionsstörungen.

Für einige Substanzen und Lösungsmittel läßt sich die Empfindlichkeit zwar durch Messung ihrer UV-Absorption erhöhen, wie es z.B. bei DNA-Sedimentationsuntersuchungen üblich ist und für Diffusionsmessungen von JACOB, FREUND und DAUNE mitgeteilt wurde [1]. Die Genauigkeit ist jedoch geringer, die Auswertung stark erschwert, will man nicht einen außerordentlich hohen Aufwand treiben. Die Konzentration ist weniger bequem zu variieren.

Es soll hier über ein einfaches Verfahren zur Erhöhung der Empfindlichkeit für die genannten Methoden um ca. eine Größenordnung berichtet werden, sowie über eine Möglichkeit zur Kontrolle auf Konvektionsfreiheit.

Die Empfindlichkeit sämtlicher optischer Methoden ist proportional der Schichtdicke der untersuchten Probe. Die effektive Schichtdicke kann nun dadurch erhöht werden, daß man das Meßlicht durch geeignete Verspiegelung der Küvettenfenster (innen oder außen) in einer Ebene konstanter Konzentration mehrmals durch die Küvette hindurchtreten läßt.

Neben der mit einigen Nachteilen behafteten halbdurchlässigen Verspiegelung hat sich vor allem die Anordnung nach Abb.1 bewährt : Durch einen schmalen unverspiegelten vertikalen Eintrittsspalt ES tritt das Licht etwas schräg in die Küvette ein und verläßt diese wieder durch den Austrittsspalt AS des gegenüberliegenden Küvettenfensters nach ein oder mehrfachem Durchtritt durch die Küvette in horizontaler Ebene.

Die Wahl der Zahl der Durchsetzungen geschieht durch geringfügige Drehung der Küvette um eine vertikale Achse oder durch geeignete Prismen (Abb.2) (notwendig bei der Sedimentation in der Ultrazentrifuge).

Zur gleichzeitigen photographischen Registrierung der Durchsetzungszahl dient der schräge Zählspalt ZS auf dem Eintrittsfenster in Abb.1, längs dem die Küvette nicht verspiegelt ist, und der gegenüber dem einfallenden Licht abgedeckt ist. Bei jeder Reflexion an der vorderen Küvettenwand geht dadurch ein Teil des Lichtes verloren, und es entstehen im Diagramm ν dunkle Streifen.

Die Abb.3 und 4 demonstrieren die beschriebene Methode. Aus der Zahl ν der dunklen Streifen folgt die Zahl der Durchgänge des Meßlichtes durch die Küvette, also der Faktor der Empfindlichkeitssteigerung, zu $(2\nu + 1)$.

Die wirkliche Küvettenlänge beträgt 2 cm, ihre Breite 6 mm. Die Küvette stammt aus der normalen Produktion und ist in ihrer optischen Qualität nicht besonders hochgezüchtet. Das zeigt auch Abb.5. Die Neigung der Basis bei 9-facher Küvettendurchsetzung rührt von dem vervielfachten Fehler der Parallelität der Fenster. Der Gradient stammt von einer 0,009 %igen DNA-Lösung (Philpotwinkel $\Theta = 60^{\circ}$). Bei Vergrößerung der wirklichen Küvettenlänge auf 6 cm entspräche dieser Gradient dem einer ca. 0,003 %igen DNA-Lösung.

Das beschriebene Verfahren ist bei Verwendung einer Doppelküvette nach Abb.1 z.B. auch bei der RAYLEIGHschen Interferenzmethode anwendbar.

Will man die durch das geschilderte Verfahren gegebenen Möglichkeiten, insbesondere mit der Interferenzmethode, voll ausschöpfen, so hat man für beste Parallelität und allenfalls Planparallelität der Küvettenfenster zu sorgen; denn Keilwinkel γ_K der Küvette sowie γ_G der Küvettenfenster werden durch die Mehrfachreflexion ebenfalls vervielfacht. Dabei zeigt die Rechnung, daß die Ablenkung $\delta^{(2\nu+1)}$ des Lichtes durch einen Küvetten-Keilwinkel γ_K je nach Lösungsmittel zwei- bis viermal stärker anwächst als die Empfindlichkeit (Abb.6). Weiterhin verursacht ein verbliebener Keilwinkel ein geringes Auswandern des Lichtes aus der Ebene konstanter Konzentration um den Betrag $z^{(2\nu+1)}$ bei $(2\nu + 1)$ facher Durchsetzung der Küvette (Abb.6), ein Fehler, der durch geringfügige Küvettenneigung auf den halben Betrag reduziert werden kann. Den damit verbundenen Nachteil hat man auch bei der UV-Absorptionsmethode[1]. Dort wird er hervorgerufen durch die Divergenz des Lichtbündels, die im wesentlichen durch die sphärischen Fehler der Optik bedingt sind.

Dieser mit ν anwachsende Versetzungsfehler macht sich am stärksten bei schmaler Gradientenzone bemerkbar. Man kann allenfalls mit unserer Küvette zu Versuchsbeginn mit einer mittleren Verstärkung arbeiten, um später bei flacher und breiter werdenden Gradienten zu höherer Vervielfachung überzugehen.

Für Sedimentationsuntersuchungen bei höheren Drehzahlen ist es aus Gründen der mechanischen Belastung erforderlich, Saphirfenster zu verwenden. Der effektive Keilwinkel, hervorgerufen durch die Kompression der Lösung und der damit verbundenen Zunahme der Brechzahl, kann (für eine bestimmte Drehzahl) durch einen Keilwinkel im Zell-Mittelstück kompensiert werden.

Nun noch einige Worte zum Konvektionsproblem :

Arbeitet man bei Diffusionsmessungen mit extrem niedrigen Konzentrationen, so ist eine Kontrolle hinsichtlich Konvektionsfreiheit wünschenswert. Dies kann durch ein zweites Diffusionsexperiment geschehen, für das folgendes gelten muß :

1. Es muß unter den gleichen thermischen Bedingungen gearbeitet werden wie bei der eigentlichen Meßsubstanz, möglichst simultan in einer zweiten Zelle einer Mehrfachküvette (s.z.B. Abb.1).
2. Der Dichteunterschied der überschichteten Flüssigkeiten des Testexperimentes soll kleiner oder höchstens gleich dem der Lösungen der eigentlichen Messung sein.
3. Der Diffusionskoeffizient der Vergleichssubstanz muß bekannt sein.

Die Forderungen 2 und 3 sind zu erfüllen, wenn man dafür zwei Lösungen einer bereits bei relativ hoher Konzentration vermessenen Substanz bei dieser und einer demgegenüber wenig verschiedenen Konzentration auswählt.

Ist Konvektion vorhanden, dann zeigt sich das in einem scheinbar erhöhten Wert für den Diffusionskoeffizienten. Ein extremes und zur Demonstration gezüchtetes Beispiel zeigt Abb.7. In allen Versuchen betrug die höhere Konzentration c_0 = 0,5 % die niedere Konzentration wurde variiert. Die Differenz beider ist als Abszisse aufgetragen. Da der wahre Diffusionskoeffizient klinischer Dextrane sich zwischen c = 0 und 0,5 % nur um wenige % ändert, ist aus der starken Zunahme der gemessenen Werte auf starken Konvektionseinfluß zu schließen.

Abschließend sei erwähnt, daß man die Empfindlichkeitssteigerung der genannten Meßmethoden auch unter dem Gesichtspunkt des geringeren Substanzbedarfs betrachten kann - das gilt besonders auch für die freie Elektrophorese - sowie ihrer prinzipiellen Anwendbarkeit bei weiteren Untersuchungen, wie z.B. mit der optischen Drehung.

L I T E R A T U R

JACOB,M., L.FREUND, M.DAUNE, J.chim.Physique 58, 521 (1961)

D I S K U S S I O N S B E M E R K U N G E N

DOSE : Haben Sie Ihre Küvette auch schon zur Messung der UV-Absorption eingesetzt ?

REINERT : Bisher noch nicht, aber es wäre insbesondere bei der freien Elektrophorese und der Sedimentation von Vorteil.

ATEN : I should like to make one remark. When you change the number of times that the light beam passes the cell, you are changing the optical pathlength essentially. A readjustment of the Philpot-Svensson-optical system might be necessary. This might explains the slope of the base line you showed.

REINERT : The readjustment of the Philpot-Svensson-optical system with changing optical pathlength in the cell is easy with our "Zeiß-Schlierengerät 80 " because of the scales, combined with the adjusting screws. In spite of the fixed entrance slit of our system no readjustment of the position of the phase plate is necessary, because the place of the cell is situated in the parallel light. Beam and therefore the slope of the base line is caused by the errors of cuvette-planparallelity.

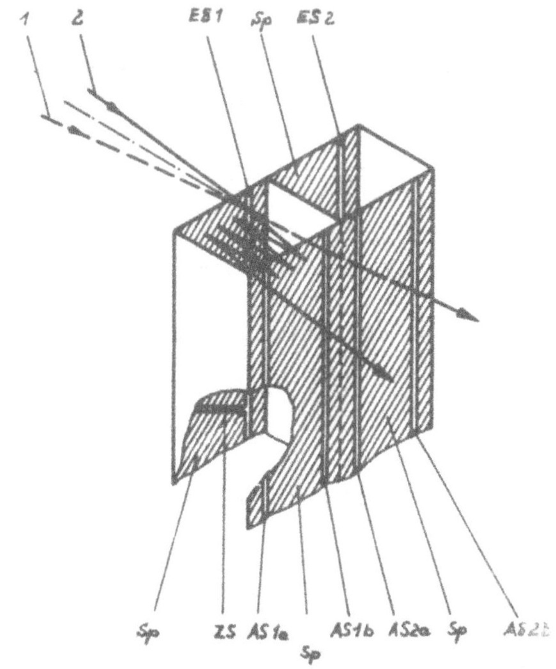

Abb.1 Doppelküvette mit verspiegelten Fenstern Sp zur Empfindlichkeitssteigerung. Eintrittsspalte: ES, Austrittsspalte: AS.
1: Richtung des einfallenden Lichtes bei einfacher Küvettendurchsetzung,
2: Einfallsrichtung bei mehrfacher Durchsetzung,
Z: Zählspalt.

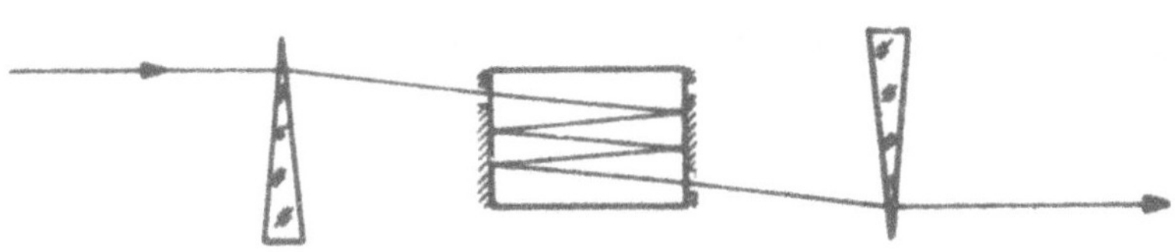

Abb.2 Erzeugung der Richtungsänderung des einfallenden Lichtes bei unveränderter Küvettenstellung durch optische Elemente mit Prismenwirkung.

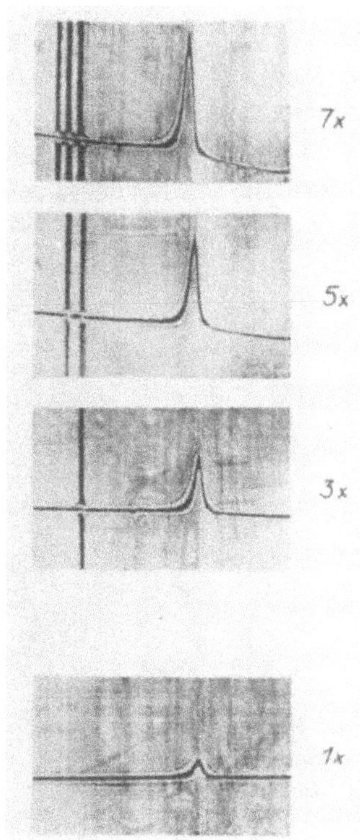

Abb. 3 Konzentrationsgradienten von 0,058 %iger Kalbsthymus-DNA, überschichtet mit dem Lösungsmittel (0,0025 m NaCl + 0,00025 m Na-Citrat), aufgezeichnet nach der PHILPOT-SVENSSON-Methode mit dem Zeiß-Schlierengerät 80. Küvettenlänge $l = 20$ mm, -breite $b = 6$ mm, Philpotwinkel $\Theta = 24°$. Aus der Zahl ν der vertikalen Streifen folgt der Faktor der Empfindlichkeitssteigerung zu $(2\nu + 1)$.

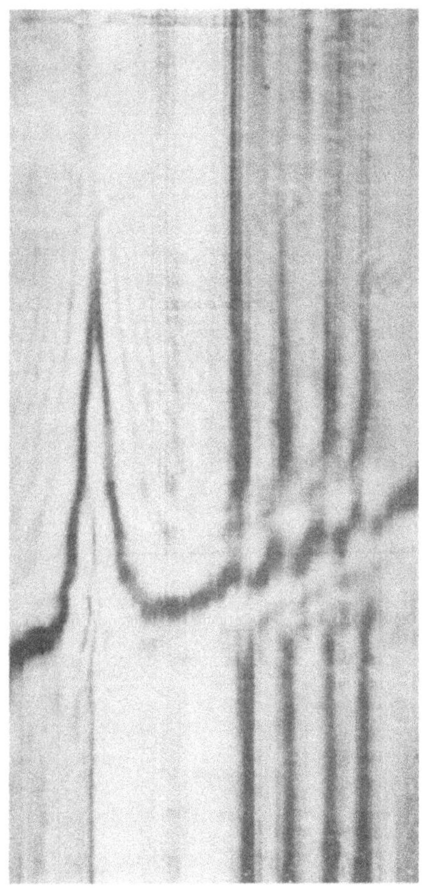

Abb. 5 DNA-Konzentrationsgradient entsprechend einer Konzentrationsdifferenz von 0,009 % bei $\Theta = 60°$ unter sonst gleichen Bedingungen wie in Abb. 3.

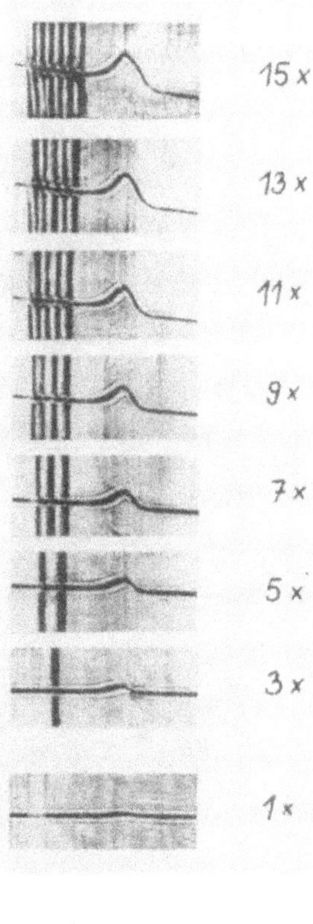

Abb. 4 Analog Abb. 3, jedoch bei einer Dextran-Konzentrationsdifferenz von 0,020 % und $\Theta = 16°$.

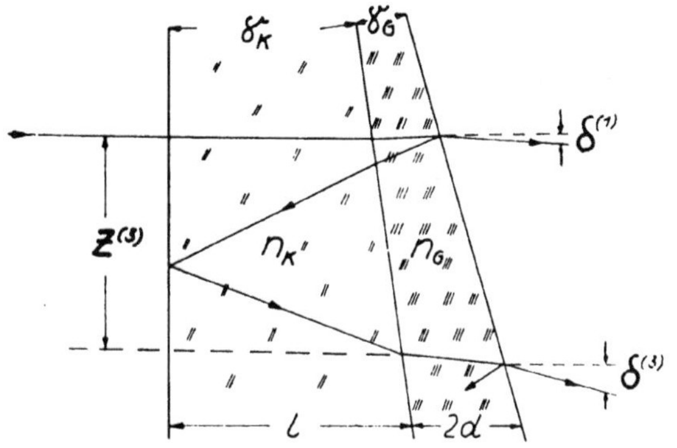

$$\delta^{(2\nu+1)} = \gamma_K\left[(2\nu+1)n_K - 1\right] + \gamma_G\left[(2\nu+1)n_G - 1\right]$$

$$z^{(2\nu+1)} = 2l\,\nu(\nu+1)\left(\gamma_K + \frac{n_G}{n_K}\gamma_G\right) + 4d\,\nu^2\left(\frac{n_K}{n_G}\gamma_K + \gamma_G\right)$$

Abb. 6 Der Einfluß geringer Keilwinkel von Küvette (γ_K) und Küvettenfenster (γ_G) auf die Ablenkung δ und Versetzung z des Lichtstrahles.

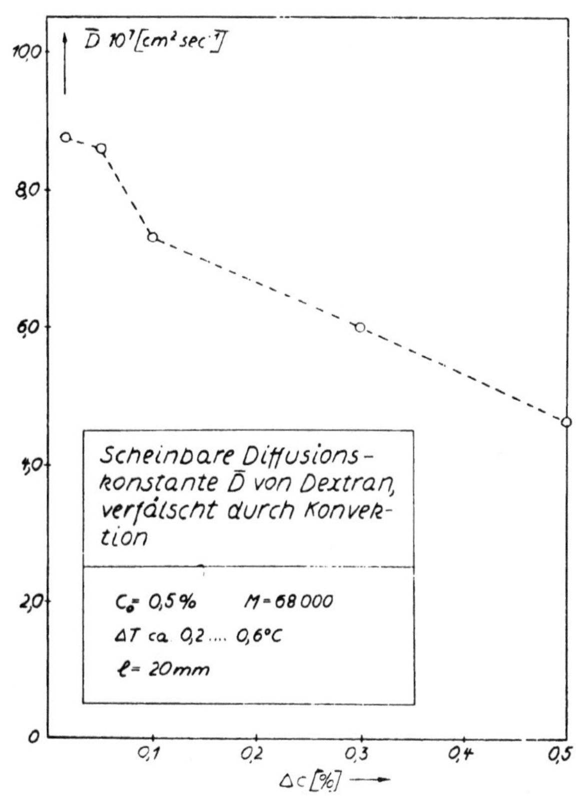

Abb. 7 (siehe auch Text)

A Allgemeine Grundlagen und Methoden

Eine Methode zur Untersuchung der Kinetik des Denaturierungsvorganges der DNA

K. E. REINERT

Institut für Mikrobiologie und experimentelle Therapie der Akademie, Jena

S U M M A R Y

A method for the measurement of the rate of DNA-helix-coil-transition is described.

Die Aufklärung des Mechanismus des Helix-Knäuel-Überganges der DNA und dessen Abhängigkeit von den verschiedensten Parametern und Einflüssen ist keineswegs abgeschlossen. Ergänzend zu den Untersuchungen mit den üblichen physikochemischen Methoden, die auf der Messung von Gleichgewichtszuständen basieren, erscheint daher eine Beschäftigung mit der Kinetik des Denaturierungsvorganges sehr nützlich. Eine dazu geeignete Untersuchungsmethode soll hier mitgeteilt werden.

Der Übergang von einer Konfiguration zur anderen kann durch Messung des zeitlichen Verlaufes einer für die zu verfolgende Änderung charakteristischen Größe - also als Relaxationsvorgang - verfolgt werden, wenn die Voraussetzungen für den Übergang in einer Zeit hergestellt werden, die kurz ist verglichen mit dem Vorgang selbst. Als solche Voraussetzungen kämen bei DNA z.B. infrage : Temperatursprung, pH-Sprung oder rascher Wechsel der Ionenstärke.

Neuerdings wurden auch Untersuchungen über die Änderung der Konformation von Proteinen mit der Temperatursprungmethode von EIGEN et.al. [4], jedoch bei spektrophotometrischer und polarimetrischer Registrierung sowie Erwärmung durch Kondensatorentladung begonnen.

Bei Nucleinsäuren ist im Prinzip zunächst zwischen dem Aufbrechen der die Sekundärstruktur der Doppelspirale fixierenden Bindungen und dem Übergang von der regelmäßigen zur statistischen Anordnung der Ketten zu unterscheiden. Daraus ergibt sich, daß Meßgrößen, die weitgehend vom Elektronenzustand des Moleküls abhängen, nicht notwendig die gleichen Ergebnisse liefern müssen wie solche, die unmittelbar durch die Gestalt des Moleküls bedingt sind.

Wir bemühen uns daher um eine hinreichend trägheitsarme Methode, die die Konfigurationsänderungen großer Makromoleküle als Relaxationsvorgang anzuzeigen gestattet. Die genannten Bedingungen erfüllt die Streulichtphotometrie, die wir für diesen Zweck anwandten. Obzwar man bisher beim DNA-Helix-Knäuel-Übergang keine Änderung des Molekulargewichtes beobachtet hat, tritt mit der Verringerung der mittleren Abmessungen des Moleküls auf Grund der verminderten innermolekularen Interferenz eine Erhöhung der Streulichtintensität bei höheren Streuwinkeln ein.

Dies wird z.T. durch Abb.1 demonstriert. Dort ist aufgetragen die Streulichtintensität einer DNA-Probe beim Streuwinkel 90° in Abhängigkeit von der Temperatur. Der T_m-Wert wurde zunächst aus experimentellen Gründen durch pH-Erniedrigung [5] vermindert. Es ergibt sich, daß die Streuintensität um einen der Messung bequem zugänglichen Betrag ansteigt, der übrigens mit zunehmendem Molekulargewicht wächst. Der zeitliche Verlauf dieser Streuintensitätsänderung z.B. bei 90° oder einem höheren Streuwinkel bis zur Erreichung des Endwertes bei sprungartiger T-Erhöhung ist der zu messende Vorgang.

In dieser vorläufigen Mitteilung soll unsere Variante der Temperatursprungmethode kurz erläutert werden.

Bei der Größe der zu messenden Übergangszeiten von z.T. vielen Sekunden müssen vor allem Störungen der Meßlösung durch gasförmige Elektrolysenprodukte, wie sie bei der Erzeugung des Temperatursprunges durch Kondensatorentladung [4] auftreten, vermieden werden. Daher wurde zwar ebenfalls eine Erhitzung durch Joule'sche Wärme vorgenommen, jedoch durch einen Hochfrequenzsender der Frequenz 800 kHz und einer Leistung von 1,5 kW, wie er zum Betrieb eines Zeiß-Ultraschallgenerators verwendet wird unter Benutzung des niederohmigen Ausganges.

Die Ausführung einer der verwendeten Meßzellen demonstriert Abb.2. Die Küvette wird nach unten und oben durch je eine ebene Platinelektrode abgeschlossen, an die die Hochfrequenzspannung angelegt wird.

Man erreicht z.Z. damit einen zeitlichen Temperaturanstieg $\Delta T/\Delta t$ von ca. 20°C/sek., der für die relativ großen Übergangszeiten bei hohen Molekulargewichten ausreichend ist und durch Verkleinerung des Probenvolumens auf über 100°/sek. vergrößert werden kann. Als Temperaturfühler dient ein Platin-Nickel-Thermoelement, dessen Pt-Komponente unmittelbar eine der Elektroden darstellt.

Ein Beispiel für eine Registrierkurve mit einer Kalbsthymus-DNA-Lösung ($M \approx 10^7$, pH 3,2) zeigt Abb.3, ebenfalls aufgenommen bei einem Streuwinkel von 90°. Der Temperatursprung wurde von Zimmertemperatur auf ca. 40°C durchgeführt. Als Lösungsmittel diente eine wässrige Lösung von 0,045 m NaCl + 0,0045 m Na-Zitrat. Diese Konzentration ergab optimale elektrische Anpassung der Meßzelle an den Hochfrequenz-Generator. Nach dem Temperatursprung geht die Streuintensität in einer meßbaren Zeit in den Wert über, der für die Anordnung des Moleküls als statistisches Knäuel charakteristisch ist.

Die Methode erfordert eine sorgfältige Vorbereitung der Meßlösung. Dazu gehört die gründliche physikalische Reinigung sowie die Entgasung zur Vermeidung von Luftbläschen nach dem Temperatursprung. Hohe Ansprüche sind auch an die chemische Reinheit der Substanz zu stellen.

Es ist beabsichtigt, Lösungen von Nucleinsäurepräparaten geringerer Heterogenität bei höherer Zeitauflösung zu untersuchen.

Schlüsse, die sich auf die Eigenschaften der DNA beziehen, wollen wir erst nach weiteren Experimenten ziehen.

Für die Anregung zur Relaxationsuntersuchung danke ich Herrn Dr.H.BERG, für vorbildliche technische Mitarbeit Frau Ch.EILHAUER.

L I T E R A T U R

1) LONGUET-HIGGINS, H.C., B.H.ZIMM, J.molecular Biol. 2, 1 (1960)
2) WOLKENSTEIN,M.V., N.GODSCHAJEW, Ju.Ja.GOTLIB, Biofisika 7, 16 (1962)
3) PTITZYN,O.B., Biofisika 7, 257 (1962)
4) EIGEN,M., L.D.MAEYER in S.L.FRIESS, E.S.LEWIS and A.WEISSBERGER eds. Technic of Organic Chemistry Vol. 8/2, Intersc.Publ. New York 1963
EIGEN,M., G.G.HAMMES in E.F.NORD ed. Advances in Enzymology Vol. XXV, Intersc.Publ. New York 1963
SPATZ,H.Ch., D.M.CROTHERS, Generaldiskussion I (Methoden), dieser Band : Inzwischen wurden im EIGENschen Labor Temperatursprungexperimente an AT-Polynucleotiden bzw. pH-Sprunguntersuchungen an DNA, jedoch bei Registrierung der Absorption durch den Elektronenübergang bei 260 nm, begonnen (nach dem Symposium ergänzt).
5) ZIMMER,Ch., H.VENNER, J.molecular Biol. 7, 603 (1963)

D I S K U S S I O N S B E M E R K U N G E N

VARSHAVSKY : Wie groß ist der Zeitbedarf für den Denaturierungsvorgang ?
REINERT : Bei einem Molekulargewicht von $M \approx 10^7$, pH 3,2 und $40^{\circ}C$ liegt die Übergangszeit bei rund 30 Sekunden.

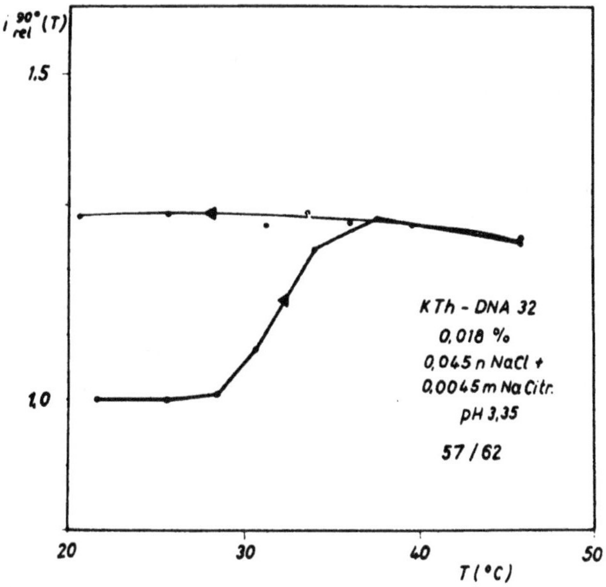

Abb.1 Streuintensität einer Kalbsthymus-DNA-Lösung in Abhängigkeit von der Temperatur beim Streuwinkel 90°

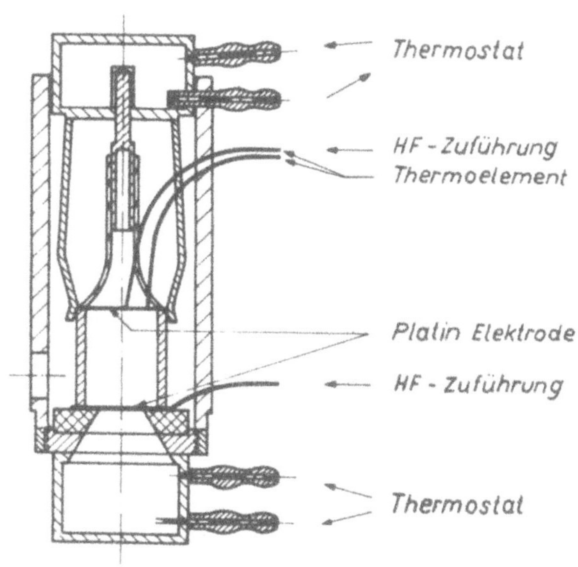

Abb.2 Ausführungsform einer Temperatursprungküvette. Das Primärlicht passiert die Küvette in einer Richtung senkrecht zur Bildebene.

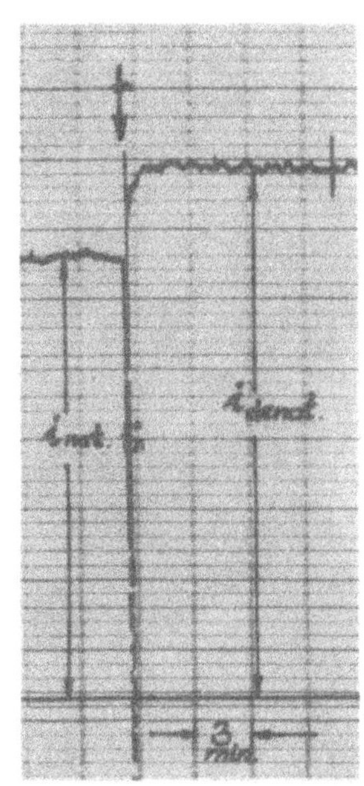

Abb.3 Beispiel für die Registrierung des Relaxationsvorganges beim Helix-Knäuel-Übergang einer Kalbsthymus-DNA Probe. Der Pfeil deutet auf den Zeitpunkt der Temperaturerhöhung.

A Allgemeine Grundlagen und Methoden

Anwendung der Röntgenkleinwinkelstreuung bei Makromolekülutersuchungen

D. UNANGST

Physikalisches Institut der Universität, Jena

S U M M A R Y

A general description of the method of small-angle scattering of x-rays will be given, followed by a critical comparison between small-angle scattering, lightscattering and electron-microscopy. The main-disadvantages of the method of small-angle scattering especially with regard to biological objects is the long registering time and the damage of sample by x-rays. Therefore some possibilities will be pointed out for weakening these disadvantages.

Die Ausdehnung der üblichen röntgenographischen Untersuchungsmethoden auf Makromoleküle verlangt wegen deren Größe besondere Untersuchungsmethoden; wegen der dabei auftretenden kleinen Streuwinkel nennt man diese Methode "Kleinwinkelstreuung".

Bei regelmäßiger Anordnung der Moleküle zu einem Kristallgitter erhält man Beugungsdiagramme mit mehr oder weniger scharfen Reflexen, wie sie von den "Weitwinkeluntersuchungen" her bekannt sind. Die Auswertung dieser "diskontinuierlichen Kleinwinkelstreuung" wird in analoger Weise wie im Weitwinkelgebiet vorgenommen. Auf diese Weise wurde z.B. von BERNAL und CROWFOOT [1)2)3)] sowie von BOYES-WATSON und PERUTZ [4)] die Struktur des Insulins bestimmt.

Eine Ansammlung von Makromolekülen, die sich nicht in einem Gitter zusammengefunden haben, streuen die Röntgenstrahlen, soweit diese Moleküle nur in einem Medium mit einer anderen mittleren Elektronendichte eingebettet sind. Infolge der mehr oder weniger regellosen Verteilung der Makromoleküle kommt es jedoch nicht mehr zu scharfen Reflexen; im allgemeinen nimmt die Streuung nach größeren Winkeln hin monoton ab. Man spricht deshalb von "kontinuierlicher Kleinwinkelstreuung", wie sie z.B. von RAMAN [5)] und KRISHNAMURTI [6)] an amorpher Kohle beobachtet wurde. Spezielle Auswertemethoden (siehe z.B. [7)8)]) ermöglichen Aussagen über Größe, Gestalt, Oberfläche, Anisotropie und Größenverteilung. Leider ist es bis jetzt nur bei speziellen Systemen möglich, alle Parameter ohne Hilfe durch andere Untersuchungsmethoden zu bestimmen.

Beim heutigen Stand der Kleinwinkeltechnik ist es möglich, Teilchen oder Gitterparameter bis etwa maximal 10 000 Å zu messen. Andere Methoden, die Aussagen in dem gleichen Größenbereich zu machen gestatten, sind die Lichtstreuung und die Elektronenmikroskopie. Diese Methoden ergänzen sich gegenseitig und die Kombination der gewonnenen Ergebnisse erleichtert oder ermöglicht überhaupt erst die erfolgreiche Untersuchung der interessierenden Systeme.

Im folgenden wird versucht, die Möglichkeiten und Schwächen der drei genannten Methoden gegeneinander abzuwägen, wobei natürlich gewisse subjektive Erfahrungen und die Blickrichtung von der Kleinwinkelstreuung her zu Einschätzungen führen, die durchaus in einzelnen Punkten oder bei besonderen Fragestellungen abweichen können. Um die Problematik der Kleinwinkeluntersuchungen besser erkennen zu können, genügt hier eine ungefähre Zuordnung und Abwägung.

Im folgenden sind die angeführten Methoden unter Gesichtspunkten angeführt, die für deren Verwendbarkeit ausschlaggebend sein können. Es bedeutet KWS Kleinwinkelstreuung, LS Lichtstreuung und EM Elektronenmikroskopie.

1. Apparativer Aufwand

KWS: Mäßig. Es werden die üblichen Röntgengeräte verwendet, an denen die speziellen Kleinwinkeleinrichtungen betrieben werden.

LS : Klein. Zur Registrierung werden Photozellen oder Sekundärelektronen-Vervielfacher benutzt.

EM : Groß. Die Elektronenmikroskope sind sehr teuer und aufwendig.

2. Einsatz von Spezialkräften

KWS: Zur Auswertung der Streudiagramme sind erfahrene Spezialkräfte erforderlich.

LS : Bei der Auswertung kann oft auf Spezialkräfte verzichtet werden.

EM : Meistens werden für die Präparation und unmittelbaren Arbeiten am Mikroskop Spezialkräfte benötigt.

3. Untersuchbare Substanzen

KWS: Alle Substanzen, bei denen die Makromoleküle in einem Gitter angeordnet oder in einem Medium mit anderer Elektronendichte eingelagert sind. Es werden etwa 1 ... 10 mm^3 Substanz benötigt.

LS : Es können nur solche Substanzen untersucht werden, die das Licht nur wenig absorbieren und örtliche Unterschiede im Brechungsindex aufweisen. Die Substanzmenge ist ebenfalls klein.

EM : Alle Substanzen, soweit sie präparierbar sind. Nur sehr kleine Substanzmengen erforderlich.

4. Ist die Untersuchung von einzelnen Objekten möglich oder werden statistische Werte ermittelt?

KWS, LS: Die Meßergebnisse sind statistische Werte. Die Bestimmung der Parameter von Einzelobjekten ist nicht möglich.

EM : Der Vorteil der Elektronenmikroskopie liegt in der Abbildung der Strukturen. Statistische Werte können erst aus einer Vielzahl von Einzelbeobachtungen ermittelt werden.

5. Die Zeit, bis zu der das Ergebnis vorliegt:

KWS: Infolge der langen Aufnahmezeit (bis zu einigen 100 Stunden) und der komplizierten Auswertung ist diese Zeit sehr lang, obgleich die Vorbereitungszeit sehr kurz ist.

LS : Kürzer als bei der KWS.
EM : Sehr unterschiedlich, je nach Präparation der Probe.

6. Einschränkungen hinsichtlich der kontinuierlichen Untersuchung veränderlicher Substanzen und lebender Materie

KWS: Es können nur sich sehr langsam verändernde Substanzen untersucht werden. Die Bearbeitung lebender Materie ist mit großem Aufwand verbunden und bisher nur in Einzelfällen gelungen.
LS : Die Untersuchung von lebender Materie ist möglich, von veränderlichen Substanzen nur in gewissen Grenzen.
EM : Die Abbildung von Veränderungen ist sehr schwierig, ebenso die von lebenden Präparaten.

7. Ungewollte Veränderungen am Objekt während der Untersuchung

KWS: Durch Strahlungsschädigung.
LS : Soweit die Substanzen lichtempfindlich sind.
EM : Bei der Präparation und während der Aufnahme infolge des Elektronenbeschusses.

In Tabelle 1 ist noch einmal in Stichworten eine Übersicht gegeben. Der Hauptnachteil der KWS-Methode ist die lange Registrierzeit. Diese lange Registrierzeit fällt bei der Untersuchung biologischer Objekte besonders stark ins Gewicht, weil das Präparat auf ungefähr 1 mm^3 Rauminhalt eine Dosis in der Größenordnung von etwa 100 r/sec. erhält und infolgedessen hoher Gefahr der Strahlungsschädigung ausgesetzt ist.

Folgende Ausführungen befassen sich mit den Möglichkeiten zur Abschwächung dieser Nachteile, wie sie u.a. auch vom Verfasser verfolgt werden. Allerdings kann hier in Kürze nur ein sehr allgemein gehaltener und summarischer Bericht gegeben werden.

Eine Verkürzung der Belichtungszeit ist durch verschiedene Vorkehrungen möglich.

1. Durch Verwenden von leistungsfähigen Strahlungsquellen wie Drehanodenröhren und Feinstfokusröhren.
2. Durch Einsatz empfindlicher Empfänger wie Zählrohr, Szintillationszähler und Halbleiter.
3. Durch Verwendung optimaler Präparatabmessungen.
4. Durch bessere Ausnutzung der verfügbaren Strahlungsleistung mit Hilfe optimaler Dimensionierung der Kollimatoren und Ausnutzung röntgenoptischer Hilfsmittel.

Bei eigenen, den Punkt 4 betreffenden Untersuchungen konnte unter Ausnutzung der Totalreflexion von Röntgenstrahlen eine Verkürzung der Belichtungszeit auf etwa 15 % erreicht werden. Weil hierüber bereits berichtet wurde [9)] und [10)], wird hier auf eine eingehende Darstellung dieses Prinzips zu Gunsten einer Abschätzung der Strahlungsbelastung des Objektes verzichtet.

Von den oben angeführten Maßnahmen können nur die unter 2. bis 4. aufgeführten zu einer Verringerung der Strahlenbelastung des Präparats führen. Steht Substanz in genügender Menge zur Verfügung, so kann durch Bewegung des Objektes bei festen Stoffen oder durch Strömung bei flüssigen Präparaten eine zusätzliche Herabsetzung der Strahlenbelastung erreicht werden.

Besonders kleine Strahlungsbelastungen des Präparates sind bei Anordnungen zu erwarten, die eine Fokussierung der Strahlung auf dem Empfänger mit Hilfe gebogener Kristalle

Tabelle 1 : Vergleich der drei Untersuchungsmethoden in Stichworten

	Röntgen-Kleinwinkelstreuung	Lichtstreuung	Elektronenmikroskopie
Geräteaufwand	mäßig	klein	groß
Spezialkräfte	Auswertung	(Auswertung)	Präparation Aufnahme
Untersuchbare Substanzen	alle mit Elektronendichtedifferenzen	schwach absorbierende mit Dichteunterschieden	soweit präparierbar
Substanzmenge	klein	klein	sehr klein
Untersuchung von Einzelobjekten	nein	nein	ja
Messung statistischer Werte	ja	ja	(ja)
Vorbereitungszeit	kurz	kurz	lang
Aufnahmezeit	lang bis sehr lang	kurz bis lang	kurz
Auswertezeit	lang bis sehr lang	lang bis sehr lang	kurz bis sehr lang bei Statistik
Gesamtzeit	lang bis sehr lang	lang	kurz bis sehr lang
Untersuchung veränderlicher Substanzen	sehr langsame Veränderungen	langsame Veränderungen	äußerst schwierig
Untersuchung lebender Substanzen	schwierig	möglich	(nicht möglich)
Schädigung der Substanz bei der Untersuchung durch	Röntgen-Quanten	Lichtquanten	Präparation und Elektronenbeschuß

bewirken. Leider sind jedoch dabei wegen des kleinen Reflexionsvermögens der Kristalle lange Belichtungszeiten erforderlich, sodaß sich Einschränkungen bei der Untersuchung veränderlicher Substanzen oder lebender Objekte (Gesichtspunkte 6) ergeben. Vergleicht man dagegen z.B. die bekannte Blendenanordnung nach HOSEMANN [11] mit Anordnungen, die die Totalreflexion ausnutzen [10], so erreicht man bei gleicher Bestrahlungsstärke des Präparates kürzere Registrierzeiten, weil wegen des kleineren gestörten Bereiches bei Ausnutzung der Totalreflexion der Abstand zwischen Präparat und Film verkleinert werden kann. Bei gleichbleibender Signalamplitude am Empfänger und Bestrahlungsstärke des

Präparates wird also die Registrierzeit und damit die Strahlungsbelastung der Probe verkleinert. Eine grobe Abschätzung ergibt, je nach Größe des kleinsten vermeßbaren Winkels, bei Rundblenden eine 4 bis 8 mal kleinere, bei Spaltblenden eine bis 4 mal kleinere Strahlungsbelastung bei Verwendung von Totalreflexionskollimatoren. Eingehendere Untersuchungen hierüber sind noch im Gange.

L I T E R A T U R

1) BERNAL,J.D., Proc.Roy.Soc.Ser.A 170, 75 (1939)
2) CROWFOOT,D., Nature 135, 591 (1935)
3) CROWFOOT,D., D.RILEY, Nature [London] 141, (1938)
4) BOYES-WATSON,J., M.PERUTZ, Nature [London] 151, 714 (1943)
5) RAMAN,C.V., P.KRISHNAMURTI, Nature [London] 124, 53 (1929)
6) KRISHNAMURTI,P., Indian J. Physics Proc. Indian Assoc.Cultivat.Sci.5, 473 (1930)
7) GUINIER,A., G.FOURNET, Small-Angle Scattering of x-Rays, New York, London 1955
8) HOSEMANN,R., Ergebn.exakt.Naturwiss. 24, 142 (1951)
9) UNANGST,D., G.DAMASCHUN, Optik und Spektroskopie aller Wellenlängen, Tagungsbericht 1960, Akad.-V., Berlin 1962
10) UNANGST,D., W.SIEGEL, Exp.Techn.Physik. 11, 277 (1963)
11) HOSEMANN,R., Z.Physik. 114, 133 (1939)

D I S K U S S I O N S B E M E R K U N G E N

DOSE : Wieviel eV Strahlenenergie werden durchschnittlich pro Molekül umgesetzt ? Durch Vergleich dieses Wertes mit den entsprechenden Ergebnissen der Strahlenchemie der Makromoleküle ließe sich dann sagen, ob die Makromoleküle während der Messung wesentlich verändert werden. Der entsprechende Schwellwert für Proteine läge z.B. zwischen 20 und 100 eV/Molekül.

UNANGST : Nimmt man eine mittlere Dichte von 1 gcm^{-3} und ein Molekulargewicht des Makromoleküls von 10^5 an, so ergeben sich bei einer Registrierzeit von 10^4 sec. (etwa 3 Std.), die bei Verwendung von Zählrohren als Empfänger erreicht werden kann, einen Wert von etwa 10 eV/Molekül für die absorbierte Energie. Benutzt man Film als Empfänger, was sich bei Untersuchung diskontinuierlicher KWS oder von Anisotropieeffekten meistens nicht vermeiden läßt, so erhält man Werte, die über 100 eV/Molekül liegen dürften.

Allerdings muß beachtet werden, daß eine Strahlenschädigung der Probe sich nicht unbedingt in einer Veränderung der Streukurve äußern muß, ein Sachverhalt, der zwar für Röntgenuntersuchungen von Nutzen sein kann, wenn das Präparat nicht danach noch mit anderen Methoden, die empfindlich für Veränderungen durch Strahlenschädigung sind, untersucht wird, der aber auch leicht zu Fehlschlüssen Anlaß geben kann.

A Allgemeine Grundlagen und Methoden

Use of Electronic Computers for Quantitative Infra-Red Analysis of Multicomponent Mixtures

L. N. DROZDOV-TIKHOMIROV

Kurchatov's Institute of Atomic Energy, Moscow

S U M M A R Y

The infrared spectrophotometric quantitative analysis of steroid hormons mixtures (number of components 2-7) and of RNA/DNA mixtures with the use of the electronic computer was performed.

Formulas are derived for calculation of the analysis mistakes due to random and systematic errors of the infrared absorbtion spectra registration.

A number of problems arising in chemical and biochemical investigations involve quantitative analysis of more or less complex mixtures.

Using the method of spectrophotometry one can in principal carry out such an analysis rather quickly and without any preliminary fractionation of the mixture.

The difficulties hampering any wide use of this method are due to the cumbersome calculating work which is necessary for the working out and the solution of the corresponding system of equations. Indeed the LAMBERT-BEER's law

$$ln \frac{I(\nu_m)}{I_0(\nu_m)} = -\sum_{i=1}^{n} k_i(\nu_m) c_i l \qquad (1)$$

if applicable, enables one to work out a system of equations which serves to determine the mixture components concentrations, c_i. The absorption spectra of the components, $k_i(\nu_m)$, and that of the mixture $ln[I(\nu_m)/I_0(\nu_m)]$ are known.

The calculation difficulties are not essential at present when electronic computers become more and more available. There are some data about the use of electronic computers in such problems [1)-4)]. The first work in this field was apparent M. ROGOFF's work (USA) [1)].

We have made an attachment to the standard infrared spectrometer UR-10 which gives the possibility to record spectra in digital code which is convenient for immediate use in the electronic computer. Using this installation and a computer we have made analysis of steroid hormone mixtures with the numbers of components from 2 to 7. To obtain infrared absorption spectra in this case the chloroform solutions of steroid hormones were used and spectra were recorded in the region 900 - 1188 cm^{-1}. After testing the method in this simplier case, the attempt has been made to perform quantitat. ı analysis of mixtures of DNA and RNA. The method of pressed KBr tablets was used to obtain the absorption spectra in this latter case and spectra were recorded in the region 700 - 1800 cm^{-1}. At the same time we have considered theoretically the errors of such a kind of the analysis. Such errors will appear because of the existence of random errors when spectrum is registered and because of the existence of systematic errors, connected with the apparatus function of the spectrometer.

The investigation of the influence of random errors of spectra recording (GAUSS distribution is supposed) on the result of quantitative analysis which is carried out with the use of minimum square deviation method shows that mean root square error of the determination of jth-component concentration may be written as

$$\overline{(\Delta\lambda_j)^2} = \frac{r^2}{(\det\|a_{ij}\|)^2}\sum_{m=1}^{N}\left[\left[\sum_{i=1}^{n}\left(\frac{\lambda_i}{S_{im}}\right)^2+\left(\frac{1}{S_m}\right)^2\right]\left[\sum_{i=1}^{n}A_{ij}\ln S_{im}\right]^2\right], \quad (2)$$

here $\lambda_j = \frac{C_j}{C_j^0}$ is the ratio of the concentration of the jth-component of the mixture to its concentration in j^{th}-standard (reference) sample (j^{th}-standard sample is the solution of pure j^{th}-component in the exactly defined concentration).

S_{im} - is the transmission of the standard sample of the ith-component in the point of spectrum ν_m,

N is the number of points in the spectrum used for the analysis,

S_m - is the mixture transmission in the spectrum point ν_m,

r - is the mean root square error of measurement of transmission (it is the parameter of apparatus and of the registration conditions)

A_{ij} - is algebraic complement of matrix $\|a_{ij}\|$, $a_{ij} = \sum_m \ln S_{im} \ln S_{jm}$

n - is the number of components in the mixture.

The expression (2) may be used for the calculation of mean root square errors of concentration determination $\sqrt{\overline{(\Delta\lambda_j)^2}}$

When N ≫ I the expression (2) may be written as

$$\overline{(\Delta\lambda_j)^2} = r^2\frac{Q_j}{N} \quad (3)$$

here Q_j does not depend on N and is a parameter characteristic for the mixture. The expression (3) shows that, all other conditions beeing equal, mean root square error of concentration determination is directly proportional to the mean root square error of measurement of transmission, r, and it is inversely proportional to the square root from the number of points in the spectrum, used for the analysis, N.

The investigation of the influence of systematic errors caused by the apparatus function of the spectrometer on the analysis results has brought the following system of equations for the calculation of systematic errors of concentration determination

$$\sum_{j=1}^{n} a_{ij}\,\Delta\lambda_j = b_i \quad (4)$$

here

$$a_{ij} = \sum_{m=1}^{N}\ln S_{im}\ln S_{jm}$$

and

$$b_i = \sum_{m=1}^{N}\ln S_{im}\left[\frac{\sigma_e^2}{6S_m}\frac{\partial^2 S_m}{\partial\nu^2} - \sum_{j=1}^{N}\lambda_j\frac{\sigma_e^2}{6S_{jm}}\frac{\partial^2 S_{jm}}{\partial\nu^2}\right]$$

σ_e is the efficient spectral width of the spectrometer slit. The system of equations (4) was worked out using the assumption that apparatus function of the spectrometer is triangular and may be written as

$$\varphi(\nu,\nu_1) = \begin{cases}\frac{1}{\sigma_e}\left(1-\frac{|\nu-\nu_1|}{\sigma_e}\right) & if\ |\nu-\nu_1|\leq\sigma_e\\ 0 & if\ |\nu-\nu_1|\geq\sigma_e\end{cases}$$

Such assumption is quite reasonable as it was shown by BRODERSEN [5]. Tables 1-6 show the results of analyses of different mixtures. The mixtures have been composed artificially and the component concentrations in them were exactly known.

Table 1

	I	III
λactual	0,557	0,443
λcalculated	0,555	0,444
$\Delta\lambda$calculated	± 0,012	± 0,008

Table 2

	V	VI
λactual	0,568	0,432
λcalculated	0,573	0,426
$\Delta\lambda$calculated	± 0,013	± 0,013

Table 3

	I	II	III	V	VI	VII	VIII
λ actual	0,813	0,000	0,000	0,000	0,000	0,000	0,000
λ calculated	0,815	0,019	0,020	- 0,058	0,016	0,058	- 0,032
$\Delta\lambda$ calculated	± 0,030	± 0,010	± 0,012	± 0,014	± 0,014	± 0,043	± 0,021

Table 4

	I	II	III	V	VI	VII	VIII
λactual	0,193	0,211	0,233	0,019	0,479	0,000	0,000
λcalculated	0,206	0,203	0,280	0,137	0,502	0,015	0,019
$\Delta\lambda$calculated	± 0,015	± 0,009	± 0,021	± 0,023	± 0,018	± 0,014	± 0,012

Table 5

	I	II	III	IV	V	VI
λ actual	0,063	0,167	0,310	0,147	0,139	0,173
λcalculated	0,055	0,176	0,307	0,136	0,164	0,161
$\Delta\lambda$calculated	± 0,016	± 0,014	± 0,015	± 0,017	± 0,021	± 0,017

Table 6

	DNA	RNA	DNA	RNA	DNA	RNA
λ actual	0,31	0,69	0,17	0,83	0,082	0,92
λ calculated	0,29	0,77	0,20	0,92	0,12	0,84
$\Delta\lambda$calculated	± 0,01	± 0,02	± 0,01	± 0,03	± 0,005	± 0,03

Thus the results of the calculation can be compared with the actual compositions of the mixtures. The tables show also the values of mean root square errors. Parameter r involved in equation (2) was determined for the conditions in which the spectrum was obtained and was found to be equal to ~0,005.

The calculated values of systematic error are not given because for the considered cases they are less than 0,002.

Roman numerals in the tables mean the following combinations:

I	-	androstendion
II	-	androstadiendion
III	-	methyltestosterine
IV	-	dianobol
V	-	cortison acetate
VI	-	Reichstein substance "S"-acetate
VII	-	cholesterine
VIII	-	lanosterine

The results of the analysis of androstendion (I) and methyltestosterine (III) mixture (Table 1) show that the calculated values of λ differ from the actual values by not more than 0,002, the value of calculated mean root square error being $\pm$ 0,012 and $\pm$ 0,008 for the first (I) and for the third (III) component respectively. Good results have been obtained also analysing two-component mixture of cortison acetate (V) and Reichstein substance "S"-acetate (VI) (Table 2).

The calculated mean root square error being equal to $\sim$ 0,013, the deviation of given and calculated values λ do not surpass 0,006.

Table 3 shows the result of the analysis of diluted solution of androstendion (I), the concentration of the rest six components being equal to zero.

One can see that the calculated values of mean root square errors correspond to the actual errors of analysis well enough. The exception is the result obtained for the component V. One can see similar picture on tables 4 and 5. Here the calculated concentrations also coincide with the actual ones in the limits of calculated mean root square error, which does not exceed in general 0,02 and also, as in the previous case, there are some exceptions (components III, V, on Table 4). The actual deviations in these cases are 2 - 2,5 times the calculated values of mean root square error. It is obvious that the possibility of such random deviation is very small and these cases have to be explained.

It seems that in this case the error might be caused by two reasons: firstly, by the non-ideality of the 100 % line of spectrometer, secondly, by the insufficiently precise fixation of frequencies, at which the measurement of transmission was carried out.

At present the check-up of these suppositions is carried out.

Table 6 shows the results of the analysis of DNA and RNA mixtures.

As one can see from the table the calculated values of the concentrations in the DNA and RNA mixtures deviate from the actual ones by the value 0,02 - 0,03 for DNA and 0,08 - 0,09 for RNA. The calculated values of mean root square errors for these cases are equal to 0,005 - 0,01 for DNA and 0,02 - 0,03 for RNA.

Great deviations of calculated concentrations from the actual ones may apparently be explained by the shortcomings of the method used when preparing KBr tablets.

At present work is carring out on the removal of these shortcomings and on the extending of accuracy of the method up to the calculated one.

L I T E R A T U R E

1) ROGOFF,M., Ann. N.Y. Acad. Sci. 69, 27 (1957)
2) STERNBERG, STILLO, SCHWENDEMAN, Analytic. Chem. 32, 84 (1960)
3) REID,J., Biochem. Biophysic. Res. Commun. 3, 337 (1961)
4) WASCHLENKO,S.K., KASCHSOLOWA,S.G., KNORRE,D.G., Biochimija 27, 142 (1962)
5) BRODERSEN,S., J. opt. Soc. America 43, 877 (1953).

A Allgemeine Grundlagen und Methoden

Tieftemperaturbeschattung als elektronenmikroskopische Präparationsmethode

M. THÜRKAUF, A. RAMEL, W. KUHN und J. RICHTER

Physikalisch-Chemisches Institut der Universität, Basel

S U M M A R Y

A method to shadow-cast in the temperature range of -160° C to -170° C is described. This means the temperature of the electronmicroscopic object during the shadow-casting process. It is the opinion to get electron-micrographs of the three dimensional shape of randomly coiled polymeric chains.

An unserem Institut interessieren wir uns für die Abbildung der räumlichen Struktur statistischer Knäuel in Lösung, z. B. der DNA-Knäuel. DNA-Fäden haben wegen ihrer Doppelhelix-Struktur eine relativ große Steifheit. Um diese Strukturen für elektronenmikroskopische Untersuchungen zu präparieren, d. h. um den Fäden den erforderlichen optischen Kontrast und die nötige mechanische Stabilität zu geben, scheint sich die Gefriertrocknung mit unmittelbar anschließender Kohle-Bedampfung am besten zu eignen. Beide Prozesse werden bei sehr tiefer Temperatur am Objekt durchgeführt, die Gefriertrocknung bei ca. -80° C, das Bedampfen mit Kohlenstoff etwas oberhalb der Temperatur von flüssiger Luft. Die Apparatur ist so gebaut, daß während des Beschattungsprozesses die Strahlungstemperatur des Raumes, in welchem sich die Objekte befinden, -160° C bis -170° C beträgt. Die Objekte können während des Beschattungsvorganges keine höhere Temperatur annehmen.

Bild 1 zeigt ein Schema der Anlage. Die Präparation umfaßt die folgenden Stufen:

1. Die mit Trägerfolie bedeckten Netze werden auf einem Schlitten (B) in den Tiefgefriertunnel (C) eingefahren. Die Kammer (D) wird auf 10^{-5} mm Hg ausgepumpt.

2. Durch Einfüllen von flüssigem Stickstoff wird das Gefäß (A) abgekühlt. Das Thermoelement (G) zeigt eine Temperatur von -185° C an.

3. Die Kammer wird nun mit trockenem Stickstoff geflutet. Die vorbereiteten Trägernetze werden durch die Öffnung (J) mit dem zu untersuchenden Material beladen, was mit Hilfe eines kleinen Zerstäubers erfolgen kann. Die entstandenen Lösungströpfchen werden augenblicklich vereist, da die Trägerfilme eine sehr niedrige Temperatur haben.

4. Die Kammer wird wieder evakuiert. Während des sich nun anschließenden Gefriertrocknens bleibt das Präparat im linken Teil (H) des Tunnels. Die Temperatur des Trocknungstunnels kann mittels eines kleinen Heizers (M) beliebig zwischen 0° C und -150° C geregelt werden, da der Wärmekontakt mit dem Kühlgefäß (A) bei (K) unterbrochen ist und nur über die Leitungsbrücke (L) verläuft.

5. Das gefriergetrocknete Präparat kann nun bedampft werden. Dazu braucht weder die Temperatur erhöht, noch das Vakuum unterbrochen zu werden. Der Trägerschlitten wird in den Tieftemperaturtunnel (B) gebracht. Von einer Kohlenstoff-Dampfquelle (N) wird durch einen Kanal (O) beschattet. (O) sorgt für die Beschränkung der aufgenommenen Strahlungsenergie.

Ein erster Test der Methode wurde mit künstlich erzeugten Nebeltröpfchen durchgeführt. Die Reihenfolge von Gefriertrocknen und Bedampfen muß dabei geändert werden. Bild 2 zeigt Kohlenstoff-Platin-Hüllen solcher Tröpfchen. Es wurde unter einem Winkel von 60° beschattet. Die meisten Partikel, mit Ausnahme der ganz großen, sind getreu nachgebildet. Die kleinsten Tröpfchen haben einen Durchmesser von ungefähr 100 Å.

Bild 3 stammt von einem Agglomerat von ca. 100 DNA-Fäden. Ein gestreckter DNA-Faden vom Molekulargewicht 5 mal 10^6, wie es die Präparatoren für unsere Substanz angegeben haben, wäre etwa fünf mal so lang wie der Durchmesser des Agglomerates. Die Torusform des Gebildes scheint eine Folge des Zerstäubens zu sein. weshalb sich diese Art der Fixierung für DNA-Suspensionen schlecht eignet. Neben die eine Schwierigkeit, eine Lösung herzustellen, in der mit guter Wahrscheinlichkeit einzelne, möglichst lange DNA-Fäden vorhanden sind, tritt eine zweite in der nächsten Phase der Präparation: das Knäuel kollabiert, und zwar ruckweise (Zeichnung Bild 4), vermutlich unter dem Einfluß der Bedampfung. Deshalb sollte die schlecht kontrollierbare Methode der Kohlebogen-Verdampfung nach BRADLEY durch eine besser regulierbare ersetzt werden (direkte elektrische Beheizung, Elektronenbeschuß).

Interessant ist noch die Feststellung, daß in einzelnen äußeren Fäden des Knäuels eine "Seele" zu sehen ist, deren Durchmesser ca. 20 Å beträgt, gleich groß wie derjenige der DNA-Fäden. Daher glauben wir annehmen zu dürfen, daß es sich hier um kohleumhüllte DNA-Stränge handelt.

Mit den Versuchen an der DNA wurden die Schwierigkeiten der auszuarbeitenden Methode offenbar. Untersuchungen an mechanisch stabileren Proteinen sollen weitere Anhaltspunkte zur Verbesserung des Verfahrens liefern.

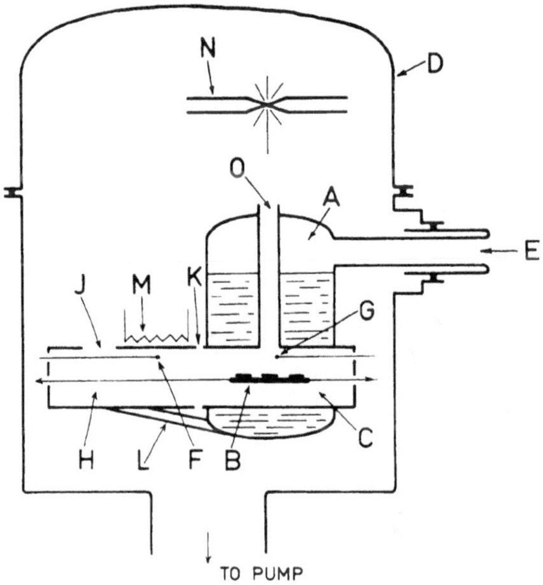

Abb. 1 Schema der Tieftemperatur-Beschattungsanlage

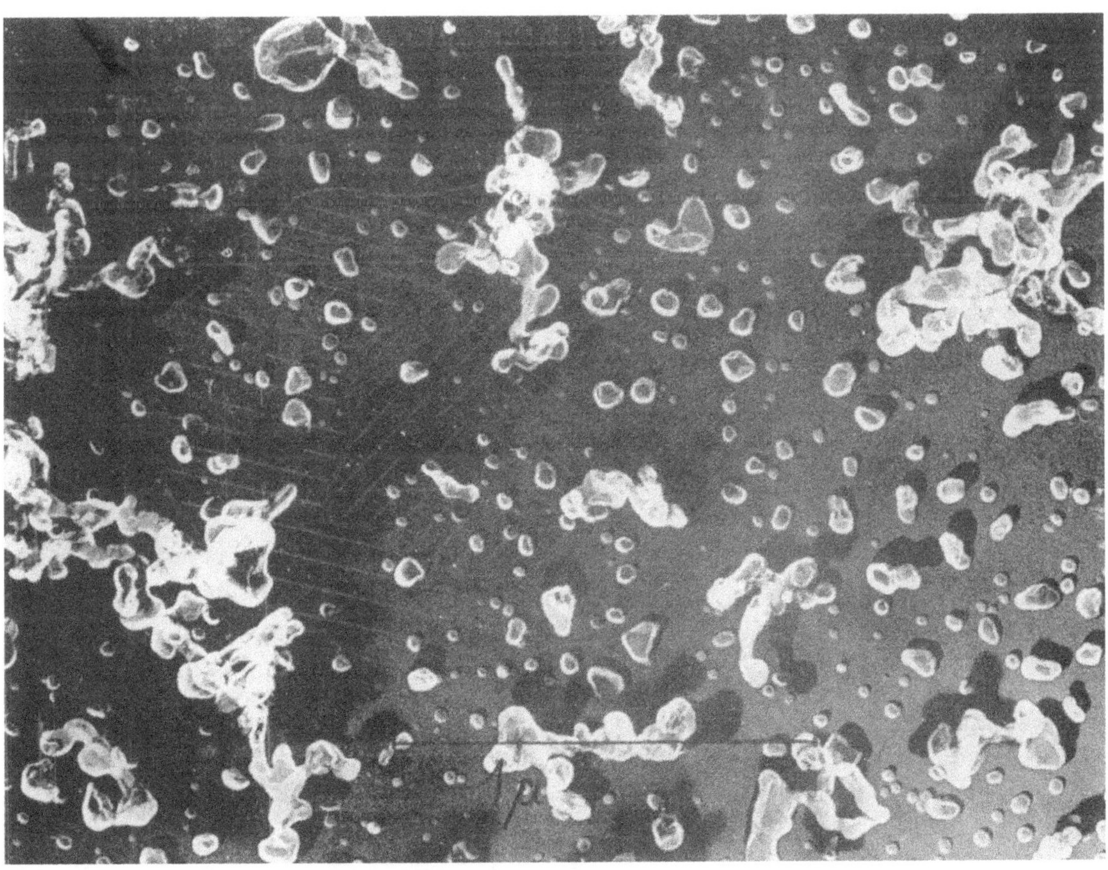

Abb. 2 Aerosol-Nebeltröpfchen, Pf-C-beschattet bei —170 °C.

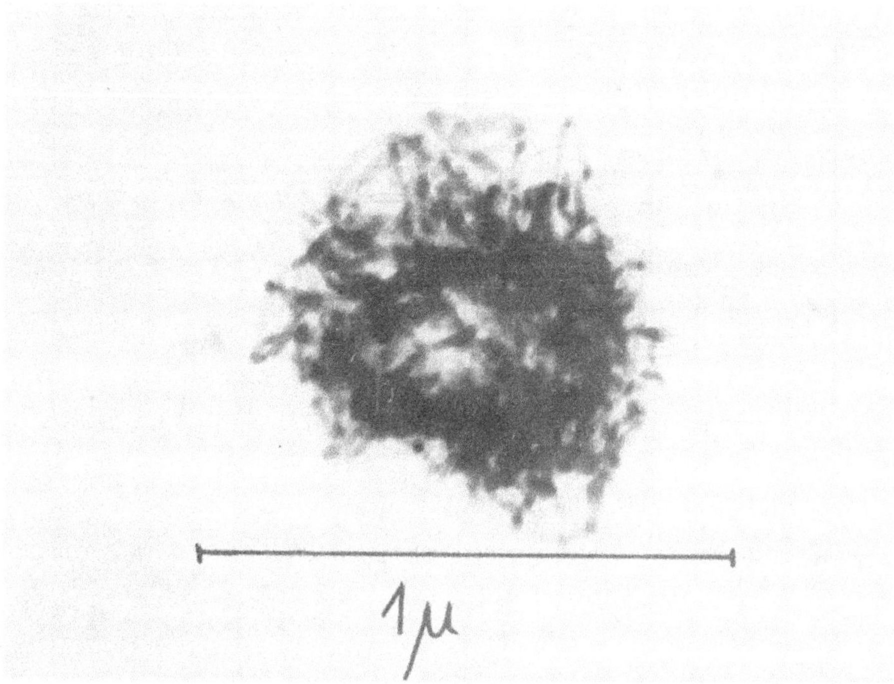

Abb. 3 DNA-Agglomerat, kollabiert, mit Kohlenstoff bedampft.

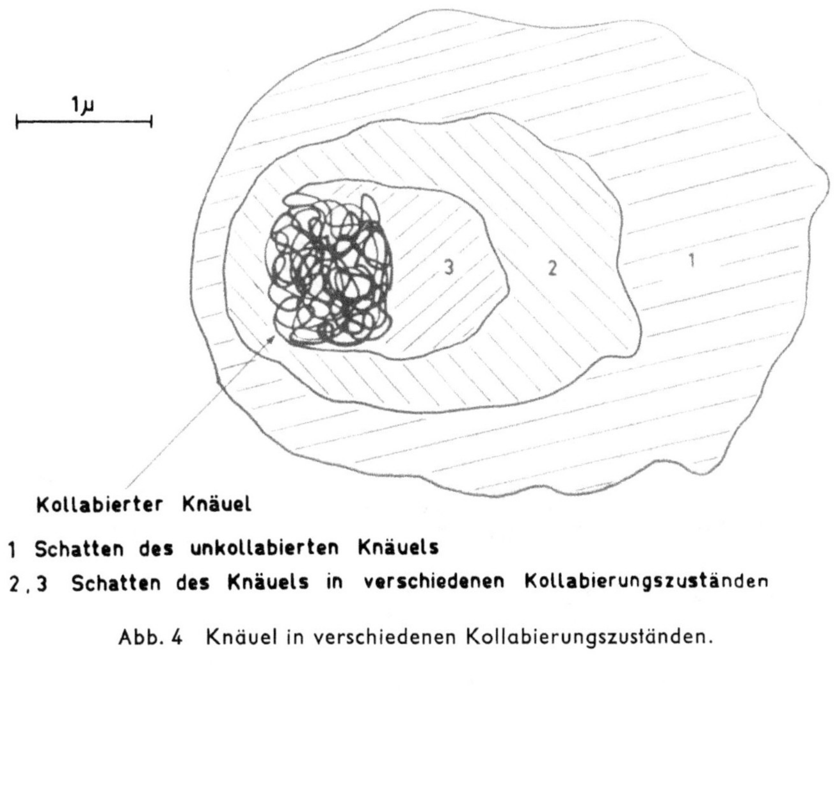

Abb. 4 Knäuel in verschiedenen Kollabierungszuständen.

A Allgemeine Grundlagen und Methoden

Über die elektrophoretische Trennung der Denaturierungsprodukte des Tropokollagens an Polyacrylamidgelen

G. REICH

Deutsches Lederinstitut, Freiberg/Sa.

S U M M A R Y

Till now the separation of components of denaturated tropocollagen was only possible by the use of the ultracentrifuge or chromatography on columns and succeeded now in a simple manner by electrophoresis on gels of polyacrylamide. The experimental foundations of this method and the results obtained in this separation of denaturated tropocollagen are described.

Tropokollagen (TC), die lösliche Form des Kollagens, zerfällt bei der Denaturierung in wenigstens 5 Komponenten. Davon können 3 (α, ß und γ) aufwendig in der Ultrazentrifuge [1], 5(α_1, α_2, β_1, β_2 und γ) bei schlechter Reproduzierbarkeit [2] durch Säulenchromatographie an CM-Cellulose [3] bzw. 2 (α und ß) in unbefriedigender Reinheit durch fraktionierte Aussalzung [4] getrennt werden. Auf der Suche nach einer einfacheren und vor allem schnellen Methode wurden mehrere Varianten der Elektrophorese zur Auftrennung der Denaturierungsprodukte des TC eingehend untersucht. Die Papierelektrophorese scheidet wegen zu hoher Affinität der Verbindungen zum Papier aus, eine Feststellung, die auch von anderer Seite bestätigt wird [5]. Dünnschichtelektrophorese und freie Elektrophorese scheitern an der Gleichheit der Äquivalentgewichte der einzelnen Denaturierungsprodukte. Mit der auf einem Molekülsiebeffekt beruhenden Stärkeelektrophorese war eine Auftrennung angedeutet. Überlegene Ergebnisse wurden jedoch bei Verwendung von Polyacrylamidgelen erhalten [6]. Polyacrylamidgel als Trägermaterial für die Elektrophorese wurde in jüngster Zeit verschiedentlich zur Klärung biochemischer Probleme angewandt [7].

Die Einfachheit der Methode und ihre Leistungsfähigkeit auch auf dem Kollagengebiet gibt Veranlassung, einige weitere Erfahrungen hier ausführlicher mitzuteilen, da die Methode ganz zweifellos eine breite Anwendung verdient und erlangen wird.

Die Elektrophoreseanordnung (Abb.1) besteht aus einem Glasröhrchen von 60 mm Länge und 4 mm Innendurchmesser (a), in dem sich ein feinporiges (b) und ein grobporiges (c) Polyacrylamidgel befinden. Letzteres ist überschichtet mit der zu trennenden Probelösung (d). Das Röhrchen befindet sich vertikal zwischen zwei aus Hart-PVC gefertigten Elektrodengefäßten (e), die mit Pufferlösung (f) gefüllt sind. Das obere Gefäß besitzt in der Grundplatte eine Bohrung, in welcher das Glasrohr mittels Gummidichtung (g) befestigt wird. In das untere Gefäß wird das Rohr durch Bohrung (h) in der Deckplatte eingeführt. Als Elektroden werden Kohlestäbe (i) benutzt.

Mit einfachen Laborhilfsmitteln ist es in analoger Weise möglich, eine Apparatur für die gleichzeitige Trennung von 8 Proteinproben zusammenzubauen (Abb.2).

Der Trenneffekt bei der gewählten Anordnung, die der von L.ORNSTEIN [7] entspricht, beruht auf folgendem Vorgang (Abb.3). Die Eiweißlösung liegt bei Beginn der Elektrophorese (a) im isoelektrischen Zustand vor. In dem Maße, wie das Pufferkation in die Probe einwandert, erniedrigt es den pH-Wert der Probelösung und nimmt an seiner Front das Protein mit. Auf diese Weise wird es zu einer schmalen Bande konzentriert (b). Im grobporigen Gel erfolgt die Auftrennung nach Äquivalentgewicht (c), im kleinporigen Hauptgel dann weiter nach Molekulargewicht und -gestalt auf Grund des Siebeffektes (d).

Daraus ergeben sich einige allgemeine Forderungen. Die pH-Werte des kleinporigen Gels und des Elektrodenpuffers sollten annähernd gleich und genügend vom IEP des Proteins entfernt liegen. Ob dabei nach der sauren oder alkalischen Seite, ist im Prinzip gleichgültig, bestimmt lediglich die Wanderungsrichtung und muß für ein bestimmtes Protein zur Erzielung optimaler Trennung erprobt werden. Für Kollagen empfiehlt sich ein saurer Puffer. Die pH-Werte im grobporigen Gel und in der Probelösung sollten dem IEP des Proteins entsprechen. Diese Forderung gilt indessen für die Probelösung nicht so streng. So konnten gute Trenneffekte auch bei direkter Verwendung von Citratpufferlösungen des Kollagens erhalten werden.

Das Porenvolumen der Polyacrylamidgele wird mehr durch die Konzentration an Polyacrylamid, als durch seinen Vernetzungsgrad bestimmt. Als Vernetzungsgrad wird dabei das Molverhältnis Methylenbisacrylamid zu Acrylamid verstanden. Je konzentrierter das Gel, desto langsamer die Wanderung. Verdünnte Gele müssen stärker vernetzt werden als konzentrierte, damit sie stabil sind und einwandfrei gehandhabt werden können. Die zu trennende Proteinmenge liegt bei den gewählten Abmessungen im Bereich von 60 - 80 µg. Die Dichte der Probelösung wird mit Glukose so erhöht, daß keine Vermischung mit dem Elektrodenpuffer erfolgen kann. Dies ist einfacher als die ursprünglich vorgeschlagene Einpolymerisation der Probe in ein grobporiges Gel [7]. Die Auslösung der Polymerisation geschieht durch Photosensibilisierung mit Riboflavin als Energieüberträger bzw. mittels Ammonpersulfat. Der Austausch von Tetramethyläthylendiamin im Gelpuffer durch leichter zugängliche Basen ist für den Polymerisationsablauf nachteilig (Tabelle 1).

Tabelle 1 : Charakteristische Daten der Polyacrylamidgelelektrophorese

	Volumen bei Röhrchengröße 4 . 60 mm	pH-Wert	Gel-konzen-tration	Vernetzungs-grad
Probelösung	0,15 ml	4-8		
Grobporiges Gel	0,15 ml	6,8	2,5 %	0,11
Feinporiges Gel	0,85 ml	4,3	7,5 %	0,012
Elektroden-pufferlösung	25 ml	4,5		

Im einzelnen gestaltete sich damit die Elektrophorese folgendermaßen :

Ein Glasröhrchen wird einseitig mit einem Gummistopfen verschlossen. Danach werden 0,85 ml der Mischung I eingefüllt, die sich aus 1 Teil Lösung 2, 2 Teilen Lösung 4, 1 Teil H_2O und 4 Teilen Lösung 7 zusammensetzt. Die Mischung wird mit etwas H_2O überschichtet, weil die Polymerisation nur unter Luftabschluß verläuft.

Nach ca. einstündigem Stehen ist die Lösung zu einem Gel erstarrt. Das überstehende Wasser wird entfernt und das Gel mit 0,15 ml der Mischung II überschichtet, die aus 1 Teil Lösung 3, 2 Teilen Lösung 5, 1 Teil Lösung 6 und 4 Teilen H_2O besteht. Wiederum wird mit H_2O von der Luft abgeschlossen.

Nach ca. 1/2 Stunde Polymerisation am Tageslicht ist auch diese Lösung zu einem Gel erstarrt. Der Verschlußstopfen wird entfernt und das Röhrchen in die beschriebenen Elektrodengefäße eingebracht. Schließlich werden 0,25 ml einer 0,5 %igen Eiweißlösung mit H_2O auf 2,25 ml verdünnt, mit einer Spatelspitze Glukose versetzt und 0,15 ml auf das obere Gel aufgetragen. Die Elektrodengefäße werden mit L ö s u n g 1 gefüllt, einem ß-Alaninpuffer vom pH-Wert 4,5, der 31,2 g ß-Alanin und 8 ml Eisessig pro Liter enthält.

Die übrigen Lösungen weisen folgende Zusammensetzung auf (alle Angaben pro 100 ml): L ö s u n g 2: 48 ml n KOH, 17,2 ml Eisessig, 4,0 ml Tetramethyläthylendiamin; L ö s u n g 3: 48 ml n KOH, 2,88 ml Eisessig, 0,46 ml Tetramethyläthylendiamin; L ö s u n g 4: 30 g Acrylamid, 0,8 g Methylenbisacrylamid; L ö s u n g 5: 10 g Acrylamid, 2,5 g Methylenbisacrylamid; L ö s u n g 6: 4 mg Riboflavin und L ö s u n g 7: 0,28 g Ammoniumpersulfat.

Bei einer angelegten Spannung von 100 V stellt sich eine Stromstärke von 5-7 mA pro Glasröhrchen ein, was zu einer Erwärmung auf etwa 30 - 35°C führt. Damit wird gleich die Renaturierung des TC verhindert. Nach 1 1/2 Stunde ist die Auftrennung beendet. Ist für spezielle Probleme diese Temperatur zu hoch, z.B. bei Untersuchung von nativem TC, so kann um die Glasröhrchen noch ein weiterer Trog gebaut werden, der mit Eiswasser versetzt ist. Dadurch gelingt es, die Temperatur im Glasröhrchen auf unter 10°C zu halten.

Von den benutzten Chemikalien sind Riboflavin (VEB Jenapharm), und ß-Alanin (VEB Berlinchemie) Handelsprodukte. Acrylamid wurde durch Verseifen von Acrylnitril (VEB Chemische Werke Buna) [8], N,N'Methylen-bis-acrylamid durch Umsetzung von Acrylamid mit Formaldehyd [9], Tetramethyläthylendiamin durch Umsetzung von Äthylendiamin mit Formaldehyd und Ameisensäure [10] hergestellt.

Zur Anfärbung wird das Gel nach Entfernen aus dem Glasröhrchen 1 Stunde mit einer 1 %igen Lösung von Amidoschwarz in 90 Teilen Methanol und 10 Teilen Eisessig behandelt. Das Auwaschen geschieht am besten über Nacht mit 7 %iger wässriger Essigsäure.

Zur Dokumentation der Versuchsergebnisse wird das so angefärbte Gel in ein schmales Reagenzglas überführt und der Zwischenraum zwischen Gel und Reagenzglaswand mit Wasser ausgefüllt. Das Gel wird im seitlich einfallenden Licht photographiert. Um reproduzierbare Aufnahmen zu erhalten, empfiehlt sich die Anwendung einer festen Anordnung aus Mikroskopierleuchte, Halterungsplatte mit weißem Karton als Reflektionswand und Photoapparat (Abb.4).

Die Anwendung der beschriebenen Methode auf Kollagenprobleme lieferte bisher folgende Resultate :

Natives Tropokollagen, gewonnen durch mehrfaches Umfällen rohen citratpufferextrahierten Prokollagens [11], liefert bei der Elektrophorese (unter Eiskühlung eine scharfe Bande. Nach Denaturierung in 1 n Essigsäure 60 Minuten bei 38°C treten drei Hauptbanden auf. Die am schnellsten wandernde liegt als Dublett vor, die am langsamsten wandernde ist von einer schmalen Vorbande begleitet. Bisweilen treten noch zwei weitere, sehr schmale Banden in der Nähe der am langsamsten wandernden Hauptbande auf. Insgesamt werden also maximal 7, mit Sicherheit reproduzierbar 5 Banden festgestellt.

Die vergrößerten Photographien der angefärbten Trennröhrchen lassen sich im Elektrophorese-Auswertegerät ERI 10 des VEB Carl Zeiß Jena im Auflicht photometrieren, jedoch läßt sich eine quantitative Bestimmung vorerst nicht darauf aufbauen, da das Auflösevermögen zu gering ist und noch nichts über die Proportionalität zwischen Eiweißmenge und Farbintensität bekannt ist. So gibt bei gleicher Proteinkonzentration natives TC eine viel weniger gefärbte Bande als im denaturierten Zustand.

Die drei Hauptbanden wurden mit a, b und c bezeichnet, das a-Dublett entsprechend a_1 und a_2, die Vorbande zur c_2-Hauptbande c_1 (Abb.5).

Besonders eingehende Untersuchungen wurden durchgeführt, um die elektrophoretisch festgestellten Denaturierungsprodukte den bereits bekannten α-, ß- und γ-Komponenten zuzuordnen. Dazu gibt es im Prinzip eine Reihe Möglichkeiten, die indessen keineswegs alle zum Ziele führten. Durch Extraktion mit Neutralsalzen gewonnenes TC besteht im Gegensatz zum citratpufferextrahierten TC vorwiegend aus α-Komponenten [12]. Tatsächlich ergab die elektrophoretische Trennung im wesentlichen nur eine Bande mit der Wanderungsgeschwindigkeit von a_1. Damit entspräche das a_1 - a_2-Dublett den α_1 - α_2-Komponenten, d.h., die Komponenten mit dem niedrigsten Molekulargewicht (125000) würden am schnellsten wandern. Dies stimmt mit den Erwartungen hinsichtlich der Siebwirkung des Gels überein. Daß jedoch auch die Molekülgestalt von Einfluß ist, geht daraus hervor, daß natives TC (Mg 350000) eine Wanderungsgeschwindigkeit aufweist, die etwa in der Mitte der von a- und b-Komponenten liegt (Abb.6a, b und c).

Die in genauer Nacharbeitung der Vorschrift zur fraktionierten Ammonsulfatfällung von denaturiertem TC [4] erhaltenen Fraktionen erwiesen sich elektrophoretisch nicht als einheitlich. Vor allem wurden zusätzlich schneller wandernde, also niedrigmolekulare sekundäre Abbauprodukte festgestellt. Daher wurde die Aussalztechnik modifiziert und in geringeren Abstufungen gefällt. Dabei gelang es, in der Reihenfolge der Fällung folgende Fraktionen zu erhalten : Niederschläge, bevorzugt bestehend aus b- und c-Banden, aus a-, b- und c-Banden gemeinsam, bevorzugt aus a-Banden. Dies legt nahe, daß es sich tatsächlich in der Reihenfolge c, b, a, entsprechend auch der Wanderungsgeschwindigkeit, um Verbindungen abnehmenden Molekulargewichtes handelt (Abb.7).

Die besonders naheliegende Untersuchung der einzelnen Fraktionen einer TC-Trennung an CM-Cellulose (CMC) brachte nicht den erwarteten Erfolg. Die Trennung ist in Abhängigkeit von der CMC-Qualität schlecht reproduzierbar und selbst, wenn Ergebnisse bei der photometrischen Auswertung erhalten wurden, die etwa den in der Literatur [3] beschriebenen Verhältnissen entsprechen, waren die einzelnen Fraktionen keineswegs elektrophoretisch einheitlich. Ähnlich verlief die Nacharbeitung der Befunde, daß durch Enzymbehandlung von TC ein großer Teil von ß- in α-Komponenten umgewandelt [13] bzw. durch Formaldehydeinwirkung γ-Komponenten in größerem Umfang gebildet würden [14].

Somit kann nach dem gegenwärtigen Stand der Untersuchungen gesagt werden, daß mit hoher Wahrscheinlichkeit die a-Banden den α-Komponenten entsprechen. b und c_2 dürften den ß-Komponenten, c_1 der γ-Komponente zuzuordnen sein. Die letztere Schlußfolgerung wird vor allem auf Grund der Intensität der Banden gezogen. Bei der Zuordnung von c_1 zu γ, die ja zunächst dem Molekülsiebeffekt widerspricht, wurde die erwähnte Feststellung mit in Betracht gezogen, daß natives TC rascher wandert als einige seiner Denaturierungsprodukte. Da auch die γ-Komponente noch aus drei Peptidketten besteht und eine ausgesprochene Tendenz zur Renaturierung besitzt [15], wäre ein nativem TC ähnliches Verhalten nicht ausgeschlossen.

Wenngleich diese Annahmen noch ihrer vollen Bestätigung bedürfen, kann die Elektrophorese bereits jetzt mit sehr gutem Erfolg für viele Probleme der Kollagenforschung eingesetzt werden. Daß sie ein wertvolles Hilfsmittel darstellt, um Trennvorgänge aller Art zu verfolgen und zu optimieren, liegt nach obigem auf der Hand.

So eignet sich die Elektrophorese an Polyacrylamidgel ausgezeichnet dazu, den Einfluß von Temperatur, Zeit und hydrotropen Substanzen auf den Abbau von TC zu verfolgen. Als Beispiel für eine Vielzahl diesbezüglicher Untersuchungen mag die elektrophoretische Auftrennung der beim Lösen von TC in H_2O bei 62°C erhaltenen Gelatine (parent gelatin) gelten. Reproduzierbar werden dabei wenigstens 10 Banden überwiegend höherer Wanderungsgeschwindigkeit erhalten (Abb.8). In diesem Zusammenhang sei lediglich darauf hingewiesen, daß sich auch Handelsgelatinen nach dem unterschiedlichen Ergebnis elektrophoretischer Auftrennung differenzieren lassen. So bestehen insbesondere deutliche Zusammenhänge zwischen Aufschlußverfahren, Ausschmelztemperatur sowie Alter des benutzten Kollagens und Anzahl und Wanderungsgeschwindigkeit der einzelnen Abbauprodukte [16].

Zusammenfassend kann festgestellt werden, daß die bereits auf andere biochemische Probleme angewandte Methode der Elektrophorese an Polyamidgelen dank ihrer Einfachheit, Schnelligkeit und guten Trennleistung auch auf dem Gebiete der Kollagenforschung, insbesondere zur Untersuchung der Denaturierungsprodukte und sonstigen Abbauprodukte des Tropokollagens erfolgreich benutzt werden kann.

Fräulein G.SCHOB ist für die verständnisvolle Mitarbeit, Herrn Professor Dr.F.STATHER für die Genehmigung zur Veröffentlichung zu danken.

L I T E R A T U R

1) DOTY,P., T.NISHIHARA, Recent Advances in Gelatin and Glue Research (G.STAINSBY ed.) Pergamon Press, London 1958, S. 92
GRASSMANN,W., K.HANNIG, I.ENGEL, Hoppe-Syler's Z.physiol.Chem. 324, 284 (1961)
Für eine Übersicht s. HANNIG,K., I.ENGEL, Leder 12, 213 (1961)
2) REICH,G., H.SIEBER, Unveröffentlicht
HÖRMANN,H., K.HANNIG, Persönliche Mitteilung
3) ROSEN,H., A.KESSLER, S.M.LEVENSON, Arch.Biochem.Biophysics 90, 167 (1960)
PIEZ,K.A., M.S.LEWIS, G.R.MARTIN, J.GROSS, Biochim.biophysica Acta Amsterdam 53, 596 (1961)
SCHLEYER,M., Hoppe-Seyler's Z.physiol.Chem. 329, 97 (1962)
PIEZ,K.A., E.A.EIGNER, M.S.LEWIS, Biochemistry 2, 58 (1963)
4) ORECHOWITSCH,V.N., V.O.SHPIKITER, Recent Advances in Gelatin and Glue Research (loc.cit. S. 87; ferner: Ber.Akad.Wiss.UdSSR 115,137 (1957)
5) DEYL,Zd., J.ROSMUS, Kozarství 12, 127 (1962); Kozarství 13, 35 (1963)
6) REICH,G., H.LEGUTKE, Z.Chem. im Druck
7) RAYMOND,S., L.S.WEINTRAUB, Science 130, 711 (1959)
DAVIS,B.J., L.ORNSTEIN, A New High Resolution Elektrophoresis Method, Vortrag vor The Society for the Study of Blood, New York Academy of Medicine, 24.3.1959
CHANG,L.O., A.M.SRB., F.C.STEWARD, Nature [London] 193, 756 (1962)
REISFELD,A., U.J.LEWIS, D.E.WILLIAMS, Nature [London] 195, 281 (1962)
RAYMOND,S., M.NAKAMICKI, B.AURELL, Nature [London] 195, 697 (1962)
RAYMOND,S., YI-IN WANG, Annal.Biochem. 1, 391 (1962)
ZELDIN,M.H., J.M.QUARD, Nature [London] 198, 389 (1963)
HJERTEN,S., J.Chromatog. 11, 66 (1963)
8) American Cyanamid Co. New York, Schweiz.P. 259 439 v. 8.11.1946 ref.C. 1950,I.917
9) MAGAT,E.E. u.a., J.Amer.chem.Soc. 73, 1028 (1951) nach HOUBEN-WEYL, Methoden der organischen Chemie, Georg-Thieme-Verlag Stuttgart 1963, Bd. 14, Teil 2, S. 393
10) MÖLLER,F., R.SCHRÖTER, in HOUBEN-WEYL loc.cit. Bd. 11, Teil 1, S. 650

11) ORECHOWITSCH,V.N., A.A.TUSTANOVSKY, K.D.ORECHOWITSCH, N.E.PLOTNIKOWA, Biochemie UdSSR 13, 55 (1948)
12) MAZUROW,V.J., V.N.ORECHOWITSCH, Biochemie UdSSR 25, 814 (1960)
13) NISHIHARA,T., T.MIYATA, Collagen-Symposium 3, 66 (1962)
vgl. aber :
HAFTER,R., H.HÖRMANN, Hoppe-Seyler's Z.physiol.Chem. 330, 169 (1963)
14) DRAKE,M.P., A.VEIS, Federat.Proc. 21, 229 (1962)
15) ALTGELD,K., A.J.HODGE, F.O.SCHMITT, Proc.nat.Acad.Sci. USA 47, 1914 (1961)
16) REICH,G., S.WALTHER, F.STATHER, In Vorbereitung

D I S K U S S I O N S B E M E R K U N G E N

DOSE : Ihre schönen Trennergebnisse haben mich sehr beeindruckt. Wir haben mit Sephadex gearbeitet und hiermit auch recht brauchbare Trennergebnisse erzielt. Der große Vorteil des Sephadex besteht darin, daß man nach der Trennung die Proteine fast quantitativ wieder eluieren kann. Ich möchte Sie fragen, ob sich die Proteine auch von den Polyacrylamiden wieder gut eluieren lassen.

REICH : Nach unseren bisherigen Ergebnissen beträgt die Eluierbarkeit wenigstens 80 % und kann als gut bezeichnet werden.

BERG : Wodurch unterscheiden sich α_1 und α_2 in der Primärstruktur ?

REICH : α_1 und α_2 unterscheiden sich in der Aminosäurezusammensetzung, wahrscheinlich auch in der Aminosäurereihenfolge, doch ist zu Letzterem noch nichts Näheres bekannt.

BERG : Erfolgt eine Renaturierung des Tropakollagens wie bei der DNA ?

REICH : Eine spontane Renaturierung erfolgt, insbesondere der γ -Komponente, die noch aus allen drei Strängen des ursprünglichen Moleküls besteht.

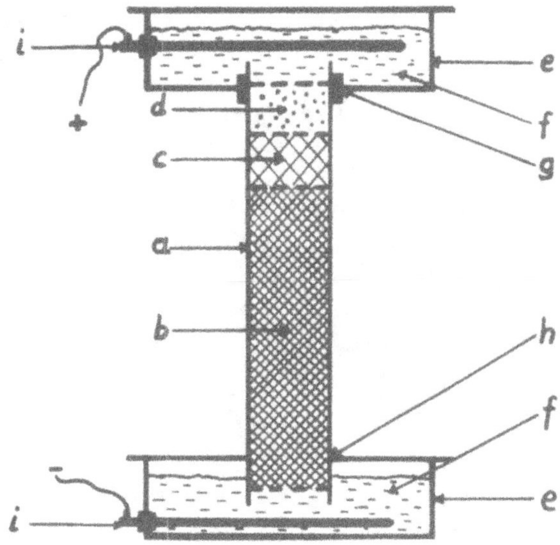

Abb.1 Prinzipieller Aufbau der elektrophoretischen Trennapparatur (a=Trennröhrchen, b = feinporiges Polyacrylamidgel, c=grobporiges Polyacrylamidgel, d = Proteinlsg., e=Elektrolytgefäße, f=Pufferlsg., g=Gummidichtung, h=Bohrung in der Deckplatte, i=Kohleelektroden).

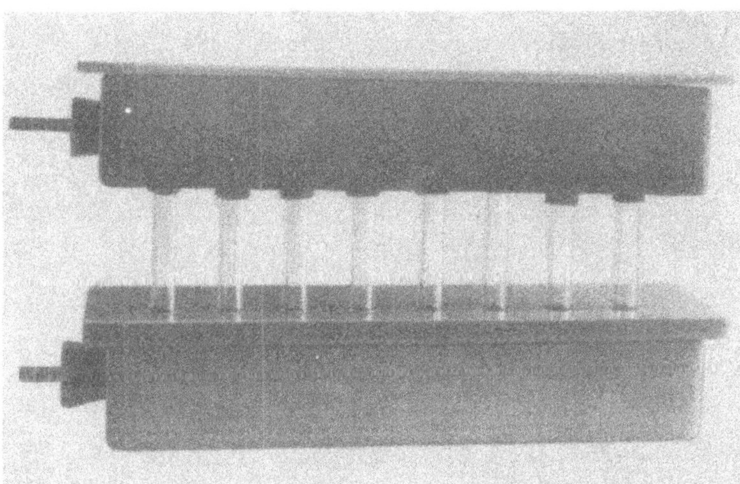

Abb.2 Trennapparatur für die Elektrophorese an Polyacrylamidgel zur gleichzeitigen Untersuchung von 8 Proben.

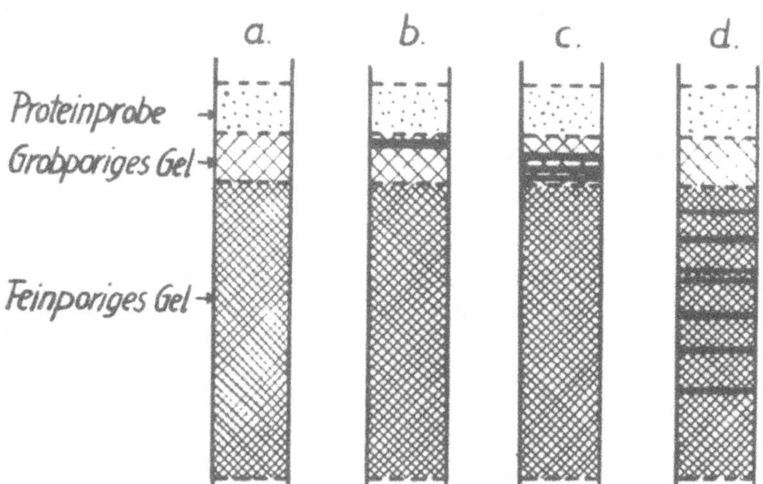

Abb.3 Schematische Darstellung des Verlaufs der Elektrophorese an Polyacrylamidgel (a=Beginn, b=Konzentrierung der Probe zu einer schmalen Bande durch pH-Gradienten, c=Trennung der Probe im grobporigen Gel nach Äquivalentgewichten, d=Trennung im feinporigen Gel nach Molekülgrösse und -gestalt).

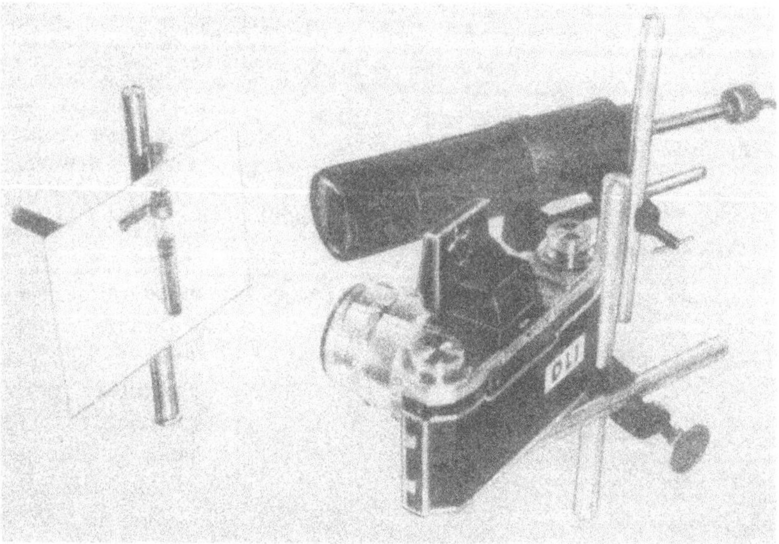

Abb.4 Einrichtung zur Dokumentation der Trennergebnisse, bestehend aus Halterung mit Reflektionswand, Mikroskopierleuchte und Fotoapparat.

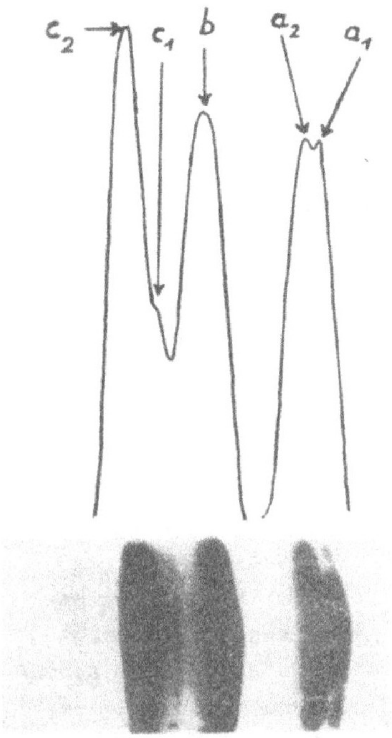

Abb.5 Elektrophoretische Auftrennung denaturierten Tropokollagens an Polyacrylamidgel, photometrische Auswertung und Nomenklatur der auftretenden Banden.

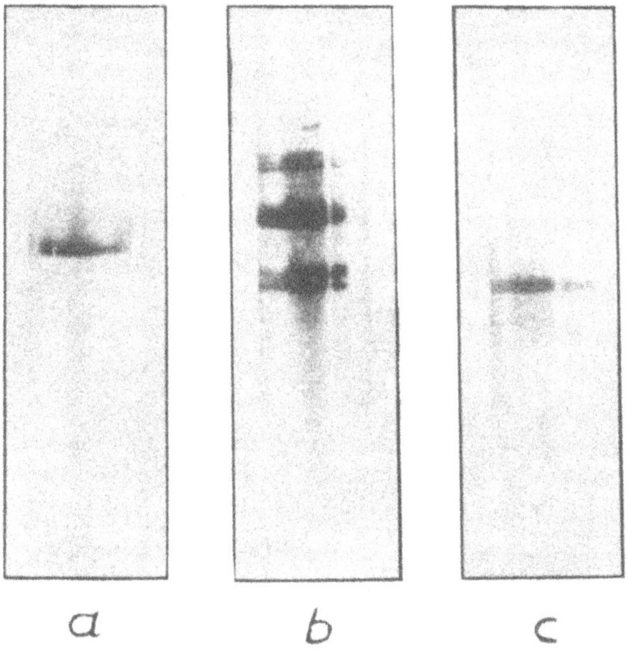

Abb.6 Wanderungsgeschwindigkeiten von nativem TC (a), denaturiertem citratpufferextrahiertem TC (b) und denaturiertem neutralsalzextrahiertem TC (c).

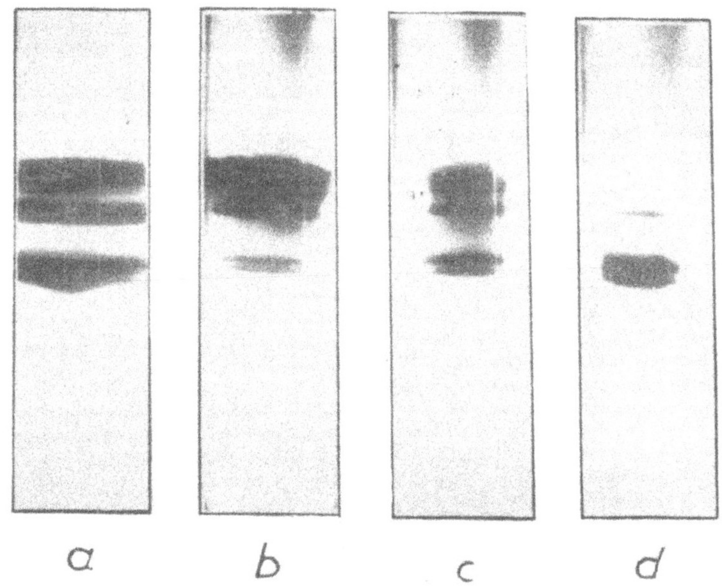

Abb.7 Elektrophoretische Trennung der durch fraktionierte Ammonsulfatfällung von denaturiertem TC erhaltenen Fraktionen (a= Originalprobe; b, c und d nach steigender Ammonsulfatzugabe erhaltene Fraktionen).

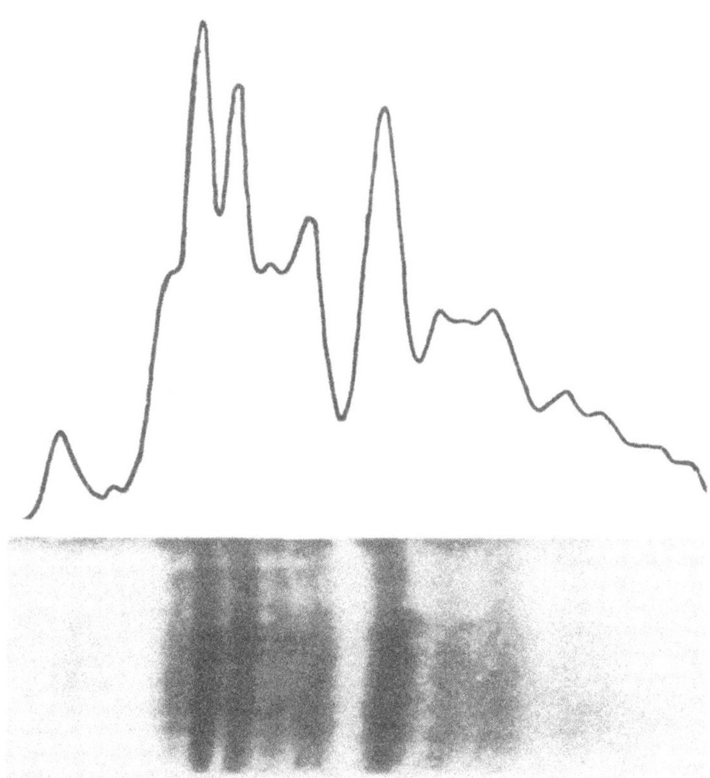

Abb.8 Elektrophoretische Auftrennung von Tropokollagen, gelöst in H_2O von 62°C (parent gelatin).

A Allgemeine Grundlagen und Methoden

Polarographic Studies of Ion Binding II. Thallous polystyrenesulphonate

S. LAPANJE, S. OMAN and C. MIKLAVC

Department of Chemistry of the University, Ljubljana

SUMMARY

The degree of ion binding in polyelectrolyte solutions was determined by means of polarographic analysis. The polyelectrolyte studied was thallous polystyrenesulphonate. Current-voltage curves of aqueous solutions were recorded without and with supporting electrolyte. In order to determine the influence of the dielectric constant on the degree of ion binding 50 % ethanol solutions of the polyelectrolyte were also studied. The extent of ion binding in both cases was obtained by assuming that only free, i.e., unbound ions contribute to the diffusion- and migration currents, respectively. For reasons of comparison, solutions of thallous p-ethylbenzenesulphonate were also investigated.

INTRODUCTION

In the first paper in this series [1] the results of the application of polarographic analysis to the study of ion binding in polyelectrolyte solutions were reported. The polyelectrolyte studied was cadmium polystyrenesulphonate. The values of the degree of ion binding obtained are in reasonable agreement with those given by another method, the measurement of ionic activity coefficients of cadmium ions in the same solutions [2]. For cadmium ions there no other data are available. However, comparing the values of the degree of ion binding for cadmium ions with the corresponding values for other counterions, obtained by different methods, we may see that they are somewhat lower. To obtain more data on the usefulness of polarographic method for the study of ion binding solutions of thallous polystyrenesulphonate were investigated. Thallous ions produce well defined waves and they display little tendency towards complex formation; in contradistinction to cadmium ions they are univalent. Furthermore nonaqueous solutions of the polyelectrolyte were studied. The solvent was a 50 % (by weight) ethanol mixture. Simultaneously, for reasons of comparison, solutions of thallous p-ethylbenzenesulphonate were studied.

E X P E R I M E N T A L

M a t e r i a l s

Thallous p-ethylbenzenesulphonate (TlEBS) was prepared from p-ethylbenzenesulphonic acid by passing it through a column containing the cation exchange resin, Dowex 50, in the thallous state. Thallous polystyrenesulphonate (TlPSS) was prepared from polystyrenesulphonic acid in the same way as the corresponding salt of the monomer. The acid was kindly supplied by DOW CHEMICAL COMPANY, Midland, Michigan. The molecular weight was approximately 10^5 and the degree of substitution 1,00.

A p p a r a t u s   a n d   p r o c e d u r e

Polarographic studies were made at 25 ± 0,5°C. A Radiometer PO 3 polarograph was used; the cells were of cylindrical shape with capillary side tubes. A pool of mercury on the bottom of the cells served as anode; electrical connection was made by means of an iron wire inserted into the side tube. Thus no salt bridge is necessary and the leakage of neutral electrolyte into solutions is eliminated. This is especially important when studying solutions without supporting electrolyte. Dissolved oxygen was removed by bubbling nitrogen or, at low concentrations, hydrogen through solutions. The capillary had the following characteristics : H = 40 cm, $m^{2/3}t^{1/6}$ = 2,69 $mg.^{2/3}sec.^{-1/2}$ at -0,8 V in a 0,1 N solution of potassium sulphate.

Current-voltage curves were first run of solutions of TlPSS and TlEBS without the supporting electrolyte. It was necessary to add a maximum suppressor to solutions of TlPSS only at concentrations above 2.10^{-3} N, but a suppressor was necessary at all concentrations for solutions of TlEBS. The supporting electrolyte was then added to solutions of TlPSS and TlEBS and polarograms were recorded. Potassium sulphate was used as the supporting electrolyte since thallous chloride has a relatively low solubility. The concentration of potassium sulphate was 50 times larger than that of the reducible salt, however, at concentrations of TlPSS and TlEBS below 2.10^{-3} N it was maintained constant at 1.10^{-1}N. Triton X 100 was added to all solutions of TlEBS and to the 6.10^{-3}N and more concentrated solutions of TlPSS.

Current-voltage curves were recorded of 1.10^{-3}N solutions of TlPSS and TlEBS, respectively, containing different amounts of potassium sulphate as well as of 1.10^{-3}N solutions of Tl_2SO_4 to which NaPSS was added.

The polarographic behaviour of both thallous salts in a 50 % ethanol mixture was studied in a similar way. Current-voltage curves were first run of solutions containing no supporting electrolyte. Triton X 100 was again used as a maximum suppressor for solutions of TlEBS, whereas to solutions of TlPSS it was not necessary to add it, though a slight maximum appeared at higher concentrations. Afterwards solutions with the supporting electrolyte were polarographed. Tetramethylammonium hydroxide was used as the supporting electrolyte and Triton X 100 as the maximum suppressor. However, no addition of suppressor to TlPSS was necessary.

R E S U L T S

The polarographic behaviour of aqueous solutions of TlEBS and TlPSS is similar to that of solutions of CdEBS and CdPSS, respectively. Therefore the relations derived in the preceeding paper [1] have been applied. However, examination of Table 1 also shows a major difference : The values of the diffusion current of thallous ions in solutions of TlPSS are studied at concentrations all about 15 % lower than the corresponding values for solutions of TlEBS. The value of the half-wave potential of thallous ions in solutions of TlPSS containing 1.10^{-1}N potassium sulphate as the supporting electrolyte is -0,52 vs. S.C.E. (sulphate agar bridge). All the results are given in Table 1.

Table 1 Diffusion current, I_d; total current in solutions containing no supporting electrolyte, I_t; diffusion current in solutions of TlPSS containing no supporting electrolyte, I_d^o (calculated from Eq. (5), Ref.[1]); degree of ion binding, 1 - f (from Eq. (4), Ref.[1]).

Concentration [eq./l.]	TlEBS I_d [μA]	TlEBS I_t [μA]	TlPSS I_d [μA]	TlPSS I_t [μA]	TlPSS I_d^o [μA]	1 - f
$1,09.10^{-2}$	87,6	183	70,5	34,4	17,2	0,810
$8,72.10^{-3}$	68,0	145	56,8	23,6	11,4	0,826
$6,54.10^{-3}$	52,5	108	42,5	18,8	9,31	0,823
$5,45.10^{-3}$	43,5	90,5	34,8	15,7	7,75	0,822
$4,36.10^{-3}$	34,6	71,9	28,4	12,0	6,25	0,819
$2,18.10^{-3}$	17,3	35,8	14,4	6,12	3,24	0,813
$1,09.10^{-3}$	8,14	16,4	7,20	3,32	1,76	0,784
$8,72.10^{-4}$	6,51	12,5	5,68	2,88	1,52	0,766
$6,54.10^{-4}$	4,70	9,05	4,10	2,21	1,17	0,751
$5,45.10^{-4}$	4,12	7,93	3,54	2,02	1,06	0,743
$4,36.10^{-4}$	3,31	6,25	2,76	1,67	0,89	0,732
$2,18.10^{-4}$	1,68	3,17	1,41	1,12	0,60	0,646
$1,09.10^{-4}$	0,82	1,58	0,69	0,56	0,30	0,634

The dependence of current on the concentration of the supporting electrolyte is seen to be the same as in the case of CdPSS and CdEBS, respectively, except for a slight maximum probably due to the change of diffusion coefficient with the concentration of the supporting electrolyte, cf. Fig.1.

The values of the current in 1.10^{-1}N solutions of Tl_2SO_4 to which NaPSS was added in gradually increasing concentrations are presented in Table 2.

Table 2 Total current, I_t, in 1.10^{-1}N solutions of Tl_2SO_4 with added NaPSS.

Concentration of NaPSS [eq./l.]	I_t [μA]
0	14,7
1.10^{-3}	6,08
5.10^{-3}	3,92
1.10^{-2}	3,86
5.10^{-2}	3,76

Solutions of TlEBS and TlPSS in the 50 % ethanol mixture offer a similar picture as the corresponding aqueous solutions. The values of the diffusion current generally are much lower than in aqueous solutions which is typical for alcohol solutions. In ethanol solutions the values of the diffusion current, at the same concentration, are app. 10 % lower in solutions of TlPSS. Owing to a poor reproducibility of current-voltage curves at low concentrations of TlPSS the 6.10^{-4}N solution was the last studied.

The results for ethanol solutions are assembled in Table 3.

Table 3 Diffusion current, I_d; total current in ethanol solutions containing no supporting electrolyte, I_t; diffusion current in ethanol solutions of TlPSS containing no supporting electrolyte, I_d^o (calculated from Eq. (5), Ref.[1]); degree of ion binding, 1 - f (from Eq. (4), Ref.[1]).

Concentration [eq./l.]	TlEBS I_d [μA]	TlEBS I_t [μA]	TlPSS I_d [μA]	TlPSS I_t [μA]	TlPSS I_d^o [μA]	1 - f
1.10^{-2}	27,4	87,6	24,3	13,8	4,55	0,834
8.10^{-3}	22,4	70,0	19,6	10,8	3,57	0,841
6.10^{-3}	17,1	51,1	15,2	7,74	2,72	0,842
4.10^{-3}	11,4	35,4	9,94	5,22	1,90	0,833
2.10^{-3}	5,69	16,5	4,69	2,62	0,98	0,826
1.10^{-3}	2,85	8,72	2,51	1,31	0,55	0,806
8.10^{-4}	2,32	6,52	2,03	1,07	0,45	0,805
6.10^{-4}	1,72	4,80	1,51	0,81	0,34	0,801

DISCUSSION

As has been mentioned before, for solutions of TlPSS the same relations are assumed to hold as for solutions of CdPSS[1]. Thus we can calculate the degree of ion binding if we know the transference numbers for thallous ions in solutions of TlPSS. Unfortunately no data are available and we again have to refer to solutions of ordinary electrolytes. In the present case however the procedure has been some different. The transference numbers for single concentrations of TlPSS have been determined from the ratio I_d/I_t for the corresponding solutions of TlEBS, i.e., solutions having the same I_t. In this way the existing situation has been approached as much as possible. The value for t_+ at the highest concentration is 0,52, decreases then and reaches the value 0,47 at the lowest concentration studied. This compares favorably with the directly measured values in solutions of Tl_2SO_4. The values of the diffusion current in solutions of TlPSS at zero concentration of the supporting electrolyte, I_d^o, are then easily obtained. For the calculation of the fraction of free ions, f, from the ratio I_d^o/I_d in solutions of CdPSS we used the values of I_d for the same solutions, since they are equal to those of solutions of CdEBS. However, in the case of solutions of TlPSS and TlEBS, respectively, the I_d values for the former are lower than for the latter, cf. Table 1. What are the reasons for this different behaviour ? One possible explanation would be in assuming a different diffusion coefficient of thallous ions in solutions of TlPSS. The fact that in solutions of CdPSS and CdEBS no such difference has been observed of course would speak against such an interpretation. Another explanation would be the surmising of an incomplete substitution of bound thallous ions by potassium ones, perhaps due to a great similarity of behaviour of the two kinds of ions. At this moment not much can be said about the role of the adsorption of polyions in lowering the diffusion current. Since the thallium maximum is a positive one, polyions at the same time act as suppressor, and, as has been mentioned above, in solutions of TlPSS with supporting electrolyte only slight maxima occur at relatively high concentrations. Current-voltage curves of solutions containing only TlPSS as a matter of fact are showing maxima at lower concentrations also but this is not surprising, since the more concentrated the solution the more stretched the curve, i.e., its slope is less. Further studies of cations having either positive or negative maxima will enlighten the nature of this phenomenon.

It is the authors belief that the second interpretation is more plausible and therefore we decided to use for I_d the values for solutions of TlEBS, i.e., f in solutions of TlPSS is set equal to I_d^o/I_d where I_d is the value for the solutions of TlEBS having the same concentration. As a matter of fact the values of the degree of ion binding are not very sensitive to the value of I_d we take. If we use the value of I_d for solutions of TlPSS the values of f are only a few per cent higher.

A comparison of the values of 1 - f in solutions of CdPSS and TlPSS shows that the existing differences are small and amount a few per cent only. In the case of TlPSS we again may notice a slight minimum when plotting f vs. concentration. In solutions of TlPSS also f is increasing with decreasing concentration which means that relatively more and more counterions are free.

The values of current in 1.10^{-3}N solutions of Tl_2SO_4 to which different quantities of NaPSS were added show the expected trend. By increasing the concentration of NaPSS more and more thallous ions are bound and the current decreases.

Let us finally consider the results obtained in the 50 % ethanol. As has been mentioned above, the values of I_d in solutions of TlPSS are lower than those in solutions of TlEBS which is in accord with observations in aqueous solutions. The arguments which have justified the use for the calculation of f from the I_d values for solutions of TlEBS presumable may be assumed to apply to ethanol solutions also. The values of transference numbers were determined in a similar way to that in aqueous solutions. The values for thallous ions reach from 0,67 at the highest concentration to 0,58 at the lowest. As far as we could ascertain there are no data on transference numbers of thallous ions in a 50 % alcohol solution, therefore no direct comparison is possible. The values for the degree of ion binding in ethanol solution are higher than those in water, cf. Fig.2, which is in accord with expectation. Since the dielectric constant of the 50 % ethanol solution is only 49,0, the electrostatic attractive forces between counterions and polyions are larger. Owing to an already large fraction of bound counterions in aqueous solutions, the increase in ethanol solutions is seen to amount to a few per cent only.

The values of 1 - f obtained for aqueous as well for alcohol solutions again are somewhat higher than those for some other univalent ions studied by different methods[3)4)].

No data on thallous ions are existing, a direct comparison is not possible.

In order to form a realistic picture of the usefulness of polarographic method, which is a simple and fast one, for the study of ion binding, it will be necessary to accumulate more experimental data, i.e., the degree of ion binding of other counterions will have to be determined with the polarographic method. Among them also should be counterions for which data are known from other methods.

The authors wish to thank Dr.P. ZUMAN, Polarographic Institute, Czechoslovak Academy of Science, for critical reading of the manuscript. They would also like to acknowledge gratefully financial support from the Boris Kidric Fund.

L I T E R A T U R E

1) LAPANJE,S., S.OMAN, Makromolekulare Chem. 53, 46 (1962)

2) OMAN,S., D.DOLAR, submitted for publication

3) RICE,S.A., M.NAGASAWA, "Polyelectrolyte Solutions", Academic Press Inc., New York 1961, Chapter IX

4) MOCK,R.A., C.A.MARSHALL, T.E.SLYKHOUSE, J.physic.Chem. 58, 498 (1954)

D I S K U S S I O N S B E M E R K U N G E N

BERG : Wird der Grenzstrom durch Tl^+ und $[Tl\ Komplex]^-$ -Abscheidung gebildet ?

LAPANJE : I would like to repeat that it has been supposed to a first approximation that only free,i.e., unbounded ions contribute to the diffusion and migration currents, respectively. Bounded thallous ions can get to the surface of mercury drops by adsorption of negative polyions only on the positive side of the electrocapillary maximum which eventually would result in a small maximum. However, in the case of proteins, which are polyampholytes, the contribution of bounded ions has to be considered.

BERG : Wenn die Leitsalzkonzentration von 0 bis 1m variiert wird, ändert sich die Struktur der Doppelschicht und damit die Adsorption oder Abstossung, was berücksichtigt werden müßte.

LAPANJE : It is well known that the structure of the double layer mercury-solution is dependent on the concentration of salt in the particular solution. One of the consequences is a change of the position of the electrocapillary maximum which in turn means that adsorption of polyanions ceases at different potentials.

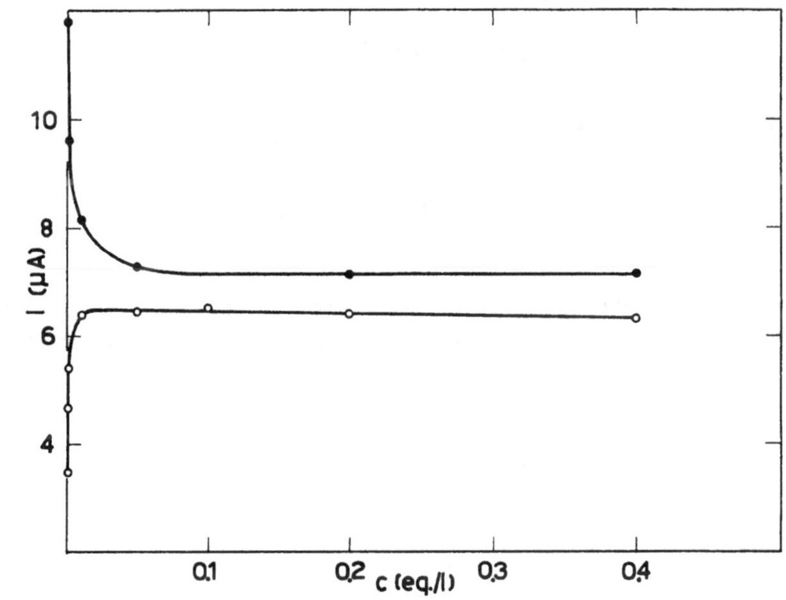

Fig.1 Influence of added potassium sulphate on the total limiting current: ●, 1.10^{-3}N TlEBS; ○, 1.10^{-3}N TlPSS.

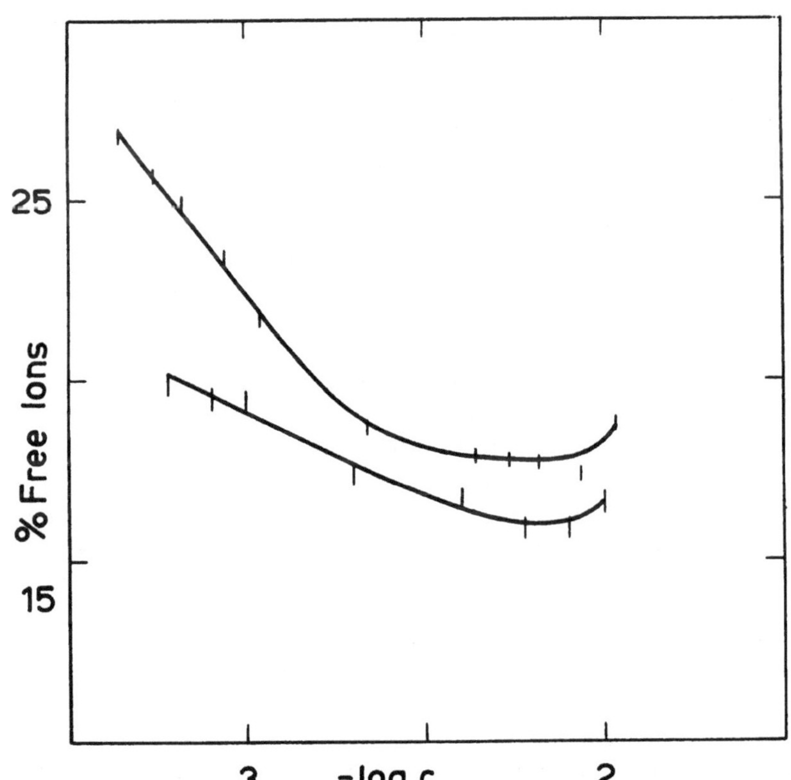

Fig.2 % Free thallous ions vs. - log of concentration of TlPSS: upper curve aqueous solutions, lower curve ethanol solutions.

Generaldiskussion I

Möglichkeiten und Grenzen der Methoden

Chairman: St. CLAESSON (Uppsala)

CLAESSON : We could start with discussion to let the biochemists tell about what would they like to know. Are there any methods which you think are good enough for your use if they would be improved sufficiently.

BERG : Auf Grund des täglichen Umgangs mit Biochemikern und Biologen will ich versuchen, etwas dazu zu sagen :

Der Biochemiker wünscht u.a. Aufklärung über Proteine und Nucleinsäuren. Insbesondere treten dabei folgende Probleme auf :

1. Die Ermittlung der Massenverteilung soll möglichst ohne vorherige Separierung durchgeführt werden können. Der Trennungsvorgang kann labile Moleküle mitunter unkontrolliert beeinflussen.
2. Wie hängt die Fermentaktivität mit der Struktur des Moleküls zusammen ? Die sterische Lage der aktiven Gruppen ist entscheidend für hohe Umsatzzahlen. Eine Methode zur Ermittlung der sterischen Anordnung bestimmter Molekülteile wird benötigt, also insbesondere zur Aufklärung von Sekundär- und Tertiärstruktur.
3. Ein Fundamentalproblem ist die Bestimmung des genetischen Codes, d.h. die Analyse der Basensequenz der DNA.

Das 4. Problem steht im Zusammenhang mit kinetischen Messungen.

Bei der Reduplikation der Nucleinsäure und dem Mechanismus der Eiweißsynthese treten Änderungen ein, die als Relaxationserscheinungen betrachtet werden können.

DOSE : Bei der Messung der Enzymaktivität in Abhängigkeit vom pH-Wert ergibt sich in manchen Fällen ein Optimum. Bei Bestrahlung oder Hitzeeinwirkung auf das Enzym kann man neben einer Inaktivierung zugleich auch eine Verschiebung des pH-Optimums beobachten. Durch Zugabe von Schutzstoffen vor der Röntgenbestrahlung von LDH aus Schweineherzen kann man erreichen, daß das Enzym nicht inaktiviert wird; die pH-Verschiebung des Optimums der Aktivität aber dennoch erhalten bleibt.

Eine Röntgenstrukturuntersuchung würde aber vermutlich keine merklichen Änderungen sichtbar machen wie z.B. auch bei den verschiedenen Myoglobinfraktionen, die sich in ihrem chromatographischen Verhalten unterscheiden.

Manchmal besitzen Enzyme bestimmte Gruppen - speziell SH-Gruppen - die mit gewissen Reagenzien besonders gut reagieren können. In der Regel ist ein erheblicher Teil dieser SH-Gruppen im Molekül maskiert (im Zentrum versteckt), sodaß er nicht mit dem Reagenz reagieren kann. Beim Studium der Kinetik einer solchen Reaktion findet man bei geringfügiger Quälung des Enzyms ein schnelleres Reagieren, was auf das Eintreten einer Entfaltung des Enzyms hinzuweisen scheint.

BERG : Weist auf die Unterschiede zwischen Struktur und Verhalten der Moleküle in Lösung gegenüber denen in der Zelle hin.

BURCHARD : Leichte Schädigungen der Enzyme führen oft nur zu einer gewissen Entfaltung. Die Enzyme sind aber meist so klein, daß man mit der Streulichtmethode kaum arbeiten kann, um diese Schädigungen feststellen zu können. Eine Kombination von Viskositäts-, Sedimentations- und Diffusionsmessungen gibt aber nach unserer Erfahrung ganz hervorragende Ergebnisse. Bei kugelförmigen Teilchen muß der Viskositätskoeffizient nach Einstein gleich 2,5 sein. Beim Vorhandensein einer Entfaltung, die zu stäbchenförmigen Teilchen führt, steigt die Viskosität rapide an. Gleichzeitig ändert sich auch die Diffusionskonstante. Da die Moleküle undurchspült sind, kann man die Ellipsoidmodelle recht gut anwenden.

BERG : Dies ist, wie ich glaube wesentlich. Die Kombination der Methoden, die die meisten Aussagen erlaubt, ist in Uppsala offenbar am besten verwirklicht.

Ich möchte noch drei weitere Fragen stellen :

1. Für Phagen-DNA können Molekulargewichte auf Grund elektronenmikroskopischer Abbildung aus der Fadenlänge ermittelt werden (Größenordnung 10^8). Bei der Isolation aus der Zelle erhält man, bedingt durch die mechanische Beanspruchung, Molekulargewichte der Größenordnung 10^7, gelegentlich aber auch höhere Werte. Wie kann man nun entscheiden, ob man Aggregate oder Einzelmoleküle vorliegen hat ?
2. Die refraktometrischen Nachweismethoden versagen meist bei niedrigen Konzentrationen. Daher hat man bei Nucleinsäuren UV-Absorptionsmethoden anzuwenden. Dabei ist nun aber eine Schädigung der Moleküle durch das Meßlicht prinzipiell denkbar. Die Frage lautet daher : Welche der Methoden sind hinreichen schonend ?

CLAESSON : I think you have brought up here an enormously important question, and that is how to work with very big molecules. If we would like for instance to determine these molecular weights by means of equilibrium sedimentation, you would need a rotor speed, I guess, on about 50 rev/min for about two years in order to reach equilibrium because the diffusion is interesting in this moment. You can see this is always not a practical method. And therefore we cannot extrapolate the methods we have been talken about to-day several powers of tens up in the molecular weight region. But there is another thing which I would like to express to ask if somebody has any comments about here. Dr.REINERT was the only one here to-day, who presented some new work on the measurements of rates. I mean really to measure something what is happening in biological systems. And I think it would be nice to hear what kind of reaction rates are you interested in biology, because after all a good physical chemist can measure reactions from half time of about 10^{-11} seconds and down to perhaps at least one year. What elementary steps are interesting in biology ?

BERG : Im Moment sind aktuell einerseits die Elementarschritte, die vor den ersten chemischen Umsetzungen bei der Photosynthese liegen : Anregungezustände, Radikalbildung; der Sehvorgang, das Zusammenwirken von Sauerstoff, Farbstoff und Licht auf Mikroorganismen, Schädigung des genetischen Materials und andererseits die Fragen der Geschwindigkeit des Helix-Knäuel-Überganges.

Weiterhin sind zu nennen die Elementarreaktionen bei Fermentumwandlungen und die Reaktionen in der Zelle gemäß CHANCE.

DOSE : Ich möchte zur Veranschaulichung noch einige Zahlen zu Fermentreaktionen ergänzen : z.B. katalysiert die LDH mit ihren drei Wirkungsgruppen die Reduktion der Brenztraubensäure oder die Oxydation der Milchsäure zu Brenztraubensäure. Ein Molekül schafft in der Minute etwa 4.10^4 Umwandlungen. Katalase schafft ca. 10^6 Umwandlungen pro Minute.

VARSHAVSKY spoke about the question: "What happens with DNA in the region of changing viscosity and optical density ?" and about problems of physical measurements for the following synthesis of messenger-RNA and protein.

BODMANN : In der Enzymchemie und allgemein bei Photoprozessen ist der eigentliche Vorgang, der der Produktbildung vorausgeht, die Enzym-Substratbindung. ALBERTY nimmt an, daß dieser Schritt in den meisten Fällen diffusionsbedingt ist. Ultrazentrifugation und Lichtstreuung geben nur Beiträge zum Verständnis der Struktur der Moleküle. Keine der Methoden läßt den Unterschied z.B. zwischen Mensch und Tier dabei erkennen.

SPATZ : Ich glaube, ich kann eine Frage von Dr.BERG beantworten. Bei uns im EIGENschen Labor ist kürzlich ermittelt worden, wie schnell sich eine Wasserstoffbrücke formiert. Am System - Caprolactam wurde die Dimerisierung untersucht. Die Bildung der Wasserstoffbrücke muß dabei schneller als in 10^{-8}sec erfolgen. Das ist außerordentlich wichtig für die Kinetik der Denaturierung von DNA. Daraus ist zu ermitteln, daß der geschwindigkeitsbestimmende Schritt des Aufbruchs etwas ganz anderes ist. Es handelt sich dabei nach unserer Auffassung offenbar eher um ein hydrodynamisches als um ein chemisches Problem.

VARSHAVSKY : Welche Methode der Geschwindigkeitsmessung wurde dabei benutzt,und welcher Prozeß wurde untersucht ?

SPATZ : I am studying the optical absorbancy and the rate of changing of the optical absorbancy of synthetic AT-polynucleotides and its kinetics of the denaturation. We have not finished these studies, but there is very much dependence on the position of the melting curve. It is approximately in 10 msec. But you see it is a very special kind of DNA.

BERG : Wird die Denaturation durch einen Temperatursprung angeregt ?

SPATZ : Yes ! I go to a lower temperature of the 260 nm melting curve of optical density and start a temperature jump which brings me up to a higher point of the melting curve when I am studying the rate at which the optical absorbancy increases. It is slow in the upper and fast in the lower part of the melting curve. But this is quite definitely more or less the special polymere that we have studied. It is a synthetic polymere and can fold back itself, it is not like natural DNA.

SAUERBIER : What happens if your temperature jump takes much more time than the transition itself ?

SPATZ : It does not. I can heat up the solution in something like 10 µsec or 100 µsec, depending on conductivity (salt concentration). I can make temperature jumps of the order of 0,5 degrees up to 5 degrees.

SAUERBIER : Some experiments were done by FREIFELDER and DAVISON in the temperature range before melting starts. They looked whether some hydrogen bonds might open. If the hydrogen bonds would be open in this range the amino groups should be accessible to formaldehyde bounded to these amino groups. So there is no indication that there are hydrogen bonds open in the temperature region before melting.

SPATZ : I studied this together with Dr.BALDWIN also.

CROTHERS : Since we did get questions talking about elementary steps of the denaturation process I have been looking actually on DNA-molecule itself. One finds now on an analogous plot - we plot the optical density vs. pH - a rise in the optical density at pH of about 11. Now I have found that if you take a solution of bacteriophage DNA (M ca. 10^8) and make an addition of alkali which raises pH from the lower part to the upper part of the curve the optical density changes and the half time of the reaction is of the order of 100 seconds. And if one takes this molecule and sheares it and reduces the molecular weight,for instance by a factor of five, then the optical density change occurs in less than ten seconds. There is a very strong dependence of this time on the length of the chain. This is why we think it is a hydrodynamic problem.

VARSHAVSKY : This time should be very sensitive to the nature of the buffer.

CROTHERS : No ! I studied the pH-transitions and these showed approximately the same rate.

BERG : The rate should depend, I think, on the ionic strength, because in pure water you have a denaturation.

CLAESSON : We are getting into details we should discuss on saturday. We have found without doubt that it is important to perform kinetic measurements in all kind of cooperation between physicists and biologists. And, I think, it is interesting to see that what has been measured up till now in almost all the cases is the change in optical density as a function of some other rapid change. And it would be certainly most interesting then to be able to measure some other variation than in optical density equally quickly.

BERG : Noch einiges zur Frage der Messung schneller Vorgänge :
Die klassischen Makromolekülmethoden haben nicht viele Möglichkeiten. Offenbar ist nur die Lichtstreuung für sehr schnelle Vorgänge verwendbar, wie z.B. für Relaxationsmessungen, sowie die Messung der Adsorption an einer Grenzfläche, z.B. einer Elektrode. Die Geschwindigkeit des letzteren Vorganges kann auch gemessen werden (KŮTA, Prag). Weiterhin ist zu nennen die Addition von Makromolekülen und niedermolekularem Farbstoff z.B. Acridinorange. Diese Adsorption ist von der Konfiguration des Makromoleküls (single oder double strand) abhängig. Prof. TUMERMAN verfolgt diesen Vorgang durch Fluoreszenzmessungen. Dabei sollte auch die Ermittlung der Einstellgeschwindigkeit des Adduktes möglich sein.

ATEN : One of the points always asked is to do the experiments with as few of material as possible. That is especially for a laboratory stammed with the bacteriophage OX 174 of SINSHEIMER. It is very difficult to have a task amount of material. OX 174 has a relatively small and single stranded DNA.
Boundary sedimentation takes about 40 µg of DNA. I'll throw your attention to the paper that newly appeared from VINOGRAD et.al. who have introduced band centrifugation, taking 30 µl of DNA solution containing about 10 µg per ml and getting the same results as in boundary centrifugation (Details in Proc.nat.Acad.Sci. USA 49, 902 (1963)).

SAMSONOW : It's possible to tell something about our methods of fractionation of the oligopeptides which we can compare with gel filtration. The fractionation of the low peptides is not very easy by gel filtration because the gel filtration is not very good in the case of molecular weights of 500 till to 1000. And so we prepared a new amphoteric resin which contains simultanously the carboxyl functional groups and tetramonium functional groups for our development of a new chromatographic procedure for the fractionation of the different groups of polypeptides. It is interesting that this type of resins, for example the phenoxyacidic group and tetramonium group, yield in general salts in the neutral solution. And after desorption of level ion one obtained the competition between indolent salts and level ion in the solution and so elutes these compounds only with water without any drop in the salt. This method is very interesting because we can obtain here in water solution the groups of peptides of different nature. This method is in comparison to the gel filtration, but the first group is never the same as in gel filtration with sephadex G 25.

CLAESSON : We have heard from the discussions to-day that the methods the physical chemists have available to-day are not precise enough, and they are not giving information enough for the work which the biochemists and biophysicists would like to do in the future and also perhaps the time resolution is too long. That means that you believe that a lot of work remains to do for the physical chemists in developing new methods.

B Polysaccharide

Lösungsverhalten und Struktur von Polysacchariden

G. MEYERHOFF

Institut für physikalische Chemie der Universität, Mainz

S U M M A R Y

For the description of the solution behaviour of threadlike molecules, knowledge of the unperturbed dimension as expressed by the end to end distance or the radius of gyration, is necessary. The unperturbed state is found in Θ-solvents, in which the intrinsic viscosity is given by

$$[\eta] = K\ M^{1/2}$$

For polysaccharides only very few Θ-solvents are known. Therefore K-values were calculated from $[\eta]$-measurements in good solvents. We used three different theoretical approaches considering the non-gaussian form of the polymer chain in a good solvent.

The K-values were used to calculate the expansion σ of the coil with different sidegroups in the unperturbed state relative to a coil without any sidegroups, where free rotation around the valence bonds can occur. This steric factor σ is a measure of the chain "stiffness". The σ-values, ranging from 1,5 to 2,6 are not greater than those of vinylpolymers. Accordingly the polysaccharides, at least in this respect, are not stiffer than other threadlike polymers, only their bond length is greater than that of vinylpolymers.

E I N L E I T U N G

Wenn wir das Lösungsverhalten von Polymeren untersuchen wollen, so stehen uns hierzu sehr verschiedenartige Verfahren zur Verfügung. Einmal die hydrodynamischen Methoden, wie Viscosität, Sedimentation, Diffusion, Strömungsdoppelbrechung und Viscoelastizität, dann die Streuung des gewöhnlichen Lichtes und der Röntgenstrahlen und schließlich die colligativen Eigenschaften. Alle liefern Aussagen über die Wechselwirkungen der gelösten Moleküle untereinander und mit dem Lösungsmittel, welche die Struktur des fadenförmigen Moleküls bestimmen.

Die Struktur des Fadenmoleküls, dessen Knäuelform sich ständig verändert, wird durch seine mittlere Ausdehnung sinnvoll erfasst. Man gibt seinen mittleren Fadenendenabstand $\langle L^2\rangle^{1/2}$, der auf lineare Fäden beschränkt bleiben muß, oder seinen mittleren Trägheitsradius $\langle R^2\rangle^{1/2}$ um den Knäuelschwerpunkt, der auch für verzweigte Moleküle ein gutes Maß bleibt, an. Wir haben hier also die Methoden zu verwenden, welche das Molekulargewicht und den Trägheitsradius möglichst hypothesenfrei zu bestimmen erlauben. Beide Größen erhält man unmittelbar aus der Messung nur durch die Lichtstreuung, wobei allerdings für M der Gewichtsdurchschnitt, für $\langle R^2 \rangle^{1/2}$ dagegen der Z-Durchschnitt erhalten wird. Leider ist der Trägheitsradius nur bis zu Werten von etwa 150 Å herab sicher bestimmbar. Das heißt, man ist für den gesamten Molekulargewichtsbereich unterhalb der Größenordnung 10^5 auf indirekte Ermittlung von $\langle R^2 \rangle^{1/2}$ angewiesen.

Mit Hilfe der Röntgenkleinwinkelstreuung misst man die Persistenzlänge q, welche dem Quotienten $\langle R^2 \rangle/_M$ proportional ist. Bei bekanntem M kann man sodann $\langle R^2 \rangle^{1/2}$ angeben. Allerdings ist der Zusammenhang zwischen q und $\langle R^2 \rangle/_M$ nur für Gauss'sche Knäuel eindeutig festgelegt.

Aus den colligativen Eigenschaften erhält man das Molekulargewicht und den zweiten Virialkoeffizienten A_2. Es ist im Prinzip möglich, aus M und A_2 den Trägheitsradius zu berechnen. Jedoch ist diese Ermittlungsart bei dem derzeitigen Stand der Theorie der makromolekularen Lösungen noch etwas ungenau.

Es bleiben also für die Bestimmung von Trägheitsradien im gesamten Molekulargewichtsbereich die hydrodynamischen Methoden übrig, von denen praktisch auch nur die Viscosität und Reibung heute genügend ausgearbeitete Meßmethoden darstellen. Mit Hilfe der Reibung kann man, falls Sedimentation und Diffusion gemessen werden, M und $\langle R^2 \rangle^{1/2}$ bestimmen. Falls M bekannt ist, läßt sich $\langle R^2 \rangle^{1/2}$ aus dem Sedimentationskoeffizienten oder aus der Viscositätszahl $[\eta]$ ermitteln. Aus dem Diffusionskoeffizienten läßt sich $\langle R^2 \rangle^{1/2}$ ohne Kenntnis von M angeben. Wie kürzlich näher ausgeführt wurde [1)], vermag jedoch die Theorie der Reibung die Experimente nicht so gut wiederzugeben wie es bei der Viscosität der Fall ist. Für diese gilt im θ-Lösungsmittel

$$[\eta]_\theta = \Phi \frac{\langle L^2 \rangle_0^{3/2}}{M} \quad (1)$$

worin $\Phi = 2{,}87 \cdot 10^{23}$ Mol^{-1}, für η in cm^3/g, eine Konstante ist, die von KIRKWOOD-RISEMAN-ZIMM [1)] berechnet wurde. Sie gibt, wie des öfteren, vgl. [2)], gezeigt wurde, das experimentelle Verhalten recht gut wieder.

In einem θ-Lösungsmittel ist außerdem

$$[\eta]_\theta = K \cdot M^{1/2} \quad (2)$$

also

$$K = \Phi \frac{\langle L^2 \rangle_0^{3/2}}{M^{3/2}} \quad (3)$$

Die entsprechenden Formeln für die Reibungszahl $[f]$ lauten

$$[f]_\theta = \Psi \langle L^2 \rangle_0^{1/2} \quad (4)$$

und

$$[f]_\theta = K_f \, M^{1/2} \quad (5)$$

also

$$K_f = \Psi \frac{\langle L^2 \rangle_0}{M^{1/2}} \quad (6)$$

Jedoch ist die Konstante $\Psi = 5{,}11$ nicht so gut geeignet, die experimentellen Fadenendenabstände richtig zu erfassen, wie es bei Φ der Fall ist. So konnte kürzlich [4)5)] gezeigt werden, daß für Vinylpolymere auf Grund von Reibungsmessungen in θ-Lösungsmittel $\Psi = 6{,}35$ zu setzen ist. Für Polysaccharide liegen derartige Reibungsmessungen jedoch nicht vor.

Daher dürfte für Polysaccharide die Viscositätszahl $[\eta]$ zusammen mit dem nach einer Absolutmethode bestimmten Molekulargewicht dasjenige Verfahren darstellen, welches z.Z. die zuverlässigsten Aussagen über die Struktur dieser Makromoleküle in verdünnter Lösung liefert. Wir wollen also vornehmlich die Viscositätszahlen benutzen, um das Lösungsverhalten der Polysaccharide zu untersuchen. Hierzu müssen wir zunächst die Grundformeln, die den Zusammenhang zwischen Knäueldimension und Molekulargewicht erfassen, zusammenstellen.

Trägheitsradius und Kettenlänge

Ein polymeres Molekül besteht im einfachsten Fall aus Grundmolekülen, deren jedes die Bindungslänge l besitzt, wobei zwei aufeinander folgende Bindungen den Valenzwinkel ϑ einschließen. Wenn sich jetzt die Grundmoleküle nicht gegenseitig beeinflussen und auch zwischen den Lösungsmittel- und den Polymermolekülen keine Wechselwirkungen auftreten, so ist die Drehbarkeit um die Valenzbindung völlig frei. Das Knäuel ordnet sich im Mittel in Form einer Gauss-Verteilung um den gemeinsamen Schwerpunkt seiner n Kettenglieder an. Es gibt für den Fadenendenabstand $\langle L^2\rangle^{1/2}$ und den Trägheitsradius $\langle R^2\rangle^{1/2}$, wenn n die Zahl der Bindungen ist,

$$\frac{1}{6}\langle L^2\rangle_{fr} = \langle R^2\rangle_{fr} = \frac{n}{6}\, l^2\, \frac{1+\cos\vartheta}{1-\cos\vartheta} = \frac{n}{6}\, b^2_{fr} \tag{7}$$

Diese Konfiguration ist allein aus sterischen Gründen nicht realisierbar. Zwei nebeneinanderliegende Grundmoleküle beeinflussen sich stets so, daß die freie Drehbarkeit um die Valenzbindung eingeschränkt ist. Dies läßt sich z.B. so ausdrücken : es steht statt eines Vollkreises um die Achse der jeweils vorhergehenden Bindung stets nur ein Teilkreis zur Verfügung, ein mittlerer Bogen der Größe φ bleibt gesperrt. Die Ausdehnung der Knäuel wird gegenüber (7) größer. Es gilt

$$\frac{1}{6}\langle L^2\rangle_0 = [R^2]_0 - \frac{n}{6}\, l^2\, \frac{1+\cos\vartheta}{1-\cos\vartheta}\;\frac{1+\cos\varphi}{1-\cos\varphi} = \frac{n}{6}\, b_0^{\;2} \tag{8}$$

Diesen Zustand, bei dem nur die direkt nebeneinanderliegenden Grundmoleküle aufeinander einwirken, bezeichnet man nach FLORY als ungestört. Der hierbei auftretende $\langle R^2\rangle_\theta^{1/2}$-Wert ist der ungestörte Trägheitsradius, wie er im allgemeinen in den θ-Lösungsmitteln (A_2-0) auftritt. Die Aufweitung des Knäuels gegenüber der üblich freien Drehbarkeit ist durch den Faktor

$$b_{fr}^{\;2}\,\frac{1+\cos\varphi}{1-\cos\varphi} = b_{fr}^{\;2}\,\sigma^2 = b_0^2 \tag{9}$$

darstellbar.

Treten jetzt noch Wechselwirkungen zwischen entfernteren Gliedern einer Kette oder Wechselwirkungen der Glieder mit Lösungsmittelmolekülen auf, so ist der sogenannte Effekt des ausgeschlossenen Volumens zu berücksichtigen. Dieser läßt sich nicht mehr, wie es zur Herleitung der Gl. (7) und (8) möglich war, durch eine Gauss-Statistik beschreiben. Man kann dann annehmen, daß das Knäuel gegenüber dem ungestörten Zustand um den Faktor α aufgeweitet ist. Es gilt

$$\frac{1}{(2+\varepsilon)\,(3+\varepsilon)}\langle L^2\rangle - \langle R^2\rangle - \alpha^2\,\langle R^2\rangle_0 = \alpha^2\,\sigma^2\,b_{fr}^{\;2}\,\frac{n^2}{6} \tag{10}$$

worin der lineare Ausdehnungsfaktor α von der Zahl der Bindungen n abhängt. Der Zusammenhang zwischen Trägheitsradius und Fadenendenabstand läßt sich durch die Abweichung ε

der Knäueldimension von der Gauss'schen Verteilung aus.

$$R^2 \sim n^{1+\varepsilon} \qquad (11)$$

ermitteln, wie PETERLIN [6] zeigte.

Bei Polysacchariden in guten Lösungsmitteln liegt eine sehr ausgeprägte Abweichung von Gauss'-Knäuel, bei dem $\varepsilon = 0$ also $R^2 \sim n$ ist, vor. Dies konnte erstmals am Verhalten der Cellulosetrinitrate in den guten Lösungsmitteln Aceton und Äthylacetat dargelegt werden [7]. Die Abb.1 zeigt aus Lichtstreuungs- und Reibungsmessungen ermittelte Trägheitsradien als Funktion der Viscositätszahl $[\eta]$, die in diesen Lösungsmitteln mit grosser Annäherung dem Molekulargewicht proportional ist. Die Neigung der eingezeichneten Geraden entspricht = 1/3. Wir können somit durch ein und dieselbe Gl. (5) mit konstantem -Wert das Verhalten der gelösten Cellulosenitrate über einen Molekulargewichtsbereich von $10^4 - 3.10^6$ gut darstellen.

Dies steht im Gegensatz zu einer von HOLTZER-BENOIT-DOTY [8], HUNT-NEUMAN-SCHERAGA-FLORY [9], HUQUES-GORING-MASON [10] und MÜNSTER-DIENER [11] vertretenen Ansicht. Danach wäre für niedrige n-Werte ε zunächst groß, um dann aber von einer gewissen Kettenlänge an null zu werden. Es soll also bei einem gewissen hohen Molekulargewicht, z.B. 300 000 die Gauss'sche Knäuelform erreicht und dann bis zu den höchsten Molekulargewichten beibehalten werden. Dieser Auffassung entspricht die in der Abb.1 gestrichelt gezeichnete Linie. Es spricht also sehr viel dafür, das gesamte Verhalten der Polysaccharide durch die Gerade der Nicht-Gauss'schen Knäuel für alle Molekulargewichte einheitlich zu erfassen. Während beim Erscheinen der Veröffentlichung [7] keine ausreichenden theoretischen Grundlagen hierzu vorlagen, ist dies inzwischen der Fall.

Diese neuen Theorien erlauben es, aus Messungen in guten Lösungsmitteln, den ungestörten Zustand im Θ-Lösungsmittel zu berechnen. Da für Polysaccharide bisher nur sehr wenig Θ-Lösungsmittel bekannt geworden sind, ist eine solche indirekte Ermittlung sehr wertvoll. Das Verhalten im ungestörten Zustand benötigen wir, um die Struktur der einzelnen Polysaccharide untereinander und mit anderen Fadenmolekülen vergleichen zu können.

Formeln zur Ermittlung von K

Unter Einführung des Expansionsfaktors α

$$[\eta] = [\eta]_0 \, \alpha^3 \qquad (12)$$

stellte FLORY [12] die auch allen späteren Theorien als Basis dienende Gleichung

$$\alpha^5 - \alpha^3 = C_F \, . \, M^{1/2} \qquad (13)$$

auf, wobei C_F allerdings nicht genau eine Konstante ist. Gl.(13) zusammen mit (12) und (2) ergibt die Beziehung

$$\frac{[\eta]^{2/3}}{M^{1/3}} = K^{2/3} + C_F \, K^{5/3} \, \frac{M}{[\eta]} \qquad (14)$$

die es im Prinzip erlaubt, K für das Θ-Lösungsmittel aus einer Auftragung von $\frac{[\eta]^{2/3}}{M^{1/3}}$ gegen $\frac{M}{[\eta]}$ von Werten in guten Lösungsmitteln aus dem Ordinatenabschnitt zu erhalten. Jedoch ist die Extrapolation nicht einfach auszuführen. Außerdem nimmt bei Polysacchariden der Exponent a in

$$[\eta] = K_\eta \, M^a \qquad (15)$$

Werte bis zu 1,0 an, während die Gl.(13) und damit auch (14) nur Werte bis a = 0,8 zu erfassen vermag, wie aus Gl.(13) folgt, wenn man α^3 gegen α^5 vernachlässigt.

Dagegen ist eine kürzlich von KURATA-STOCKMAYER-ROIG [13] vorgeschlagene Näherungsgleichung

$$\alpha^3 - \alpha = C_{KSR}\, g\, (\alpha)\, M^{1/2} \tag{16}$$

mit

$$g\,(\alpha) = \frac{8\,\alpha^3}{(3\alpha^2+1)^{3/2}} \tag{17}$$

viel weitgehender gültig. Aus ihr folgt

$$\frac{[\eta]^{2/3}}{M^{1/3}} = K^{2/3} + C_{KSR}\, g\,(\alpha)\, \frac{M^{2/3}}{[\eta]^{1/3}} \tag{18}$$

Eine Kombination von Gl.(16), (17) und (18) zeigt, daß hiermit Exponenten bis a = 1,0 der Gl.(15) erfaßt werden können. Wenn auch a-Werte $>1{,}0$ durchaus möglich sind [14], so ist die Gl.(18) für uns hier ausreichend anpassungsfähig, da bei den hier untersuchten Polymeren nur Exponenten bis a = 1,0 auftreten.

Die praktische Anwendung von Gl.(18) geschieht graphisch durch Auftragung von $\frac{[\eta]^{2/3}}{M^{1/3}}$ gegen g (α) $\frac{M^{2/3}}{[\eta]^{1/3}}$. Sie erfordert also eine Kenntnis von g (α). Hierbei kann man sich mit dem nicht exakten Wert $\alpha_\eta^3 = [\eta]/[\eta]_0$ begnügen, wobei η_0 aus $K^{2/3}$ für α = g (α) = 1 als erste Näherung verwandt wird. Eine zweite oder höchstens dritte Näherung ergibt dann den endgültigen K-Wert. Von PTITSYN-EJZNER [15] wurde das Nicht-Gauss'sche-Verhalten der Fadenmoleküle ebenfalls, aber auf anderem Wege, erfaßt. Jedoch ist eine Ermittlung von K nach ihrer Methode auch nicht einfacher als nach Gl.(18). Das experimentelle Material über die verschiedenen Typen von hochpolymeren Fadenmolekülen wurde von in allerletzter Zeit KURATA und STOCKMAYER [16] entsprechend der Gl.(18) ausgewertet. Inzwischen erschien jedoch eine weitere Veröffentlichung von STOCKMAYER-FIXMAN [17], die von einem Ansatz von FIXMAN [18]

$$\alpha^3 - 1 = C_{Fi}\, M^{1/2} \tag{19}$$

ausgehen und eine Gleichung von KURATA-YAMAKAWA [19] benutzen, um die einfache Beziehung

$$[\eta] = K\, M^{1/2} + 0{,}51\,\phi\, B\, M \tag{20}$$

zu erhalten. Hierin werden die kurzreichenden Wechselwirkungskräfte durch das erste Glied (K) und die weitreichenden Kräfte, der Effekt des ausgeschlossenen Volumens, durch das zweite Glied (B) getrennt erfaßt. B ist eine Funktion des FLORY'schen Parameters X, der mit der örtlichen freien Energie der Polymer-Lösungsmittel Wechselwirkungen zusammenhängt.

Da wir uns hier vorwiegend für den Einfluß der kurzreichenden Wechselwirkungen interessieren, gibt also

$$\frac{[\eta]}{M^{1/2}} = K + C_{SF}\, .\, M^{1/2} \tag{21}$$

eine weitere, aber wesentlich einfachere Möglichkeit, K aus $[\eta]$-Werten in guten Lösungsmitteln zu bestimmen.

Eine Beziehung wie Gl.(21) haben bereits mehrere andere Autoren wie KRIGBAUM [20][21] und BURCHARD [22] früher benutzt. Deren Vorgehen ist somit durch STOCKMAYER-FIXMAN [17] lediglich theoretisch besser fundiert worden.

Wir verwenden im folgenden beide Methoden, um K graphisch für Cellulose, Amylose und einige ihrer Derivate zu ermitteln.

Daneben benutzen wir ein anderes Verfahren, das auf theoretischen Grundlagen von PTITSYN [23] und PETERLIN [6] beruht. Es wurde, bei etwas anderer Herleitung, von H.LÜTJE[24] zur Bestimmung von b_0 für Polymethacrylate benutzt. Der Proportionalitätsfaktor in Gl.(11) ist $b_0^2/(2+\varepsilon)(3+\varepsilon)$. Damit wird der FLORY'sche Expansionskoeffizient

$$\alpha^2 = \frac{\langle R^2\rangle}{\langle R^2\rangle_0} = \frac{(2+\varepsilon)(3+\varepsilon)}{6} n^{\varepsilon} \qquad (22)$$

Nach PTITSYN [23] ist $[\eta]$ mit variablem $\phi = \phi(\varepsilon)$ darstellbar zu

$$[\eta] = \phi(\varepsilon) \frac{\langle R^2\rangle^{3/2}}{M} \qquad (23)$$

wobei $\phi(\varepsilon)$ graphisch und tabellarisch [23] gegeben ist. Die Beziehung (23) gibt das Experiment gut wieder, wie früher [3] gezeigt werden konnte. Für kleine ε gilt

$$\phi(\varepsilon) = 2{,}87.10^{23}(1 - 2{,}63\varepsilon + 2{,}86\,\varepsilon^2) \qquad (24)$$

Zwischen ε und a besteht die Beziehung $\varepsilon = 1/3\,(2a-1)$
Mit den Gln.(2), (15), (22) und (23) können wir also schreiben

$$\frac{K_\eta M^a}{K\,M^{1/2}} = \frac{\phi(\varepsilon)}{\phi} \frac{\langle R^2\rangle^{3/2}}{\langle R^2\rangle_0^{3/2}} = \frac{\phi(\varepsilon)}{\phi}\left[\frac{6}{(2+\varepsilon)(3+\varepsilon)} n^{\varepsilon}\right]^{3/2}$$

oder

$$K = K_\eta \frac{\phi}{\phi(\varepsilon)} \left[\frac{(2+\varepsilon)(3+\varepsilon)}{6} M_0^{\varepsilon}\right]^{3/2} \qquad (25)$$

Diese Beziehung wollen wir ebenfalls verwenden um K für den ungestörten Zustand zu ermitteln. Wir benötigen hierzu nur die Konstanten K_η und a der Viscositätsmolekulargewichtsbeziehung (15), die wir zum größten Teil aus Tabellen [3][16] entnehmen können.

Dimensionen bei freier Drehbarkeit

Die Berechnung der Dimension von Fadenmolekülen bei völlig freier Drehbarkeit soll hier nicht näher beschrieben werden. Es wird lediglich auf die Originalliteratur von WALL [25], BENOIT [26], MARKOVITZ [27], ELIEZER und HAYMAN [28] und für die Amylose von COWIE [29] und BURCHARD [30] hingewiesen. Bei den Polysacchariden ist angenommen, daß alle Glukoseringe die sogenannte Sesselform annehmen, obwohl man für die Amylose die Wannenform nicht ganz ausschließen kann. In der Tabelle I sind die ungestörten Dimensionen, wie sie sich für Vinylpolymere, Polypeptide, Polyisoprene und Amylose- und Glukoseketten ergeben, aufgeführt.

Wir sehen, daß die Größe $\langle L^2\rangle_{fr}^{1/2}/n^{1/2}$ für die Cellulosekette den größten Wert annimmt, für Amylose liegt er dagegen noch unter dem Wert für die trans-Formen des Peptids und des Isoprens. Daneben ist die Bindungslänge **l**, wie sie sich im Mittel aus den Einzelbindungen und den Winkeln zwischen diesen im Grundmolekül ergibt, aufgeführt. Man kann nach KURATA-STOCKMAYER [16] die Steifheit der verschiedenen Ketten charakterisieren, wenn man die Anzahl ν der Bindungen, die l ergeben, berücksichtigt. Bei den Glukosen sind dann für den Ring 3 Bindungen zu zählen, so daß $\nu = 5$ wird. In der letzter Spalte sind dementsprechend die durch $\nu^{1/2}$ dividierten Werte von $\langle L^2\rangle^{1/2}/n^{1/2}$ aufgeführt. Wieder ergibt sich für die Cellulosekette den höchsten Wert, während die Amylosekette in dieser Betrachtungsweise praktisch den Vinylpolymeren gleich zu setzen ist.

Tabelle 1 : Kettendimensionen bei freier Drehbarkeit

Kettentyp	Bindungen ν	Bindungslänge l [Å]	$\frac{\langle L^2\rangle_{fr}^{1/2}}{n^{1/2}}$	$\frac{\langle L^2\rangle_{fr}^{1/2}}{(n.\nu)^{1/2}}$
Vinyl	2	2,52	3,08	2,18
Peptid cis	3	2,75	2,92	1,68
Peptid trans	3	3,80	5,28	3,05
Isopren cis	4	4,55	4,02	2,01
Isopren trans	4	5,12	5,80	2,90
Glukose cis	5	4,6	4,95	2,21
Glukose trans	5	5,62	7,85	3,51

Ermittlung von K aus dem Experiment

Für Polysaccharide stehen sehr viele Bestimmungen der Viscositäts-Molekulargewichtsbeziehung (15) zur Verfügung. Wir wollen hier nur solche Molekulargewichte verwenden, die direkt nach einer Absolutmethode bestimmt wurden. Auf osmotisch gemessene Molekulargewichte, die stark von der Uneinheitlichkeit der Präparate beeinflußt werden, greifen wir nur zurück, falls keine Werte von M aus der Lichtstreuung oder aus Ultrazentrifugenmessungen bekannt geworden sind.

Während es für Amylose (cis-Glukose) relativ zahlreiche Lösungsmittel, die eine optische Beobachtung erlauben, gibt (vgl.Abb.4), waren für Cellulose bis vor kurzem nur fast undurchsichtige Lösungsmittel bekannt. Jedoch liegen jetzt für Cellulose zwei Untersuchungen von CLAESSON-BERGMANN-JAYME [31)] und von HENLEY [32)] in den neuen Lösungsmitteln, einem Eisen-Weinsäure-Natriumkomplex (EWNN) und Triäthylendiamin-Cadmiumhydroxyd (Cadoxen) vor. In der Abb.2a ist die graphische Bestimmung von K nach Gl.(18) und in der Abb.2b nach Gl.(21) ausgeführt. Auch die beiden Molekulargewichte, die IMMERGUT-RANBY-MARK [33)] in Kupferäthylendiamin (CED) osmotisch bestimmten, sind eingezeichnet. Sie lassen sich durch die jeweils für EWNN eingezeichnete Gerade noch ziemlich gut mit erfassen.

Für die zahlreichen Untersuchungen an Cellulosederivaten zeigen die Abb.3a und 3b als Beispiel die Auswertung von Mainzer Messungen an Cellulosenitraten in Aceton[7)34)].

Messungen an Amylose führten EVERETT-FOSTER[35)] und COWIE[29)] aus, die für die Viscositätsmolekulargewichtsbeziehung ein einigermaßen übereinstimmendes Bild ergeben, Abb.4. In den Abbildungen 5a, 5b und 5c wird der K-Wert ermittelt. Hieraus und noch deutlicher aus den K-Werten nach Gl.(25) ergibt sich eine beträchtliche Diskrepanz zwischen den beiden Untersuchungen, vgl.Tab.2. Während EVERETT-FOSTER [35)] K = 0,113 in 1/3 n KCl als Θ-Lösungsmittel bestimmten und die berechneten K-Werte nur wenig tiefer liegen, sie entsprechen der oberen gestrichelten Linie der Θ-LM in der Abb.5, müßte nach COWIE [29)] die untere gestrichelte Linie dieser Abb. das Θ-Lösungsmittel repräsentieren. Diese untere Linie (K = 0,055) erscheint sehr tief. Der wahre K-Wert dürfte daher näher bei dem EVERETT-FOSTER'schen [35)] Wert liegen.

In der Tab.2 sind sämtliche nach den drei Methoden ermittelten K-Werte für Cellulose, Amylose und deren Derivate zusammengestellt. Die Übereinstimmung mit den direkt im Θ-Lösungsmittel gefundenen K-Werten ist recht gut. Allerdings liegen derartige Messungen nur für wenige Polysaccharide vor.

Tabelle 2 : Ermittlung der K-Werte von Polysacchariden

Polysaccharid	Methode+++	Temperatur	K_{KSR} Gl.(18)	K_{SF} Gl.(21)	K (ε) Gl.(25)	K gemessen	Literatur
Cellulose	SD	25	0,260	0,300	0,230	-	31,32, 33
- triacetat	SD	20	0,108+	0,10	0,13	-	16,38
- tributyrat	OS	30	0,097+	0,105	0,135	(0,085)	39
- tricarbanilat	LS	20	0,120	0,100	0,098	-	37
- tricaproat	LS	35	0,240+	0,250	0,255	0,245;41°C	40
- tricaprylat	OS	30	0,127+	0,100	0,090	0,129;48°C++	39
- trinitrat	LS	20	0,100	0,130	0,160	-	7,34
Äthylcellulose	LS	25	0,232+	0,180	0,175	-	41
Hydroxyäthylcellulose	LS	25	0,620+	0,81	0,52	-	42
Äthylhydroxyäthyl-cellulose	SD	25	0,550+	0,55	0,36	-	43
Methylcellulose	SD	20	0,940	0,95	0,78	-	44
Amylose	LS	25	0,115	0,125	0,075	(0,113;25°C)	35
	LS	25	0,060	0,055	0,040	-	29
- triacetat	LS	25	0,047+	0,055	0,03	-	36
- tricarbanilat	LS	20	0,043	0,037	0,03	-	37

\+ Werte von KURATA-STOCKMAYER [16]   ++ bei 140°C K = 0,113

+++ SD = M aus Sedimentation und Diffusion, OS = osmotisch, LS = Lichtstreuung

Auch der Vergleich der nach den Gln. (18), (21) und (25) berechneten K-Werte untereinander ist noch befriedigend. Man muß dabei beachten, daß bei den graphischen Extrapolationen, die nach $M \to 0$ und $[\eta] \to 0$ erfolgen, leicht die niedrigen Molekulargewichte den K-Wert zu stark beeinflussen. Dabei sind die niedrigen Molekulargewichte oft und besonders bei $M_{osm.}$ weniger genau als die höheren. Dementsprechend ist die Berechnung von K aus K_η und a nach Gl.(25) häufig günstiger. Sollten sich aber K_η und a nur schwierig festlegen lassen, so ist den graphischen Methoden mehr Wert beizumessen. Dies trifft z.B. bei den Amyloseacetaten [36] und den Amylosecarbanilaten [37] zu. Deren Darstellung $[\eta] = f(M)$ in doppeltlogarithmischen Koordinaten ergibt eine etwas nach oben gekrümmte Kurve, läßt sich also durch ein konstantes a nur mit etwas größerer Näherung erfassen.

Bei den Celluloseäthern ist zu beachten, daß nur ein sehr schmaler Molekulargewichtsbereich, teilweise nur eine halbe Größenordnung, zur Verfügung stand. Eine genauere Extrapolation war hier daher schwierig.

In der Tabelle 3 werden die K-Werte benutzt, um aus dem Grundmolekulargewicht M_0 und nach Gl.(3) $\frac{\langle L^2\rangle_0^{1/2}}{n^{1/2}}$, den ungestörten Fadenendenabstand, dividiert durch die Wurzel aus dem Polymerisationsgrad n, zu berechnen. In der letzten Spalte ist das Verhältnis $\sigma = \langle L^2\rangle_0^{1/2} / \langle L^2\rangle_{fr}^{1/2}$ aufgeführt, welches angibt, wie stark das Knäuel durch kurzreichende Wechselwirkungskräfte gegenüber dem Zustand bei völlig freier Drehbarkeit aufgeweitet ist.

Tabelle 3 : Ungestörte Dimensionen $\frac{\langle L^2\rangle_0^{1/2}}{n^{1/2}}$ der Polysaccharidmoleküle

	M_0	K	$\frac{\langle L^2\rangle_0^{1/2}}{n^{1/2}}$	A	$\frac{\langle L^2\rangle_0^{1/2}}{\langle L^2\rangle_{fr}^{1/2}}$
Cellulose	162	0,25	12,1		1,5
- triacetat	288	0,11	12,3		1,6
- tributyrat	372	0,11	14,0		1,8
- tricarbanilat	519	0,10	16,0		2,0
- tricaproat	456	0,24	20,1		2,6
Äthylcellulose	232	0,18	13,0		1,7
Hydroxyäthylcellulose	236	0,60	19,7		2,5
Äthylhydroxyäthylc.	210	0,50	16,3		2,1
Methylcellulose	187	0,90	20,0		2,5
Amylose	162	0,10	8,9		1,8
- triacetat	288	0,045	9,2		1,9
- tricarbanilat	519	0,037	11,5		2,3

Wir vergleichen zunächst die Cellulose mit der Amylose und deren Triacetate und Tricarbanilate untereinander. Es ist recht plausibel, daß das Celluloseskelett durch die Seitenglieder weniger stark aufgeweitet wird als das infolge der cis-Stellung erheblich enger geknäuelte Amyloseskelett. Die Aufweitung infolge der Rotationsbehinderung ist in den drei Fällen etwa gleich groß, jeweils $\Delta\sigma \approx 0{,}3$. Außerdem läuft bei diesen drei Typen von Seitengliedern M_0 mit σ symbat.

Bei den übrigen Celluloseestern und den Celluloseäthern liegt jedoch kein eindeutiger Zusammenhang zwischen der Größe der Seitenglieder, wie sie durch M_0 ausgedrückt wird, und der Knäuelaufweitung σ vor.

Interessant ist jedoch insbesondere der Absolutwert von σ. Er liegt zwischen 1,5 und 2,6, also in demselben Bereich wie er für andere Polymere gefunden wurde. Vgl. z.B. [45)] Polydimethylsiloxan $\sigma \approx 1{,}4$, Polymethylmethacrylat $\sigma \approx 1{,}8$, Polystyrol $\sigma \approx 2{,}4$. Für Polyvinylcaproat ergibt sich nach [16)] mit $\sigma \approx 2{,}7$ der höchste bei Vinylpolymeren ermittelte Wert. Für Cellulosederivate findet man genau so beim Caproat den höchsten Wert für σ.

Insgesamt gesehen liegen unsere σ-Werte für Cellulosemoleküle wesentlich niedriger als sie z.B. von FLORY [9)] angenommen wurden. Dieser berechnete aus Lichtstreuungsmessungen an Cellulosetrinitrat in Äthylacetat $\sigma = 4{,}5$. Dieser Wert wurde unter der Annahme ermittelt, daß sowohl im guten - wie im Θ-Lösungsmittel ein Gauss'sches Knäuel vorliegt. Als Expansionskoeffizienten α ermittelten HUNT-NEWMANN-SCHERAGE-FLORY [9)] für den Molekulargewichtsbereich von M_w = 41 000 - 573 000 ein mittleres $\alpha = 1{,}04$. Der Quotient gemessene Knäueldimension im guten Lösungsmittel zur Dimension bei freier Drehbarkeit

$$\frac{\langle L^2\rangle_{fr}^{1/2}}{\langle R^2\rangle^{1/2}} = \alpha$$

nach Gl.(10) wurde also in 1,04 . 4,5 aufgespalten. Dagegen dürfte unsere Aufspaltung in 2,75 . 1,7 mit $\alpha = 2{,}75$ sinnvoller sein.

Wie bereits KURATA-STOCKMAYER [16] darlegten, sprechen auch die neuesten Messungen der Röntgenkleinwinkelstreuung gegen die Annahme eines hohen σ-Wertes wie $\sigma = 4,5$. Dieser stand einigermaßen in Übereinstimmung mit dem Wert von KRATKY-SEMBACH [46], die eine Persistenzlänge q = 114 Å sein sollte. Nach den letzten Ergebnissen von HEINE-KRATKY-PORODSCHMITZ [47] liegt der q-Wert von Cellulosenitraten jedoch wesentlich niedriger und zwar bei q ≈ 22 Å. Unserem Cellulosenitratwert, der aus $[\eta]$ und M berechnet wurde, entspricht eine Persistenzlänge q ≈ 24 Å. Die Messungen der Röntgenkleinwinkelstreuung stützen also unsere Modellannahme für das Cellulosemolekül.

Hohe σ-Werte, wie $\sigma = 4,5$ wurden als Charakteristikum für eine besondere "Steifheit" des Fadenmoleküls benutzt. Da nun aber für Cellulose und Amylose sich nach unserer Berechnungsart durchaus normale σ-Werte ergeben, erscheint die Bezeichnung als sehr "steif" für die Polysaccharide nicht mehr begründet.

Es ist zwar so, daß die Bindungslänge l, bedingt durch den Aufbau des Glukosemoleküls, groß wird. Darüber hinausgehende Unterschiede gegenüber den Vinylpolymeren sind, wenn man ein Nicht-Gauss'sches Knäuel zugrundelegt, jedoch zur Deutung des Lösungsverhaltens nicht mehr erforderlich.

L I T E R A T U R

1) KIRKWOOD,J.G., J.RISEMAN, J.Chem.Phys. 16, 565 (1948)
2) ZIMM,B.H., J.Chem.Phys. 24, 269 (1956)
3) MEYERHOFF,G., Fortschr.Hochpol.Forsch. 3, 59 (1961)
4) LÜTJE,H., G.MEYERHOFF, Makromol.Chem. im Druck
5) LÜTJE,H., G.MEYERHOFF, Z.physik.Chem. NF im Druck
6) PETERLIN,A., J.Chem.Phys. 23, 2464 (1955)
7) MEYERHOFF,G., JUPAC-Symposium Prag 1957, J.Pol.Sci. 29, 399 (1958)
8) HOLTZER,A.M., H.BENOIT, P.DOTY, J.Phys.Chem. 58, 624 (1954)
9) HUNT,M.L., S.NEWMANN, H.A.SCHERAGA, P.J.FLORY, J.Phys.Chem. 60, 1278 (1956)
10) HUQUE,H.M., D.A.GORING, S.C.MASON, Can.J.Chem. 53, 952 (1958)
11) MÜNSTER,A., H.DIENER, JUPAC-Symposium Wiesbaden 1959 II A2
12) FLORY,P.J., Principles of Polymer Chemistry Cornell Univ.Press 1953
13) KURATA,M., W.H.STOCKMAYER, A.ROIG, J.Chem.Phys. 33, 151 (1960)
14) MEYERHOFF,G., J.Pol.Sci. 43, 269 (1960)
15) PTITSYN,O.B., Y.E.EJZNER, Vysomol.Soed. 1, 966, 1200 (1959)
16) KURATA,M., W.H.STOCKMAYER, Fortschr.Hochpol.Forsch. III, 196 (1963)
17) STOCKMAYER,W.H., M.FIXMAN, J.Pol.Sci. C I, 137 (1963)
18) FIXMAN,M., J.Chem.Phys. 23, 1656 (1955)
19) KURATA,M., H.YAMAKAWA, J.Chem.Phys. 29, 311 (1958)
20) KRIGBAUM,W.R., J.Pol.Sci. 18, 316 (1955)
21) KRIGBAUM,W.R., J.Pol.Sci. 28, 213 (1958)
22) BURCHARD,W., Makromol.Chem. 50, 20 (1961)
23) PTITSYN,O.B., Usp.Fiz.Nauk 49,371 (1959), Soviet Physics Usp. 2,797(1960), (engl. Übersetzung)
24) LÜTJE,H., Dissertation Mainz 1962
25) WALL,F.T., J.Chem.Phys. 11, 67 (1943)
26) BENOIT,H., J.Pol.Sci. 3, 376 (1948)
27) MARKOVITZ,H., J.Chem.Phys. 20, 868 (1952)
28) ELIEZER,J., H.J.G.HAYMAN, J.Pol.Sci. 23, 387 (1957)
29) COWIE,J.M.G., Makromol.Chem. 42, 230 (1960)

30) BURCHARD,W., Makromol.Chem. 42, 151 (1960)
31) CLAESSON,S., W.BERGMANN, G.JAYME, Svensk Paperstidn. 62, 141 (1959)
32) HENLEY,D., Ark.Kemi 18, 327 (1961)
33) IMMERGUT,E.H., G.G.RANBY, M.MARK, Ind.Eng.Chem. 45, 2383 (1953)
34) MEYERHOFF,G., Naturw. 41, 13 (1954)
35) EVERETT,W.W., J.F.FOSTER, J.Am.Chem. Soc. 81, 3464 (1959)
36) COWIE,J.M.G., J.Pol.Sci. 49, 455 (1961)
37) BURCHARD,W., E.HUSEMANN, Makromol.Chem. 44-46, 358 (1961)
38) HOLMES,F.H., D.I.SMITH, Tr.Far.Soc. 53, 67 (1957)
39) MANDELKERN,L., P.J.FLORY, J.Am.Chem.Soc. 24, 2517 (1952)
40) KRIGBAUM,W.R., L.H.SPERLING, J.Phys.Chem. 64, 99 (1960)
41) SCHERER,P.C., A.TANENBAUM, D.W.LEVI, J.Pol.Sci. 43, 531 (1961)
42) BROWN,W., Ark.Kemi 18, 227 (1961)
43) MANLEY,R.St.J., Ark.Kemi 9, 519 (1956)
44) UDA,K., G.MEYERHOFF, Makromol.Chem. 47, 168 (1961)
45) HAUG,A., G.MEYERHOFF, Makromol.Chem. 53, 91 (1962)
46) KRATKY,O., H.SEMBACH, Makromol.Chem. 16, 115 (1959)
47) HEINE,S., O.KRATKY, G.POROD, P.J.SCHMITZ, Makromol.Chem. 47, 682 (1961)

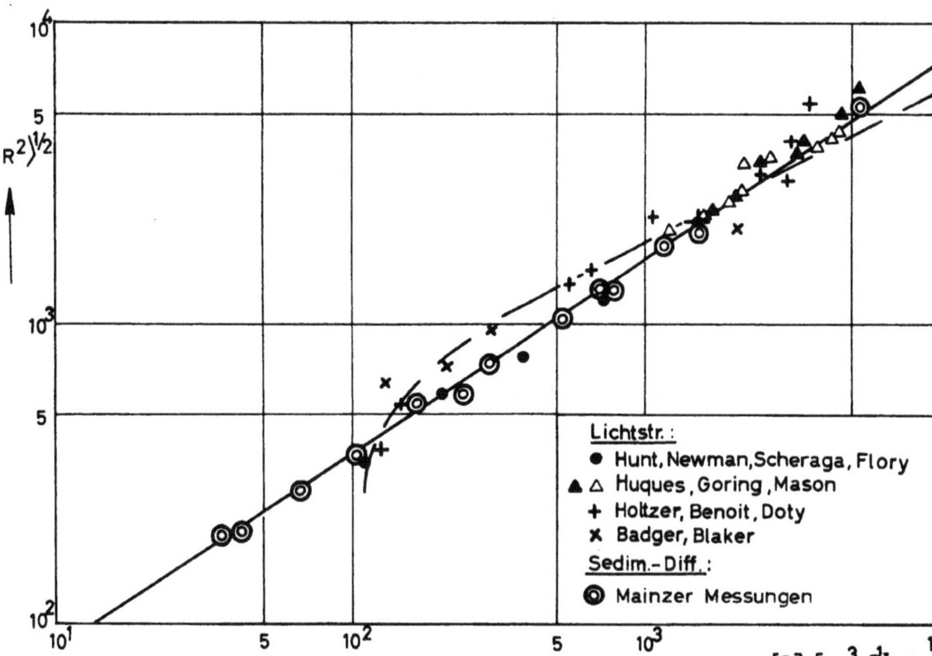

Abb.1 Trägheitsradien von Cellulosenitraten[7] in Aceton (◉,▲,+,x) und Äthylacetat (●,▲). Ausgezogene Gerade $R^2 \sim M^{1+\epsilon}$, gestrichelte Kurve - Gauß-Knäuel für M > 500 000.

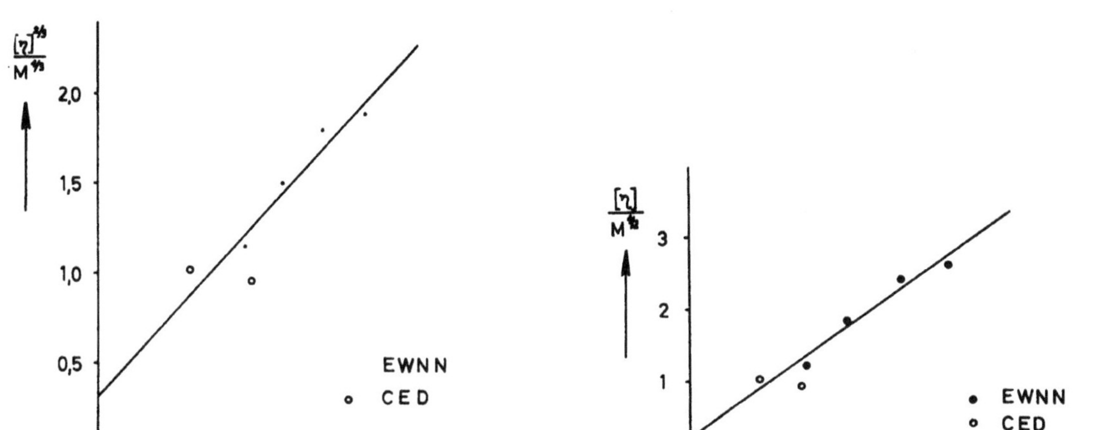

Abb.2a und b Ermittlung von K für Cellulose in EWNN[31] und Cadoxen[32] (M aus Sedimentation und Diffusion) sowie Cuen[33] (M_{osm}).

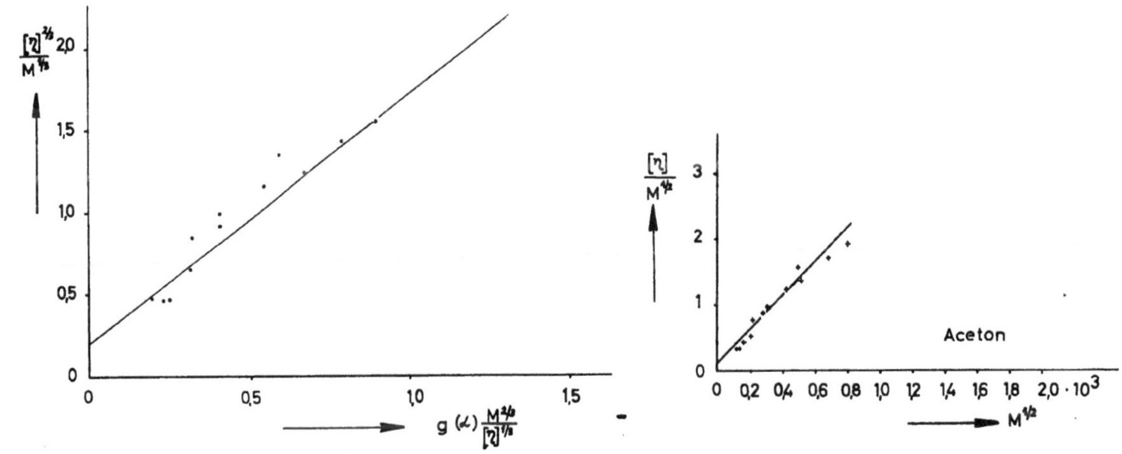

Abb.3a und b Ermittlung von K für Cellulosenitrate in Aceton (M_{SD}) [7].

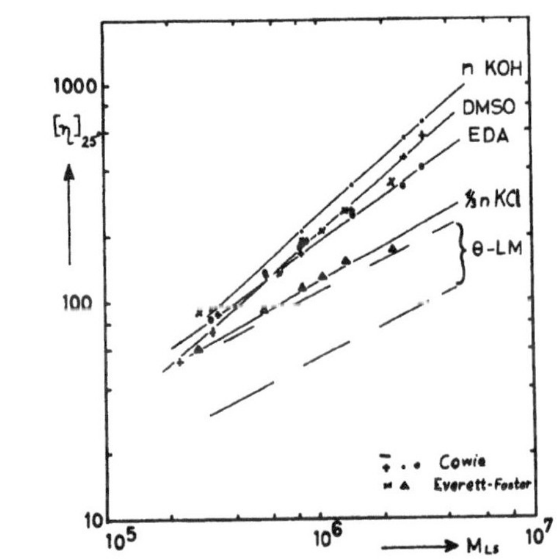

Abb.4 Viscositäts-Molekulargewichtsbeziehungen für Amylose in verschiedenen Lösungsmitteln [35)36)].

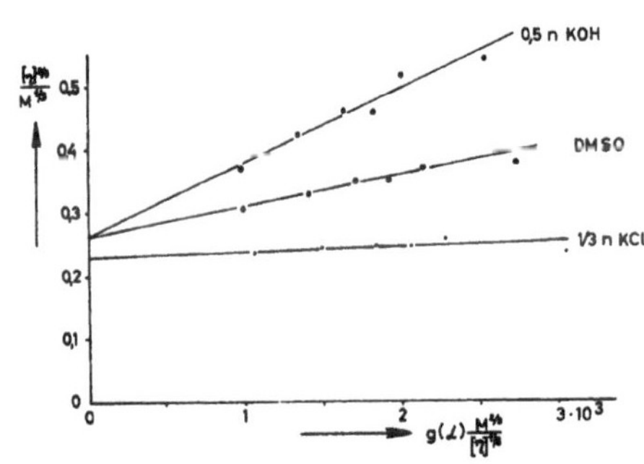

Abb. 5a

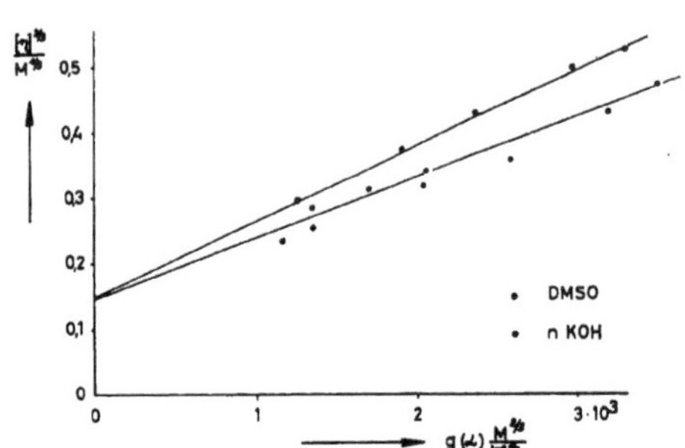

Abb. 5b

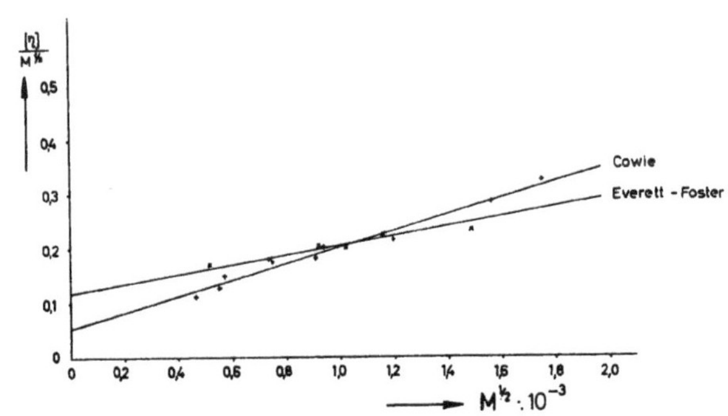

Abb. 5c

Abb.5a, b und c Ermittlung von K für Amylose in verschiedenen Lösungsmitteln [35)36)].

B Polysaccharide

Gestaltsbestimmung von Polysaccharidderivaten mit Hilfe von Streulicht- und Diffusionsmessungen

W. BURCHARD

Institut für makromolekulare Chemie der Universität, Freiburg/Br.

S U M M A R Y

The carbanilate derivates of amylose and cellulose has been studied in different Θ - solvents by means of light scattering. The imperturbated dimensions of the amylosetricarbanilate has been found to be strongly dependent from the composition of the Θ - solvents; on the other side the imperturbated dimensions of the cellulosetricarbanilate in different Θ-solvents remained unchanged. From this behavior it is concluded, that in the case of amylosetricarbanilate the ring of the anhydroglucose can posess two different stable conformations and in the case of cellulosetricarbanilate only one. This is confirmed by the ratio of the imperturbated radii of gyration $\langle R_0^2 \rangle^{1/2}$ to the radii of free rotation R_f, which is nearly the same in the amylose- and cellulosecarbanilatemolecule. A comparison with other cellulose derivates shows the high stiff properties of the two polysaccharide carbanilates.

Von den vielen im Laufe der Zeit bekannt gewordenen Polysacchariden verdienen zwei eine besondere Beachtung. Es sind dies die Amylose und die Cellulose, beides unverzweigte stereoregulierte Fadenmoleküle, die beide aus derselben monomeren Einheit, nämlich der Glucose aufgebaut sind. Der einzige Unterschied besteht in der Bindungsart : In der Amylose sind die Glucoseeinheiten in C_1 und C_4 α-glycosidisch, in der Cellulose dagegen ß-glycosidisch gebunden. Welche Bedeutung die Bindungsart für die Gestalt der Makromoleküle hat, verdeutlicht die Abb.1. Hier sind die sterischen Verhältnisse dreier aufeinanderfolgender Kettenglieder dargestellt. Wie man sieht, steht man bei der Amylose vor der Schwierigkeit, welche Konformation die Glucose in der Kette einnimmt, da neben der wahrscheinlichsten Sesselkonformation C1 auch die 3B-Konformation gut denkbar ist. Bei der Cellulose besteht diese Schwierigkeit nicht, da die Wannenform der ß-Glucose instabil ist. Unter jedem Bild ist die Größe des mittleren Fadenendenabstandes bei freier Drehbarkeit der Glucoseeinheiten um die Sauerstoffvalenzen angegeben [1].

Danach sollten also Cellulose und Amylose mit der 3B-Konformation etwa gleiche Trägheitsradien haben, während Amylose mit C1-Konformation um gut 30 % kleinere Trägheitsradien haben sollten.

Man kann nun diesen Sachverhalt experimentell nur schlecht an den Naturstoffen selbst prüfen, weil die Lösungseigenschaften der beiden Polysaccharide extrem verschieden sind, so daß bisher kein geeignetes gemeinsames Lösungsmittel bekannt geworden ist. Wir haben uns daher in Freiburg nach einem geeigneten Derivat umgesehen und fanden in dem Umsetzun produkt von Phenylisocyanat mit den Polysacchariden ein Derivat mit den gewünschten Eigenschaften [2]. Diese sogenannten Carbanilate haben gegenüber den bisher verwendeten Derivaten eine Reihe entscheidender Vorteile :

1. Die Umsetzung läßt sich ohne Abbau der Kette leicht durchführen, wobei sich stets tri substituierte Produkte erreichen lassen.
2. Die Carbanilate sind außerordentlich beständig gegen Abbau.
3. Sie lösen sich in einer Fülle der üblichen organischen Lösungsmittel, vor allem Ester Ketone und Äther.
4. Man erhält infolge der großen Seitengruppen eine Substanzzunahme um den Faktor 3.

Wir haben nun zunächst an einer Reihe von Amylose- und Cellulosetricarbanilaten Streu lichtmessungen in Aceton durchgeführt [2]. Das Ergebnis zeigt Abb.2. Man findet also in der Tat, daß die mittleren quadratischen Trägheitsradien der Cellulosecarbanilate etwa um den Faktor 2,5 größer sind als bei den Amylosecarbanilaten, so daß angenommen werden kann, daß in beiden Ketten die Glucose C1-Konformation hat. Jedoch haben die Cellulosederivate fast fünfmal so hohe Virialkoeffizienten, Abb.3, und daher bestand die Möglichkeit, daß die Unterschiede in den Trägheitsradien auf einen verschieden starken Lösungsmitteleinfluss zurückgehen. Um dieses zu entscheiden ist es notwendig, Messungen in Θ -Lösungsmitteln auszuführen, in denen der 2.Virialkoeffizient Null ist und in denen das Lösungsmittel keinen Einfluß auf die Gestalt ausüben sollte. Um derartige Θ-Lösungsmittel zu finden, habe ich verschiedene Mischungen von Lösungsmittel mit Fällungsmittel hergestellt und Trägheitsradius und 2. Virialkoeffizient durch Lichtstreuung bestimmt. Nach den Messungen an Polystyrol [3], Polymethylmethacrylat [4] und unsubstituierter Amylose [5] war zu erwarten, daß zwischen Trägheitsradien und 2. Virialkoeffizienten bei einem festen Molekulargewicht ein eindeutiger Zusammenhang besteht. Statt dessen ergab sich beim Amylosecarbanilat folgender Sachverhalt (Abb.4). Bei Mischungen von Methyläthylketon mit Methanol ergibt sich eine andere Linie als bei Mischungen von Diäthylketon mit Methanol, und beide sind verschieden von der bei Dioxan/Methanol. Das Besondere an diesen Messungen ist nun, daß die Kurven verschiedene Ordinatenabschnitte haben, die Trägheitsradien sind also in den drei Θ-Lösungsmitteln verschieden. Dieses widerspricht der Hypothese von FLORY, nach der dieser Trägheitsradius unabhängig von der Lösungsmittelzusammensetzung sein soll. Man hat diese These mathematisch in die folgende Form gebracht :

Wenn $\langle R^2\rangle$ der Trägheitsradius in einem guten Lösungsmittel ist und $\langle R_0^2\rangle$ der in einem Θ-Lösungsmittel, so sollte gelten [6)7)]

$$\langle R^2\rangle = \alpha^2 \cdot \langle R_0^2\rangle \qquad \text{(I)}$$

heißt Expansionskoeffizient und soll allein den Lösungsmitteleinfluß beschreiben.
Der 2. Virialkoeffizient sollte der Beziehung

$$A_2 = \frac{4\pi^{3/2}}{M^2} \cdot N_0 \cdot \langle R_0^2\rangle^{3/2} \cdot z \cdot h\,(z/\alpha^3) \qquad \text{(II)}$$

genügen [8], wobei $z \cdot h\,(z/\alpha^3)$ eine Wechselwirkungsfunktion ist, die nur vom Expansionskoeffizienten abhängt.

Aus der ersten Gleichung läßt sich der Expansionskoeffizient berechnen und aus der zweiten die Funktion $z.h(z/\alpha)$. Trägt man diese aus den drei Kurven gewonnenen Werte gegeneinander auf, so erhält man jetzt nur noch einen einzigen Kurvenzug. Die beiden erwähnten Gleichungen sind also für jede der drei Kurven erfüllt. Dies macht nun deutlich, daß die "ungestörten" Trägheitsradien $\langle R_0^2\rangle$ von dem Fällungsmittel Methanol nicht beeinflußt werden. Die verschiedenen Trägheitsradien $\langle R_0^2 \rangle$ müssen sich daher schon während des Lösungsvorganges in Dioxan bzw. in den Ketonen einstellen. Es liegt nun nahe, diese verschiedenen "ungestörten" Trägheitsradien durch verschiedene Glucosekonformationen in der Amylosekette zu erklären. Wie ich etwas später zeigen werde, gelingt auch in zwei der drei Fälle eine Zuordnung. Zuvor möchte ich mich aber den Messungen am Cellulosetricarbanilat zuwenden.

Das Erscheinungsbild ist hier (Abb.6) noch wesentlich undurchsichtiger. Das Molekülknäuel verhält sich nur in den Gemischen von Dioxan/Methanol normal. Bei den Ketonen dagegen tritt der eigentümliche Effekt auf, daß der Trägheitsradius bei Zugabe von nur 5 % Wasser oder 10 % Methanol sprunghaft ansteigt, und im Falle von Aceton und Äthylmethylketon in die Kurve von Dioxan/Methanol einmündet. Die beiden höheren Ketone Diäthylketon und Dipropylketon zeigen ein noch ausgeprägteres Verhalten, bilden aber einen getrennten Kurvenzug, der ebenfalls für $A_2 = 0$ in denselben Ordinatenabschnitt einmündet. Aus den Messungen ist nur eines mit Sicherheit zu entnehmen : Der Trägheitsradius von Cellulosecarbanilat am θ-Punkt ist unabhängig von der Zusammensetzung des θ-Lösungsmittels. Das war aber auch zu erwarten, da aus energetischen Gründen die ß-Glucose nur in der C1-Konformation stabil ist. Damit wird nun auch für das Amylosecarbanilat eine Zuordnung der Glucosekonformationen möglich. Bildet man nämlich das Verhältnis von R_0/R_f und legt beim Amylosecarbanilat die beiden fraglichen Glucosekonformationen vor, so erhält man die Ergebnisse der folgenden Tabelle (Tab.1). Dieses Verhältnis ist ein Mass für die Rotationsbehinderung. Wie man sieht, würde das Amylosecarbanilat in Diäthylketon mit C1-Konformation der Glucose die gleiche Rotationsbehinderung haben wie in Dioxan bei Vorliegen der 3B-Konformation, und diese Rotationsbehinderung ist nahezu die gleiche wie im Cellulosecarbanilat.

Tabelle 1

Substanz	θ-Lösungsmittel bei 20°C	$\langle R_0^2 \rangle$ $[Å^2] \cdot 10^4$	$\langle R_0^2 \rangle^{1/2} / \langle R_f^2 \rangle^{1/2}$ Konformation C1	3B
AC P_w=4200	Methyläthylketon/ 34% Methanol	39,0	4,60	3,07
	Diäthylketon/ 33% Methanol	46,0	5,00	3,33
	Dioxan/ 51% Methanol	90,0	7,00	4,66
CC P_w=1620	Dioxan/ 57,5% Methanol	38,5	4,71	-

Die Werte von Amylosecarbanilat in Methyläthylketon ordnen sich zunächst nicht in dieses Schema ein, jedoch erscheint es möglich, daß im Molekül ein Gemisch von beiden Glucosekonformationen vorliegt, wie es bei unsubstituierter Amylose von REEVES [9] schon vor längerer Zeit gefordert wurde. Die Ergebnisse lassen sich allerdings auch anders deuten. So ist es denkbar, daß die glycosidische Bindung durch Dioxan, Methanol und Wasser stark solvatisiert wird, wodurch eine wesentlich größere Rotationsbehinderung zustandekommen kann als in reinen Ketonen, die vermutlich nur die Seitengruppen solvatisieren und damit nicht zur Rotationsbehinderung beitragen. Mit dieser Vorstellung wäre das sprunghafte

Ansteigen des Trägheitsradius vom Cellulosecarbanilat bei Zugabe von Methanol bzw. Wasser einfach zu erklären. Für das Amylosecarbanilat würde das aber heissen, daß sie Glucose in der Kette auch bei Dioxanlösungen C1-Konformation hätte. Die Rotationsbehinderung wäre dann mit R_o/R_f = 7,0 immens hoch. Mir erscheint dies wenig wahrscheinlich.

Tabelle 2

Polymere Substanz	$6(R_o^2/P)w$ A^2	A_m A	$(R_o^2/R_f^2)^{1/2}$	Bestimmungs-methode	Autor
Amylosetricarbanilat in Diäthylketon C1 in Dioxan 3B	 530 1230	 128 (31) 202 (33)	 4,60 4,66	 Θ-Lösungsmittel Θ-Lösungsmittel	 BURCHARD BURCHARD
Cellulosetricarbanilat	1330	259 (50)	4,71	Θ-Lösungsmittel	BURCHARD
Cellulosetrinitrat in Äthylacetat	1200	233 (45)	4,47	aus A_2 nach OROFINO u. FLORY	HUNT, NEWMAN, SCHERAGA u. FLORY
Hydroxyäthylcellulose in Wasser in Cadoxen	 780 320	 152 (30) 62 (12)	 3,60 2,30	aus A_2 nach OROFINO u. FLORY	BROWN, HENLEY u. ÖHMAN
Na-Carboxymethylcellulose in Cadoxen	324	63 (12)	2,32	aus A_2 nach OROFINO u. FLORY	BROWN, ÖHMAN u. HENLEY
Cellulose in Cadoxen	362	70 (14)	2,46	aus A_2 nach OROFINO u. FLORY	BROWN, ÖHMAN u. HENLEY
Amylose	79	19 (5)	1,78	Θ-Lösungsmittel	BURCHARD
Polymethylmethacrylat	30	12±2 (5)	1,82	Θ-Lösungsmittel	SCHULZ u. KIRSTE

Es ist nun aufschlußreich, die ungestörten Knäueldimensionen mit denen anderer Cellulosederivate zu vergleichen. In Tab.2 ist als erstes das Verhältnis von R_o^2/P angegeben, und daraus ist die Länge des Kuhn'schen Segmentes berechnet worden. Diese Größe berechnet sich aus $\langle R_o^2 \rangle/P$ zu [10]

$$A_m = \frac{6 \cdot \langle R_o^2 \rangle}{P} \cdot \frac{L}{P} \qquad \text{(III)}$$

wobei P der Polymerisationsgrad ist und L die maximale Länge des Fadenmoleküls, die sich ohne Deformation der Valenzwinkel ergibt. Die Werte in den Klammern bedeuten die Anzahl der monomeren Einheiten pro Segment. Dann folgt eine Spalte mit den R_o/R_f-Werten und schließlich die Methode und der Bearbeiter.

Man entnimmt dieser Tabelle, daß die Carbanilate von den aufgeführten Polymeren die steifsten Fadenmoleküle sind, dann folgen die Cellulosenitrate, die Hydroxyäthylcellulosen, die Carboxymethylcellulosen, und schließlich die Cellulose und Amylose selbst. Als letztes ist zum Vergleich Polymethylmethacrylat als ein Vertreter der Vinylpolymeren angegeben. Die unsubstituierten Polysaccharide sind also keineswegs, wie bisher angenommen war, besonders steif, sondern entsprechen in der Flexibilität durchaus den Vinylpolymeren. Auffällig ist der Unterschied in der Rotationsbehinderung zwischen der Cellulose und der Amylose, der im Gegensatz zu den Ergebnissen an den Carbanilaten steht. Dazu ist nun zu sagen, daß der Trägheitsradius bei der Cellulose nicht direkt am θ-Punkt gemessen wurde, sondern über die Theorie von OROFINO und FLORY aus dem 2. Virialkoeffizienten berechnet wurde [14]. Gegen dieses Verfahren sind aber einige Bedenken anzuführen. Dieses verdeutlicht die Abb.7. Hier ist für verschiedene Substanzen die Wechselwirkungsfunktion z.h (z/α^2) gegen α^2 aufgetragen, so wie man sie aus den Messungen erhält. Aus dieser Gegenüberstellung ist zu entnehmen, daß die Funktion z.h (z/α^3) entgegen der herrschenden Auffassung nicht nur von dem Expansionskoeffizienten abhängt, sondern auch von der Art des Polymeren, oder, wie ich vermute, von der Steifheit der Kette. Die gestrichelte Kurve gibt die Theorie von FLORY und OROFINO wieder in der Gestalt, wie sie von HUNT, NEWMAN, SCHERAGA und FLORY [11], sowie von BROWN, HENLEY und ÖHMAN [12)13] angewendet wurde. Man sieht, daß die Theorie die experimentellen Ergebnisse in keinem Falle richtig wiedergibt. Es erscheint mir daher durchaus denkbar, daß auch bei den unsubstituierten Amylosen und Cellulosen in der Rotationsbehinderung keine nennenswerten Unterschiede bestehen.

Zum Schluß möchte ich noch ein Ergebnis unserer Diffusionsmessungen an den Amylosetricarbanilaten anführen. Die Diffusionskonstante hängt mit dem Trägheitsradius nach der Theorie von KURATA und YAMAKAWA [15] wie folgt zusammen.

$$D \cdot \eta_0 = \frac{RT}{\psi(x) \cdot \langle R_0^2 \rangle^{1/2} \cdot \alpha^{0,654}} \qquad \text{(IV)}$$

Hierin ist R die Gaskonstante, ψ (x) ein Durchspülungsfaktor und η_0 die Viskosität des Lösungsmittels. Die Werte D. η_0 werden danach umso kleiner, je größer der Trägheitsradius ist. Trägt man also die $D\eta_0$-Werte gegen den Polymerisationsgrad auf, so muß sich beim Vergleich mit anderen Polymeren entscheiden lassen, welches das steifste Molekül ist. In Abb.8 habe ich eine solche Auftragung vorgenommen, und daraus entnimmt man, daß Cellulosenitrate und Amylosecarbanilate etwa gleich große Trägheitsradien besitzen in Übereinstimmung mit den direkten Messungen des Trägheitsradius, während die Trägheitsradien von PMMA wesentlich kleiner sind und dementsprechend die $D\eta_0$-Werte wesentlich größer. Das bestätigt erneut, daß die Amylosetricabanilate mit zu den steifsten Fadenmolekülen gehören.

L I T E R A T U R

1) BURCHARD,W., Makromolekulare Chem. 42, 151 (1960)
2) BURCHARD,W., E.HUSEMANN, Makromolekulare Chem. 44-46, 358 (1961)
3) OUTER,P., C.I.CARR, B.H.ZIMM, J.chem.Physics 18, 830 (1950)
4) SCHULZ,G.V., R.KIRSTE, Z.physik.Chem. (Frankfurt/M.) 30, 171 (1961)
5) BURCHARD,W., Makromolekulare Chem. 59, 16 (1963)
6) FLORY,P.J., J.chem.Physics 17, 303 (1949)
7) FOX,T.G., FLORY,P.J., J.physic.Colloid Chem. 53, 187 (1949);J.Polymer Sci.5,745(1950)
8) STOCKMAYER,W.H., Makromolekulare Chem. 35, 54 (1959), dort weitere Literatur

9) REEVES,R.E., J.Amer.chem.Soc. 76, 4595 (1954)
10) vgl. STUART, Physik der Hochpolymeren Bd.1, Springer-Verlag, Heidelberg
11) HUNT,M.L., S.NEWMAN, A.SCHERAGA, P.J.FLORY, J.phys.Chem. 60, 1278 (1956)
12) BROWN,W., D.HENLEY, J.ÖHMAN, Makromolekulare Chem. 64, 49 (1963)
13) BROWN,W., J.ÖHMAN, D.HENLEY, Makromolekulare Chem. 62, 164 (1963)
14) OROFINO,T.A., P.J.FLORY, J.chem.Physics 26, 1067 (1957)
15) KURATA,M., H.YAMAKAWA, J.chem.Physics 29, 311 (1958)
16) MEYERHOFF,G., J.Polymer Sci. 29, 399 (1958)
17) CANTOW,H.J., Z.physik.Chem. (Frankfurt/M.) 1, 390 (1964)

DISKUSSIONSBEMERKUNGEN

BERG : Ihre Substanzen eignen sich wegen ihrer Stabilität und relativen Einheitlichkeit zu Präzisionsmessungen und damit zur Überprüfung der Theorien. Wie hoch ist die Einheitlichkeit Ihrer Präparate ?

BURCHARD : Bei unseren ersten physikalischen Messungen haben wir aus Kartoffelstärke isolierte natürliche Amylose verwendet, die mit M_w/M_n ca. 2 eine breite molekulare Verteilung hat. Nachdem wir aber durch ausführliche Messungen nachweisen konnten, daß aus Glucose-1-phosphat (Cori-Ester) durch Phosphorilase synthetisierte Amylosen dieselben physikalischen Eigenschaften besitzen, haben wir später nur noch synthetische Amylosen verwendet. Der Mechanismus dieser Synthese verläuft nach einem Poissonmechanismus, d.h. es entstehen im allgemeinen Produkte mit hoher Einheitlichkeit von M_w/M_n ca.1,04. Die Substanzen weisen allerdings in manchen Fällen eine breitere Verteilung auf, die sich vor allem in einem langen Ausläufer zu kleineren Molekulargewichten äußert. Diese Verbreiterung entsteht vermutlich nachträglich beim Aufarbeiten der Substanzen durch Einwirkung von α-Amylase. So erhielten wir in einem Falle eine Verteilung mit $M_w/M_n = 1,24$ und $M_z/M_w = 1,05$.

Bei den Cellulosecarbanilaten handelt es sich in jedem Falle um sorgfältig fraktionierte Mittelfraktionen,deren Einheitlichkeiten in der Größenordnung von M_w/M_n=1,1 liegen.

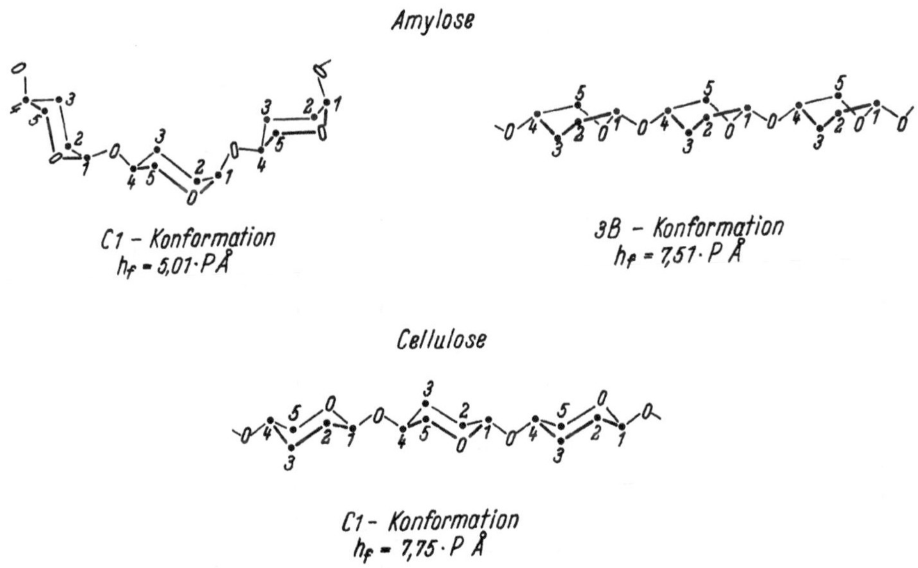

Abb.1 Die sterischen Verhältnisse bei Amylose und Cellulose (h_f=Fadenendenabstand bei freier Drehbarkeit der Glucoseeinheiten).

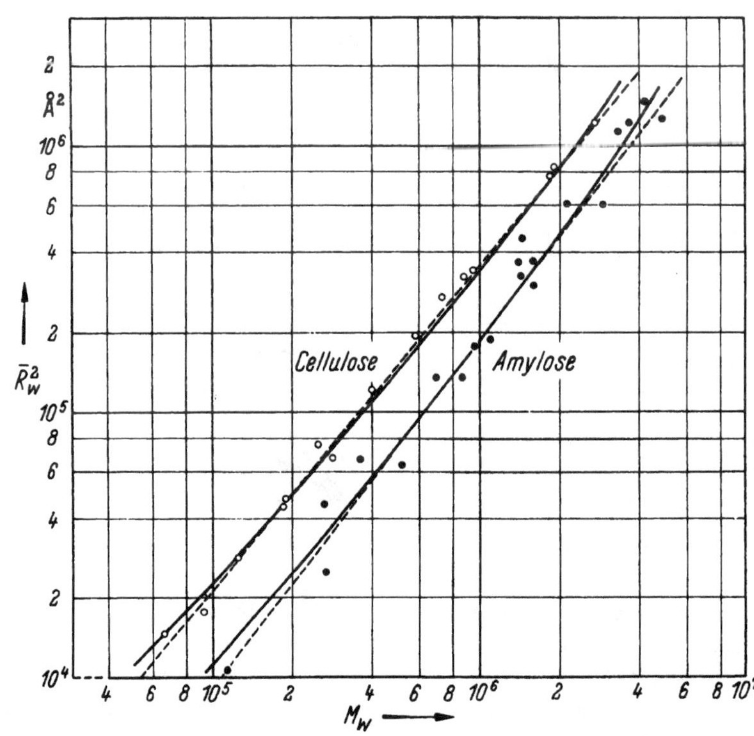

Abb.2 Die Trägheitsradienquadrate bei Amylose- und Cellulosetricarbanilaten in Aceton in Abhängigkeit von Molekulargewicht bei 20°C.

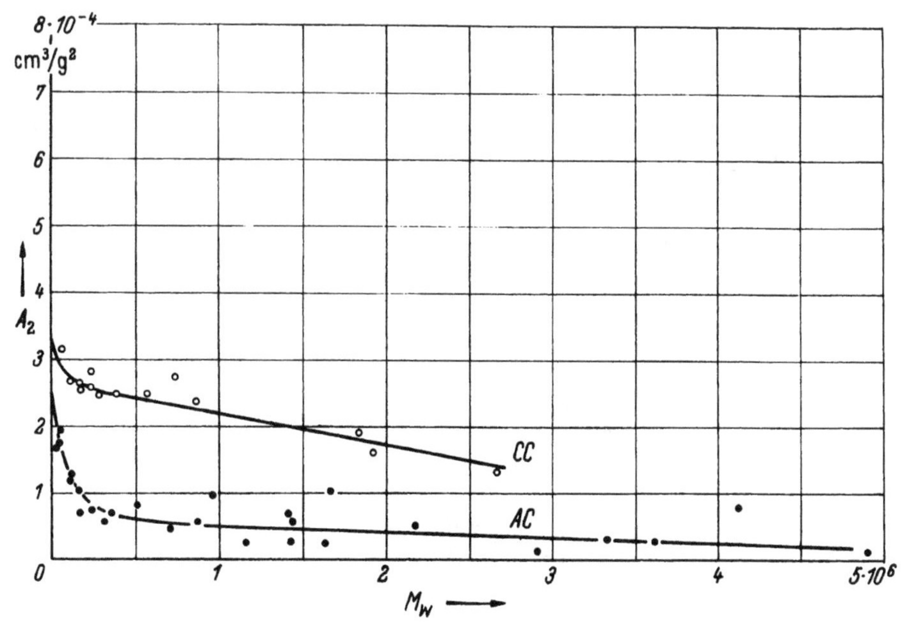

Abb.3 Abhängigkeit des 2.Virialkoeffizienten vom Molekulargewicht bei Amylose- (AC) und Cellulosetricarbanilat (CC) in Aceton bei 20°C.

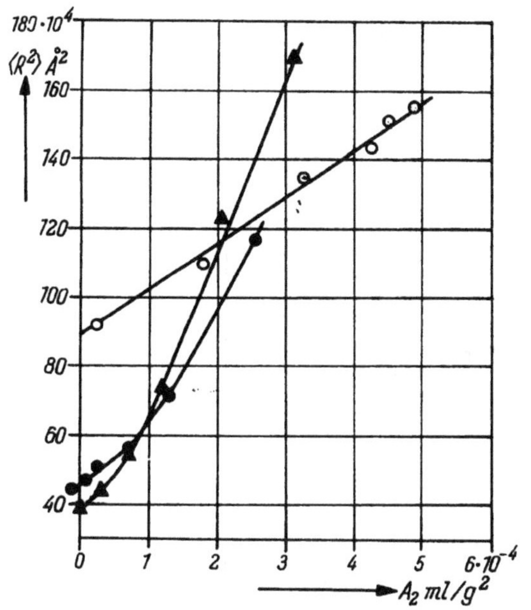

Abb.4 Beziehung zwischen Trägheitsradius und 2.Virialkoeffizienten bei Amylosetricarbanilat (h_w=2,18.10^6) in - o - Dioxan/Methanol,-•- Methyläthylketon/Methanol,-▲- Diäthylketon/Methanol,bei 20°C.

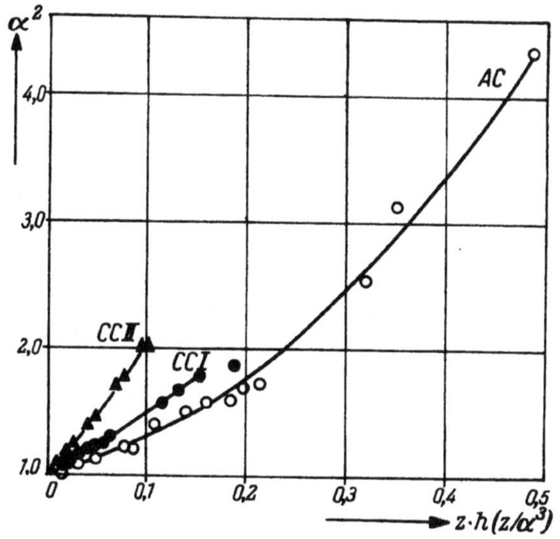

Abb.5 Beziehung zwischen dem Expansionskoeffizienten α^2 und der Wechselwirkungsfunktion Z.h(Z/α^3) bei Amylosetricarbanilat (AC) und Cellulosetricarbanilat (CC).

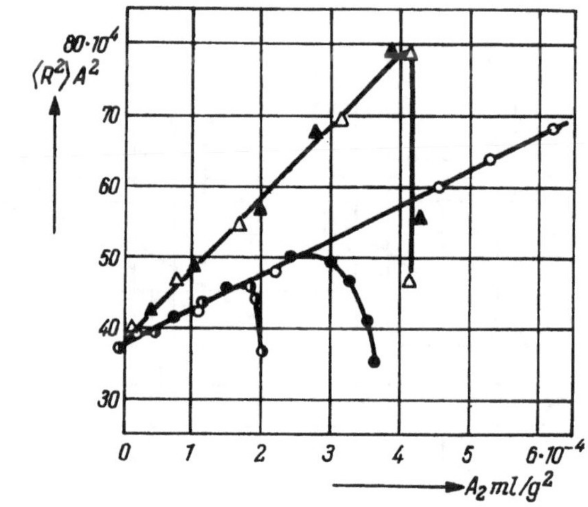

Abb.6 Beziehung zwischen Trägheitsradius und 2.Virialkoeffizienten bei Cellulosetricarbanilat CM_w=8,4.10^5) in - o - Dioxan/Methanol, -◐- Aceton/Wasser, -●- Methyläthylketon/Methanol, -▲- Diäthylketon/Methanol, -△- Dipropylketon/Methanol bei 20°C.

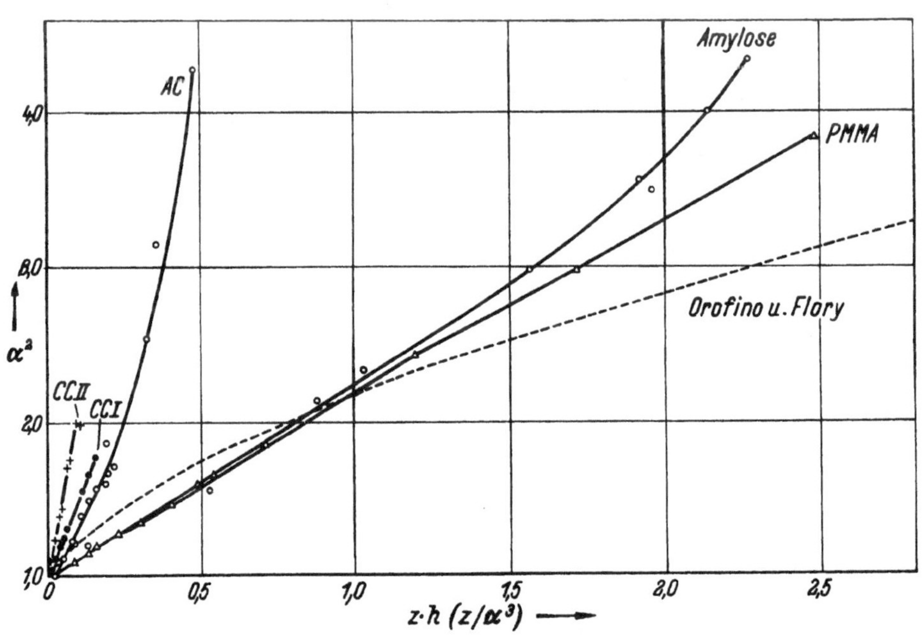

Abb.7 Verlauf der Wechselwirkungsfunktion Z.h (Z/α^3) nach Messungen an Polymethylmethacrylat (PMMA), Amylose, Amylosetricarbanilat (AC) und Cellulosetricarbanilat (CC I und CC II) und nach der Theorie von OROFINER und FLORY.

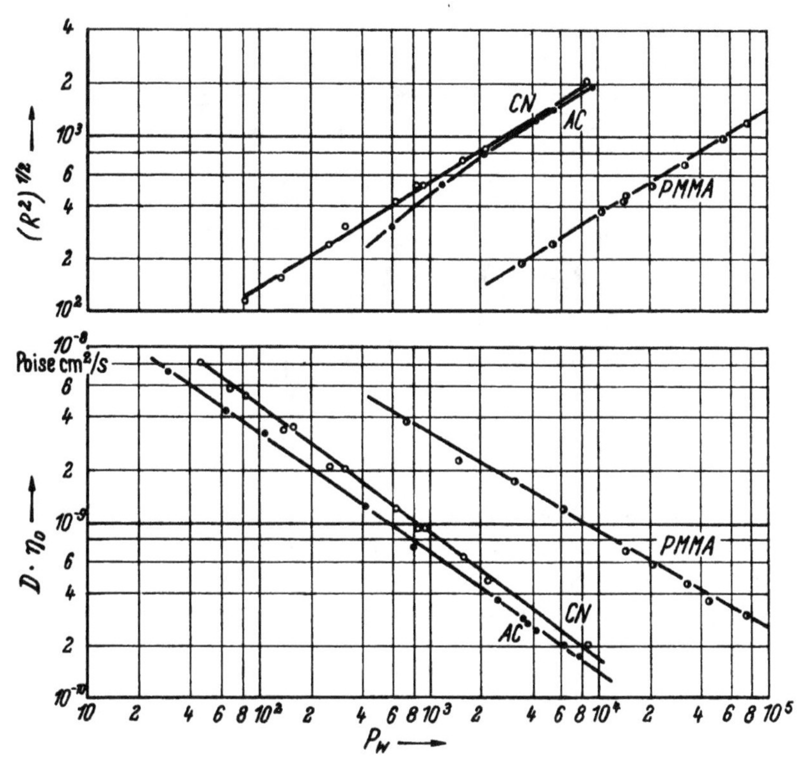

Abb.8 Vergleich der Trägheitsradien und der $D.\eta_0$-Werte von Cellulosetrinitro (CN) in Aceton, Amylosetricarbanilat (AC)in Dioxan und Polymethylmethacrylat (PMMA) in Aceton.

B Polysaccharide

Fraktionierung von Polysacchariden mit Gelfiltration

K. A. GRANATH

AB Pharmacia, Uppsala

S U M M A R Y

In a few years gel filtration has become an important tool in the purification of biological materials and the separation or fractionation of high molecular substances. A brief theory of the method is given below together with a review of its applications in the polysaccharide field.

B e s c h r e i b u n g d e r M e t h o d e

Unter Gelfiltration versteht man eine Art von Chromatographie, die sich dadurch auszeichnet, daß sie die Trennung von Molekülen auf der Basis von Molekulargewichtsunterschieden erlaubt. Als Trägermaterial werden Körnchen aus vernetzten Makromolekülen benutzt. Die im Lösungsmittel gequollene Gelmatrix besteht aus Poren verschiedener Größe. Wenn sich in dem umgebenden Lösungsmittel (meistens Wasser oder Pufferlösung) Substanzen befinden, deren Moleküldurchmesser von etwa der gleichen Größe ist wie die Poren im Gel, so durchdringen Moleküle bis zu einer bestimmten Größe das Wasser in den Gelpartikeln (Innenvolumen). Den etwas größeren Molekülen ist nur ein Teil der Poren zugänglich, während den größten Molekülen gegebenenfalls nur das Lösungsmittel außerhalb der Gelkörnchen zur Verfügung steht. Füllt man ein solches gekörntes Gel zusammen mit dem Quellmittel in ein senkrechtes Glasrohr, bringt obenauf ein Substanzgemisch der oben beschriebenen Art und eluiert mit dem gleichen Lösungsmittel, so wandern die größten Moleküle an den Gelpartikeln vorbei. Sie verlassen die Säule zuerst, und zwar mit einer dem Lösungsmittelvolumen außerhalb der Gelpartikel entsprechenden Flüssigkeitsmenge (Außenvolumen, void volume). Die kleinsten Moleküle, die das Gelinnere völlig penetrieren, werden erst nach Durchgang einer weitaus größeren, dem äußeren und inneren Lösungsmittelvolumen entsprechenden Flüssigkeitsmenge, eluiert. Stoffe mit mittleren Molekulargewichten erscheinen dazwischen in der Reihenfolge der Abnahme ihres Molekulargewichtes.

Eine mit Gelpartikeln im Quellmittel gefüllte Säule kann mit folgenden Kennzeichen beschrieben werden:

$$V_t = V_o + V_i + V_g, \text{ d. h.}$$

das Totalvolumen, das sich aus der Geometrie des Gelbettes ergibt, setzt sich zusammen aus dem Lösungsmittelvolumen außerhalb der Gelpartikel (V_o), dem Volumen innerhalb der Gelpartikel (V_i) und dem vergleichsweise geringen Volumen der Gelmatrix (V_g).

Alle diese Volumina sind "Materialkonstanten" für jede Gelsäule und brauchen nur einmal bestimmt zu werden. V_o ist identisch mit dem Elutionsvolumen einer völlig von der Gelmatrix ausgeschlossenen Substanz oder Molekülgröße. Dementsprechend kann V_o mit Tusche oder Hämoglobin bestimmt werden. V_i hängt von der Quellbarkeit des verwendeten Gels ab. Beim Einwiegen von a Gramm trockenen Gelpulvers errechnet sich das innere Volumen gemäß:

$$V_i = a.W_r$$

Dabei ist W_r = water regain, d. h. Wasseraufnahme in g/g Trockengel.

Wenn das Trockengewicht der Säulenfüllung nicht bekannt ist, läßt sich V_i aus der Dichte des Gels im nassen Zustand ermitteln:

$$V_i = \frac{W_r.d}{W_r + 1} (V_t - V_o)$$

Für Dextrangel ist $d = \frac{1 + W_r}{0,61 + W_r}$, also ebenso eine Materialkonstante.

Eine interessante experimentelle Bestimmungsmethode des inneren Volumens V_i basiert auf der Tatsache, daß ein kleines Molekül ein dem inneren und äußeren Volumen entsprechendes Elutionsvolumen hat, d. h. $V_e = V_o + V_i$. Geeignete Substanzen dafür sind z. B. mit Tritium markiertes Wasser und KCl.

Jede Fraktionierung mit Gelfiltration basiert auf einer unterschiedlichen Beeinträchtigung der Diffusion je nach Molekülgröße und Größe der Gelporen. Deshalb werden die Moleküle beim Durchlaufen der Säule in verschiedenem Grad verzögert. Das Elutionsvolumen summiert sich dann zu

$$V_e = V_o + K_d.V_i, \text{ wobei}$$

$K_d.V_i$ den Bruchteil des inneren Volumens angibt, der gerade für die bestimmte Molekülgröße zugänglich ist. Für die beiden Extremfälle (unbehinderte Diffusion $K_d = 1$; totaler Ausschluß, $K_d = 0$) ist K_d ein wahrer Verteilungskoeffizient. K_d kann ebenfalls als Stoffkonstante für eine gewisse Molekülgröße in einem Gel mit bestimmtem Vernetzungsgrad betrachtet werden, weitgehend unabhängig von pH, Ionenstärke oder Konzentration der Lösung. Man hat gegebenenfalls eine lineare Abhängigkeit zwischen $K_d^{1/3}$ und $MG^{1/2}$ gefunden.

Die Elutionsvolumina für zwei Stoffe mit verschiedenen K_d-Werten $V'_e = V_o + K'_d.V_i$ und $V''_e = V_o + K''_d.V_i$ unterscheiden sich um einen Bruchteil des inneren Volumens wie

$$V_s = (K'_d - K''_d) V_i.$$

Mehrere hydrophile Gele, z. B. Stärkepartikel, Agar und Polyacrylamidgel, sind für die Gelchromatographie in Wasserlösung benutzt worden. Das weitaus am meisten gebräuchliche Trägermaterial ist dabei das Dextrangel Sephadex[R] (Pharmacia, Uppsala, Schweden). Es besteht aus Dextranketten, die mit Epichlorhydrin an den Hydroxylgruppen quervernetzt und dadurch unlöslich gemacht worden sind. Starke Vernetzung liefert ein Gel mit geringem Quellungsvermögen, geringer Vernetzungsgrad bewirkt hingegen eine starke Quellung im Wasser. Die Porosität wird durch die Angabe von "water regain" definiert. Für die vorliegenden Typen variiert W_r von 2.5 bis 20 und die entsprechenden oberen Ausschlußgrenzen liegen, anhand von Prüfungen mit Dextranfraktionen, bei Molekulargewichten von 5000 bis 200 000. Nach der Perjodatoxydation sind noch 40 % bzw. 80 % von den Glucosemolekülen in diesen Sephadextypen unvernetzt. Glucosidbindungen bestimmen die chemische Stabilität des Gels. Starke Mineralsäuren (stärker als 0.1 N) führen mit der Zeit zur Hydrolyse von diesen. Bei langem Stehen vermeidet man eventuelles Bakterienwachstum durch Zusatz von Spuren von Konservierungsmittel zur Elutionsflüssigkeit. Sephadex kann in gequollenem Zustand bei 110° C sterilisiert werden. Eventuell kann man gequollenes Sephadex nach vorherigem Entquellen mit ansteigenden Äthanolzusätzen trocknen.

Experimentelle Technik

Um ein optimales Separationsvermögen und eine befriedigende Durchflußgeschwindigkeit zu erzielen, muß die Chromatographiesäule sorgfältig gepackt werden. Einzelheiten über die Quellung des Gels, Packung der Säule und Auftragen der Probelösung sind in der Literatur [1] ausführlich beschrieben.

Für die Handhabung einer Dextrangel-Säule gelten die üblichen Regeln der Säulenchromatographie. Die Wahl des Gels, die Dimensionen der Säule und die Analyse des Eluates hängen weitgehend von dem Zweck des Experimentes ab. Der Trenneffekt ist von der Konzentration der Probe weitgehend unabhängig, wenn nicht die Diffusion durch zu hohe Viskosität behindert wird. Das Volumen der Probe dagegen muß beachtet werden. In Gruppenseparationen mit großem K_d-Unterschied kann das Probevolumen 20 - 30 % vom Volumen der Gelpackung ausmachen. Bei Fraktionierung von Polymerhomologen empfiehlt sich eine möglichst lange Säule und ein kleines Probevolumen. Eine Säulengeometrie von 1:20 bis 1:100 und Probevolumina von 5 - 10 % sind üblich. Wenn auch für feine und gleichmäßige Körnung des Gels und geringe Durchflußgeschwindigkeit (1 - 2 ml/cm^2 per Std.) gesorgt wird, dürfte man eine optimale Trennleistung erwarten.

Abbildung 1 zeigt eine Anordnung der Gelfiltration für präparative Zwecke. Besonders wirkungsvoll läßt sich bei kleinem K_d-Unterschied, wie z. B. beim Trennen von Oligomeren, eine Gelpackung durch zyclische Wiederholung des Trennvorgangs ausnutzen [2]. Beim Fraktionieren breiter Molekulargewichtsbereiche empfiehlt sich weiterhin eine Reihenschaltung von Säulen, die mit Gel von verschiedener Porosität gefüllt sind.

Fraktionierung von Polysacchariden

Seit der Entdeckung des Trennvermögens des Sephadex [3] hat die Gelfiltration in kurzer Zeit interessante Anwendungsbereiche gefunden, vor allem bei der Separation biologischen Materials. Die Fraktionierung der hochmolekularen Substanzen ist erst kürzlich durch die Einführung der hochporösen Gelsorten möglich geworden, und deshalb ist die Anzahl der Veröffentlichungen noch beschränkt. Man hat diese Methode vorerst für die Entsalzung der Polysaccharide und für Gruppenseparationen benützt. So entfernt man in klinischen Analysen durch Gelfiltration die Glucose aus dem Serum vor der Bestimmung der Polysaccharide wie Inulin oder Dextran [4][5]. Für die Entfernung des überschüssigen Jodates nach vollendeter Perjodatoxydation der Polysaccharide wurde Sephadex vorgeschlagen [6], und in unserem Laboratorium hat man für diesen Zweck bei der Präparation von oxydierten Polysacchariden eine semikontinuierliche Technik entwickelt.

Selbstverständlich wurde Sephadexgel frühzeitig für die Fraktionierung der niedermolekularen Homologen der Dextranhydrolysate benützt. Abb. 2 zeigt einen Teil eines solchen Versuches,um die Zusammensetzung des Dextranhydrolysates in Intervallen zu verfolgen. Im Fall a) war das mittlere Molekulargewicht ($\bar{M}_n$) der hydrolysierten Probe 320, und im Fall b) war $\bar{M}_n$ 190. Die Hydrolyse lief in 0.33 N HCl bei 100° C während 3 bzw. 9 Stunden. Nach dem Neutralisieren wurde die Probe mit einem kombinierten Ionenaustauscher entsalzt und die Fraktionierung mit Sephadexgel (W_r = 2.5) in einer Säule von 4 x 56 cm ausgeführt. Das Volumen der Probe war 5 ml und die Konzentration variierte von 2 bis 6 %. Die Fraktionierung nahm 7 bis 10 Stunden in Anspruch. Die Konzentration der eluierten Fraktionen wurde mit Anthron gemessen [7] und die einzelnen Bestimmungen des Polymerisationsgrades durch Endgruppenanalyse nach SOMOGYIs Methode [8] durchgeführt. Die einzelnen Oligomeren wurden weiterhin papierchromatographisch identifiziert. Keine der Komponenten wurde in diesem Versuch völlig rein gewonnen wie auch die gemessenen Molekulargewichte andeuten.

Durch Anwendung von Gelen verschiedener Quellbarkeit sollte es möglich sein, den Verlauf eines Polysaccharidabbaus innerhalb weiter Grenzen quantitativ verfolgen zu können.

Eine weitere Anwendung der Gelfiltration ist die präparative Separation der Oligosaccharide. Cellodextrin-Oligosaccharide wurden bis zum Hexameren erfolgreich voneinander getrennt [9] und die Dextranhydrolysate werden bei uns zu einzelnen Isomaltose-Oligosacchariden präparativ aufgearbeitet.

Geladene Polysaccharide lassen sich gut trennen, wenn für einen erforderlichen Elektrolytzusatz im Elutionsmittel gesorgt wird [10].

Eine vollständige Trennung der sukzessiven Individuen einer homologen Reihe erfordert eine unrealistisch große Gelsäule, sobald das Molekulargewicht der Homologen nur wenig differiert. Eine gewisse Fraktionierung nach Moleküldimensionen ist dagegen innerhalb des ganzen Bereiches möglich falls die verschiedenen Molekülgrößen unterschiedlich in das verwendete Gel hineindiffundieren können, d. h., K_d-Werte von 0 bis 1 haben.

Die stark vernetzten Gelsorten (W_r 2.5 bis 7.5) haben einen guten Trenneffekt im Molekulargewichtsbereich von etwa 1000 bis 50 000 [11]. Die Schärfe der eluierten Fraktionen wurde mehrfalls durch M_w/M_n Messungen kontrolliert. Abb. 3 zeigt eine Trennung von zwei Dextranfraktionen ($\bar{M}_n$ 2600 und 17 000). Diese wurden teils separat teils gemischt durch Sephadex eluiert. Die Probemenge war 50 ml 6 % Lösung bei der Fraktionierung der Einzelfraktionen. Die gemischte Probe bestand aus 100 ml, darin 3 % von jeder Fraktion. Die Probe passierte zuerst eine Gelsäule vom hochporösen Geltyp (W_r = 10.6) und wurde danach mittels Kapillarschlauch auf die zweite Säule übergeleitet. Diese war mit stark vernetztem Gel gefüllt (W_r = 2.6). Die Säulevolumina waren 1800 ml bzw. 3400 ml. Die Elutionsgeschwindigkeit war 50 ml/Std. und wurde mit einer peristaltischen Pumpe geregelt.

Mit den höchstporösen Geltypen (W_r 10 - 20) kann man den Molekulargewichtsbereich der klinischen Dextranpräparate erreichen. Die Gelfiltration ist tatsächlich heute als Routinemethode für die Analyse der Molekulargewichtsverteilung des klinischen Rheomacrodex^R ($\bar{M}_w$ 40 000, $\bar{M}_n$ 22 000) eingeführt.

Abbildung 4 zeigt die Molekulargewichtsverteilung eines Dextranpräparates, fraktioniert in zwei sukzessiven Gelsäulen, gefüllt mit Sephadex von W_r 20 bzw. von W_r 10. Die Gelvolumina waren 3000 ml und 1800 ml. Die Elutionsgeschwindigkeit war 40 ml/Std. Dabei nahm die Fraktionierung 120 Stunden in Anspruch. Die eluierten Fraktionen wurden auf ihre Grenzviskosität geprüft.

Abbildung 5 zeigt die Elutionskurve von Inulin. Die Fraktionierung wurde mit demselben Säulenpaar wie die bei der Dextranfraktionierung in Abb. 3 ausgeführt. Die eingebrachte Probe bestand aus 10 g Inulin (Gurr Ltd.) in 120 ml Wasser. Die fraktionierte Substanz erschien in 1375 ml Eluat aufgeteilt in Fraktionen von 55 ml. Die Inulinkonzentration in den Fraktionen wurde mit Anthron bestimmt. Nachher wurden die Fraktionen im Vakuum konzentriert und getrocknet. Die Molekulargewichte, die in dem Diagramm eingeführt sind, wurden mittels Lichtstreuung gemessen. Die unfraktionierte Substanz hatte $\bar{M}_w$ = 6800.

Betreffend andere Polysaccharide ist vor kurzem die Trennung der Amylose von Amylodextrin mit Sephadexgel beschrieben worden [12]. Dabei erschien Amylodextrin (D.P. 25) gut fraktioniert, Amylose dagegen zu hochmolekular, um in die Poren des verwendeten Gels (W_r = 7.5) einzudringen.

Die obigen Beispiele dürften den Nutzen der Gelfiltration für die Fraktionierung der Polysaccharide beleuchtet haben. Außer ihren präparativen Möglichkeiten sollte die Methode auch geeignet sein, bei entsprechender Kalibrierung und genauer Reproduktion der Versuchsbedingungen zur Darstellung der Molekulargewichtsverteilung auf der Basis der Elutionskurve zu gelangen.

L I T E R A T U R

1. FLODIN, P., Dextran gels and their applications in gel filtration. AB Pharmacia, Uppsala, 1962
2. PORATH, J., H. BENNICH, Arch. Biochem. Biophys. Suppl. 1, 319 (1962)
3. PORATH, J., P. FLODIN, Nature (London) 183, 1657 (1959)
4. JACOBSSON, L., Clin. Chim. Acta 7, 180 (1962)
5. ÖSTLING, G., Acta Soc. Med. Upsaliensis 65, 222 (1960)
6. CHATTERJEE, A.K., G.J. DURANT, H. HENDRICKSON, Y.C. LEE, R. MONTGOMERY, Biochem. Biophys. Res. Comm. 4, 425 (1961)
7. SCOTT, T., E. MELVIN, Anal. Chem. 25, 1656 (1953)
8. ISBELL, H., C. SNYDER, N. HOLT, M. DRYDEN, J. Research Natl. Bur. Standards 50, 81 (1953)
9. FLODIN, P., K. ASPBERG, Biological Structure and Function. Vol. 1, p. 345 Academic Press, London (1961)
10. GREGORY, J., L. RODÊN, Biochem. Biophys. Res. Comm. 5, 430 (1961)
11. GRANATH, K., P. FLODIN, Makromol. Chem. 48, 160 (1961)
12. NORDIN, P., Arch. Biochem. Biophys. 99, 101 (1962)
13. JOHANSSON, B., L. RYMO, Acta Chem. Scand. 16, 2067 (1962)

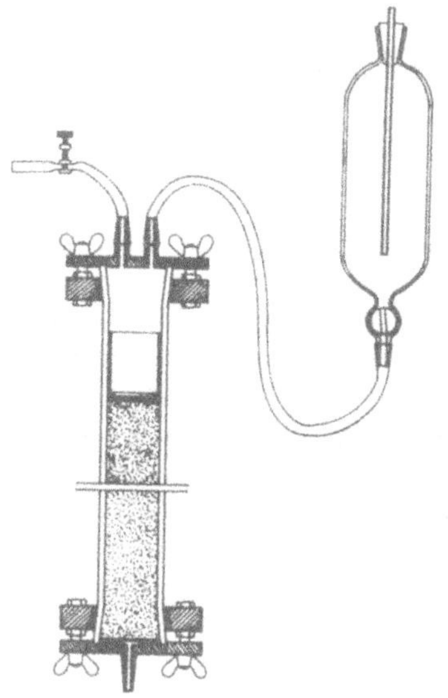

Abb.1 Säulenanordnung für Gelfiltration.

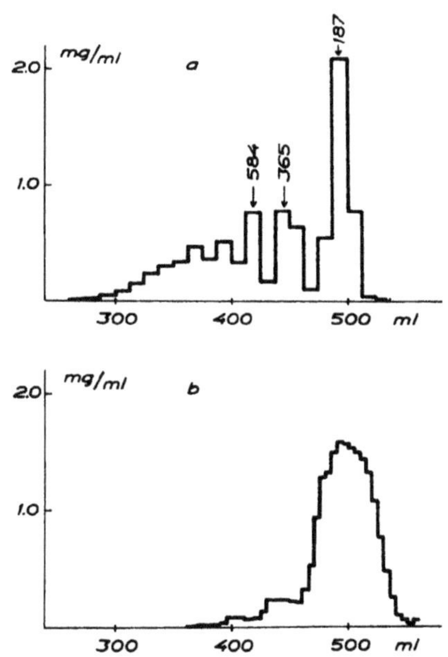

Abb.2 Elutionsdiagramme der Dextranhydrolysate, fraktioniert mit Sephadex (W_r 2.5). a) Zeit der Hydrolyse 3 Std.; $\overline{M}_n$=320. b) Zeit der Hydrolyse 9 Std.; $\overline{M}_n$=190.

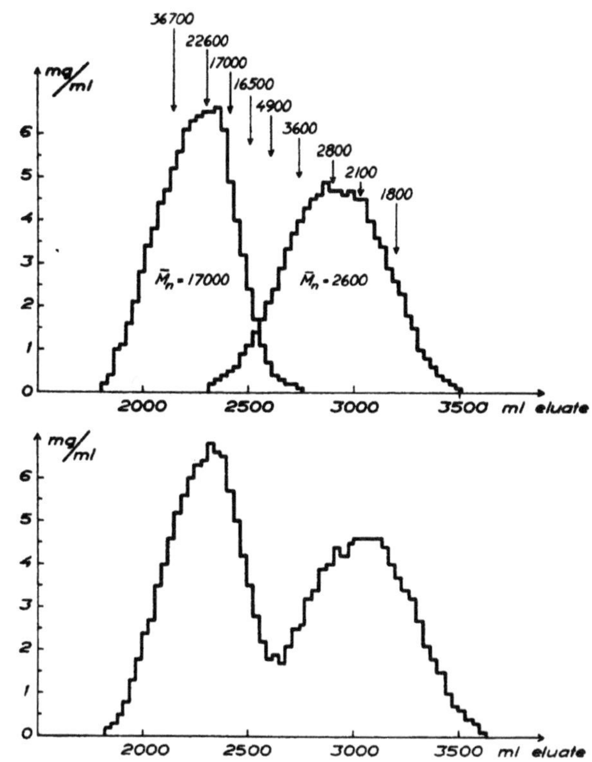

Abb.3 Gelfiltration von zwei Dextranfraktionen ($\overline{M}_n$ 17 000 und 2 600). Das obere Diagramm zeigt die Elutionskurven der Einzelfraktionen, das untere die Elutionskurve der Mischung derselben.

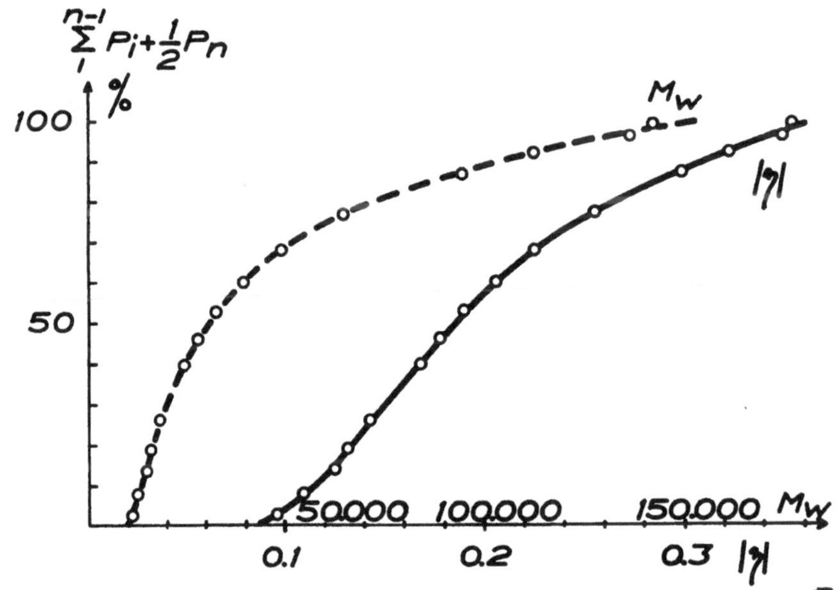

Abb.4 Die Molekulargewichtsverteilung eines Dextranpräparates ($\overline{M}_w$ = 45 000; $\overline{M}_n$=14 000), basiert auf Fraktionierung mit Sephadex. Die η-Kurve wurde auf Basis der Grenzviskositätsmessungen und Fraktionsgewichte gezeichnet. Die $\overline{M}_w$-Kurve wurde nach der Gl. $/\eta/=2,43.10^{-3}M_w^{0,42}$ kalkuliert. $\overline{M}_w$ einiger Fraktionen nach Lichtstreuungsmethode in guter Übereinstimmung mit den kalkulierten Werten.

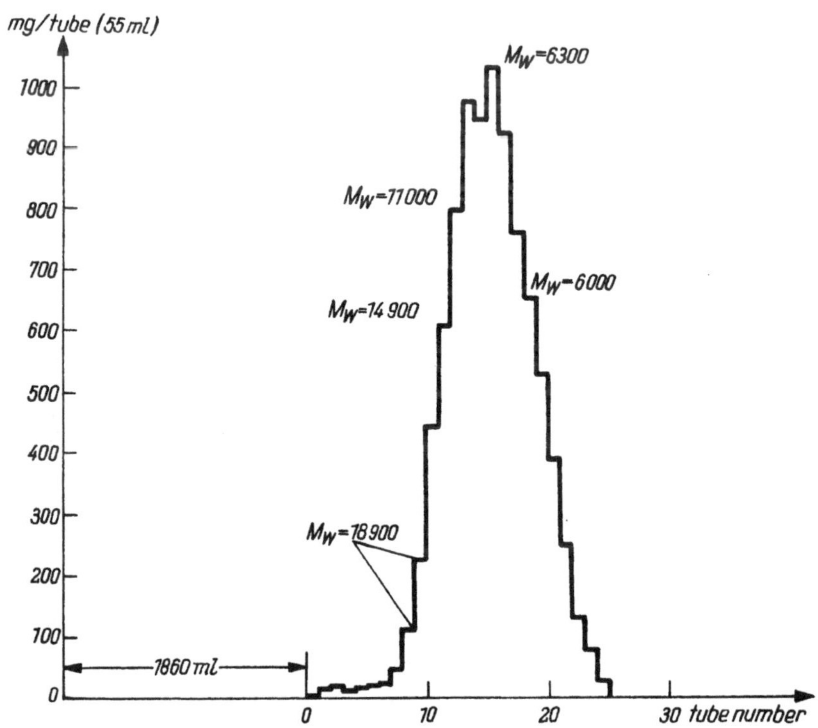

Abb.5 Das Elutionsdiagramm von Inulin. Die eingezeichneten Molekulargewichte wurden mit der Lichtstreuungsmethode gemessen.

B Polysaccharide

Beiträge zur Fraktionierung durch Gelfiltration

H. SEIDE und G. LANGHAMMER

Institut für Physik und physikalische Chemie der Hochpolymeren der Akademie, Leipzig

S U M M A R Y

It is shown that the fractionation of synthetic polymers in a suitable range of molecular weights can be achived by gel filtration and that the resulting molecular weight distribution functions are in excellent agreement with those by other methods. Also for the characterisation of a gel which may be applied for gel filtration, expecially a dextran gel, the gel filtration of a synthetic (and water soluble) high polymer with a well known molecular weight distribution may be used with advantage. Polyvinylpyrrolidone (Kollidon) has proved to be well suited for this purpose. From our experiments emerges that the so called value of "water-regain" alone is not sufficient for the characterisation of such a gel. Gels which have the same water regain may exhibit very different selectivities. It is proposed to supply the usual representation of an elution curve as a plot of concentration versus elution volume by a plot of intrinsic viscosity (or molecular weight) versus the elution volume. In a double logarithmic plot one gets two straight lines which are connected by a short transition curve, which may degenerate to a distinct break. The same picture is found with polystyrene gels.

E I N L E I T U N G

Das Grundsätzliche zur Gelfiltrationsmethodik hat meine Vorrednerin, Frau Dr.GRANATH, bereits in ihrem Vortrag geschildert, so daß ich auf eine Wiederholung verzichten darf. Da in der analytischen und evtl. technischen Praxis die Methode bisher wohl ausschließlich auf Trennungen in wässrigen Lösungen angewandt wurde, hat sich Frau Dr.GRANATH auf die Erörterung solcher Fälle beschränkt, insbesondere auf die Anwendung des vernetzten Dextrans als Gelsubstanz, an dessen Erprobung sie selbst je großen Anteil hat [1][2][3]. Auch die übrigen, neben den von der Firma Pharmacia kommerziell zur Verfügung gestellten vernetzten Dextranen herangezogenen Gelsubstanzen, wie Stärke- [4], Polyacrylamid- [5], Agar-Agar- [5][6], und Agarose-Gele [5][7] sind nur für wässrige Lösungen geeignet. Obwohl unsere Interessen, von ursprünlich ganz anderen Gesichtspunkten ausgehend, hauptsächlich bei den synthetischen Polymeren und nichtwässrigen Lösungsmitteln liegen, haben wir uns ebenfalls etwas mit Dextrangel beschäftigt. Dies einmal, weil im Institut von Herrn Prof. LEIBNITZ (Institut für Verfahrenstechnik der organischen Chemie der DAW) durch Herrn Dr. BEHRENS und Herrn Dr. GABERT ebenfalls vernetzte Dextrane hergestellt worden sind, die zu prüfen wir gebeten wurden, zum anderen, weil wir hofften, mit einem uns aus Thermodiffusionsversuchen [15] wohlbekannten wasserlöslichen Polymeren (Polyvinylpyrrolidon) Aufschlüsse über die Wirkungsweise der Gelfiltration zu erhalten. Ein Nebenziel der Untersuchungen war folgendes : Die biologischen Substanzen, zu deren Trennung die Gelfiltration bisher fast ausschließlich eingesetzt wurde, besitzen gewöhnlich ein diskretes Molekulargewichtsspektrum mit teilweise weit auseinanderliegenden Komponenten. Es erschien uns günstig, zur Testung eines Gels ein Polymeres mit wohlbekannter und praktisch kontinuierlicher Molekulargewichtsverteilung zu verwenden. Im folgenden möchten wir über unsere Ergebnisse berichten und damit einen kleinen Beitrag zur Kenntnis und Weiterentwicklung dieser gerade für die biogenen Makromoleküle so wertvollen Methode zu liefern.

1. Versuchsdurchführung

1.1 Eingesetzte Produkte

Es standen uns je eine Probe Sephadex G 100 und G 200 (Labormuster) zur Verfügung, die uns freundlicherweise von der Pharmacia AB Uppsala übersandt worden waren. Die im folgenden als Epidex bezeichneten Produkte wurden uns von Herrn Dr.GABERT und Herrn Dr.BEHRENS überlassen. Derartige vernetzte Dextrane werden unter der angeführten Bezeichnung in Kürze von Seiten des VEB Serumwerke Bernburg als Labormuster im 100-g-Maßstab zur Verfügung stehen. Die als sehr geeignete Testsubstanz herangezogenen Polyvinylpyrrolidone der BASF waren Kollidone mit den K-Werten 25 und 60, im folgenden kurz als K 25 und K 60 bezeichnet.

Allen Genannten möchten wir an dieser Stelle nochmals für die Bereitwilligkeit danken, mit der sie uns unterstützt haben.

1.2 Vorbereitung der Kolonnen

Besondere Sorfalt mußte auf die Vorbereitung des chromatographischen Betts verwandt werden. Das getrocknete und gesiebte Produkte (Korngrößen von 0,06...0,5 mm Ø) wurde unter Rühren in 0,5 m Natriumchloridlösung mindestens 24 Stunden gequollen. Die feinsten Gelpartikel wurden durch mehrmaliges Dekantieren entfernt. Wir verwendeten Glaskolonnen, die am unteren Ende mit einem Hahn versehen waren. Das Gelbett lagerte auf einem Stopfen aus Glaswatte, bei den späteren Versuchen auf Glasfritten (G 3). Der Durchmesser der von uns benutzten Kolonnen betrug 2,8 cm. Versuche mit engeren Kolonnen (1,2 cm Ø) ergaben schlechtere Ergebnisse. Wahrscheinlich traten schon Randeffekte auf. Die Kolonnen müssen sorgfältig gefüllt werden, da bei Kanalbildung oder Einschluß von Luftblasen die Lösung nicht in der gewünschten Weise gleichmäßig an den Gelkörnern verbeifließt, worunter die Trennwirkung leidet. Bei uns hat sich am besten das Einsedimentieren der Gelkörner unter ihrem eigenen Gewicht bewährt. Zu diesem Zweck wurden die Kolonnen genau senkrecht montiert und mit Wasser von Raumtemperatur gefüllt. Temperaturunterschiede dürfen nicht auftreten, da dies zu unerwünschter Blasenbildung führen kann. Eine Aufschwemmung des Gels wird durch einen Trichter in die Kolonne gebracht. Die Gelkörner sedimentieren langsam in die Kolonne und bei genau senkrechter Montage wächst die Kolonnenfüllung ganz gleichmäßig über den gesamten Querschnitt der Kolonne. Nach dem Sedimentieren wird das Natriumchlorid mit Wasser ausgewaschen. Wenn kein Chlorid mehr nachweisbar war und auch interferometrisch keine Verunreinigungen mehr nachgewiesen werden konnten, war die Kolonne versuchsfertig. Die Oberfläche wurde durch ein aufgelegtes Filterblättchen vor Aufwirbelung geschützt. Die Substanz wurde mit einer Kapillare an die Glaswand gegeben, um eine Kanalbildung durch Auftropfen zu vermeiden.

1.3 Bestimmung des Quellwertes und des "Water regain"

Die Gele sollten nicht allein durch die mit ihnen erhaltenen Verteilungskurven der Testsubstanz, sondern auch durch andere, bisher allein verwendete Kenngrößen charakterisiert werden. Dies war notwendig, um die günstigste Arbeitsweise bei der Vernetzung festzulegen und die in verschiedenen Chargen vernetzten Dextrane möglichst schnell miteinander vergleichen zu können. Da man die Quellung der Substanz ausnutzt, liegt es nahe, den Quellwert zu bestimmen. Die Quellwerte geben gleichzeitig Aufschluß über die Mengen, die man für eine vorgegebene Kolonne benutzen muß. Die Zeitabhängigkeit des Quellwertes ist bei allen Trockenprodukten in einer Apparatur nach ENSLIN [9] gemessen worden. Die Ergebnisse sind in Tabelle 1 gezeigt.

Die Bestimmung des Quellwertes, d.h. des von der Trockensubstanz im ganzen aufgenommenen Wassers, reicht nicht aus zur eindeutigen Kennzeichnung der Gele. Ein Teil des Wassers befindet sich im Gelkorn, der zweite zwischen den einzelnen Körnern. Dieser zweite Teil ist abhängig von den Zwischenräumen zwischen den Körnern und somit von der Korngröße. Damit wird auch der Gesamtwassergehalt des gequollenen Gels von der Korngröße abhängig. Diese Abhängigkeit konnte experimentell bestätigt werden, wie sich aus den Versuchen mit der ENSLIN-Apparatur ergab, bei denen man unmittelbar den Gesamtwasserinhalt des Gels erhält. Der Wasseranteil, der sich innerhalb der Gelkörner befindet, sollte in Beziehung stehen zum Grad der Vernetzung, also charakteristisch für das Gel und unabhängig von der Korngröße sein. Nach den bisherigen Vorstellungen (z.B. [8]) sollte dieser Wasseranteil ein direktes Maß für die Maschengröße sein. Dieser Anteil wird im folgenden immer als "Water regain" (Wr-Wert) bezeichnet. Anhand dieses Wertes kann gegebenenfalls entschieden werden, ob bei der Vernetzung verschiedener Chargen das gleiche Produkt entsteht. Fernerhin müßte, je größer dieser Wert ist, die Maschenweite größer sein und damit müßten größere Moleküle in das Gel eindringen können. Die experimentelle Bestimmung des Wr-Wertes erfolgte nach einer von PEPPER vorgeschlagenen und von FLODIN [8] modifizierten Weise. Dabei werden etwa 10 ml des Gels in einen Plastbehälter, der unten mit einem kleinen Perlonnetz verschlossen ist, gegeben. Der Behälter kann in ein Zentrifugenglas gehängt werden, hat aber nur die halbe Länge des Glases, so daß der untere Teil des Glases frei bleibt. Beim Zentrifugieren kann das Wasser aus den Zwischenräumen zwischen den Körnern nach unten austreten. Der Anteil in den einzelnen Körnern verbleibt im Gel. Der Inhalt des Behälters wird gewogen und bei 105°C bis zur Gewichtskonstanz getrocknet. Der Wr-Wert wird in Gramm Wasser je Gramm Trockensubstanz ausgedrückt. Die Naßdichte der Gele ist mit dem zentrifugierten Gel im Pyknometer bestimmt worden. Die Ergebnisse sind ebenfalls in Tabelle 1 dargestellt.

1.4 Die verwendeten Gele

Es wurden in der Hauptsache drei Gele getestet, die im folgenden als Epidex A, B, C bezeichnet werden. Sie sind unter stets gleichen Bedingungen mit Epichlorhydrin vernetzt worden. Die unter gleichen Bedingungen vernetzten Gele ergaben jeweils gut reproduzierbare Wr-Werte. In Tabelle 1 sind noch einige Gele, die mit den entsprechenden Mengen Epichlorhydrin unter anderen Bedingungen vernetzt worden sind, aufgeführt (Epidex a, b). Die Proben d und e sind unter den gleichen Bedingungen wie Epidex a und b, aber mit geringeren Mengen an Epichlorhydrin vernetzt worden. Unter den gleichen Bedingungen wie die Standardprodukte A, B und C wurden zwei Proben Nativdextran vernetzt. Sie tragen die Bezeichnungen Epidex E und Epidex F.

Tabelle 1

Bezeichnung:	Wr-Wert g/g	Quellwert ml/g	Naßdichte g/ml
Sephadex G 200	20,2	40	-
Epidex A	13,3	18,2	1,01
Epidex B	6,7	10,9	1,05
Epidex C	4,6	8,0	1,07
Epidex a	4,1	-	1,08
Epidex b	6,6	-	1,06
Epidex d	(nicht bestimmbar) sehr groß	-	-
Epidex e	8,1	-	-
Epidex E	19,3	-	-
Epidex F	14,6	-	-

Die große Bedeutung der Vernetzungsbedingungen ergibt sich aus Tabelle 1. Die Produkte Epidex A und Epidex a sind mit gleichen Mengen an Epichlorhydrin vernetzt worden. Der Wr-Wert beträgt einmal 13,3 und einmal 4,1. Ein anderes Beispiel ist Epidex b, das mit mehr Epichlorhydrin vernetzt worden ist als Epidex a, jedoch einen größeren Wr-Wert (6,6) aufweist als jenes (4,1). Die Vernetzungsbedingungen haben auch, wie die folgenden Versuche zeigen, entscheidenden Einfluß auf die Selektivität der Gele. Auf den großen Effekt der Vernetzungsbedingungen hat schon FLODIN [8] aufmerksam gemacht, wobei die Beeinflussung des Wr-Wertes im Vordergrund stand.

2. ERGEBNISSE

2.1 Die erhaltenen Verteilungskurven und ihre Abhängigkeit vom Wr-Wert

Zuerst sollen die Produkte mit verschiedenen Wr-Werten betrachtet werden, die unter gleichen Bedingungen vernetzt worden sind. Zum zweiten ist von Interesse, wie sich die Selektivität ändert, wenn man Gele mit gleichen Wr-Werten, die aber unter verschiedenen Bedingungen vernetzt worden sind, zur Trennung benutzt.

In Abb.1 sind die mit Epidex A, B und C erhaltenen Verteilungskurven für K 25 dargestellt. Man sieht, daß im niedermolekularen Teil der Verteilungskurve Übereinstimmung herrscht. Die Verteilungskurven mit Gelen von kleinerem Wr-Wert zeigen bei größeren Molekulargewichten einen steileren Anstieg. In Abb.2 sind mit verschiedenen Durchlaufgeschwindigkeiten aufgenommene Verteilungskurven von K 25 an Epidex A dargestellt. Man sieht, daß die Kurven im untersuchten Bereich zwischen 1,4 ml/h und 25 ml/h gut reproduzierbar sind. Von großer Bedeutung für eine präparative Anwendung der Methode, ebenso wie für die eigentliche Fraktionierung, ist dabei noch, daß die Kurven mit 5 %-iger Ausgangslösung gewonnen worden sind. Auf diese Weise sind Mengen von 1 g an etwa 100 ml Gel fraktioniert worden. Auch noch höhere Ausgangskonzentrationen (bis 20 %) können in diesem Molekulargewichtsbereich noch mit guten Ergebnissen angewendet werden. Einzelne Fraktionen aus diesen Versuchen sind durch Trübungstitration untersucht worden. Deren Ergebnisse nach einer von SCHOLTAN [10] angegebenen Methode können als recht zuverlässig angesehen werden. Diese Messungen waren u.a. deshalb von Interesse, weil zu entscheiden war, ob die mit 5 %-iger Ausgangskonzentration durchgeführten Versuche wirklich Fraktionen mit enger Verteilung ergeben. Bekanntlich muß man bei der fraktionierten Fällung mit sehr viel geringerer Konzentration arbeiten, um ausreichend scharfe Fraktionen zu erhalten. Es kann festgestellt werden, daß die mit der Gelfiltration bei der relativ hohen Ausgangskonzentration gewonnenen Verteilungen der Fraktionen durchaus den Ergebnissen einer guten Fällungsfraktionierung gleichwertig sind. Als Beispiele sind in Abb.3 die Verteilungskurven einiger Fraktionen dargestellt. Die Fraktionen haben, vor allem bei kleinen Molekulargewichten, sehr enge Verteilungen ($U < 0,1$), bei mittleren Molekulargewichten ist $U < 0,2$ und bei den höchsten $U < 0,4$. Dieses Ergebnis ist bemerkenswert, da an derartigen Gelen das Dextran selbst fraktioniert werden kann. Nach Mitteilung der Firma Pharmacia und Angaben von FLODIN [8] sollte es möglich sein, die Kolonnendimensionen erheblich zu vergrößern. Kolonnen mit etwa 35 Liter Gelvolumen sind von FLODIN zur Trennung von Dextranfraktionen in halbkontinuierlicher Verfahrensweise betrieben worden. Dies könnte evtl. zur Gewinnung von klinischem Dextran von Bedeutung sein. Wenn man die Ergebnisse an Kollidon übernehmen darf, was sehr wahrscheinlich ist, so müßten die Mittelfraktionen in ihrer Verteilung sehr eng zusammenliegen und etwa 60 % der gesamten Substanz enthalten.

Die mit Kollidon K 25 durchgeführten Versuche ergaben, wie Abb. 2 zeigt, eine fast als vorzüglich anzusprechende Verteilungskurve. Aus der Gelfiltration erhält man folgende M_w- und M_n-Paare: 23 600/17 900; 23 200/16 500; 24800/17 200.

Mit den Uneinheitlichkeiten $U_1 = 0,32$, $U_2 = 0,41$, $U_3 = 0,39$ und den Mittelwerten $\bar{M}_w = 23\,900$ und $\bar{M}_n$ 17 200, $U = 0,39$. Die Meßpunkte liegen sehr gut auf der aus dem k-Wert [10] berechneten Verteilungskurve, die identisch ist mit der durch Trübungstitration erhaltenen (ausgezogene Kurve). Die Kurve entspricht einem $\bar{M}_w$ von 25 000 und einem $\bar{M}_n$ von 18 000 sowie einem U von 0,39.

Wir können also zeigen, daß am Dextrangel, außer dem Dextran selbst, auch andere wasserlösliche Polymere mit kontinuierlicher Molekulargewichtsverteilung fraktioniert werden können und das Ergebnis, in Übereinstimmung mit dem Befund von Frau Dr. GRANATH am Dextran, im geeigneten Molekulargewichtsbereich vorzüglich ist. Bei höheren mittleren Molekulargewichten der Testsubstanz (Kollidon K 60) wurde noch keine so gute Übereinstimmung zwischen der wirklichen Verteilungskurve und der mittels Gelfiltration erhaltenen erreicht.

Einen weiteren Effekt der Vernetzungsbedingungen zeigt Abb.4. Hier werden zwei Verteilungskurven von K 25 wiedergegeben, die mit Epidex C und Epidex a erhalten worden sind. Die Wr-Werte beider Gele sind etwa gleich (Epidex C 4,6; Epidex a 4,1). Die Verteilungskurven zeigen aber unter sonst gleichen Bedingungen einen grundsätzlich anderen Verlauf. Bei Epidex C ergibt sich eine Fraktionierung über einen weiten Molekulargewichtsbereich. Epidex a gibt dagegen nur Trennungen bei sehr niederen Molekulargewichten. Der mengenmäßig größere Teil des Polymeren wird vom Gel ausgeschlossen, und es ergibt sich ein steiler Anstieg der Verteilungskurve. Der Befund und seine Deutung werden bestätigt durch Versuche mit K 60. Mit Epidex a konnte keine Fraktionierung nachgewiesen werden, d.h., das Polymere wird völlig vom Gel ausgeschlossen. Bei Epidex C ergab sich ein Fraktioniereffekt zwischen $[\eta] = 0,34$ und $[\eta] = 0,76$. Das bedeutet, daß Epidex a trotz des fast gleichen Wr-Wertes viel selektiver wirkt.

Der gleiche Effekt zeigt sich bei Epidex d, einem nach dem gleichen Verfahren sehr schwach vernetzten Gel. Die mit diesem Produkt für K 60 erhaltene Verteilungskurve zeigt Abb.5. Epidex d war leider mechanisch wenig stabil, so daß noch kein Wr-Wert bestimmt werden konnte. Desto erstaunlicher ist, daß man mit diesem Produkt eine ziemlich scharfe Ausschließungsgrenze erhält, die in dieser Weise weder bei Sephadex G 200, noch bei Epidex E und Epidex F auftrat. Auch dieses Produkt ist also trotz großer Quellfähigkeit sehr selektiv wirksam.

Auf Grund der vorliegenden Messungen ist es fraglich, ob der Wr-Wert, wie vorher aufgeführt, tatsächlich mit dem Vernetzungsgrad eindeutig in Zusammenhang steht. Es dürfte aber der Nachweis erbracht sein, daß der Wr-Wert a l l e i n noch nichts über die zu erwartenden Trennungen, die Ausschließungsgrenzen und die Selektivität aussagen kann.

2.2 Darstellung der Elutionskurven

In Abb.6 sind einige Elutionskurven in der herkömmlichen Weise als Konzentrationsangabe über dem Elutionsvolumen dargestellt. Es ergeben sich bei Epidex A und B sowie Epidex E und F etwa symmetrische Kurven von der Art einer Gauss'schen Fehlerkurve. Bei Epidex C zeigt sich nach dem Maximum eine deutlich ausgeprägte Schwelle, so daß der rechte Teil der Kurve verzerrt erscheint. Diese Verzerrung wird sowohl bei Arbeiten mit K 25 als auch mit K 60 beobachtet; sie bleibt hier außer Betracht.

Aus den Elutionskurven können nur bedingt Rückschlüsse auf die Verteilungskurven gezogen werden. Bei festgelegten Kolonnenabmessungen, Probemengen und Konzentrationen wird auch die Elutionskurve umso breiter sein, je breiter die Verteilung des Polymeren ist. So ergeben sich bei der Gelfraktion von Fraktionen (enge Verteilung) sehr schmale Elutionskurven.

Instruktiver sind nach unserer Ansicht die Elutionskurven, wenn man anstelle der Konzentration, die Grenzviskositätszahl oder das Molekulargewicht über dem Elutionsvolumen aufträgt. In Abb.7 sind so zwei Kurven von K 60 an Epidex E gezeichnet. Jede Kurve läßt sich im doppeltlogarithmischen Netz ziemlich gut als zwei Geraden mit verschiedener Steigung, verbunden durch ein kurzes Übergangsstück, darstellen. Weitere Diagramme dieser Art sind in Abb.8 und 9 gezeigt. Deutlich unterscheiden sich auch hier wieder die Kurven, die mit Epidex C und mit Epidex a erhalten werden. Die mit Epidex C für K 25 erhaltenen Kurven können näherungsweise durch zwei Geraden mit Verbindungsstück dargestellt werden. Die Kurven für K 25 mit Epidex a zeigen dagegen einen gleichartigen Verlauf, wie die Elutionskurven von K 60 an Epidex E oder F. Dieser Befund führt zu der Annahme, daß diese Kurven etwas über das Trennverhalten der Gele aussagen können und daß damit jedem Teil der Kurve eine physikalische Bedeutung zugeschrieben werden kann. Nach unserer Auffassung wäre dies so möglich, daß der untere Ast (größte Steigung) demjenigen Teil der Moleküle entspricht, der die Gelkörner frei passieren kann. Das Verbindungsstück gibt den Bereich wieder, in dem teilweiser Ausschluß der Moleküle stattfindet. Der Rest der Moleküle sollte nicht mehr oder nur noch mit sehr geringen Anteilen in die Gelkörner eintreten (flacher Teil der Kurve). Man kann sich vorstellen, daß die Steigung der Geraden noch mehr Aufschluß über die Wirksamkeit der Gele geben könnte. Das vorliegende Material ist aber noch zu gering, um hierüber schon Aussagen machen zu können.

Der steile Anstieg bei K 25 an Epidex A, B und C kann in Übereinstimmung mit den sich ergebenden Verteilungskurven vielleicht so gedeutet werden, daß die Vernetzung nicht gleichmäßig stattgefunden hat. Die Vernetzungsbrücken sind nicht in statistisch gleichem Abstand eingebaut worden, sondern es haben sich Bereiche starker Vernetzung gebildet, zwischen denen sehr große Zwischenräume bestehen. Dies würde bedeuten, daß in diese Zwischenräume noch verhältnismäßig große Moleküle eintreten können.

Diese Vorstellung würde das kleiner werdende Übergangsstück erklären, wo die Moleküle teilweise ausgeschlossen werden und die andere Form der Elutionskurven bei K 60 an den gleichen Gelen. Bei größerem Gehalt an Epichlorhydrin werden die Poren in den Inseln gleichmäßiger, wenn auch noch große Zwischenräume zwischen diesen vorhanden sind. Damit muß aber der Bereich des teilweisen Ausschlusses kleiner werden. Anders liegen die Verhältnisse bei K 60. Auf Grund des größeren Molekulargewichtes können die großen Moleküle dieses Produktes nicht mehr in die Zwischenräume zwischen den Vernetzungsinseln eintreten. In diesem Bereich erfolgt dann teilweiser Ausschluß. Überträgt man diese Vorstellung auf Epidex a, so bedeutet die Elutionskurve, daß bei diesem Gel die Vernetzung sehr gleichmäßig stattgefunden hat und die Porengrößen nur in einem kleinen Bereich schwanken. Die Vorstellung wird durch die mit Epidex E für K 60 erhaltenen Ergebnisse gestützt. Epidex E war ein schwach vernetztes Nativdextran und sehr weich. Mit der frisch gefüllten Kolonne hatte die Gerade, die den teilweisen Ausschluß darstellen soll, eine größere Steigung. Beim zweiten Versuch an der gleichen Kolonne war das Gel zusammengedrückt worden, damit wurden die Poren kleiner und die Gerade verläuft flacher. Die Elutionskurve war unabhängig von der Füllhöhe (17,5 bzw. 30 cm). Die Elutionskurven sind identisch.

Diese Versuche sind zunächst nur orientierender Art, die vorgeschlagene Deutung ist noch mehr oder weniger spekulativ. Doch dürfte gezeigt sein, daß hinter diesen Kurven wahrscheinlich ein physikalischer, das Verhalten des Gels bestimmender Sachverhalt steht und daß sie zur Charakterisierung der Gele verwendbar erscheinen. Ähnliche Kurven wurden übrigens auch bei der Gelfiltration von Polystyrol über vernetztes Polystyrol erhalten.

Z U S A M M E N F A S S U N G

Unsere Versuche haben gezeigt, daß Fraktionierungen von wasserlöslichen Substanzen mit kontinuierlicher Molekulargewichtsverteilung an Dextrangelen bei geeignetem Molekulargewichtsbereich mit sehr guten Ergebnissen, auch bei hohen Ausgangskonzentrationen, durchgeführt werden können. Allerdings sind Dextrangele zur Bestimmung der vollständigen Verteilungskurve bis jetzt leider nur bei verhältnismäßig niedermolekularen Substanzen anwendbar; hier aber mit sehr guten Ergebnissen, die auch die präparativ-technische Anwendung in einigen Fällen nahelegen. Zur Charakterisierung der Gele wird, neben der Bestimmung des Wr-Wertes, die Testung durch ein Polymeres bekannter Verteilung und Auswertung in der beschriebenen Weise empfohlen. Dies erlaubt, mehr über das allgemeine Trennverhalten und über die Selektivität auszusagen als die bisher benutzte Elutionskurve. Eine Verbesserung der Selektivität der Gele erscheint durch systematische Untersuchung der Vernetzungsbedingungen möglich. Es kann noch nicht entschieden werden, welche Vernetzungsmethode bei Substanzen mit einem Molekulargewichtsspektrum die günstigsten Ergebnisse liefern wird.

Die Übertragung der Gelfiltrationsmethode auf nichtwässrige Lösungen erscheint recht aussichtsreich [11] [12] [13] [14]. Über die Besonderheiten in diesem Falle und diesbezügliche Ergebnisse soll an anderer Stelle berichtet werden.

Wir danken auch an dieser Stelle den Herren Prof. HOLZMÜLLER und Prof. LEIBNITZ dafür, daß sie uns die Möglichkeit gaben, diese Untersuchungen durchzuführen und für ihr förderndes Interesse, ebenso Herrn Dr. GABERT und seinen Mitarbeitern für die Herstellung der Gele. Ferner möchten wir unseren Mitarbeiterinnen, Frau A.BUHE und Fräulein K.LEYSER für ihre wertvolle Hilfe danken.

L I T E R A T U R

1) PORATH,J., P.FLODIN, Nature 183, 1657 (1959)
2) FLODIN,P., K.GRANATH, Symposium über Makromoleküle, Verlag Chemie Wiesbaden 1959, II C 6
3) GRANATH,K., P.FLODIN, Makromolekulare Chem. 48, 160 (1961)
4) LATHE,G.H., C.R.J.RUTHVEN, Biochem.J. 62, 655 (1956)
5) HJERTEN,S., R.MOSBACH, Anal.Biochem. 3, 109 (1962)

6) POLSON,A., Biochem.Biophys.Acta 50, 565 (1961)
7) HJERTÉN,S., Archiv.Biochem.Biophys. 99, 466 (1962)
8) FLODIN,P., Dextran Gels and their Application in Gel Filtration Uppsala 1962 Pharmacia
9) ENSLIN,O., Chemische Fabrik 6, 147 (1933)
10) SCHOLTAN,W., Makromolekulare Chem. 24, 104 (1957)
11) VAUGHAN,M.F., Nature 188, 55 (1960)
12) VAUGHAN,M.F., Nature 195, 801 (1962)
13) LANGHAMMER,G., K.QUITZSCH, Makromolekulare Chem. 43, 160 (1961)
14) Chem.Eng.News 40, 43 (1962)
15) SEIDE,H., G.LANGHAMMER, Plaste u. Kautschuk 9, 12, 573 (1962)

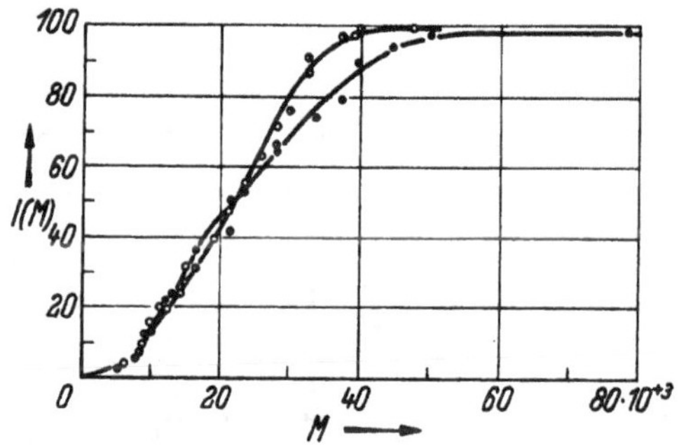

Abb.1 Molekulargewichtsverteilungskurven für Kollidon K 25 durch Gelfiltration an verschiedenen Epidex-Präparaten (● Epidex A; O Epidex B; ◐ Epidex C).

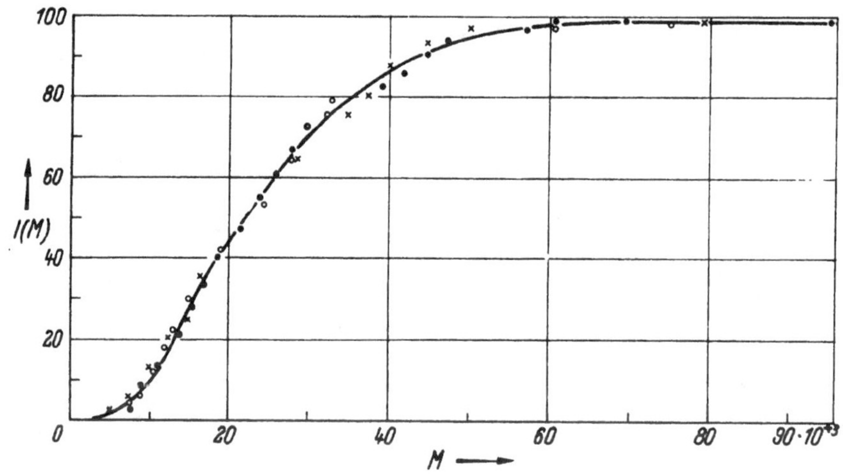

Abb.2 Molekulargewichtsverteilung von Kollidon K 25 durch Gelfiltration an Epidex A bei verschiedenen Durchflußgeschwindigkeiten (O 1,4 ml/h; × 8,1 ml/h; ● 25 ml/h). Aufgabe 20 ml 5 %ige Lsg. Ausgezogene Kurve: Verteilungskurve aus k-Wert berechnet u. aus Trübungstitration (identisch). Einzelangaben über Molekulargewichtsmittelwerte im Text.

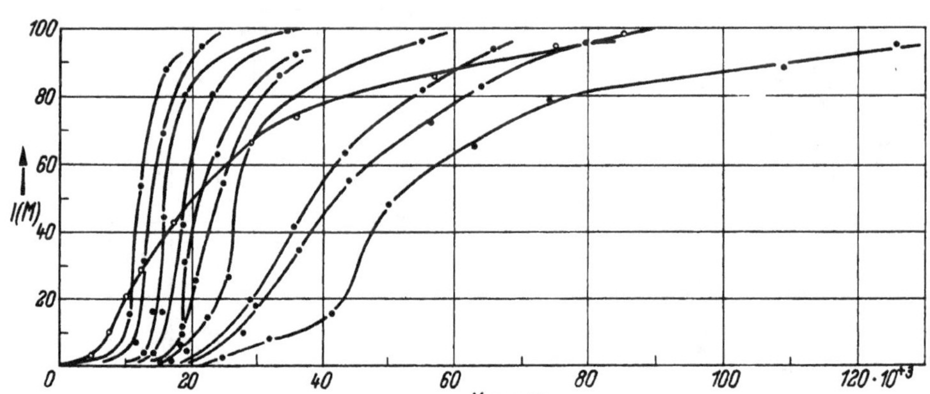

Abb.3 Molekulargewichtsverteilungen einiger mittels Gelfiltration gewonnener Fraktionen von K 25 (aus Messungen der Trübungstitration) (● Trübungsmessungen an Fraktionen; O unfraktioniertes Produkt).

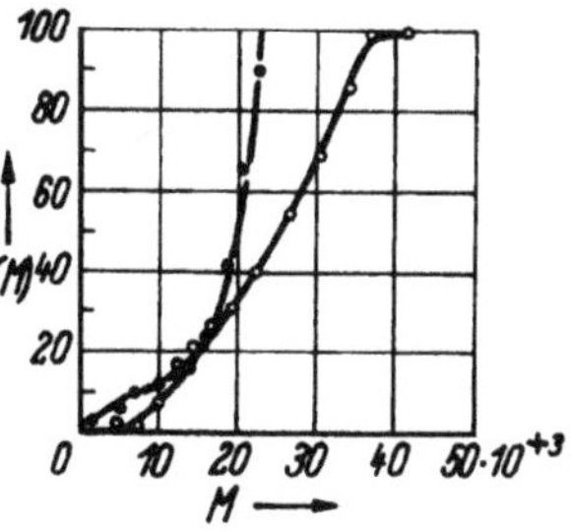

Abb.4 Molekulargewichtsverteilung von Kollidon K 25 mittels Gelfiltration an Epidex a (Wr-Wert 4,1 (●)) und Epidex C (Wr-Wert 4,6 (○)).

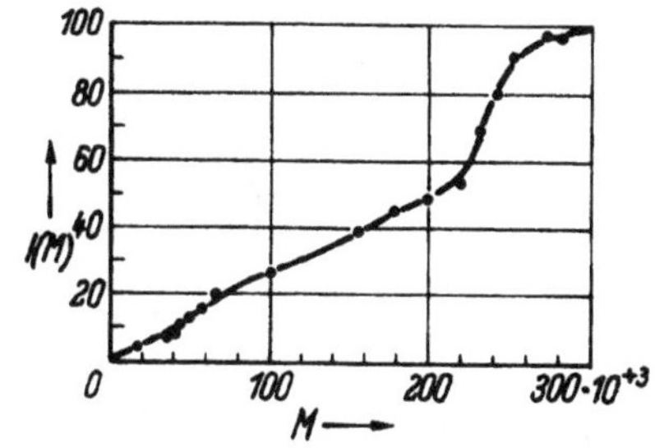

Abb.5 Molekulargewichtsverteilung von Kollidon K 60 aus Gelfiltration an Epidex d (Wr-Wert groß, aber nicht bestimmbar). Gelfiltration: M_w=171000; M_n=113000; U=0,51 zum Vergl. Trübungstit.: M_w=205000; M_n=73000; U=1,81

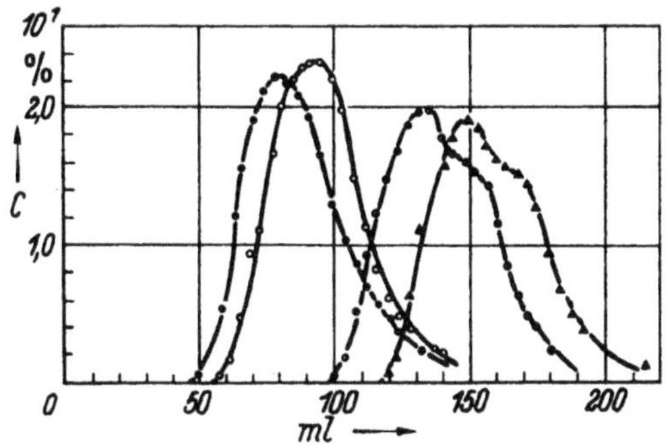

Abb.6 Elutionskurven von K 25 an verschiedenen Mustern von Epidex. Epidex B(Wr-Wert 6,7): ● 58 ml/h; ○ 10 ml/h; Epidex C (Wr-Wert 4,6): ◑ 18 ml/h; △ 12 ml/h.

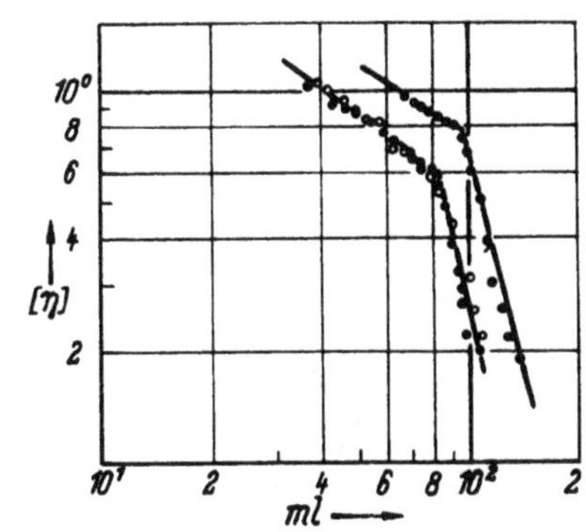

Abb.7 Elutionskurven von K 60 an Epidex E. ● 20 ml 2,5 % 8,2ml/h
◑ 10 ml 5,0 % 10,4 ml/h
○ 5 ml 10,0 % 6,8 ml/h.

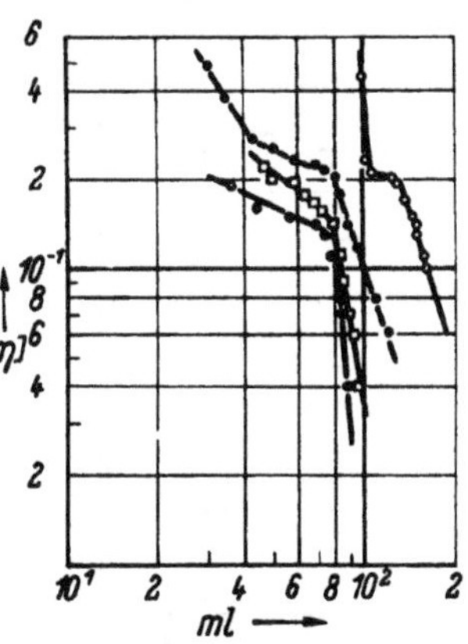

Abb.8 Elutionskurven von K 25 an versch. Epidexmustern. ● Epidex A, ○ Epidex C, ◓ Epidex a, □ Epidex b.

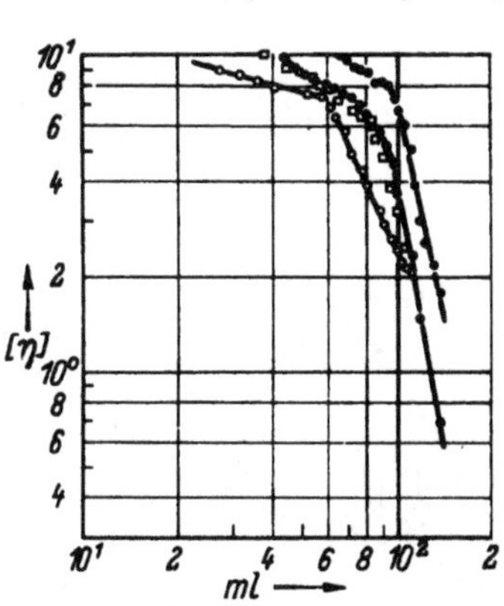

Abb.9 Elutionskurven von K 60 an versch. Epidexmustern. ● Epidex E (frisch gefüllt), □◓ Epidex E (nach Absetzen der Säule), ○ Epidex d.

B Polysaccharide

Zum Ultraschallabbau von Dextran

H. LÜCKE

Institut für Mikrobiologie und experimentelle Therapie der Akademie, Jena

S U M M A R Y

Ultrasonic depolymerisation of dextran molecules in air and CO_2, respectively, was studied at various solution concentrations. By addition of CO_2 during the sonic treatment the degradation happened more slowly than by bubbling with air. Comparison with theoretical curves calculated by DOTY and STEINER proved, that the degraded dextran molecules are monodisperse coils. The dependence of the second virial coefficient on the molecular weight was indicated.

Die Depolymerisationswirkung des Ultraschalls wurde an einem Dextran von Pharmacia, Uppsala, das dem hiesigen Institut zu Versuchszwecken überlassen wurde, verfolgt. Das mittlere Molekulargewicht des unbeschallten Dextrans betrug nach Angaben der Herstellerfirma $M_w = 2,18.10^6$. Unsere Lichtstreuungsmessungen lieferten ein Gewichtsmittel von $M_w = 2,3.10^6$. Abbildung 1 zeigt ein Zimm-Diagramm des Dextrans. Nach der Methode der doppelten graphischen Extrapolation von ZIMM wurden alle Lichtsteuungsmessungen ausgewertet.

Aufgabe war es, zwischen mechanischem und chemischem Abbau zu unterscheiden. Dazu wurden wäßrige Lösungen von Dextran in Luft- oder CO_2-Atmosphäre beschallt. Nach verschiedenen Beschallungszeiten erfolgten Lichtstreuungsmessungen.

Ein Industrie-Generator (RFT Erfurt) mit einer maximalen Leistung von 1,5 kW in Verbindung mit einem 800 kHz-Ultraschall-Topf vom VEB Carl ZEISS, Jena, stand für die Abbauversuche zur Verfügung. Für die praktische Verfolgung des Abbaues fand im Institut für Mikrobiologie Jena entwickeltes Lichtstreugerät Verwendung.

E R G E B N I S S E

Da für die Molekulargewichtsbestimmung durch die Lichtstreuungsmethode Verdünnungsreihen der zu untersuchenden Lösungen notwendig sind, wurde durch Relativmessungen an einem Bernburger Dextran der Ultraschallabbau auf eine Volumenabhängigkeit hin untersucht. Die experimentellen Werte genügen der Beziehung $A_v = \frac{1}{V}$ (Abb.2), wenn A_v ein relatives Maß der Abbauwirkung darstellt.

$$A_v = \frac{i_{90}^{o} - i_{90}^{30}}{i_{90}^{o}} \qquad (i_{90}^{t} = 90^{o} \text{ -Intensität bei Beschallungszeit t})$$

Daher wurde bei sämtlichen Abbauversuchen auf konstantes Lösungsvolumen geachtet.

Abbildung 3 gibt die Ergebnisse unserer Abbauversuche an luftgesättigten Dextranlösungen wieder, und zwar sind die aus den Zimm-Diagrammen berechneten Molekulargewichte als Funktion der Beschallungsdauer t dargestellt.

Parameter der Kurvenschar ist die Dextrankonzentration der Lösungen während der Beschallung. Das Ausgangsmolekulargewicht war bei allen 4 Versuchsreihen das gleiche ($2{,}3 \cdot 10^6$). Man erkennt, daß die Depolymerisation der Dextranmoleküle in den ersten 60' sehr rasch verläuft und dann allmählich abklingt. Nach einer Beschallung von 360' ist das Molekulargewicht auf weniger als 10 % des Ausgangswertes abgesunken. Das Endmolekulargewicht M_e, das durch sehr langzeitige Beschallung nicht mehr unterschritten werden kann, wurde bei unseren Versuchen nicht erreicht; die Gestalt unserer Abbaukurven deutet aber darauf hin, daß ein solcher Grenzwert existiert, dem sich die 4 Abbaukurven gemeinsam nähern. Die Depolymerisationsgeschwindigkeit hängt merklich von der Konzentration der beschallten Lösung ab. Aus Abb.3 ($c = 5{,}4;\ 2{,}7;\ 3{,}4;\ 1{,}4 \cdot 10^{-3}\,g/cm^3$) und noch deutlicher aus Abb.4 ist ersichtlich, daß in dem von uns untersuchten Konzentrationsbereich bei konstanter Beschallungsdauer die Abbauwirkung des Ultraschalls mit steigender Konzentration zunimmt. Offenbar tritt mit wachsender Konzentration in zunehmendem Maße eine gegenseitige Behinderung und Vernetzung der Dextranmoleküle ein, die eine Vergrößerung der Relativbewegung zwischen Lösungsmittel und Gelöstem und damit eine erhöhte Depolymerisation zur Folge hat.

Die bei der Beschallung CO_2-gesättigter Dextranlösungen gewonnenen Abbauergebnisse sind in Abb.5 zusammengestellt. Qualitativ ergeben sich die gleichen Abhängigkeiten von der Beschallungsdauer sowie von der Lösungskonzentration wie bei der Beschallung in Luftatmosphäre (Abb.6). Der Abbau unter CO_2-Zufuhr geht langsamer vor sich als bei luftgesättigten Lösungen (Abb.7).

Wir nehmen an, daß der Unterschied der Depolymerisation darauf beruht, daß der Molekülabbau in Luft nicht allein auf eine mechanische Depolymerisationswirkung des Ultraschalls zurückgeführt werden kann, sondern daß zusätzlich ein chemischer Abbau durch während der Beschallung in der Lösung erzeugte Radikale erfolgt. Bei Anwesenheit von Luft (CO_2-gesättigte Lösung) wird die Bildung dieser Radikale offenbar gehemmt.

Für eine Reihe unserer Abbauversuche haben wir das Verhältnis der charakteristischen Molekülgröße zur benutzten Wellenlänge $\hat{d}/\lambda$ aus den "Nullkonzentrationsgeraden" unserer Zimm-Diagramme berechnet. Die Werte sind nach fallendem Molekulargewicht absinkende Grössen. Abb.8 und Abb.9 zeigen den Zusammenhang zwischen $\hat{d}/\lambda$ und der Unsymmetriezahl $[z]$ bzw. der reziproken Streufunktion $P(90)^{-1}$. Ein Vergleich mit den eingezeichneten theoretischen Kurven von DOTY und STEINER [1] erlaubt Rückschlüsse auf die Gestalt der Dextranmoleküle. Aus dem chemischen Aufbau ist bekannt, daß die Glukosebausteine zu langkettigen Makromolekülen polymerisiert sind. Eine kompakte Kugelform kommt als Modell für Dextranmoleküle nicht in Betracht. Nach Abb.8 kann eine klare Entscheidung zwischen poly- und monodispersem Knäuel wegen der Streuung der Meßpunkte nicht getroffen werden. In Abb.9 hingegen scharen sich die experimentellen Werte vorwiegend um die für monodisperse Knäuelmoleküle gültige theoretische Kurve, d.h. es handelt sich um ein Dextran relativ geringer Uneinheitlichkeit.

Abbildung 10 gibt die Änderung der 2.Virialkoeffizienten mit dem Molekulargewicht des abgebauten Dextrans wieder. (B aus der Anfangssteigung der Streuwinkelnullkurven.) Trotz der außerordentlich starken Streuung der Werte,besonders bei hohen Molekulargewichten, zeichnet sich eine Abhängigkeit des Virialkoeffizienten vom reziproken Molekulargewicht ab, wie sie MÜNSTER [2] für hinreichend hohe Polymerisationsgrade fand, während GRIMLEY[3] für das gleiche Gebiet zu $B \sim 1/\sqrt{M}$ gelangt.

L I T E R A T U R

1) DOTY,P., R.F.STEINER, J.Chem.Physics 18, 1211 (1950)
2) MÜNSTER,A., Makromol.Chem. 2, 27 (1948)
3) GRIMLEY,T.B., Proc.Roy.Soc. [London] A 212, 339 (1952)

D I S K U S S I O N S B E M E R K U N G E N

BODMANN : Wie groß ist die Uneinheitlichkeit des Dextrans ?

LÜCKE : Da zu den Untersuchungen keine andere Methode als die Lichtstreuung herangezogen wurde, sind Aussagen über den Grad der Uneinheitlichkeit nur qualitativ möglich (vgl.Abb.9). Die Lichtstreuung liefert nur das Gewichtsmittel des Molekulargewichts. Zur Bestimmung der Molekulargewichtsverteilung wurde bei ähnlichen Proben früher die Ultrazentrifuge verwendet (Pharmazie 14, 380 (1959)).

GRANATH : Wie wurde die Heterogenität bestimmt ?

LÜCKE : Spezielle Untersuchungen in dieser Richtung wurden von mir nicht durchgeführt. Mithilfe der Trübungstitration sind einige Anhaltspunkte gewonnen worden (Pharmazie 13, 72 (1958)).

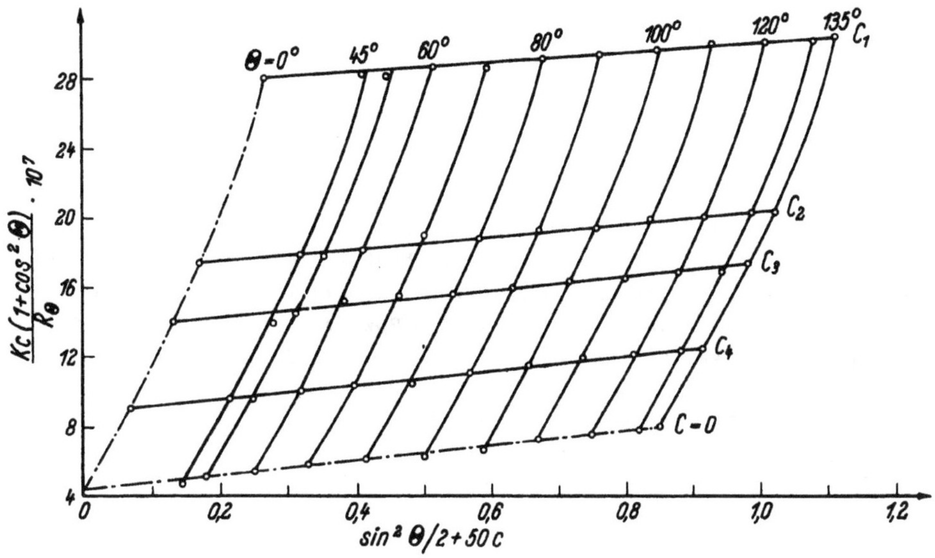

Abb.1 Zimm-Diagramms des unbeschallten Dextrans

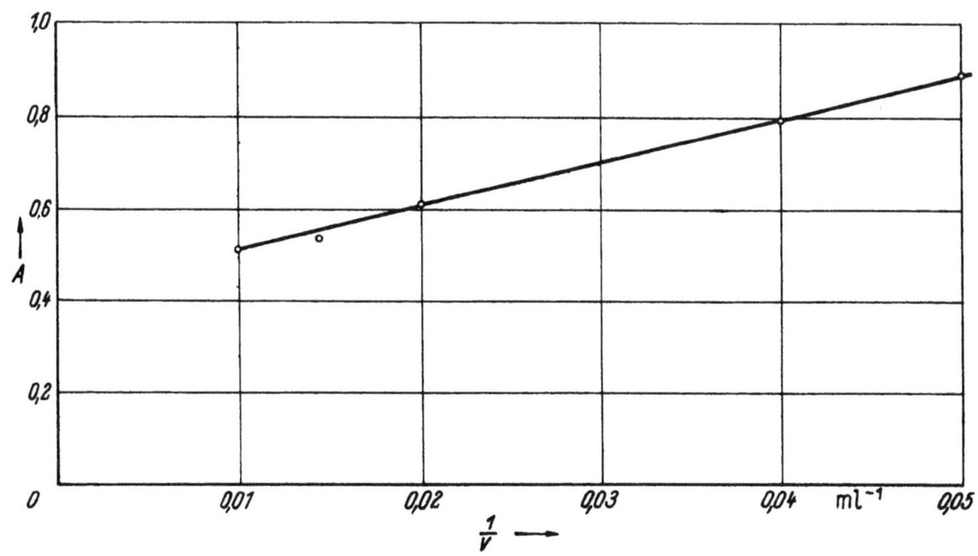

Abb.2 Volumenabhängigkeit des Abbaues

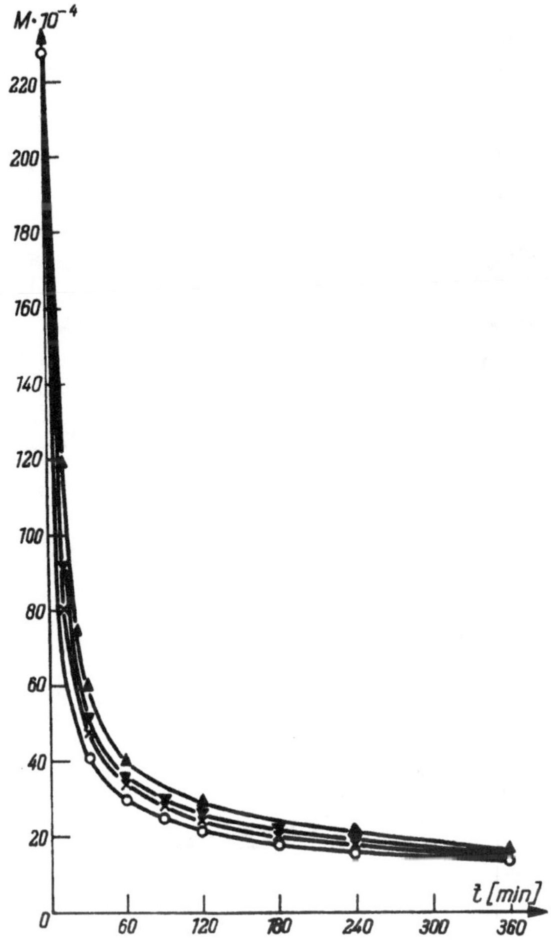

Abb.3 Abbaukurven des Dextrans bei Beschallung in Luftatmosphäre. Lösungskonzentrationen: o–o c = 5,4 ; ▼–▼ c = 3,4 ; x – x c = 2,7 ; Δ–Δ c = 1,4 $\cdot 10^{-3}$ g/cm^3

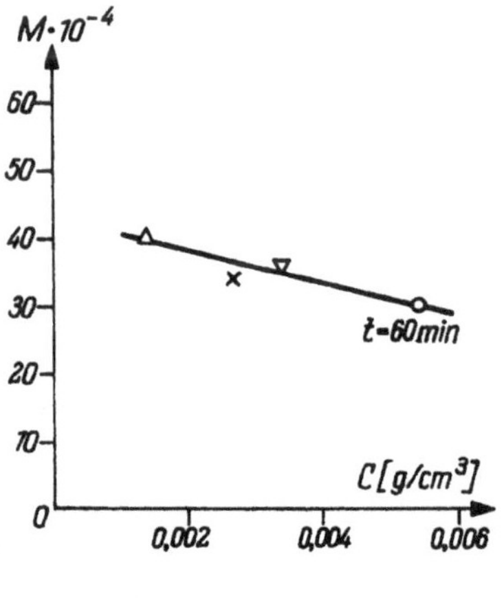

Abb.4 Konzentrationsabhängigkeit des Ultraschallabbaues in Luftatmosphäre. t=60 min

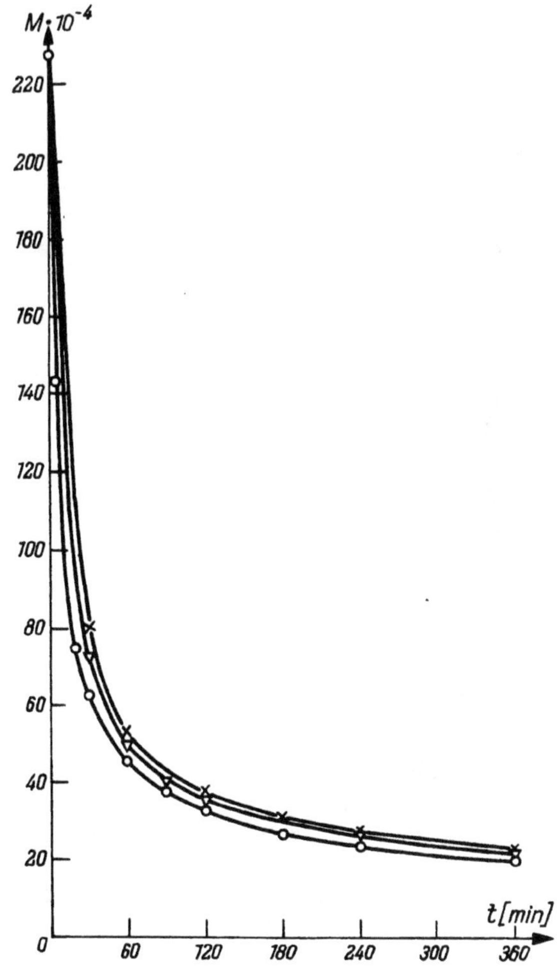

Abb.5 Abbaukurven verschieden konzentrierter Dextranlösungen bei Beschallung unter CO_2-Zufuhr. Lösungskonzentrationen : ×-× c= 5,4, ▽-▽ c=3,4, o-o c=2,7 $.10^{-3}$g/cm^3

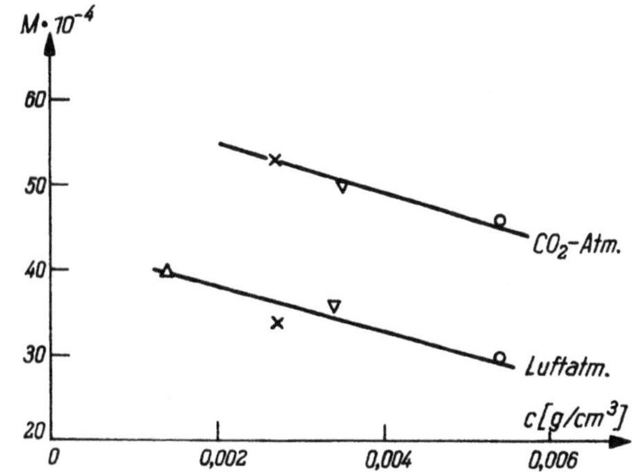

Abb.6 Abhängigkeit des Molekulargwichtes von der Lösungskonzentration bei konstanter Beschallungszeit t = 60'

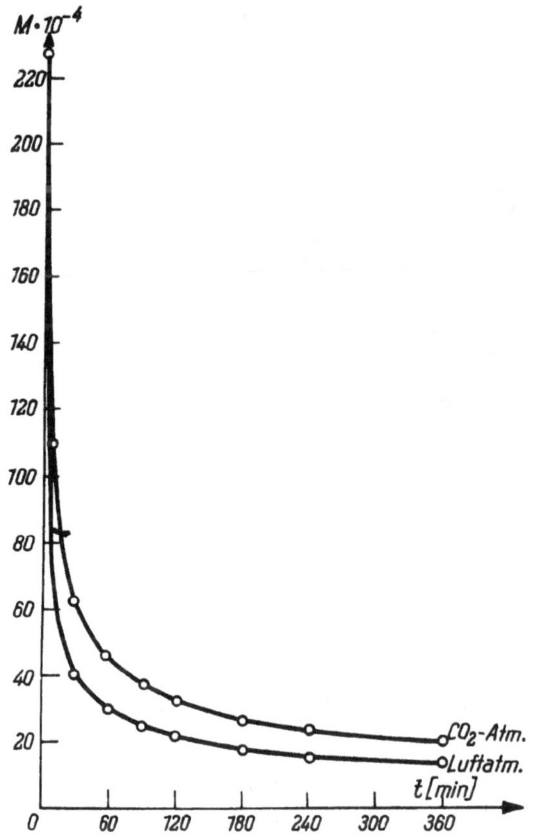

Abb.7 Vergleich zweier Abbaukurven für Luft- und CO_2-Atmosphäre. Lösungskonzentration der beschallten Proben: $c = 5{,}4.10^{-3} g/cm^3$

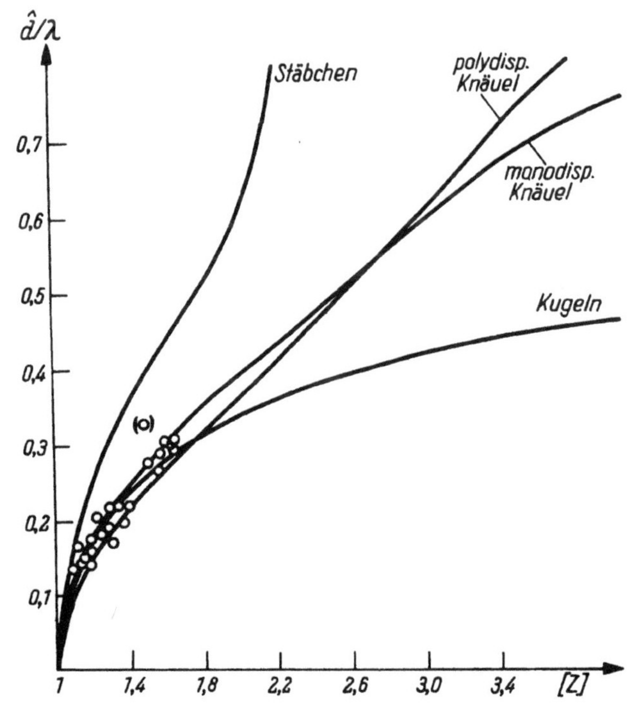

Abb.8 Verhältnis der charakteristischen Molekülgröße $\hat{d}$ zur Wellenlänge des verwendeten Lichts in Abhängigkeit von der Unsymmetriezahl [z]. Eingezeichnet sind die theoretischen Kurven nach DOTY und STEINER.

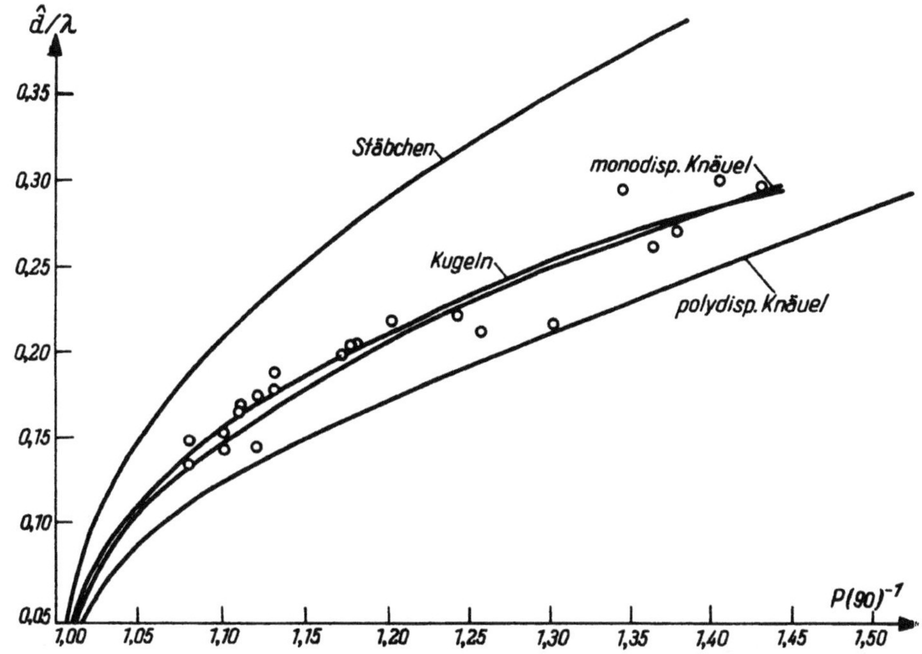

Abb.9 Zusammenhang zwischen $\hat{d}/\lambda$ und der reziproken 90°-Streuintensität. Theoretische Kurven nach DOTY und STEINER mit den **experimentellen** Werten der abgebauten Dextranproben.

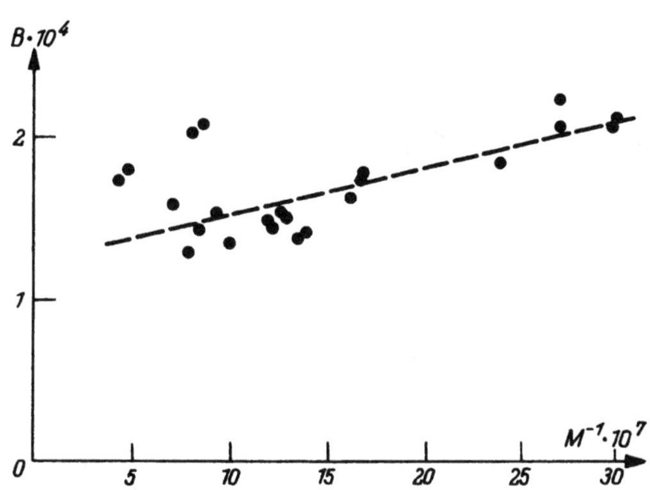

Abb.10 Änderung des 2.Virialkoeffizienten mit 1/M

B Polysaccharide

Rheologische Untersuchungen an Cellulosexanthogenatlösungen

B. PHILIPP und K. WULF

Institut für Faserstoff-Forschung der Akademie, Teltow-Seehof

S U M M A R Y

Rheological studies on technical solutions of cellulose xanthate are reported which have been carried out by means of an UMSTÄTTER structural viscosimeter using capillaries of different length. Although there are differences in initial viscosity of solution provoked by the various viscose parameters, the viscosity at high shear stress shows the nearly unchanged minimal value of 1 poise. Further, the flow behaviour is governed by the initial viscosity rather than by the various viscose parameters. The results presented here indicate that, as suggested by HOPKINS and WHATLEY, the flow curve does not exhibit a symmetrical shape with inflection point.

Während Viskositätsmessungen an verdünnten Polymerlösungen vorwiegend analytischen Zwecken, nämlich einer Bestimmung von Molekulargewicht und Knäueldimensionen dienen, und hierbei die gegenseitige Wechselwirkung der Polymermolekeln (HUGGINSsche Konstante k) und der Einfluß des Schergefälles nur die Rolle von Korrekturgrößen spielen, steht für rheologische Untersuchungen an konzentrierten Polymerlösungen eben diese Abhängigkeit der scheinbaren Viskosität vom Schergefälle im Mittelpunkt des Interesses. Analytisches Ziel solcher Untersuchungen ist für den Polymerchemiker natürlich eine Charakterisierung der Struktur dieser Lösungen und der Änderung dieser Lösungsstruktur im Scherfeld mit Hilfe rheologischer Parameter. Allerdings sind wir von diesem Ziel heute zumeist noch weit entfernt, nicht zuletzt infolge der noch im vollen Fluß befindlichen Entwicklung der Methodik und eines Mangels an wirklich vergleichbarem Versuchsmaterial. Verhältnismäßig eingehend untersucht wurden aufgrund ihres Interesses zur Herstellung von Spinnlösungen verschiedene Cellulosederivate. Bekanntlich ist Cellulose nur in Form ihrer Derivate löslich, wobei sich rheologisch wesentliche Unterschiede ergeben zwischen nichtionogenen Derivaten (Nitrat, Acetat) in organischen Lösungsmitteln und wässrigen Lösungen, in denen die Cellulose etwa als CMC oder als Xanthogenat in Form eines Polyelektrolyten vorliegt. So zeigt das System Cellulosenitrat - Butylacetat ein relativ übersichtliches Fließverhalten, nämlich entsprechend den Untersuchungen von EDELMANN [1] (Abb.1) die nach UMSTÄTTER [2] zu erwartende symmetrische Fließkurve bei Auftragung von log D gegen log σ mit einem wohl definierten Wendepunkt. Hingegen ist das Fließverhalten bei den in wässriger Lösung als Polyelektrolyten vorliegenden Cellulosederivaten komplizierter. Wie Abb.2 für eine Lösung des Cellulose-Eisen-Weinsäure-Nitrat-Komplexes zeigt, verläuft hier die Fließkurve unsymmetrisch [3]. Für das in Zusammenhang mit dem technischen Viskoseverfahren besonders interessante System Cellulosexanthogenat - wässrige NaOH liegen neben einigen älteren rheologischen Arbeiten von PHILIPPOFF [4] und BREDEE [5] eine Reihe neuerer Veröffentlichungen vor, die sich jedoch z.T. dahingehend widersprechen, als von SCHURZ [6] und von PAKSVER [7] analog zum System Cellulosenitrat-Butylacetat ein Wendepunkt der Fließkurve festgestellt und im Hinblick auf Molekulargewicht und Molekulargewichtsverteilung ausgewertet wird, während HOPKINS und WHATLEY [8] einen in weiten Grenzen linearen Verlauf der Fließkurve fanden, der keinen Wendepunkt erkennen ließ.

Ziel der nun kurz zu referierenden eigenen Untersuchungen war es, zur Klärung dieser für das rheologische Verhalten von Viskose recht wesentlichen Frage beizutragen, daneben den Einfluß einiger Viskosierparameter auf die Fließkurvenform zu untersuchen und schließlich deren Änderung mit abnehmendem Längen-Radius-Verhältnis der Meßkapillare, also bei schrittweiser Annäherung an die Dimensionen der technischen Spinndüse zu prüfen.

M E T H O D I S C H E S

Zunächst ein paar Worte zur Methodik ! Zur Verfügung stand ein Strukturviskosimeter nach UMSTÄTTER [9], bei dem die Lösung in einem Vorratsgefäß stufenweise unter einen immer höheren, genau meßbaren Druck zwischen 0,1 und 100 atü gesetzt wird und nach Passieren der Meßkapillare in eine Meßbürette läuft, die eine Ermittlung der jeweiligen Volumengeschwindigkeit gestattet. Das Gerät wurde von uns dahingehend verbessert, daß

1. durch eine Umkonstruktion des Verschlußkopfes entsprechend Abb.3 alle zusätzlichen Kapillarwege soweit als möglich in ihrer Länge reduziert und in ihrem Durchmesser möglichst groß gehalten wurden, um den hierdurch bedingten Fehler auch bei Verwendung kurzer Meßkapillaren in erträglichen Grenzen zu halten;
2. wurde die Verbindung der Kapillaren mit dem Verschlußkopf und die Abdichtung dieser Verbindung besonders einfach und sicher mit Hilfe eines auf die Kapillaren aufgeschrumpften Kunststoffkonus vorgenommen, der zusätzlich verklebt wird (Abb.4). Verwendet wurden Kapillaren mit einem Durchmesser von 0,5 mm und den Längen 15 cm, 10 cm, 1 cm und 0,1 cm, d.h. Längen-Radien-Verhältnissen von 600, 400, 40 und 4.

Außerdem wurden die Cellulosexanthogenatlösungen auch im Freiflußviskosimeter nach UMSTÄTTER [10] gemessen, um den Bereich der konstanten Anfangsviskosität η_0 näher zu charakterisieren.

Experimentell wird beim Strukturviskosimeter der Druck P vorgegeben und das in der Zeit t aus der Kapillare fließende Volumen V bzw. die Volumgeschwindigkeit Q = V/t ermittelt. Berechnet wird hiermit die scheinbare Schergeschwindigkeit D :

$$D = \frac{4\,Q}{\pi R^3} \left[\mathrm{sec}^{-1} \right]$$

und die scheinbare bzw. unkorrigierte Schubspannung σ_u :

$$\sigma_u = \frac{R.P}{2\,L} \left[\mathrm{dyn\ cm}^{-2} \right]$$

Auf eine Berechnung der wahren Schergeschwindigkeit $\dot{\gamma}$ nach EISENSCHITZ, RABINOWITSCH und WEISSENBERG [11] wurde verzichtet. Hier interessierte vor allem die Korrektur der scheinbaren Schubspannung, die formal mit Hilfe einer fiktiven Anlauflänge l, die zur Kapillarlänge L addiert wird, erfolgt. Diese Korrektur ist erforderlich, da die dem Gesamtdruck P entsprechende Energie nicht nur in der Kapillare als Wärme dissipiert wird, sondern sich nach HERZOG und WEISSENBERG [12] prinzipiell auf einen elastischen Anteil (= potentielle Energie) und die miteinander verkoppelten Anteile der kinetischen und der Wärmeenergie verteilt. Der Gesamtdruck ist damit aufgespalten in den Druckanteil für die Energiedissipation in der Kapillare, der der Länge L der Kapillare proportional ist, in den Druckanteil für die potentielle bzw. elastische Energie, der einer fiktiven Verlängerung der Kapillare um die "Anlauflänge" l porportional ist und als COUETTE-Korrektur in Erscheinung tritt, und in den Druckanteil für die kinetische Energie, die der HAGENBACH-Korrektur H entspricht.

$$P = B.L + B.l + H$$

Die Bestimmung der Anlauflänge ist unter Verwendung mindestens zweier verschieden langer Kapillaren möglich, wie aus Abb.5 ersichtlich ist. Bei jeweils konstanter Schergeschwindigkeit wird der um die HAGENBACH-Korrektur H verminderte Gesamtdruck P gegen die Kapillarlänge aufgetragen. Der negative Abszissenabschnitt stellt die Anlauflänge dar. Bei hinreichend großem Längen-Radien-Verhältnis gilt $L \gg l$ und die COUETTE-Korrektur kann vernachlässigt werden.

Die Cellulosexanthogenatlösungen wurden nach dem klassischen Viskoseverfahren hergestellt, wobei wir folgende Parameter variierten : Den Zellstofftyp hinsichtlich seines übermolekularen Faserbaus (Sulfit - Sulfat), die Alkali- und Cellulosekonzentration, den DP durch verschieden weit getriebenen Abbau der Alkalicellulose und den Substitutionsgrad, sowohl durch verschieden hohes Aufxanthogenieren mit unterschiedlichen CS_2- als auch durch verschieden weit getriebene Abspaltung der ionogenen Gruppen im Nachreifeprozeß. Zwischen technischen, also sulfid- und trithiocarbonathaltigen Viskosen und den entsprechenden Reinxanthogenatlösungen wurde kein Unterschied in der Fließkurvenform gefunden, so daß bei den nun zu behandelnden systematischen Untersuchungen auf diese umständliche und langwierige Reinigung verzichtet werden konnte.

E R G E B N I S S E   U N D   D I S K U S S I O N

Der folgenden Betrachtung soll zunächst die recht übersichtliche Darstellung: scheinbare Viskosität $\eta' = \sigma_u/D$ gegen Logarithmus der Schubspannung σ_u (Abb.6) zugrunde gelegt werden. Die im Freiflußviskosimeter gemessenen scheinbaren Viskositäten liegen unter Berücksichtigung der Fehlerbreite der Methode noch im Bereich konstanter Anfangsviskosität η_o. Diese Anfangsviskosität kann durch entsprechende Wahl der Viskosierparameter wie DP, Cellulose- und Alkali-Konzentration und Substitutionsgrad, sowie durch die Meßtemperatur in weiten Grenzen variiert werden. Mit zunehmender Schubspannung sinkt η' um ein bis drei Zehnerpotenzen ab und erreicht unabhängig von der Anfangsviskosität stets annähernd die gleiche Minimalviskosität von $1 \pm 0{,}5$ Poise. Es tritt also bei hoher Schubspannung eine ausgesprochene Nivellierung zwischen den einzelnen Lösungen unterschiedlicher Anfangsviskosität ein. Der Abfall von η' erfolgt sehr steil und praktisch linear mit $\log \sigma$ im Bereich 4.10^3 bis 4.10^4 dyn cm^{-2} für die 10 cm-Kapillare. Es sei bemerkt, daß die 10 cm lange und die 15 cm lange Kapillare identische Fließkurven ergaben; in beiden Fällen ist also die COUETTE-Korrektur kleiner als die Streuung der Meßpunkte und kann vernachlässigt werden. Auf die abweichenden Kurven bei kurzen Kapillaren komme ich später zurück. Zur Kennzeichnung des Verlaufs von η' mit $\log \sigma$ eignet sich neben η_o auch die Neigung m des linearen Kurvenstücks. Da bei allen Lösungen eine annähernd gleiche Minimalviskosität erreicht wird, ist diese Neigung primär abhängig von der Anfangsviskosität η_o. Daneben ist ein Einfluß einzelner Viskosierparameter insofern zu erkennen, als die Neigung m etwas stärker zunimmt, wenn die Erhöhung der Anfangsviskosität durch Temperaturerniedrigung vorgenommen wird, als wenn ein Anstieg von η_o durch fortschreitende Nachreife oder durch Verringerung der Alkalikonzentration erfolgt (Abb.7). Der Zellstofftyp wirkt sich weder auf diese Neigung der Fließkurve m noch auf andere hier ausgewertete rheologische Parameter wie etwa die ca. 7 kcal/Mol betragende Fließaktivierungsenergie aus. Obwohl aus anderen Untersuchungen - eigenen wie fremden - geschlossen werden muß, daß der übermolekulare Bau der Ausgangscellulose die Struktur der Lösung mitbestimmt und sich sogar in der Regeneratfaser manifestiert, scheint offenbar die Fließkurve der Lösung, zumindest unter den hier angewandten Meßbedingungen, auf diesen Einfluß nicht anzusprechen.

Aus den Versuchsergebnissen kann gefolgert werden, daß bei sehr hoher Schubspannung ($\sigma > 10^5$ dyn cm^{-2}) unabhängig von der Anfangsviskosität und den diese bestimmenden Viskosierparametern die Wechselwirkung zwischen den xanthogenierten Celluloseketten in einem für alle Viskosen gleichen Ausmaße durch den Fließvorgang aufgehoben ist. Bei niedriger Schubspannung dagegen scheint die Art der zwischenmolekularen Bindungen eine Rolle zu spielen, wobei die durch Temperaturerniedrigung hervorgerufene zusätzliche und reversible Wechselwirkung in einem engeren Schubspannungsbereich überwunden wird als die wahrscheinlich stärkeren Haftpunkte infolge fortschreitender Nachreife.

Der Einfluß einer abnehmenden Kapillarlänge auf die Fließkurve ist ebenfalls aus Abb.6 ersichtlich. In der η' - log σ -Darstellung sind die Kurven der 1 cm- und 0,1 cm-Kapillaren gegenüber der 10 cm-Kapillare nach rechts verschoben, wobei zumindest die linearen Abschnitte der Kurven annähernd parallel verlaufen. Die mit den kurzen Kapillaren erreichten "Minimalviskositäten" liegen ebenfalls bei etwa 1 Poise, werden allerdings erst bei einer unkorrigierten Schubspannung von etwa 5.10^5 bis 1.10^6 dyn cm^{-2} erreicht.

Die meist benutzte Fließkurvendarstellung, d.h. die Auftragung von log D gegen log σ zeigt Abb.8. Die Fließkurven für die drei Kapillaren fächern mit wachsender Schubspannung auseinander, am stärksten ist dies bei der kürzesten Kapillare der Fall. Damit ist die "Anlauflänge l" abhängig von der Schergeschwindigkeit D. Entsprechende Ergebnisse hat auch KIRSCHKE [13] mit strukturviskosen Ölen erhalten. Zum Vergleich der einzelnen Lösungen hinsichtlich ihrer Anlauflänge wurden diese für ein konstantes Geschwindigkeitsgefälle D = 10^3 sec^{-1} berechnet. Den relativ geringen Wert des Geschwindigkeitsgefälles wählten wir aus Gründen der Meßgenauigkeit. In Abb.9 sind die Anlauflängen gegen die Anfangsviskosität aufgetragen. Sie nehmen mit letzterer allgemein etwa linear zu, ohne daß ein gesicherter Einfluß einzelner Viskosierparameter zu erkennen wäre.

Aus der log D - log σ -Darstellung ist ersichtlich (Abb.8), daß die Fließkurven für die 10 cm-Kapillare keinen Wendepunkt aufweist, bei höheren Schubspannungen in dieser Auftragung annähernd linear verläuft und " η_∞" experimentell nicht erreicht wird.

Zusammenfassend können wir feststellen, daß erstens technische Cellulosexanthogenatlösungen unterschiedlichster Anfangsviskosität bei hoher Schubspannung praktisch stets die gleiche Minimalviskosität von etwa 1 Poise aufweisen, daß zweitens das Fließverhalten in erster Linie von der Anfangsviskosität bestimmt wird und der Einfluß der einzelnen Viskosierparameter demgegenüber zurücktritt, und daß drittens unsere Ergebnisse in Übereinstimmung mit HOPKINS und WHATLEY [8] gegen eine symmetrische Fließkurve mit Wendepunkt sprechen.

Für die verantwortliche Durchführung des viskosetechnischen Teils der vorliegenden Arbeit danken wir Herrn Chemie-Ingenieur Horst DAUTZENBERG.

L I T E R A T U R

1) EDELMANN,K., E.HORN, Faserforsch. u. Textiltechn. 10, 531 (1959)

2) UMSTÄTTER,H., Z.Elektrochem. 54, 374 (1950)
UMSTÄTTER,H., Strukturmechanik, Dresden und Leipzig, Steinkopff 1948

3) EDELMANN,K., E.HORN, Faserforsch.u.Textiltechn. 12, 190 (1961)

4) PHILIPPOFF,W., H.E.KRÜGER, Kolloid-Z. 88, 215 (1939)

5) BREDÉE, H.L., I.DE BOOYS, Kolloid-Z. 96, 24 (1941)

6) SCHURZ,J., Kolloid-Z. 154, 97 (1957)
Kolloid-Z. 155, 45 (1957)

7) VINOGRADOV,G.V., A.A.KONSTANTINOV, E.A PAKSVER, A.P.FROLOVA, Chim.Volokna (Chem. Fasern) 5, 33 (1963)
PAKSVER,E.A., V.G.VINOGRADOV, A.A.KONSTANTINOV, A.P.FROLOVA, Chim. Volokna (Chem. Fasern) 5, 38 (1963)

8) HOPKINS,T.E., J.W.WHATLEY, J.Applied Polymer Sci. 6, 600 (1962)

9) UMSTÄTTER,H., Einführung in die Viskosimetrie und Rheometrie, Berlin-Göttingen-Heidelberg: Springer-V. 1952, S.96

10) UMSTÄTTER,H., Einführung in die Viskosimetrie und Rheometrie, Berlin-Göttingen-Heidelberg: Springer-V. 1952, S.87

11) EISENSCHITZ,R., B.RABINOWITSCH, K.WEISSENBERG, Mitt.Dtsch.Mat.-Prüf.-Anst.Sonderh.9 (1929),S.91; Zitat bei Meißner,J.:Materialprüfung 5,1 (1963)

12) HERZOG,R.O., K.WEISSENBERG, Kolloid-Z. 46,277 (1928)
RABINOWITSCH,B., Z.Physik.Chem. A 145, 1 (1929)

13) KIRSCHKE,K., Rheologica Acta 2, 147 (1962)

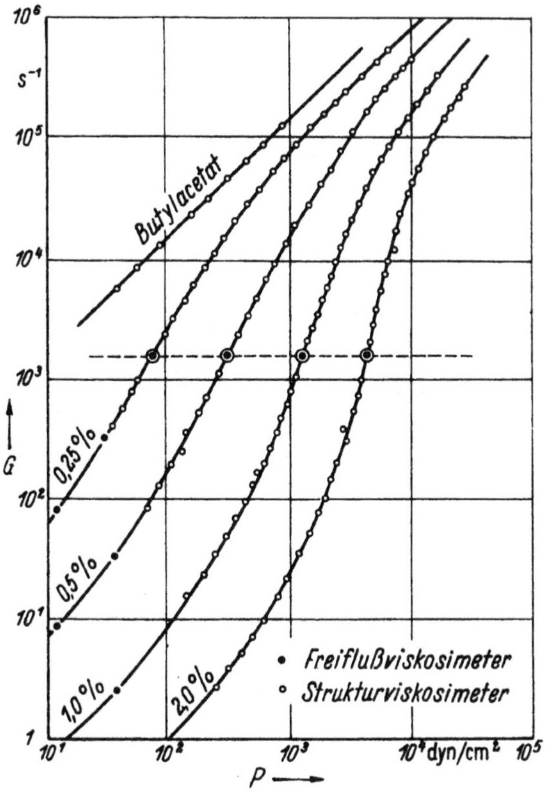

Abb.1 Mechanisches Zustandsbild eines nitrierten Zellstoffs, Nitrat-DP=1060, in Butylacetat (1)

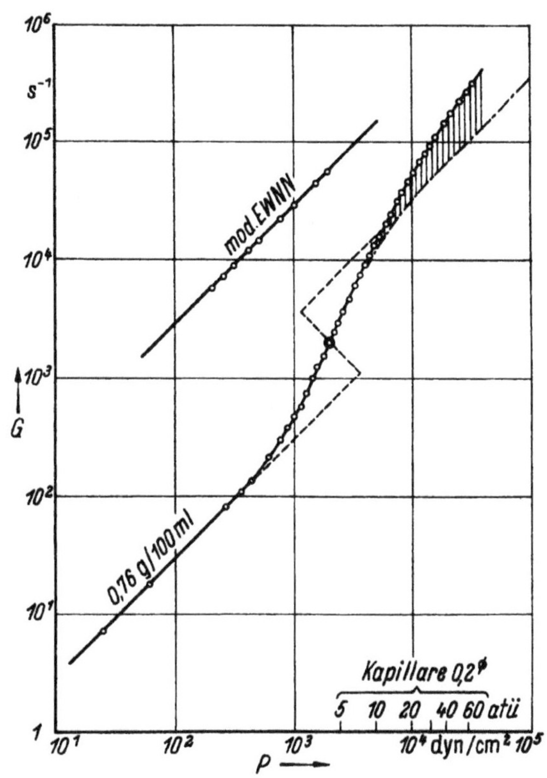

Abb.2 Fließkurve einer 0,76%igen Lösung des Zellstoffs (DP=1060) in dem modifizierten EWNN. Die Abweichung von der symmetrischen Kurve ist schraffiert eingezeichnet. ——— gemessene Kurve —·— Verlauf der symmetrischen Kurve (2)

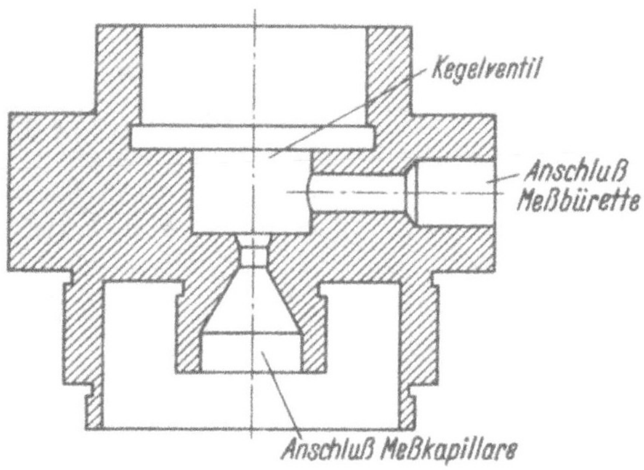

Abb.3 Verbesserter Verschlußkopf für das Strukturviskosimeter nach UMSTÄTTER (9)

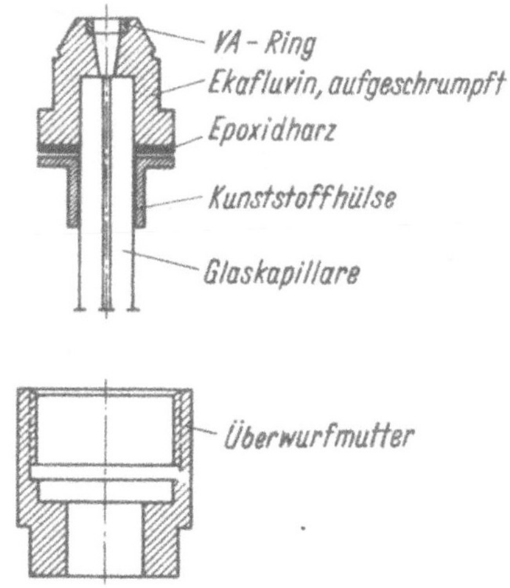

Abb.4 Anschluß und Abdichtung der Meßkapillare mit dem Verschlußkopf

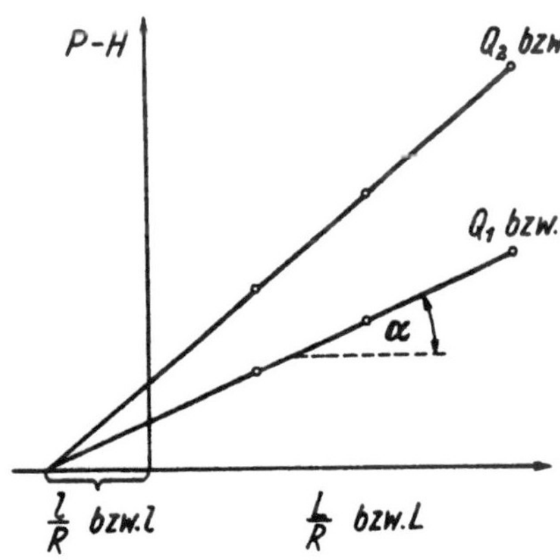

Abb.5 Bestimmung der "Anlauflänge l"

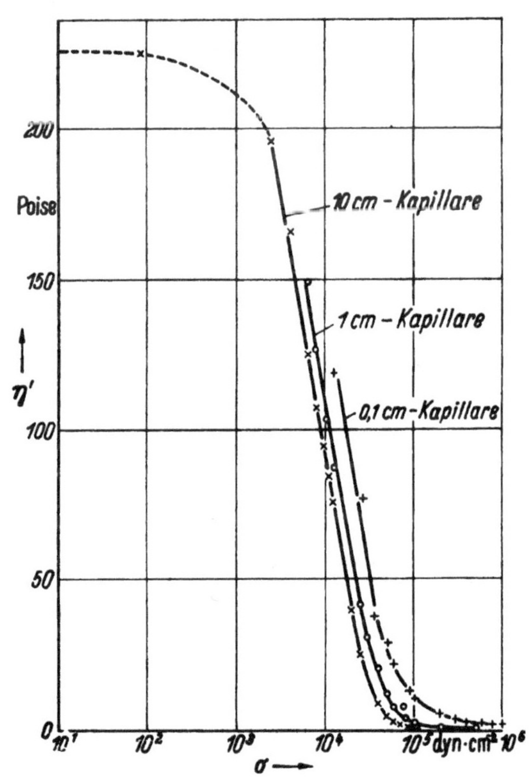

Abb.6 Scheinbare Viskosität η' als Funktion des Logarithmus der Schubspannung

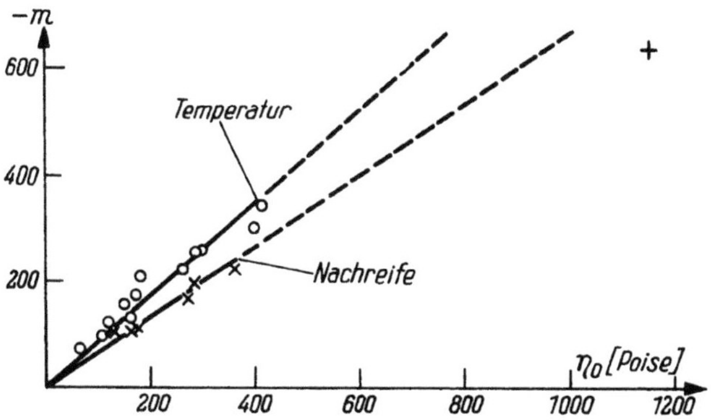

Abb.7 Neigung m des linearen Teils der Kurve aus Abb.6 in Abhängigkeit von der Anfangsviskosität η_0
Parameter: Temperatur und Nachreife
+ : m-Wert für Lösung mit nur 3,26 % NaOH gegenüber 8% NaOH der anderen Lösungen

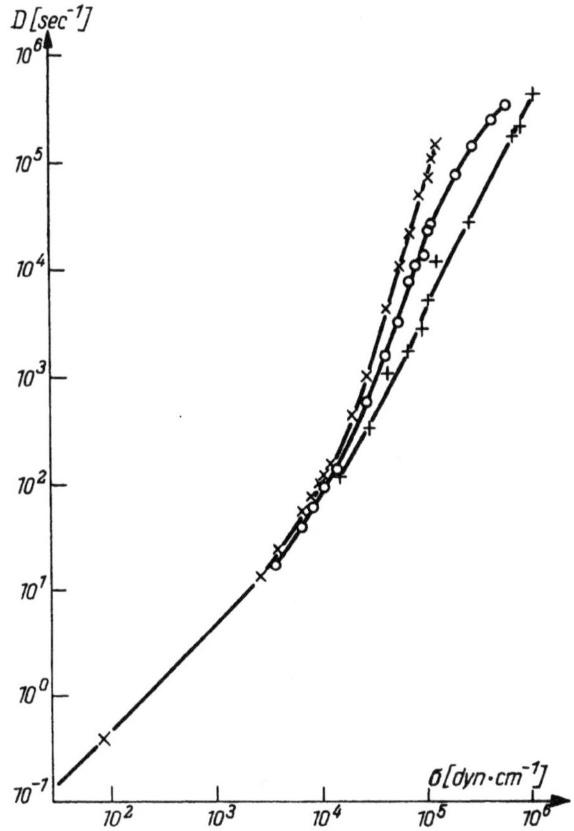

Abb.8 Logarithmus der Schergeschwindigkeit D als Funktion des Logarithmus der Schubspannung
× 10 cm-Kapillare
○ 1 cm-Kapillare
+ 0,1 cm-Kapillare

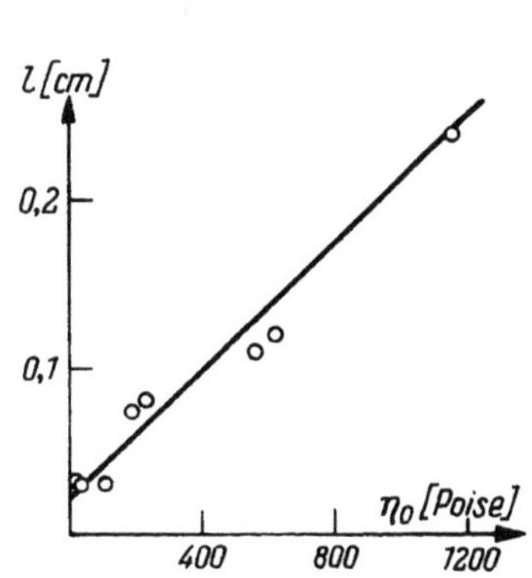

Abb.9 "Anlauflänge l" als Funktion der Anfangsviskosität η_0

C Proteine

Substrate Control of Protein Conformation

R. LUMRY

Laboratory for Biophysical Chemistry of the University, Minneapolis, Minnesota

S U M M A R Y

As presently known the special physiological reactions of proteins are due to 1) the passive positioning of reactants near ionizable groups of the protein; 2) the effective local dielectric constant; and 3) the distortion of primary bonds of the substrate and of the protein produced in folding the protein (the "rack" effect). The relative importance of (2) and (3) for heme protein is discussed and the mechanism of the participation of (3) in hemoglobin and cytochrome c is presented. The present status of attempts to detect conformation changes in chymotrypsin during its catalytic reaction is discussed. At present there is no evidence for drastic changes at normal pH values for the reaction. Optical rotation and reversible thermal denaturation studies indicate significant changes in conformation at extremes of pH and these have proved valuable in studying the changes in the conformation necessary to support catalysis. A new kind of conformation change is suggested in which only small displacements in the protein structure take place but convert the protein from catalytically inactive to a catalytically active form. It is possible that this change is associated with the freezing out of side chain rotation.

This contribution to the symposium has to do with the more mysterious features of the relationship which exists between the chemical function of a globular protein and the conformation of that protein. Many authors have suggested that proteins provide more than a passive binding site for substrates and functional groups, although none of the evidence presented has suggested any very specific hypothesis as to the basis of unique protein mechanisms. Increasingly good evidence has suggested that some mechanisms are n o t involved and has allowed some reasonable assessments of the importance of those mechanisms not excluded by available evidence. For example, it is almost certain that proteins do not have special low-lying electronic states which function uniquely in catalysis. Other suggestions are less easily excluded by available evidence but there remain at the present time three features of globular proteins which probably play the significant roles in enzymic catalysis, oxygen binding, oxidation-reduction reactions and the like.

First among these three though not necessarily first in importance is still the passive role of the protein as a matrix in or on which are positioned the substrates and necessary acidic or basic functional groups of the protein in a way conducive to a low free energy of activation for one or more slowest steps of the overall catenary chain of catalysis.

Simple association of two reactants in such a matrix may lead to remarkable rate acceleration and specificity even without the aid of other factors. For example in Fig. 1 is shown the oxidation-reduction reaction between cytochrome c and the corresponding hemichrome-hemochrome system. Cytochrome c to a first approximation consists of the hemichrome--an octahedral iron complex--embedded in a protein matrix. The effect of the reaction is to transfer one positive charge from a water environment to a protein environment. The net free energy change in this transfer is considerable as shown in the figure. However if an intelligent guess of about 20 is made for the effective dielectric constant, nearly all the free energy change can be explained by charge transfer alone. A variety of studies of the factors influencing the oxidation-reduction potential of this protein [1)] tell us that this explanation is quantitatively incorrect and that a large part of the observed free-energy change must be explained in other ways. Nevertheless the argument is qualitatively sound and it must be concluded that the local electrostatic environment provided by the protein is a major factor in determining the behavior of specific protein reactions. Furthermore the nearest dipolar groups and occasional buried charges will dominate the electrostatic situation. These are determined in position and type by the amino-acid composition of the protein since it alone determines conformation. It is thus entirely possible for Nature to modify the electrostatic environment and thus the characteristics of a given protein reaction by modification of the amino-acid composition.

Thus far two protein roles have been mentioned. One is the orientation of reactants and directly involved functional groups. The other is the effective electrostatic field at this cluster. Both roles are undoubtedly important but we have been forced by the major part of the results we have obtained on protein systems during the last fifteen years to conclude that there is at least one more equally fundamental and equally important role that protein conformation plays in protein reactions [2)]. Perhaps the easiest way to describe this role is to discuss the case of hemoglobin for which the favorable evidence is the most impressive.

Hemoglobin consists of four separate globular sub-proteins strongly bound to each other, very similar in structure, and each containing a heme group to which oxygen is bound at the iron atom in a reaction thought to displace a water molecule from the iron atom. There are many puzzling features of this apparently simple reaction which to many workers in the field [3)4)] have suggested an intimate involvement of protein conformation involvement which can be developed pictorially with several figures. Fig. 2 shows the octahedral hemochrome complex buried in the protein. The sixth ligand position is occupied by water or oxygen, the base plane of the complex ion by the four porphyrin nitrogen atoms and the fifth ligand position by the nitrogen atom of an imidazole group also bound to the protein. It was proposed that the protein retains some control over the orientation of this imidazole group so as to distort it from the normal geometry found in the corresponding simple unstrained hemochrome free of the protein. In Fig. 3 are shown some of the ways in which this orientation can and probably does influence all the 3d orbitals of the iron atom. Twisting, stretching, compression or bending--all can change the orientation of these orbitals with respect to the porphyrin plane and to the remainder of the protein and, in so doing, change orbital hybridization. The $3d_{z^2}$ orbital is the most influenced but this is also the orbital used to bind oxygen or water at the sixth position. Hence it was postulated that the detailed electronic properties of the iron atom and thus its reactions at the sixth position were in a very important way controlled by the conformation of the protein. Thus far, this mechanism is essentially a static one. On folding, the protein produces a net distortion of the imidazole-iron bond but no further change occurs during a reaction at the oxygen position. However such a picture is unreasonable since if changes in imidazole position can alter the electronic situation at the oxygen position, changes in the kind of ligand at the oxygen position must also alter the electronic situation at the imidazole position. Such changes can upset the balance of forces between protein and iron so that the protein might change conformation or at least undergo some readjustment of atom positions as a result of reaction at the oxygen position. At this point we were pleased to note that HAUROWITZ [6] and WYMAN and ALLEN [3] had advanced good though indirect evidence for gross changes in the protein occurring as a result of oxygenation. TAKASHIMA using dielectric-dispersion methods set out to find more direct evidence and our initial results are shown in Fig. 4 [7]. Without going into a detailed description of the results it can be said that this figure is now reasonably well understood as indicating in its maxima and minimum large changes in protein conformation and in charge distribution. MATSUMIYA [8] working under identical solution conditions found no change in sedimentation constant with oxygenation. Hence the binding of oxygen produces a progression of changes in protein conformational properties. These experiments were carried out at very low salt--about 10^{-4} M.--where the effects are very large. At physiological salt concentrations the effects are much smaller and this kind of observation no longer was adequate to show the changes in the protein on oxygenation. A variety of indirect evidence from experiments at higher salt concentrations provided further evidence for oxygenation -linked conformation changes but there was no direct evidence until very recently that changes in conformation occur at higher salt concentrations. Very recently PERUTZ and coworkers [9] have shown that the ß sub-proteins do undergo changes in relative position during oxygenation. These changes are considerable and can o n l y result from changes in the fine details of the internal conformations of the sub-proteins whether or not changes in charge distribution also occur. Nevertheless the resolution of the x-ray diffraction studies of hemoglobin has not yet been sufficient to reveal these inner changes. This is an important point since it demonstrates that x-ray diffraction studies may not always have sufficiently high resolution to detect the all-important conformational shifts of the protein which occur in and support physiological reactions. It also demonstrates that these conformational shifts are probably not large in most cases although at present the x-ray method for hemoglobin does not set a very tight limit as its resolution is only about 5.5 angstrom units.

Additional experiments in our laboratory and by ANTONINI et al. [10][11] in Professor ROSSI-FANELLI's laboratory in Rome provide very strong supporting evidence for the mechanism of hemoglobin which has been presented. Alterations in the protein do alter the electronic properties of heme iron and do appear to do so primarily in the way which has been outlined. It may be illuminating to explain briefly one of the more puzzling features of the oxygenation of hemoglobin, which is the so-called heme-heme interaction in which addition of one oxygen molecule to one sub-protein increases the affinity for oxygen at the heme iron of a second sub-protein. Fig. 5 is a very rough description of this mechanism. It is seen in this figure that the rearrangement of conformation produced in one sub-protein by oxygenation is spread from the imidazole region of a second sub-protein. In the second sub-protein the resulting small conformation changes alter the imidazole orientation relative to its iron atom thus increasing the affinity for oxygen to establish the heme-heme linkage. There are many complexities to this picture and certainly additional ways in which the protein is involved. These complexities are well explained by the mechanism presented, so well in fact that it is no longer advisable to seek evidence favorable to the mechanism but rather necessary to construct experiments designed to show that the mechanism is not correct--a critical point in the development of any new theory.

There is fortunately no electrostatic problem in hemoglobin to complicate its mechanism. Such a problem does exist in cytochrome c but about the same conclusions as to mechanism can be reached for this protein [1][12] as for hemoglobin, although the study is less well advanced and in some ways more complicated. It does appear that in cytochrome c, as in hemoglobin, the electronic properties of heme iron, notably the oxidation-reduction potential, are controlled to a major extent by the orientation of the imidazole ligands at iron positions five and six and that this orientation is in turn controlled by protein folding. In view of the interest of many of the participants in this symposium in DNA, it may be useful to employ cytochrome c to explain the way in which the functional electronic properties of at least some proteins are evolved by changes in amino-acid composition.

As in the case of hemoglobin where the affinity for oxygen is determined by protein folding, the amino-acid composition of cytochrome c determines the way this protein is folded. The details of the folding in turn govern the orientation of the ligands around the iron atom and thus control the electronic properties of the 3d orbitals. No particular magic is involved. As LUMRY and EYRING [5] pointed out many years ago, it is all a matter of free energy. The folding and the orientation of the iron ligands are determined by the amino-acid composition and the normal tendency in Nature for the total protein to fold to the state of lowest free energy. Hence replacement of amino-acid residues at key points in the protein by other residues with different side chains alters details of folding and thus alters electronic properties. Evolution of electronic properties, then, is just as simple and as fundamental as the evolution of other characteristics of living tissue. Different oxidation-reduction potentials, different oxygen affinities, different catalytic efficiencies and the like are thus determined by mutational replacements of one residue by another and a large number of modified electronic states are available to Nature in this way to allow functional as well as morphological evolution.

If the mechanisms which have been discussed are of general significance, it should be possible to detect static or dynamic conformational involvement in the catalytic reactions of enzymes. There has been considerable effort expended in a search for such involvement in recent years. Thus far nothing definite has appeared. Our own work has been concentrated on chymotrypsin, but even after five years we have found no positive evidence that conformational changes occur during the function of this enzyme under normal conditions of salt and pH. Some of the work has been encouraging but little definite evidence for or against such changes has appeared. It would be inappropriate to discuss fully the work done in our laboratory and in that of HESS but a few of the more interesting results may be summarized.

Figure 6 presents the results of studies of optical rotation changes in α-chymotrypsin. We use the MOFFITT a_o parameter to scale these changes. At present it is thought that changes in this parameter without corresponding changes in the helix-measuring parameter, b_o, measure changes in the freedom of the groups vicinal to assymetric carbon atoms as well as any changes in the polarizability of the local environment at such carbon atoms. There is very little α-helix in chymotrypsin--about 18 percent--and it has been found that this helix-content as measured by b_o does not change under any conditions except on complete unfolding in urea. As the figure shows there are large changes in optical rotation in chymotrypsin itself on varying pH [13]. However it is possible to make derivatives of the protein with reagents such as di-isopropylfluorophosphate (DFP) which produces the di-isopropyl derivative (DIPCT) or p-nitrophenyl trimethylacetate which produces the trimethylacetylenzyme (TMACT). The latter compound is a typical "acyl enzyme" [14]. It has been shown that these reagents produce acylation of the serine hydroxyl group which appears to be intimately involved in the catalytic process. Except at the most extreme pH values, where outright denaturation takes place, formation of the derivatives completely restores the optical rotation value to that of free chymotrypsin at its most active pH which is 7.8. In the case of DIPCT there is an overcompensation showing that more drastic changes in structure occur with DFP than with other reagents. It should be pointed out that HESS and coworkers [15] have carried out many similar experiments although the quantitative agreement between their findings and ours is often less than satisfactory. In any event all workers agree qualitatively on the crucial points and from these we may draw the conclusion that there is a change in conformation, or at least a change in tightness of folding, rate of opening up, etc. which can be produced in chymotrypsin by adding acid or by adding base. The effect is to alter the enzyme from its active form at pH 7.8 but the resulting inactive forms are restored to the active conformation by the specific reagents in the course of acetylation even though the pH is still far removed from 7.8. One then observed optical-rotation changes in reactions with these reagents at pH values other than 7.8 but these are accidental and not a characteristic of the normal catalytic reaction. We have carried out a number of experiments to detect conformation changes at pH 7.8 and although very small changes in optical rotation can be observed on deacylation of TMACT at this pH there is little reason to interpret these as indicating any significant conformation change. Of the several methods we use only the measurement of indole fluorescence yield [16] shows peculiar properties for the acy enzyme at pH 7.8 and we are not at present inclined to believe that these imply any deep-seated change in the conformation of the protein. The active site of chymotrypsin is very much like a binding site on serum albumen--it has very little specificity and will bind almost anything [17]--but most of the inhibitor binding reactions in chymotrypsin appear to be much slower than the catalytic reactions.

The optical-rotation results by themselves are not sufficient to convince us that any major change in conformational characteristics occur when chymotrypsin reacts with substrates or specific acylating or phosphorylating reagents. Stronger evidence is supplied by studies of the reversible denaturation of the protein in its free and derivative forms. Fig. 7 shows the behavior of free chymotrypsin as measured by ultraviolet difference spectra. On heating the protein undergoes denaturation reversibly in a mechanism which strictly follows mass-law behavior, although in this case as in all other protein cases studied carefully, there are large heat capacity effects. In Fig. 8 the behavior of DIPCT is the same kind of experiment as shown [18]. The protein now denatures in at least two distinct steps and the thermodynamic quantities involved are quite different from those obtained in the study of free chymotrypsin. Starch-gel electrophoresis shows that only one form of protein is present. Such behavior can only result from a fundamental change in the character of the protein and it must be remembered that this change is produced by the binding of a single DIP group. There is thus considerable evidence that the phosphorylation reaction is not minor and there is evidence of other types from both HESS's laboratory and ours to give additional confidence to this interpretation. It is also not a totally trivial process from the point of view of the study of chymotrypsin kinetics since the pH dependence of the kinetic parameters of the catalytic reaction are related directly to changes in conformation rather than to simple changes in the ionization of a few acidic or basic groups.

Table 1 [13][19] shows that substrates and even carboxylic-acid products of substrate cleavage also produce readily detectable changes in the optical rotation of chymotrypsin at pH 2.5 or, indeed, at any pH other than about 7.8. A rather good case has been made by BENDER et al. [20] for the normal occurrence of primary-bonded acyl-enzyme intermediate forms in the catalysis of normal chymotryptic substrates. Our evidence in some ways supports this view and suggests in agreement with BENDER that changes in optical rotation and the like occur as a result of acyl-enzyme formation. On the other hand, the carboxylic acids lack sufficient free energy relative to hydrolysis to form acyl enzymes and yet they too cause rotation changes. Tentatively we are led to the idea that at acid or basic pH values both the weakly-bound MICHAELIS-MENTEN complex and the acyl enzyme have altered conformational characteristics. The work is however in a very early stage and no definite conclusions are yet possible.

Table 1

Added compound or forms	$-a_o$
CT	490
DIP-CT	405
TMA-CT	420
t-CINNAMOYL-CT	425
ACETYL-L-TYROSINE	460
ACETYL-L-TRYPTOPHAN	450
β-PHENYLPROPIONIC ACID	460
ACETYL-L-PHENYLALANINE-METHYL ESTER	445
ACETYL-L-TYROSINE-ETHYL ESTER	435
ACETYL-D-TYROSINE-ETHYL ESTER	495 - 515
ACETYL-L-TYROSINE AMIDE	515
L-PHENYLALANINE-ETHYL ESTER	485

The changes in optical-rotatory-dispersion parameter a_o of chymotrypsin in solutions containing virtual substrates, real substrates and enantomorphic isomers of real substrates. The pH is 2.5.

Thus far we have emphasized the absence of any substrate-produced change in conformation or conformational characteristics of the protein at its normal operating pH. But the fact of the matter is that we are still convinced that such changes do occur and result in the distortion of protein or substrate primary bonds so that these bonds become sufficiently reactive to explain the high values of enzymic rate parameters. In hemoglobin at higher salt concentrations the changes in internal folding are relatively small and probably consist of small shifts in the positions of a number of atoms. We are inclined to believe that similar small shifts--undetectable in most optical rotation work--occur in chymotrypsin and are equally important in determining the electronic properties of both the protein and substrate bonds at which the chemical reactions occur. It is perhaps misleading to call such changes conformation changes and we have preferred to call them changes in conformational characteristics. Indeed we have come to refer to this kind of protein process as a subtle change, meaning, among other things, that we do not at all understand it. Fortunately this term is not a sterile one since our recent studies of plasma albumin [21)], chymotrypsinogen [22)], enolase [23)] and hemoglobin have demonstrated equally mysterious small changes often detectable only in refined proton-exchange measurements, solubility changes, or by large entropy changes, although almost never in the other physical parameters we use. We suggest that such changes involve the folded integrity of the protein. Quite possibly in such changes side-chain motion which is highly restricted under more "native" conditions becomes less hindered, the protein opens up more readily and water penetrates more frequently. We are just at the beginning of the investigation of this kind of process and indeed of the whole aggregate of processes which have for many years loosely been called reversible denaturation. As this report indicates we are also very little advanced toward any real understanding of the conformational basis of enzymic catalysis. Evidence favoring the mechanism presented for hemoglobin is formidable and it is primarily on the basis of these findings that one may retain an optimism toward the possibility of a similar conformational basis for enzymic catalysis. It is not at all certain that any such basis exists except in the sense of the first two of the proteins roles presented at an early point in this report. Nevertheless it should be emphasized that the additional protein role based on internal strain and distortion caused by folding is at the present still the best possibility for the explanation of the unique phenomenon of enzymic catalysis.

Acknowledgements: This report is based on work supported by the U. S. Department of Defense through the Office of Naval Research and by the U. S. Public Health Service, National Institutes of Health.

R E F E R E N C E S

1. SULLIVAN, J., unpublished observations from this laboratory.
2. LUMRY, R., J. Biophysics (Japan) 1, No. 3, 1 (1961)
3. WYMAN, J., Jr., D. W. ALLEN, J. Polymer Sci. 7, 499 (1951)
4. MANWELL, C., Ann. Revs. Physiology 22, 191 (1960)
5. LUMRY, R., E. EYRING, J. Phys. Chem. 58, 110 (1954)
6. HAUROWITZ, F., Hoppe-Seyler's Z. Physiol. Chem. 254, 266 (1938)
7. TAKASHIMA, S., J. Am. Chem. Soc. 78, 541 (1956). TAKASHIMA, S., R. LUMRY, ibid. 80, 4238 (1958)
8. MATSUMIYA, H., R. LUMRY, R. BURNS, American Chemical Society Abstracts, 136th meeting, Atlantic City, N. J., 1959, 3C.
9. PERUTZ, M. F. et al., Nature [London], in press.
10. ROSSI-FANELLI, A., E. ANTONINI, Arch. Biochem. Biophys. 80, 308 (1959). ROSSI-FANELLI, A., E. ANTONINI, A. CAPUTO, ibid. 85, 37 (1959)
11. ANTONINI, E., J. WYMAN, Jr., P. MORETTI, A. ROSSI-FANELLI, Biochim. Biophys. Acta 71, 124 (1963)
12. LUMRY, R., A. SOLBAKKEN, J. SULLIVAN, L. REYERSON, J. Am. Chem. Soc. 84, 142 (1962)
13. PARKER, H., R. LUMRY, J. Am. Chem. Soc. 85, 483 (1963)
14. BENDER, M. L., J. Am. Chem. Soc. 84, 2582 (1962)
15. WOOTON, J. F., G. P. HESS, J. Am. Chem. Soc. 84, 440 (1962). HAVSTEEN, B. H., G. P. HESS, J. Am. Chem. Soc. 84, 491 (1962). LABOUESSE, B., B. H. HAVSTEEN, G. P. HESS, Proc. Nat. Acad. Sci. 48, 2137 (1962). HAVSTEEN, B., B. LABOUESSE, G. P. HESS, J. Am. Chem. Soc. 85, 796 (1963)
16. LUMRY, R., S. YANARI, F. BOVEY, to be published.
17. HEIN, G., C. NIEMANN, Proc. Nat. Acad. Sci. 47, 1341 (1961). HEIN, G., C. NIEMANN, J. Am. Chem. Soc. 84, 4495 (1962). WALLACE, R. A., A. N. KURTZ, C. NIEMANN, Biochem. 2, 824 (1963)

18. BILTONEN, R., unpublished observations from this laboratory.
19. PARKER, H., unpublished observations from this laboratory.
20. BENDER, M. L., G. E. CLEMENT, F. J. KÉZDY, B. ZERNER, J. Am. Chem. Soc. 85, 358 (1962)
21. BENSON, E., B. HOLLOWAY, R. LUMRY, J. Biol. Chem., to be published November (1963)
22. BRANDTS, J., R. LUMRY, J. Phys. Chem. 67, 1484 (1963)
23. ROSENBERG, A., R. LUMRY, to be published.

D I S C U S S I O N - Q U E S T I O N S

ZAHRADNIK : I feel that it would be of some value to carry out quantum-chemical calculations concerning the electronic structure of some components of the systems under study. I would like to know whether you have performed some calculations of this kind.

LUMRY : Of course, quantum-mechanical calculations cannot give any really quantitative information. They can however tell us whether or not significant changes in electronic state of a complexed metal ion can occur through rearrangement of ligand geometry. In our own studies we have considered only the simplest approach to rehybridization which suggests that angular distortions in particular have large effects on the orbital situation in the complex ion. More detailed studies, in some cases very detailed, are being carried out in two laboratories in Japan but as yet nothing has been reported. Experimentally the "trans effect" is reasonably well known and is the current model for ligand interaction. In this effect there is a clear connection between the ligands trans to each other across the central metal ion of an octahedral complex so that changes of one ligand either in type of position force changes in the electronic at the ligand trans to it. Unfortunately there is not yet much theoretical consideration of this well known effect.

BLUMENFELD : Can you tell about the changes in electronic state of iron due to the oxygenation in more detailes ? And why these changes take place ?

LUMRY : According to our hypothesis the imidazole bond to iron in hemoglobin is not broken during oxygenation but rather the bond is altered in its properties for a rearrangement of the imidazole group and the way the 3d orbitals interact with imidazole nitrogen and with the porphyrin nitrogen atoms. This is a quantitative change due to twisting, bending, stretching of the bonds. We expect even that iron may be well displaced from the heme plane. In this way the electronic properties of iron and the bonds it makes are partially controlled by protein folding.

BLUMENFELD : What can you say about the nature of N-Fe bond in reduced hemoglobin ?

LUMRY : I can only say that the iron nitrogen oonds in reduced hemoglobin are quantitatively different from the same bonds in model hemochromes. I do not think that the porphyrin plane will allow change of coordination number from four to six in the oxygenation or deoxygenation process.

JORDAN : Could you elaborate the details of the renaturation of chymotrypsin. Does renaturation occur on cooling quickly or has you carried out slowly the cooling ?

LUMRY : Chymotrypsinogen shows a relatively simple, totally reversible, thermal denaturation. The reaction is reasonably fast but has a very large temperature coefficient for the rate. **Just above T_m** the reaction occurs in about 15 sec. No problems of reversibility are encountered except at long standing of the denatured form which than aggregates irreversibly. The reaction follows the simply mass-law expectations for a two-state equilibrium. However in contrast to unfolding processes like that of polyglutamic acid or DNA, the protein is not very extensively unfolded in the process though the entropy of the process is remarkably high. On the whole the process is in most ways simpler than DNA denaturation.

BEHLKE : Die Struktur H_2O - Fe - Imidazol trifft nur für das Hämoglobin zu. Im Methämoglobin muß die 5. und 6.Koordination des Eisens durch H_2O-Moleküle besetzt sein. Wir konnten das aus spektralphotometrischen und magnetischen Messungen entnehmen. Wir nehmen an, daß ein Unterschied zwischen der Struktur von PERUTZ und KENDREW besteht. Sie haben am Kristall gearbeitet, wir hingegen in wässriger Lösung. Durch Zugabe von hohen Salzkonzentrationen muß man mit einer Verdrängung des am Eisen gebundenen Wassers rechnen. (Näheres siehe Vortrag: W.SCHELER:Symposium für molekulare Zellphysiologie. Vortrag: J.BLANCK: Symposium für molekulare Zellphysiologie, Berlin,Aug.1963, Akad.V. (im Druck)).

LUMRY : This is a most remarkable conclusion which I do not understand. It is very important if true and though it seems unreasonable to me, I cannot really exclude it on the basis of what I now know. In fact some experiments on water binding by hemoglobin carried our by Mr. HNOJWEJY in REYERSON's laboratory at Minnesota are consistent with this picture you have presented. It will be most interesting to see the details of this work and to hear of the results of future experiments of this sort.

ELECTROSTATIC CONTRIBUTION TO THE FREE ENERGY OF THE OXIDATION-REDUCTION REACTION

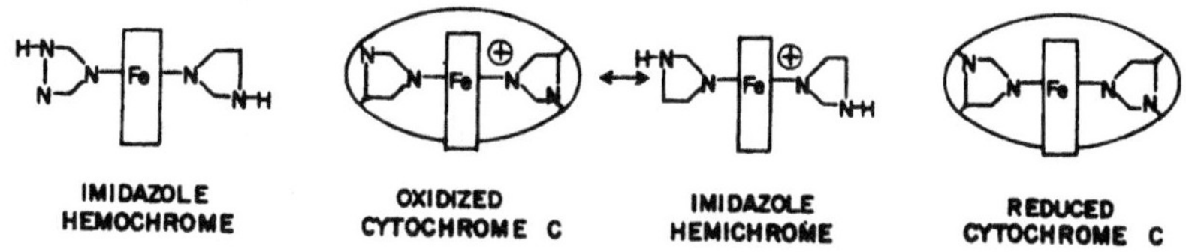

$\Delta E_0 = 0.365$ V

$\Delta F^0 = -8.3$ kcal. mole^{-1}

Fig.1 The reaction between oxidized cytochrome c and the di-imidazole hemochrome complex in water.

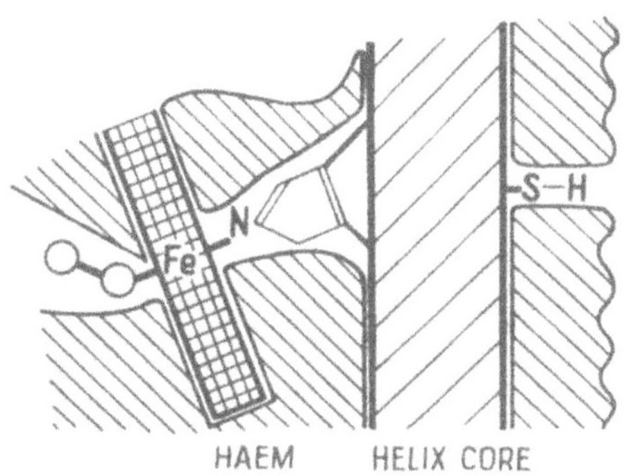

Fig.2 A Schematic detail of the heme complex ion embedded in myoglobin. A protein imidazole groups is bound at the fifth ligand position of heme iron and an oxygen molecule is shown bound in the sixth position.

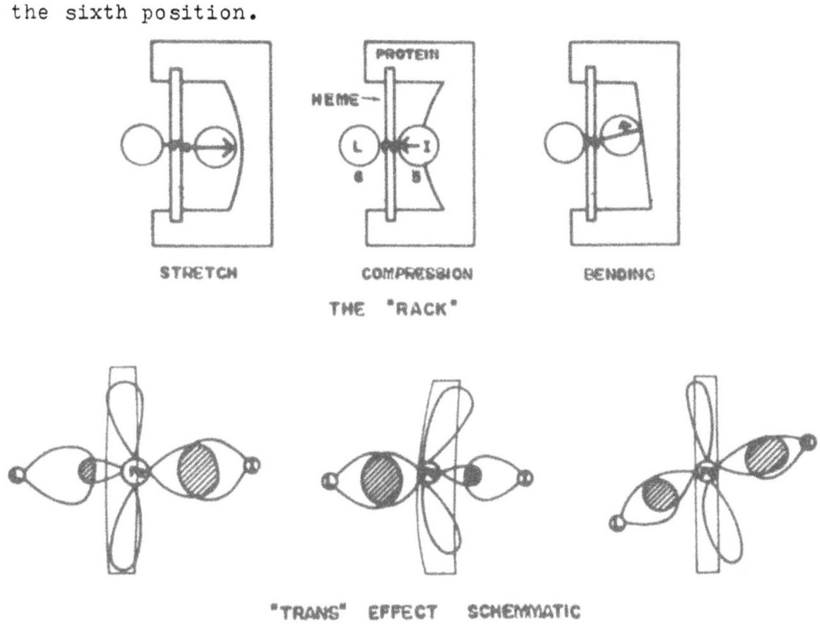

Fig.3 Some possible orbital rearrangements in the sigma bonding of heme to protein showing the effect of changes in imidazole position.

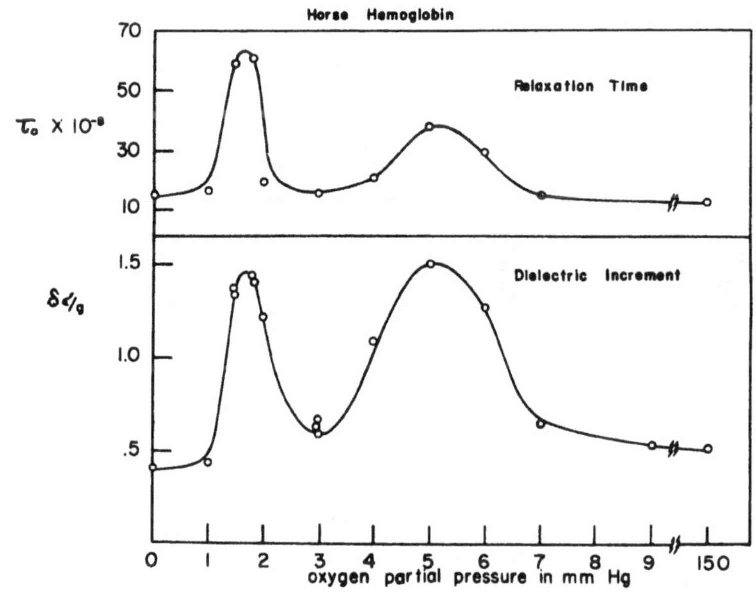

Fig.4 The dielectric relaxation time, T_0, and dielectric increment of horse hemoglobin solutions as a function of oxygen pressure showing large changes in these conformational parameters during the several steps of oxygenation. Low salt concentration.

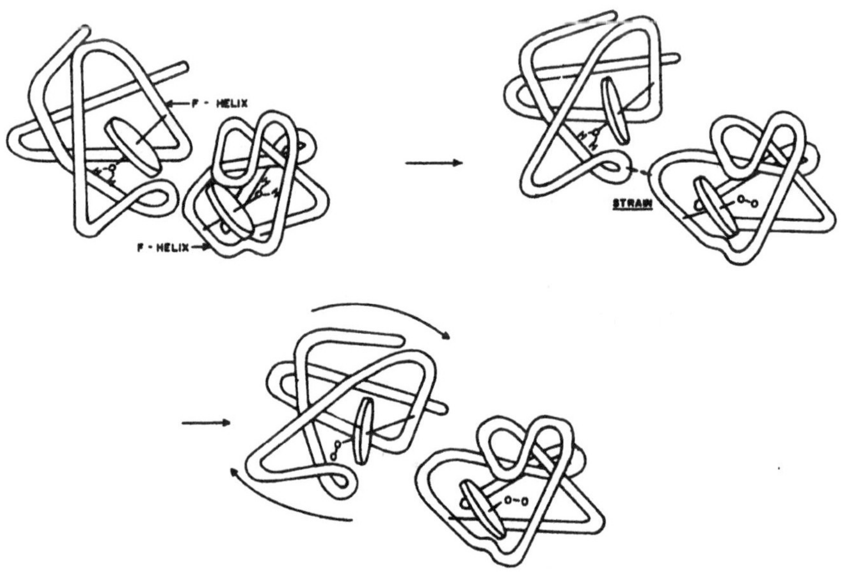

Fig.5 A rough description of the heme-heme interaction process in hemoglobin. Oxygenation of one heme produces distortions of structure passing from imidazole group of one sub-protein to another and changing oxygen affinity at the second heme group.

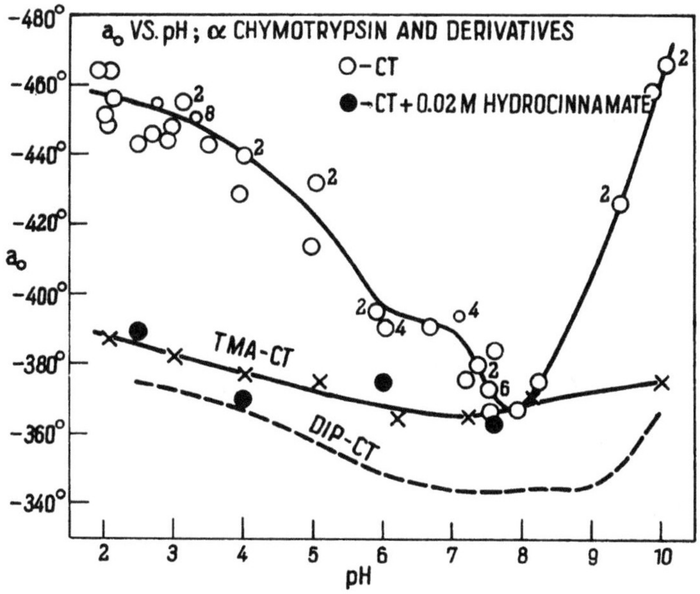

Fig.6 The behavior of the optical-rotatory-dispersion parameter a_o as a function of pH for α-chymotrypsin and two derivatives. 25°.

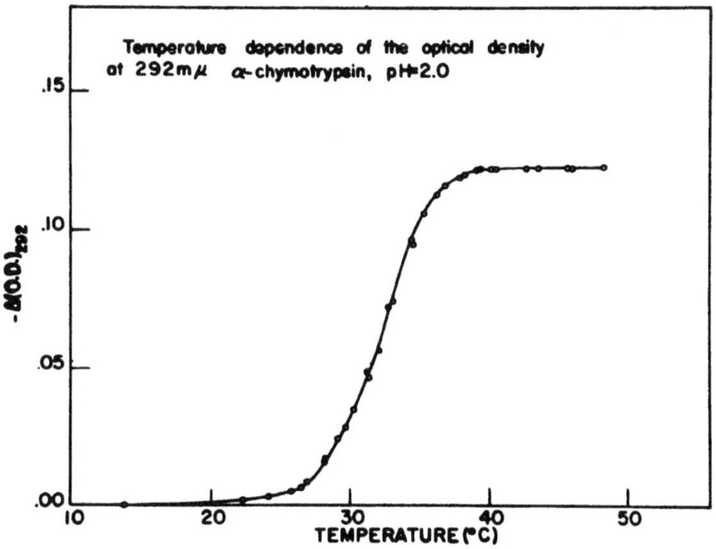

Fig.7 The reversible thermal denaturation of α-chymotrypsin as measured with indole absorbance changes.

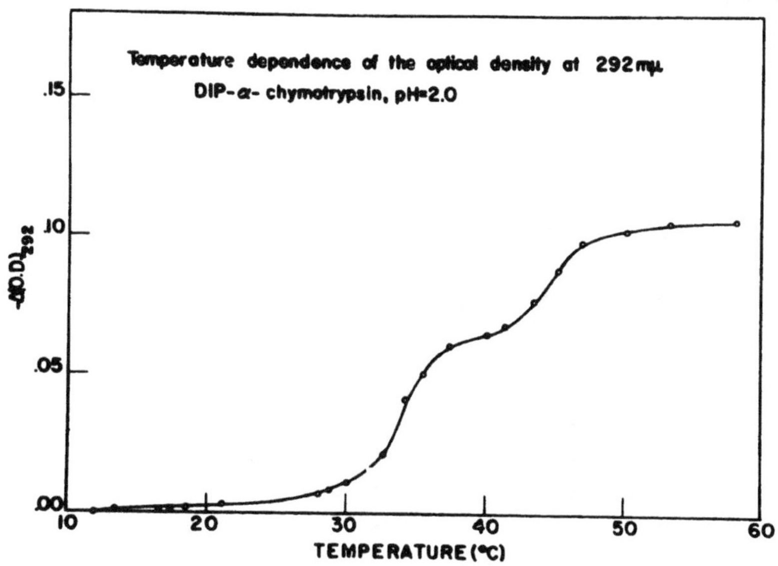

Fig.8 The reversible thermal denaturation of di-isopropylphosphoryl chymotrypsin as measured with indole absorbance changes.

C Proteine

Magnetochemische Aufschlüsse über das System Metall/Porphyrin/Protein

W. HABERDITZL

Institut für physikalische Chemie der Humboldt-Universität, Berlin

S U M M A R Y

In the isolated state the components of the above combination are magnetochemical marked in the following sense.

1. Ions of heavy metals: Spinparamagnetism of the $d_{\varepsilon}^{n}\ d_{\gamma}^{m}$ - electrons; strong dependence of term sequence and term distances (also orbital part) from symmetry and strength of the ligand, according to this a great variability of valency and of ability of binding.

2. Porphyrine system: The largest of known diamagnetisms of organic compounds due to the ringcurrent of π-electrons.

3. Protein: Normal additiv amino acid diamagnetism. Because of the sterical structure a particular variability in the symmetry of the ligand field (changes of the tertiaer structure).

The magnetical behaviour of the hem derivatives, obtained by combination of the components 1 + 2, 1 + 2 + 3 illustrates on the basis of the properties 1 - 3 the origin of a system, the state of binding of wich is influenced in a large extent by the smallest changes in the ligand field.

In agreement with this behaviour a MO-scheme of hem protein is proposed.

Das System Metall/Porphyrin/Protein gehört zu den am intensivsten erforschten biochemischen Untersuchungsobjekten. Trotz des Einsatzes zahlreicher moderner physikalisch-chemischer Untersuchungsmethoden sind zwar viele Einzelheiten der biochemischen Funktionen dieses Systems (z. B. Photosynthese, Atmung, O_2-Transport) aufgeklärt worden, die meisten grundlegenden Probleme der atomaren Primärreaktionen dieser Funktionen und die mit ihnen zusammenhängenden Besonderheiten der chemischen Bindung sowohl am Metall-Zentralatom (z. B. Oxyhämoglobin) als auch am Porphyrinsystem (Photolyt der CO_2-Assimilation ?) sind noch ungelöst. Seit den grundlegenden Arbeiten PAULINGs haben magnetische Untersuchungsmethoden (Messungen der statischen Suszeptibilität und der Spinresonanz) wesentlichen Anteil an der Aufklärung chemischer Bindungen und Reaktionen von Porphyrinkomplexen. Deren Komponenten sind in isoliertem Zustand magnetochemisch folgendermaßen aufgezeichnet: (In diesem Vortrag werden nur die Methoden der statischen Suszeptibilitätsmessung sowie der ESR behandelt. Auf die Anwendung der magnetischen Kernresonanzspektroskopie sowie auf Mößbauereffekt-Messungen wird nicht eingegangen.

1. Schwermetallionen

Diese besitzen Spinparamagnetismus der $d_\varepsilon^n\ d_\gamma^m$ - Elektronen mit oder ohne Bahnanteil und zeigen starke Abhängigkeit der Termfolge und des Termabstandes sowie des Bahnanteils von Symmetrie und Stärke des Ligandenfeldes, was große Variabilität von Wertigkeit und Bindungsfähigkeit bedingt. Als Beispiel sei das Fe^{+++} ($d_\varepsilon^3\ d_\gamma^2$), wie es z. B. im Methämoglobin vorliegt, betrachtet. Das freie Ion besitzt einen $^6S_{5/2}$-Grundzustand, dessen resultierendes Bahnmoment gleich Null ist. Trotzdem findet man auch bei schwachen Kristallfeldern eine geringe Nullfeldaufspaltung. Ihre Ursache konnte für Kristallfelder kubischer Symmetrie durch VAN VLECK und PENNEY [1] und Felder tri- und tetragonaler Symmetrie durch ABRAGAM und PRYCE [2] aufgeklärt werden. Sie hängt vom Energieabstand zwischen dem ^6S-Grundterm und dem $3\ d^4\ 4\ s\ ^6D$- bzw. dem ^4P-Term des freien Ions (ca. $2{,}5 \cdot 10^4\ cm^{-1}$) sowie von der Spin-Bahn-Kopplungskonstanten λ(ca. 300 cm^{-1}) und den Kristallfeldpotentialen ab. Die Abspaltung berechnet sich zu ca. 0,1 cm^{-1}, was mit den experimentellen ESR-Daten übereinstimmt.

Die grundlegende Untersuchung der Aufspaltung von d^5-Konfigurationen in starken Kristallfeldern, wie sie bei zahlreichen Fe^{+++}-Komplexen vorliegen (z. B. Cyaniden) stammt von HOWARD [3]. Sie wurde von STEVENS [4] ergänzt und erweitert. Die Einbeziehung des MO-Verfahrens in die Kristallfeldtheorie oktaedrischer d^5-Komplexe wurde weiterhin von OWEN [5] durchgeführt. Mit ihr ist die bisher wohl zweckmäßigste Methode der Untersuchung der Komplexbindungsverhältnisse auch in komplizierteren Ligandenfeldern gegeben worden. In einem MX_6-Komplex z. B. werden die für die d_ε/d_γ -Aufspaltung (Δ = 10 Dq) und damit für das "Spinkriterium" wichtigen d-MO's durch folgende Linearkombinationen zwischen d-Metall- und p_σ -Liganden-AO's gebildet:

$$\sigma_{z^2}^* = \alpha\, d_{z^2} - \sqrt{1-\alpha^2}\ \frac{1}{\sqrt{12}}\ (2p_6 - 2p_3 + p_1 + p_2 - p_4 - p_5)_\sigma$$

$$\sigma_{x^2-y^2}^* = \alpha\, d_{x^2-y^2} - \sqrt{1-\alpha^2}\ \frac{1}{2}\ (p_2 + p_4 - p_1 - p_5)_\sigma$$

$$\sigma_{z^2} = \sqrt{1-\alpha^2}\ d_{z^2} + \frac{\alpha}{\sqrt{12}}\ (2p_6 - 2p_3 + p_1 + p_2 - p_4 - p_5)_\sigma$$

$$\sigma_{x^2-y^2} = \sqrt{1-\alpha^2}\ d_{x^2-y^2} + \alpha\frac{1}{2}\ (p_2 + p_4 - p_1 - p_5)_\sigma$$

Die Indices 1-6 bezeichnen die Ligandenatome auf den Achsen x, y, z, -x, -y, -z. Bei "reiner" Ionenbindung ist der Überlappungskoeffizient $\alpha = 1$, bei "reiner" Kovalenz ist $\alpha^2 = 0{,}5$. α läßt sich aus ESR-Messungen ermitteln, falls Δ und λ bekannt sind. Nach OWEN gilt

$$g = 2{,}0023 - \alpha^2 \cdot \frac{8\,\lambda}{\Delta}$$

Für die Größe von Δ ist weiterhin auch die π-Bindung durch Linearkombination von d_ε -Metall- und p_π -Liganden-AO's wichtig:

$$\pi_{xy}^* = \beta d_{xy} - \sqrt{1-\beta^2}\ \frac{1}{2}\ (p_1 + p_2 - p_4 - p_5)_\pi$$

$$\pi_{xy} = 1 - \beta^2 d_{xy} + \beta\,\frac{1}{2}\ (p_1 + p_2 - p_4 - p_5)_\pi$$

Die π_{yz}, π_{yz}^*, π_{zx}, π_{zx}^* -MO's erhält man entsprechend durch zyklische Vertauschung der Indices. Bei fehlender π-Bindung ist $\beta = 1$. Das bisherige experimentelle Material läßt sich im Fall des Fe^{+++} in zwei Gruppen zusammenfassen: Entweder ist Δ noch so klein, daß der $d_\varepsilon^3\ d_\gamma^2\ ^6S_{5/2}$ -Term Grundterm bleibt oder Δ ist so groß, daß ein d_ε^5-Grundterm mit L' = 1 und S' = 1/2 resultiert. Der erste Fall liegt bei den meisten hydratisierten Fe^{+++}-Salzen vor, der zweite z. B. bei $K_3[Fe(CN)_6]$. Im ersten Fall beträgt $\Delta \approx 10\,000\ cm^{-1}$, im zweiten Fall ist $\Delta \approx 50\,000\ cm^{-1}$. Im zweiten Fall ist zu beachten, daß durch die asymmetrische d_ε-Besetzung und die daraus resultierende 3-fache Bahnentartung des Grundzustandes ein erhebliches magnetisches Bahnmoment und durch den Jahn-Teller-Effekt Symmetrieerniedrigung auftritt. Sie nimmt bei zunehmender Stabilisierung durch d_ε-π-Bindung zu.

Die Wechselwirkung der in 3 Kramers-Dubletts aufspaltenden d_ε^5-MO's mit einem statischen Magnetfeld wurde in quantitativer Hinsicht neuerdings vor allem von KAMIMURA [6] behandelt. Sie ist die Grundlage für eine Analyse der experimentellen ESR- und Suszeptibilitäts-Daten. Die Lage der Kramers-Dubletts hängt auch hier vor allem von λ und Δ ab, wobei Δ den Abstand zwischen den d_ε-Niveaus angibt. Beide Größen werden durch Kovalenzbildungen beeinflußt, λ durch die Potentialänderung am Metallion infolge dativer π-Bindung. Das effektive magnetische Moment unterscheidet sich vom reinen Spinwert (μ_s) wie folgt:

$$\mu_{eff} = \mu_s \left(1 - \frac{\alpha\lambda}{\Delta}\right) \quad \begin{array}{l} (\alpha = 2 \text{ für D-Terme}) \\ (\alpha = 4 \text{ für F-Terme}) \\ (\alpha = 0 \text{ für S-Terme}) \end{array}$$

2. Porphyrinsystem

Das Porphyrinsystem ist in magnetochemischer Hinsicht in besonderer Weise ausgezeichnet. Sein Diamagnetismus ist der größte bekannte aller organischen Substanzen. Der Diamagnetismusanteil,der dem London-Diamagnetismus des Ringstromes des konjugierten π-Elektronensystems entspricht, konnte auf zwei Wegen unabhängig voneinander ermittelt werden [7]. Einmal läßt sich mit Hilfe zuverlässiger Diamagnetismusinkremente aus den gemessenen Suszeptibilitäten dieser Anteil bestimmen, andererseits läßt er sich direkt experimentell bestimmen durch Vergleich mit dem analog gebauten Bilirubin, bei dem der große konjugierte π-Elektronenring unterbrochen ist. Auf beiden Wegen erhält man einen Ringstromdiamagnetismus, aus dem sich mit Hilfe folgender Beziehung der Radius des Stromringes bestimmen läßt:

$$\chi_{Ring} = \frac{N_L e^2}{4\ mc^2} \sum_n \overline{r^2} = 4{,}248\ n \cdot \overline{r}^2 \cdot 10^{10} \quad (n = \text{Zahl der } \pi\text{-Elektronen})$$

Man erhält r = 3,21 Å. Es zeigt sich, daß ein Stromring dieser Abmessung sich mit dem Areal der konjugierten Doppelbindungen des Porphyrinmoleküls deckt. Dieses Ergebnis stützt ebenfalls die Vorstellungen über die Ursache des anormal hohen Diamagnetismus des Porphyrins. Auch die Anisotropie des Diamagnetismus des Porphyrinmoleküls geht auf den Ringstromdiamagnetismus zurück. Interessante Beziehungen ergeben sich aus der Tatsache, daß der Diamagnetismus der π-Elektronen-Kollektive solcher Makrozyklen eine Art Mittelstellung einnimmt zwischen dem Langevin-Diamagnetismus lokalisierter Elektronen und dem Landau-Diamagnetismus des Elektronengases. Auf die besondere Rolle des Porphyrinsystems als Ligandenfeld wird sicher auch die Wirkung des vom π-Elektronen-Kreisstrom induzierten sekundären magnetischen Momentes auf die im Zentrum des Porphyrinmoleküls gebundenen paramagnetischen Metallionen einen Einfluß haben (s. Abb. 1).

3. Protein

Der makromolekulare Anteil der Hämproteine, z. B. das Globin, verhält sich magnetisch normal. Es konnte gezeigt werden, daß der Globindiamagnetismus sich additiv aus dem Diamagnetismus der das Protein aufbauenden Aminosäuren zusammensetzt [8]. Durch die besondere sterische Variabilität der Tertiärstruktur des Proteins kann jedoch das Ligandenfeld des wahrscheinlich über Imidazol-N an das Metall gebundenen Proteinanteils variiert werden.

Durch magnetische Messungen läßt sich z. B. die Kinetik der H_2O-Diffusion bei der reversiblen Reaktion Methämoglobin ⇌ Anhydro-Methämoglobin verfolgen, da ersteres paramagnetisch, letzteres diamagnetisch ist (bzw. nur den Spinparamagnetismus nur eines Elektrones besitzt) [9].

Besonders wertvoll erwiesen sich die ESR-Messungen der Southamptoner Schule in Gemeinschaft mit den struktur-analytischen Arbeiten von KENDREW und PERUTZ im Hinblick auf die räumliche Orientierung der Porphyrinebene relativ zur Polypeptidkettenrichtung im Myoglobin sowie auf die Orientierung der 4 Hämgruppen relativ zu den Kristallachsen des Hämoglobins. Solche Aussagen sind auf Grund der Anisotropie des g-Faktors in Hämproteinen mit ^6S-Grundzuständen mit Hilfe von ESR-Messungen bei verschiedenen Ausrichtungen der Kristallachsen von Hämprotein-Einkristallen relativ zum Magnetfeld möglich [10]. Durch die Zusammenarbeit von Röntgenstrukturanalyse und ESR-Spektroskopie gelingt also ein Einblick in die Primärstruktur in unmittelbarer räumlicher Nachbarschaft der funktionell wirksamen Atomgruppierung im Protein wie sie von keiner anderen Methode erreicht wird.

Das System 1. + 2. bzw. 1. + 2. + 3.

Auch hier soll die Behandlung auf Fe^{+++}-Komplexe beschränkt bleiben. Unter 1. wurde bereits angeführt, daß sich aus dem bisherigen experimentellen Material die Einteilung der Fe^{+++}-Komplexe in die beiden Gruppen "starkes" ($\Delta \approx 50\,000$ cm^{-1}) und "schwaches" ($\Delta \approx 10\,000$ cm^{-1}) Ligandenfeld ergibt. Die für die Spinverteilung wichtige Differenz zwischen der (kristallfeldtheoretisch so bezeichneten) Kristallfeldstabilisierungsenergie Δ = 10 Dq und der Energie der Spinaustauschwechselwirkung ist in beiden Fällen wesentlich größer als kT bei Zimmertemperatur (ca. 200 cm^{-1}).

Die Besonderheit des Porphyrin/Protein-Ligandenfeldes liegt nun neben der Symmetrieerniedrigung gegenüber der kubischen MX_6-Symmetrie und der damit verknüpften zusätzlichen Aufspaltung und Anisotropie der magnetischen Eigenschaften vor allem darin, daß hier die oben erwähnte Differenz in den Bereich von ca. 200 cm^{-1} fällt (wie vom Verfasser bereits 1957 vermutet wurde [11]) und damit folgende experimentell überprüfbare Eigenschaften auftreten:

1. Geringe Variationen des Ligandenfeldes bedingen verschiedene Grundzustände verschiedener Entartung und mit verschiedenem Spin ($S \longrightarrow S'$).

2. Temperaturabhängigkeit der Multiplizität des Grundzustandes und damit des Absorptionsspektrums sowie des magnetischen Momentes bei bestimmten Liganden, bei denen $\Delta_{SS'} \approx kT$ ist.

3. Geringe Variationen des Ligandenfeldes bedingen durch verschiedene Spin-Bahn-Kopplungskonstanten und verschiedene Δ-Werte verschiedene g-Faktoren der ESR.

1. konnte vom Verfasser [12] sowie von SCHELER, SCHOFFA und JUNG [13] an Hand der Methämoglobin-Komplexe mit N_3', NO_2', SCN', CNO' usw. nachgewiesen werden.

2. wurde vom Verfasser 1957 postuliert und konnte später im Falle der Absorptionsspektren von GEORGE, BETTLESTONE und GRIFFITH [14] sowie von SCHELER, BLANCK und GRAF [15] nachgewiesen werden.

3. wurde von BENNETT, INGRAM, GEORGE und GRIFFITH [16] bzw. GIBSON, INGRAM und SCHONLAND [17] 1955 bzw. 1958 gefunden.

Abb. 2 zeigt ein qualitatives MO-Niveauschema für Hämin und Methämoglobin-Komplexe, das die oben entwickelten Vorstellungen verdeutlichen soll. Die absolute Lage der Niveaus ist willkürlich, die Ligandenorbitals und -MO's sind weggelassen. Die relative Zuordnung der MO's beruht im wesentlichen auf den Arbeiten von J. S. GRIFFITH [18] und GIBSON, INGRAM und SCHONLAND [17]. Neuerdings ist auch von

SPANJAARD und BERTHIER [19] mit Hilfe der halbempirischen LCAO-MO-Methode für Ferro- und Ferriporphyrin ein quantitatives Niveauschema angegeben worden. Die relative Zuordnung der magnetisch wirksamen MO's entspricht bei allen Fe-Porphyrinkomplexen dem Schema der Abb. 2. Was sich jedoch durch kleine Variationen von Stärke und Symmetrie des Ligandenfeldes ändert, ist vor allem der Wert von Δ, der Energieabstand zwischen dem untersten und obersten besetzten magnetisch wirksamen Niveau. Der Übergang vom Hämin zum Methämoglobin (s. Abb. 2) hat offenbar zwei Auswirkungen:

a) Durch die neue N-Bindung (Imidazol-N) wird Δ vergrößert und kommt in den Bereich, wo $\Delta - \zeta \approx kT$ und die Voraussetzung der oben genannten Eigenschaften 1.-3. gegeben ist.

b) Das Protein-Ligandenfeld ist sterisch so variabel, daß zusätzlich die oben unter 1.-3. genannten Variationsmöglichkeiten begünstigt werden.

Offenbar sind die Voraussetzungen für 1.-3. bei den eiweißfreien Fe-Porphyrinen noch nicht voll erfüllt. So zeigen z. B. die Protohäminkomplexe mit Cl', Br', J', SCN', COO' alle den theoretischen Spinmagnetismus der $d_\varepsilon^3\ d_\gamma^2$-Besetzung ohne Bahnanteil [20]. Bei bestimmten Methämoglobinverbindungen (z. B. mit NO_2' oder OH') ist jedoch vor allem die Voraussetzung a) erfüllt. Damit läßt sich für diese Δ abschätzen. Da sich nach den Untersuchungen von GRIFFITH die Austauschenergie der "Spinpaarung" für den S = 1/2-Grundzustand zu ζ = 5 B + 8 C (B, C = Racahparameter, $B \approx C/4 \approx 1\,000$ cm^{-1} für Fe^{+++}) berechnet, beträgt also $\zeta \approx 37\,000\ cm^{-1}$ und damit $\Delta \approx \zeta \pm kT_{Zimmer} \approx 37\,000 \pm 200\ cm^{-1}$.

Dies ist ein Beispiel für Hinweise auf die Energiewerte molekularer Orbitals auf Grund von Messungen der statischen Suszeptibilität.

Noch detailliertere Hinweise gibt die Analyse der ESR-Spektren. So konnte GRIFFITH mit Hilfe der g-Faktoren von Methämoglobinazid den Energieabstand zwischen den t_{2g}-Kramers-Dubletts d_{xy}, d_{xz}, d_{yz} in dieser Verbindung berechnen.

Zum Abschluß sei noch auf die Porphyrinkomplexe mit Fe^{++} kurz hingewiesen: Am evidentesten tritt vielleicht die Bedeutung magnetischer Messungen bei der Erklärung der einzigartigen Sonderstellung der O_2-Fe-Bindung im Oxyhämoglobin zutage. Der ganz und gar unerwartete Diamagnetismus dieser Verbindung war der Ausgangspunkt für eine eingehende Analyse [18] der besonderen Valenzzustände des hämoglobingebundenen, ursprünglich paramagnetischen O_2-Moleküls, die zu dem Ergebnis führte, daß die Molekülachse des O_2 parallel und nicht senkrecht zur Porphyrinebene ausgerichtet ist.

L I T E R A T U R

1) VAN VLECK und PENNEY, Phil. Mag. 17, 961 (1934)
2) ABRAGAM und PRYCE, Proc. Royal Soc. A 205, 135 (1951)
3) HOWARD, J. chem. Phys. 3, 207 (1935)
4) STEVENS, Proc. Royal Soc. A 219, 542 (1953)
5) OWEN, Proc. Royal Soc. A 227, 183 (1955)
6) KAMIMURA, J. Phys. Soc. Jap. 11, 1171 (1956)
7) HAVEMANN, HABERDITZL und GRZEGORZEWSKI, Z. phys. Chem. 217, 217 (1961)
8) HAVEMANN, HABERDITZL und RABE, Z. phys. Chem. 218, 417 (1961)
9) HAVEMANN, HABERDITZL, Z. phys. Chem. 210, 267 (1959)
10) INGRAM, KENDREW, INGRAM, GIBSON, PERUTZ, Nat. 178, 905 (1956)
11) HAVEMANN und HABERDITZL, Z. phys. Chem. 209, 135 (1958)
12) HAVEMANN und HABERDITZL, Chem. Technik 8, 418 (1956)
13) SCHELER, SCHOFFA und JUNG, Naturwiss. 43, 159 (1956)

14) GEORGE, BETTLESTONE und GRIFFITH, Haematin Enzymes, S. 105, Oxford 1961
15) SCHELER, BLANCK und GRAF, Naturwiss. 50, 500 (1963)
16) BENNETT, INGRAM, GEORGE und GRIFFITH, Nat. 176, 394 (1955)
17) GIBSON, INGRAM und SCHONLAND, Discuss. Faraday Soc. 26, 81 (1958)
18) GRIFFITH, Proc. Roy. Soc. A 235, 23 (1956)
19) SPANJAARD und BERTHIER, J. Chim. Phys. 1961, 169
20) HAVEMANN, HABERDITZL und MADER, Z. phys. Chem. 218 71 (1961)

D I S K U S S I O N S B E M E R K U N G E N

BLUMENFELD.- Have you made any experiments on the ESR of free porphiryns?

HABERDITZL.- Wir haben solche Messungen noch nicht durchgeführt. Meines Wissens sind wohl von INGRAM ESR-Signale von metallfreien Porphyrinen unter verschiedenen Bedingungen analysiert worden.

BLUMENFELD.- What are the exact differences between various proteins that lead to their different properties in influencing the electronic state of iron in heme?

HABERDITZL.- In erster Linie ist es der verschiedene Abstand zum Imidazol-N bei verschiedener Proteinkonfiguration, der verschiedene Überlappungszustände bei Linearkombinationen von Fe-N-Orbitals und damit verschiedene Δ -Werte bedingt.

BEHLKE.- Die Ergebnisse, die Sie auf magnetochemischem Wege am Hämoglobin gefunden haben, können wir auf Grund von Temperaturdifferenzspektren ebenfalls ermitteln. Wir haben darüber hinaus noch die Myoglobinkomplexe untersucht. Die Ergebnisse sind etwas unterschiedlich, wie durch die veränderte Bindung des Proteins am Hämin zu erwarten war. (s. Vortrag: J. BLANCK, Symposium für Molekulare Zellphysiologie, Berlin, 13. - 17. August 1963).

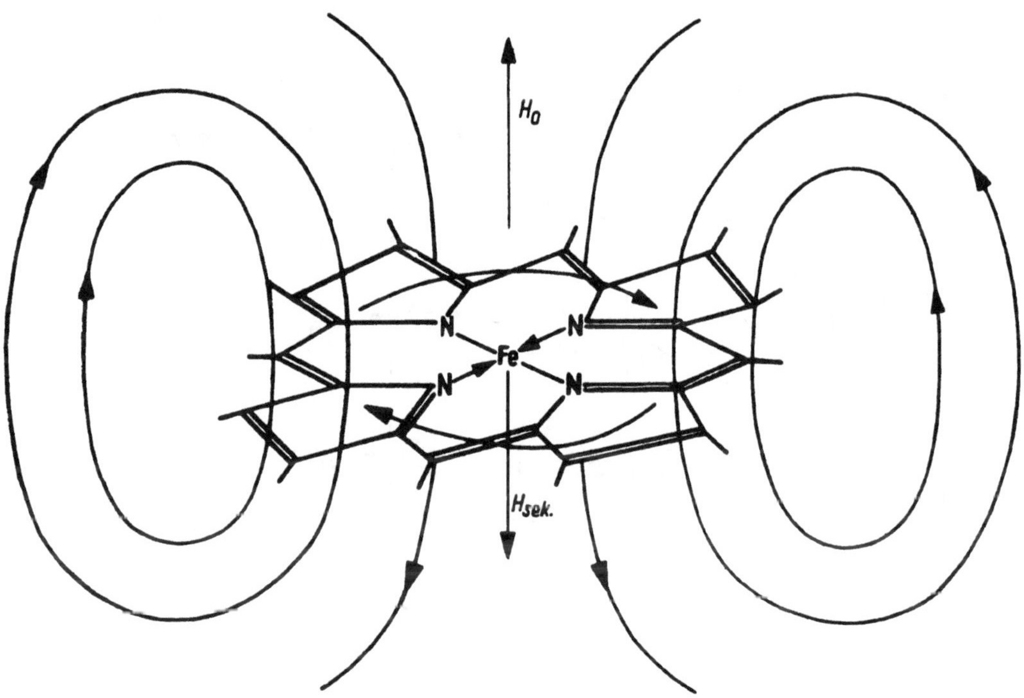

Abb.1 Induziertes Magnetfeld im Porphyrinligandensystem

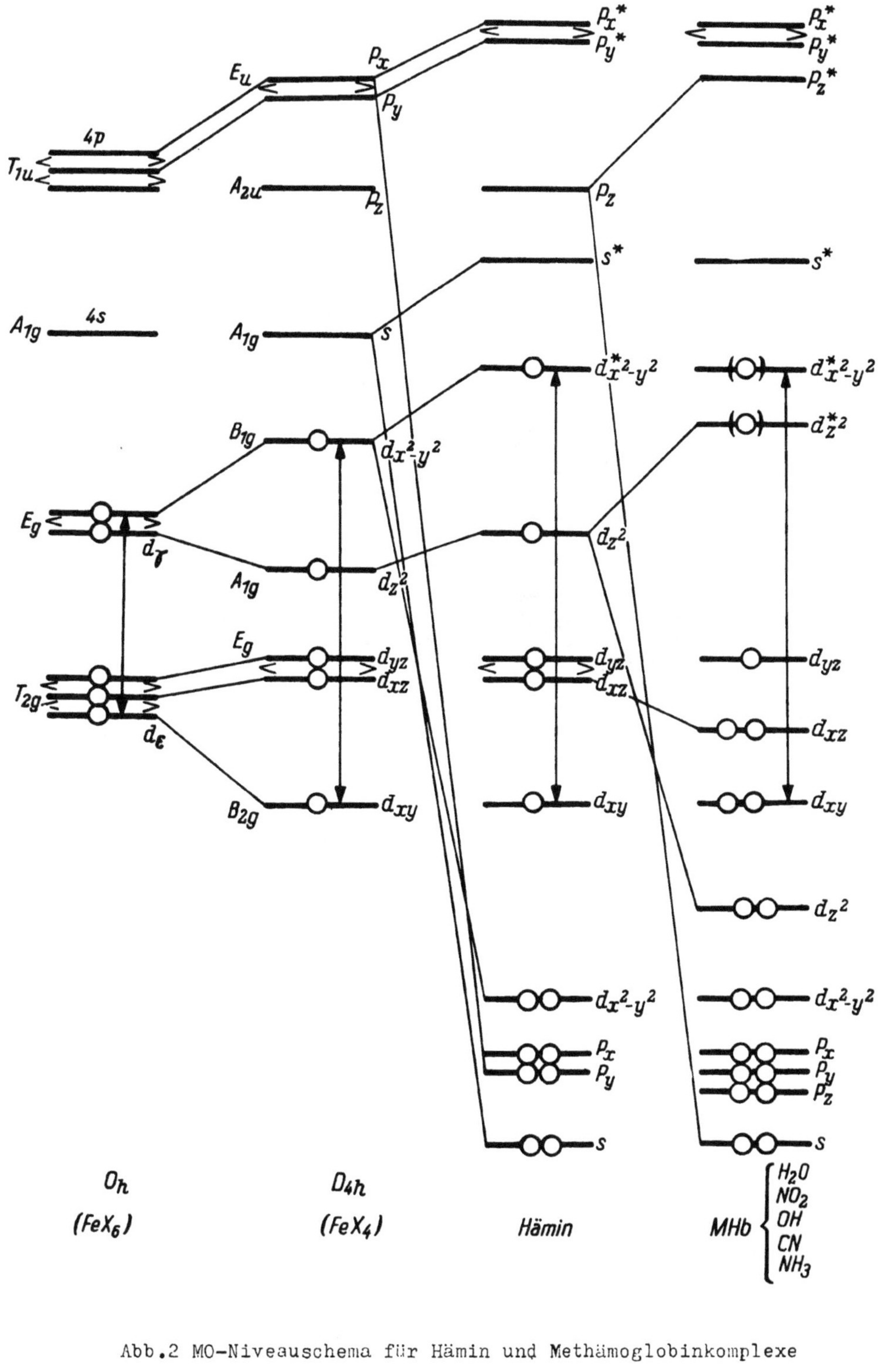

Abb.2 MO-Niveauschema für Hämin und Methämoglobinkomplexe

C Proteine

Optische Drehung und Heterogenität des menschlichen Serumalbumins

I. FRIČ, J. ŠPONAR, S. STOKROVÁ und J. JANATOVÁ

Institut für organische Chemie und Biochemie der Akademie, Prag

S U M M A R Y

The heat denaturation of human serum albumin was studied by means of measurements of optical rotation. The heterogeneity in heat stability of albumin components found previously was confirmed. The results of further experiments on albumin components revealed differencies in optical rotation and in changes of optical rotation on denaturation as well. The same value of the rotatory dispersion constant λ_c for all the albumin components was obtained. Because of the possibility that albumin splits into components on heating, measurements on albumin fractions isolated by chromatography on DEAE-Cellulose were carried out. Conclusions concerning the structural heterogeneity of human serum albumin can be drawn on the basis of these results.

Im Zusammenhang mit den Untersuchungen der Wärmedenaturierung des menschlichen Serumalbumins, die von STOKROVÁ und SPONAR [1) 2)] mit Hilfe der Elektrophorese und Sedimentationsanalyse vorgenommen wurden, beschäftigten wir uns mit diesem Problem auch unter dem Gesichtspunkt der Veränderungen der optischen Drehung. Auf Grund der Analyse der kinetischen Messungen der Drehungsveränderungen einerseits und des durch Fällung mit Ammoniumsulfat bestimmten Anteils des denaturierten Albumins andererseits [3)], konnten die Schlußfolgerungen der erwähnten Verfasser über die Heterogenität des Albumins bestätigt werden : das Serumalbumin verhält sich als eine Mischung von Komponenten mit unterschiedlicher Beständigkeit gegen Wärme.

Aus der Abhängigkeit der optischen Drehung vom Denaturierungsgrad f_d der Albuminlösung ließ sich überdies eine weitere Unterschiedlichkeit der vorausgesetzten Albumin-Molekülarten voraussagen. Der Denaturierungsgrad f_d wird als Verhältnis der Konzentration des denaturierten Albumins zur Gesamtkonzentration des Albumins in der Lösung definiert. Diese Abhängigkeiten sollten unter Annahme eines "all ore none" Vorganges und gleicher Drehungseigenschaften aller Albumin-Molekülarten einen linearen Verlauf haben. Die experimentellen Kurven weisen aber eine Krümmung auf, ohne Rücksicht darauf, ob es sich um kinetische oder scheinbare Gleichgewichtsdaten handelte. Als Beispiel solcher Abhängigkeit kann die obere Kurve (schwarze Punkte) auf Abb.2 dienen. Diese Krümmung kann nur dann gedeutet werden, wenn den beständigeren Molekülarten eine niedrigere spezifische Drehung und / oder eine größere Drehungsveränderung beim Übergang vom nativen in den denaturierten Zustand zugehört [3)]. Ein direkter Beweis dafür ergab sich aus der

Messung der spezifischen Drehung des bei gegebenen Bedingungen beständigen Albuminanteils, der nach Entfernung (durch Fällung) des denaturierten Anteils in der Lösung verblieb. Wie aus der Abb.1 ersichtlich, fällt die Drehung dieses resistenten Rückstandes mit ansteigendem Denaturierungsgrad ab.

Um die Eigenschaften des resistenten Albumins eingehender untersuchen zu können, bereiteten wir es in einer größeren Menge vor. Eine fünfprozentige Albuminlösung (0,2 m Phosphatpuffer, pH 6,8) wurde 2 Stunden auf 65°C erwärmt. Unter diesen Bedingungen denaturiert ungefähr 75 % des in Lösung enthaltenen Albumins. Das denaturierte Protein wurde dann mit 2 m Ammoniumsulfat ausgefällt und der resistente Anteil aus der Lösung zurückgewonnen und auf einer Sephadexkolonne entsalzt. Das resultierende Präparat besaß einige gleiche Eigenschaften wie das ursprüngliche gesamte Albumin : gleiche Farbe und Löslichkeit, gleiche Sedimentationskonstante ($s_{20,w}$ = 4,0 S), elektrophoretische Beweglichkeit ($U = -5{,}10 . 10^5$ $cm^2 sec^{-1} V^{-1}$) und auch Rotationsdispersionskonstanten λ_c = 269 mμ und $b_0 = -230°$. Die spezifische Drehung des resistenten Albumins wurde aber um 5° niedriger gefunden. Das ist auch aus Abb.2 ersichtlich, wo das Verhalten des Gesamtalbumins beim Erwärmen mit dem seines resistenten Anteils verglichen wird. Die mit gleichen Zahlen bezeichneten Punkte entsprechen gleichen Denaturierungsbedingungen. Nach zweistündigem Erwärmen auf 65° (unter den Bedingungen des vorher beschriebenen präparativen Versuchs) ist die optische Drehung des gesamten Albumins um 7° und sein Denaturierungsgrad auf 75 % gestiegen, während vom resistenten Albumin unter gleichen Bedingungen erwartungsgemäß nur 4 % gefällt werden konnten und seine Drehung (und auch Rotationsdispersionskonstanten) praktisch unverändert geblieben sind. Die weiteren Punkte entsprechen dann dem zweistündigen Erwärmen bei 75° (2; 2') und 80° (3; 3'). Die optische Drehung beider verglichenen Präparate besitzt beim Denaturierungsgrad $f_d = 1$ wahrscheinlich einen gleichen Drehungswert. Das unterstützt die Annahme, die jedoch durch direkte Messung nicht bewiesen werden konnte, daß alle Albuminkomponenten im denaturierten Zustand die gleiche Drehung besitzen.

Das unbehandelte Albumin und sein untersuchter resistenter Anteil unterscheiden sich offensichtlich nicht im Gehalt der α-helikalen Konformation, wie es aus der Übereinstimmung ihrer Rotationsdispersionskonstanten hervorgeht. Das folgt auch aus dem übereinstimmenden Verlauf der Veränderungen der λ_c-Werte mit f_d (Abb.3).

Kurz zusammengefaßt wird die Heterogenität des Serumalbumins dadurch gekennzeichnet, daß mit steigender Denaturierungsbeständigkeit die spezifische Drehung der Komponenten abfällt und ihre Drehungsveränderung beim Übergang vom nativen in den denaturierten Zustand ansteigt. Mit Rücksicht darauf, daß die stufenweise Denaturierung einer Albuminlösung eine Art Fraktionierungsverfahren vorstellt, kann auf Grund ähnlicher Angaben, wie sie auf Abb.1 gezeigt wurden, die Form der Verteilungskurve der Albuminkomponenten hinsichtlich ihres Drehungsvermögens abgeschätzt werden. Aus solchen Berechnungen folgt eine stark asymmetrische differentielle Verteilungskurve mit einem Maximum im Bereiche der größten Drehungswerte, d.h. im Bereiche der am leichtesten denaturierbaren Komponenten. Diese Ergebnisse stehen in guter Übereinstimmung mit der auf andere Weise abgeleiteten Stabilitätsverteilung der Albuminkomponenten [2)].

Die Denaturierungsversuche allein ermöglichen keine Entscheidung darüber, ob das Albumin bereits im nativen Zustande heterogen ist, oder ob seine resistenten Komponenten erst während des Denaturierungsvorganges aus ursprünglich homogenem Stoffe entstehen. Prinzipiell könnte die Denaturierungsreaktion im ersten Falle durch eine Folge der Reaktionen dargestellt werden, die gleichzeitig mit unterschiedlicher Reaktionsgeschwindigkeit verlaufen

$$N_1 \xrightarrow{k_1} D_1 \; ; \; N_2 \xrightarrow{k_2} D_2 \; ; \; N_3 \xrightarrow{k_3} D_3 \; ; \; \ldots\ldots\ldots N_n \xrightarrow{k_n} D_n \; .$$

Im zweiten Falle handelte es sich um einen Komplex von Folge- und Zweigreaktionen gemäß dem Schema

$$\begin{array}{ccccccccc}
N & \xrightarrow{k'_1} & R_1 & \xrightarrow{k'_2} & R_2 & \xrightarrow{k'_3} & R_3 \cdots\cdots\cdots & \xrightarrow{k'_{n-1}} & R_{n-1} \\
\downarrow k_1 & & \downarrow k_2 & & \downarrow k_3 & & \downarrow k_4 & & \downarrow k_n \\
D_1 & & D_2 & & D_3 & & D_4 & & D_n
\end{array}$$

Obwohl wir die letzte Möglichkeit als unwahrscheinlich betrachten, versuchten wir die Entscheidung in dieser Frage durch Untersuchung von chromatographisch auf DEAE-Zellulose hergestellten Fraktionen des Albumins herbeizuführen. Die Ergebnisse eines bereits veröffentlichten Versuchs werden in der Tabelle 1 wiedergegeben [4].

Tabelle1 Phosphatpuffer 0,02 m, pH 6,8, mit steigendem Gehalt an NaCl (0,05 bis 0,3 m)

Fraktion	$[\alpha]_D^{25^0}$	λ_c, mμ	c,g/100 ml
30	-70,3	275	0,405
35	-71,3	272	0,331
38	-69,2	270	0,294
40	-	274	0,273
45	-68,0	270	0,203
50	-65,2	271	0,148
55	-63,0	271	0,111
60	-59,8	276	0,072
64	-56,6	275	0,054

Die Albuminlösungen wurden in dem Zustande, wie sie die chromatographische Kolonne verlassen haben, gemessen. Man sieht, daß die spezifische Drehung mit der Reihenzahl der Fraktionen abfällt. Die ersten Fraktionen sind im Albumin am häufigsten vertreten, wie es aus der Form des entsprechenden Elutionsdiagramms folgt. Die Werte der Rotationsdispersionskonstanten λ_c sind innerhalb der Fehlergrenzen bei allen Fraktionen gleich, wenn auch etwas größer,als es üblich ist.

Die Daten, die in der Tabelle 2 angeführt sind, beziehen sich auf Albuminfraktionen, die aus dem Eluat nach Entsalzen durch Vakuumsublimation im festen Zustande hergestellt wurden. Im Versuche 10 wurde das Albumin in zwei Fraktionen, im Versuche 14 in drei und im Versuche 16 in 5 Fraktionen aufgeteilt. In allen drei Versuchen sieht man wieder den Abfall der spezifischen Drehung mit der Reihenzahl der Fraktionen. Die λ_c-Konstanten sind in allen Fällen gleich, vielleicht mit Ausnahme der Fraktion 16-5. Von Fraktionen 10-1 und 10-2 wurden die Anfangsgeschwindigkeitskonstanten der Denaturierungsreaktion bei 65^0 ermittelt. Tatsächlich ist die Fraktion, die den niedrigeren Drehungswert besitzt, gegen die Wärmeeinwirkung mehr stabil, wie vorausgesetzt. Die immunoelektropho-

Tabelle 2

Fraktion	$[\alpha]_D^{25^\circ}$		λ_c, mμ	Immunoelektrophorese	Lösungsmittel	Bemerkung
10-1	-64,1	-	270	rein	Phosphat 0,2 m	$k^*=5,3.10^{-4}sec^{-1}$
10-2	-60,2	-58,2	267	α_1- Globulin	pH 6,8	$k^*=2,9.10^{-4}sec^{-1}$
14-1	-64,8	-62,6	267	rein	Phosphat 0,2 m	
14-2	-60,8	-60,8	269	rein	pH 6,8	
16-1	-61,8	-64,2	266	rein	NaCl 0,1 m	
16-2	-65,9	-64,4	269	rein	pH 5,6	
16-3	-64,0	-61,6	269	rein		
16-4	-59,3	-61,1	266	rein		
16-5	-56,0	-58,8	264	α_1- Globulin		

*Geschwindigkeitskonstante der Denaturierungsreaktion bei 65°

retische Prüfung der Fraktionen zeigt, daß sie mit einer Ausnahme immunologisch rein sind und daß die Anwesenheit des einen niedrigeren Drehungswert besitzenden α_1-Globulins für den Abfall der Drehungswerte der Fraktionen nicht verantwortlich ist. In der zweiten und dritten Spalte der Tab.2 werden Drehungswerte verglichen, die auf spektrophotometrischer (zweite Spalte) bzw. anderer, z.B. refraktometrischer (dritte Spalte) Konzentrationsbestimmung beruhen. Aus dem Vergleich ergeben sich Unterschiede zwischen beiden Wertereihen, besonders auffallend ist ein steilerer Abfall der auf spektroskopischer Konzentrationsbestimmung beruhenden Werte mit der Fraktionsreihenfolge. Es besteht also eine Möglichkeit, daß die Albuminkomponenten sich hinsichtlich des Extinktionskoeffizienten unterscheiden. Wir versuchten auch dieses Problem aufzuklären, aber da uns wenig Material zur Verfügung stand, konnten wir nicht zu einer eindeutigen Antwort gelangen. Die Lage des Absorptionsmaximums der Fraktionen im Vergleich mit dem unbehandelten Albumin bleibt unverändert. Bei der differentiellen Spektrophotometrie der Fraktionen gegen Albumin (ΔD_{280}= 0) wurde aber ein Peak bei 296 mμ beobachtet (Abb.4). Gleiches Verhalten haben wir bei allen untersuchten Fraktionen beobachtet. Die relative Intensität dieses Peaks scheint mit der Reihenzahl der Fraktion anzusteigen. Es ist wahrscheinlich, daß diese Erhöhung der Absorption bei 296 mμ den chemischen oder Ladungsveränderungen in naher Umgebung eines Anteils der Tyrosinreste, die aber auch erst während des Adsorptions- und Desorptionsprozesses an der DEAE-Zellulose stattfinden können, zuzuschreiben ist.

Die Untersuchung der chromatographischen Fraktionen des Albumins scheint also im wesentlichen die Heterogenität des Albumins bereits im nativen Zustande zu beweisen. Das Verhalten des Albumins bei der Wärmedenaturierung wird dann durch die Eigenschaften und relative Vertretung seiner im nativen Zustande anwesenden Komponenten bestimmt.

Mit Rücksicht darauf, daß bei bereits untersuchten Proteinen keine Unterschiede in der primären Struktur gefunden wurden, sind die Ursachen der Unterschiedlichkeit der Albumin-Molekülarten wahrscheinlich in ihrer sekundären oder tertiären Struktur zu suchen. Die experimentellen Daten schließen Unterschiede im α-Helixgehalt aus. Wir konnten schon früher auf Grund von Denaturierungsversuchen zeigen [3], daß, obgleich die Eigenschaften der Albuminkomponenten, z.B. die optische Drehung oder die Denaturierungsgeschwindigkeit, vom pH und von der Ionenstärke abhängig sind, ihre Verteilung dagegen durch diese Parameter nicht beeinflußt wird. Es ist deshalb wahrscheinlich, daß für die Heterogenität des Albumins eher feste kovalente Bindungen als die elektrostatischen Kräfte, Wasserstoffbrücken usw. verantwortlich sind. Hier können die Disulfidbrücken, die die einzelnen Segmente der Peptidkette des Albumins zusammenbinden, in Erwägung gezogen werden. Eine mögliche Erklärung der Rolle dieser Bindungen liegt darin, daß die Molekülarten des Albumins sich bei gleicher Primärstruktur durch die Lage der Disulfidbrücken unterscheiden. Die relativ wahrscheinlichste und deshalb häufigste Anordnung der Disulfidbrücken sollte nach KAUZMANN [5] die Verbindung der benachbarten Halbcystinreste sein. Nehmen wir an, daß eine solche Anordnung den Segmenten der Peptidkette die größte Bewegungsfreiheit ermöglicht und sogleich die Einwirkung anderer die tertiäre und sekundäre Struktur stabilisierender Kräfte erschwert. Je mehr das Molekül solche durch Verbindung entfernterer Halbcystinreste entstandene Disulfidbrücken enthält, desto kompakter ist das Molekül und desto größer seine Stabilität. Die niedrigere spezifische Drehung der beständigeren Molekülarten kann dadurch erklärt werden, daß in kompakteren Molekülen ein größerer Anteil der Peptidbindungen sich im hydrofoben Milieu innerhalb des Moleküls befindet. Die ungeheure Anzahl der möglichen Verbindungen der Halbcystinreste [5] wird sicher durch sterische Hinderungen und andere Einflüsse stark herabgesetzt. Trotzdem können irgendwann in der Historie des Albumins günstige Bedingungen zur Bildung einer genügend großen Anzahl Albuminisomere entstehen, um dadurch das Verhalten des Albumins bei den Denaturierungs- und chromatographischen Versuchen erklären zu können. Man kann voraussetzen, daß solche Bedingungen bereits bei der Biosynthese des Proteins an der Matrix der Nucleinsäure existieren. Es ist bekannt, daß bei der Biosynthese nur Cystein aktiviert und eingebaut wird. Dagegen wurde bisher in dem linearen Molekül der Ribonucleinsäure kein Kode entdeckt, der die dreidimensionale Struktur des synthetisierten Proteins bestimmen könnte. Es besteht also bei Proteinen vom größeren Molekulargewicht und Cystingehalt gewisse Freiheit im Kombinieren der Cysteinreste zur Disulfidbindung.

Das vorgeschlagene Modell der Albuminheterogenität scheint auch durch die Befunde von HARTLEY, PETERSON und SOBER [6] unterstützt zu werden. Diese Autoren befaßten sich eingehend mit der Chromatographie des Albumins und sind zu der Ansicht gekommen, daß die Heterogenität des Albumins hinsichtlich der Adsorption an DEAE-Zellulose größtenteils mit dem Molekulargewicht und mit der Reaktivität der SH-Gruppe verknüpft ist. Die unterschiedliche Reaktivität der SH-Gruppe kann dadurch bedingt sein, daß diese Gruppe sich in einzelnen Molekülarten des Albumins an unterschiedlichen Stellen der Peptidkette und in unterschiedlicher Umgebung befindet.

L I T E R A T U R

1) STOKROVÁ,S., J.ŠPONAR, Collect. czechoslov.chem.Commun. 27, 2516 (1962)
2) STOKROVÁ,S., J.ŠPONAR, Collect. czechoslov.chem.Commun. 28, 659 (1963)
3) FRIČ,I., Dissertation, Tschechoslow.Akad.der Wissenschaften, Prag 1962
4) ŠPONAR,J., I.FRIČ, S.STOKROVÁ, J.KOVÁŘIKOVÁ, Collect. czechoslov.chem.Commun. 28, 1831 (1963)
5) KAUZMANN,W., in "Sulphur in Proteins", Kap.II/2, S.93, Academic Press,New York 1959
6) HARTLEY,R.W., E.A.PETERSON, H.A.SOBER, Biochemistry 1, 60 (1962)

D I S K U S S I O N S B E M E R K U N G E N

LUMRY : Recent experiments of FOSTER and his coworkers on acid BSA support the Prague group's results in finding heterogeneity in the protein. This heterogeneity is unlikely to be due to changes in amino-acid composition. Recently BENSON, HOLLOWAY and I have detected very substantial evidence for heterogeneity in BSA at higher pH values. Studies of proton exchange from pH 5,5 to 8,5 can only be explained on the basis of such heterogeneity which appears as differences in rate of proton exchange for the different forms. This heterogeneity might be a kinetic phenomenon- the so-called kinetic stability of LUMRY and Eyring-depending on a slow interconversion of conformations due to large activation free energies for such processes. On the other hand it may be a consequence of a different distribution of disulfide bond compositions. In the latter case the heterogeneity need not be attributed to kinetic effects so long as disulfide bond rearrangement-HUGGINS crossing- does not take place. At present our results indicate reasonably well that the distribution of disulfide bonds is different or at least that is the better of the two explanations. However experiments with catalysts for the disulfide crossing reaction did not fully substantiate this possibility and we are still highly confused about the true state of affairs.

FRIČ : Unser Modell der Heterogenität des menschlichen Serumalbumins, das auf unterschiedlicher Verteilung der disulfidischen Brücken beruht, hat selbstverständlich einen rein spekulativen Charakter. Obwohl es die Eigenschaften der Albuminkomponenten besonders hinsichtlich ihrer Verteilung, Wärmebeständigkeit und optischen Drehung gut erklärt, haben wir für seine Gültigkeit keine direkten Beweise. Es ist deswegen für uns sehr anregend, daß auch Dr. LUMRY die experimentellen Befunde zu ähnlichen Schlüssen geführt hat. Die Möglichkeit, daß die Albuminkomponenten erst z.B. bei Erwärmung durch Umgruppierung der Disulfidbrücken entstehen, betrachten wir als weniger wahrscheinlich auch aus dem Grunde, daß in diesem Falle diese Reaktionen bereits in nativen Molekülen stattfinden müßten. Es ist außerdem nicht wahrscheinlich, daß die relative Vertretung der auf diese Weise entstandenen Molekülarten in relativ breitem, von uns untersuchten pH-Bereichen (6,0 bis 8,5) gleich wäre.

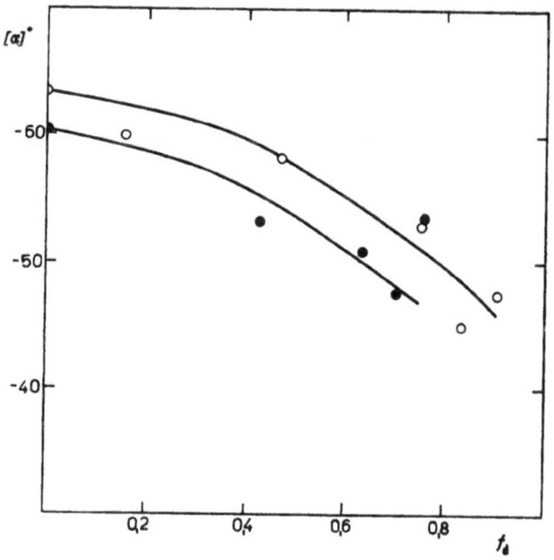

Abb.1 Abhängigkeit der optischen Drehung des beständigen Albuminrückstands vom Denaturierungsgrad f_d der Ausgangslösung. Denaturierungsbedingungen: O c=2,29 g/100 ml, Phosphatpuffer μ=0,1, pH=7, $t_{den.}$=76°C; ● c=0,860 g/100ml, Boratpuffer μ=0,1, pH=8,8, $t_{den.}$ = 67°

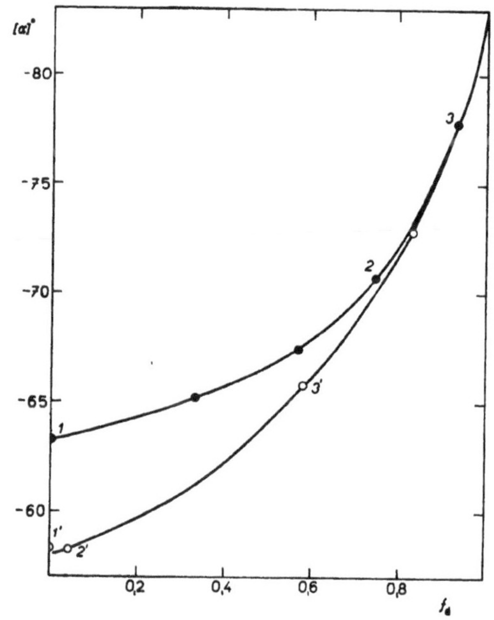

Abb.2 Abhängigkeit der optischen Drehung vom Denaturierungsgrad f_d für die Lösungen des unbehandelten Albumins ● und des isolierten beständigen Albuminanteils O. c~1,0 g/100ml, Phosphatpuffer μ=0,2, pH=6,76. Punkte 1,1' nichterwärmte Lsgn., 2,2' nach Erwärmen bei 65°C; 3,3' bei 75°C.

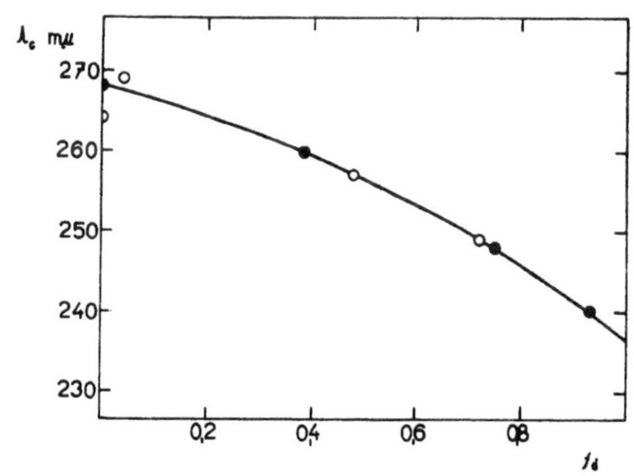

Abb.3

Abhängigkeit der λ_c-Werte vom Denaturierungsgrad f_d für die Lösungen des unbehandelten Albumins ● und des isolierten beständigen Albuminanteils O .Versuchsbedingungen s.Abb.2.

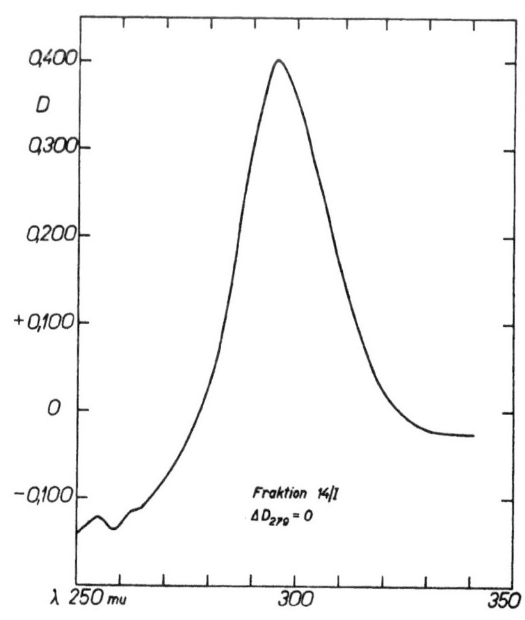

Abb.4 Differenzielles Absorptionsspektrum der chromatographisch hergestellten Fraktion 14-1 gegen das unbehandelte Albumin.

C Proteine

The Study of the Structure and Transformations of Proteins by the Methods of Sorption and Catalysis on Ion Exchange Resins

G. V. SAMSONOV

Institute of Macromolecular Compounds of the Academy, Leningrad

S U M M A R Y

A method for the analysis of local dipolarity of ions in solution has been developed on the basis of sorption of aminoacids and polypeptides on ion exchange resins. This method has been used for the analysis of the protein electrolytic structure, in particular for determining the distance between the ionogen groups carrying opposite charges. A method has been developed for the analysis of the variability of the protein macromolecule size by studying their sorption on ion exchange resins. A degree of the changes in size of the macromolecules of ribonuclease, trypsin, lysozyme and serum albumin in disulfide bond breaking has been investigated.

As a result of the investigation of catalytic transformations of protein hydrolysates on ion exchange carboxyl resins both the specific hydrolysis and the synthesis of peptide bonds have been realized.

The interaction of dipole ions (that is, ions carrying both a positive and a negative charge) with ion exchange resins should differ from the sorption of ions carrying one or several charges of the same sign. The dipole ion sorption should be characterized by both the mutual attraction and the mutual repulsion of the sorbate and the sorbent. The particular nature of the phenomenon of dipole ion sorption can be easily revealed if we compare the sorption capacity of the simplest dipole ions (amino acids) on the salt and on the hydrogen forms of cation exchange resins (table 1).

As the table shows, amino acids have a great sorption capacity on the cation exchange resins in the hydrogen form. The sorption of monoaminomonocarboxylic amino acids does not take place on the salt forms of the resins, since sodium ions cannot neutralize the carboxyl group of the zwitterion. Hence, for the salt forms of the resin the electrostatic repulsion of the sorbate and the sorbent should be maintained and low sorption capacity for the zwitterions should be observed. In contrast to this, the amino acid zwitterions on resins in the hydrogen form have a greater sorbtion capacity, since under these conditions the transfer of hydrogen ions to the carboxyl group of the amino acid takes place. As a result the electrostatic repulsion disappears.

Several methods can be used for reducing the repulsion of the zwitterion and the salt form of the resin. It can be reduced by the increase of the ionic strength of the solution, by using the aqueous-acetonic medium instead of the aqueous one and also by increasing the distance between the charges of the zwitterion.

table 1 Capacity of sorption of aminoacids on sulforesins

Resin	Aminoacids	Capacity H-form	$\frac{mg\text{-}eq}{ml}$ Na-form
	glycine	2,2	0,02
SDW-3	alanine	1,75	0,011
	leucine	1,95	0,08
SBS-2	glycine	1,2	0,044

table 2 Capacity of sorption of aminoacids by sulforesin SNF (Na-form) and carboxylic resin KFU (Na-form) from water and water-acetone mixtures

Aminoacids	Resin	Capacity water	$\frac{mg\text{-}eq}{gr}$ water-aceton
glycine	KFU	0,14	0,82
alanine	-	0,086	0,67
glycine	SNF	0,098	0,62
alanine	-	0,054	0,286

Figure 1 shows the sorption capacity for alanine on sulforesin SDW-3 in the sodium form as a function of the ionic strength of the solution. The abscissa gives the concentration of sodium chloride in solution. We can see that at first, when the ionic strength of the solution increases, the sorption capacity for alanine grows. This is due to the screening of the second zwitterion charge in accordance with the DEBYE-HÜCKEL theory of strong electrolytes. A further drop in the sorption capacity is caused by the competing action of sodium ions. The second method of increasing the sorption capacity of cation exchange resins in the salt form for amino acids is based on the use of a solvent giving a lower degree of the ionization of the zwitterion carboxyl group. Acetone is such a solvent.

As table 2 shows, in changing from aqueous to aqueousacetonic solution a greater amount of amino acids can be sorbed on the salt forms of cation exchange resins. Finally, the experimental data show that the sorption capacity of the above mentioned sorbents increases in the series amino acids-dipeptides-tripeptides (table 3).

table 3 Bipolar qualities of aminoacids, polypeptides and proteins (sorption on sulforesins)

Substance	M	Resin	m_H/m_{Na}
glycine	75	SDW-3	106
glycyl-glycine	132	-	19,4
leucyl-leucyl-glycine	304	-	6,3
serum albumin	69000	SBS	
serum globulin	176000	-	

table 4 Capacity of sorption of proteins on SBS and Dowex-50 resins

Protein	Capacity mg/gr H-form	Na-form
SBS		
γ-globulin	171	0
serum albumin	145	0
ovalbumin	96	0
insulin	125	0
AKTU	117	63
Myogen	47	26
Chymotrypsinogen	90	105
Dowex 50 x 1		
RN-ase	1120	750
Trypsin	900	1320
Chymotrypsin	1150	1640
serum albumin	1420	120

It is very interesting to draw the analogy between the peculiarities of the sorption of the simplest zwitterions amino acids and peptides and those of proteins. This could serve as a new method of studying intermolecular interaction in systems containing both protein ions and other ions. It should also aid the study of the structural peculiarities of proteins in connection with the distribution of functional groups in the protein globule.

First of all the question arises whether proteins have the ability to be sorbed according to the laws of the sorption of the simplest zwitterions-amino acids and peptides. The data in table 4 show that certain proteins have a strongly pronounced sorption capacity according to the mechanism of zwitterion sorption, for example serum albumin and globulin and egg albumin. These proteins are sorbed only on the hydrogen form of the resins and are not sorbed on the salt form of sulforesins. Thus the globules of these proteins carry very closely spaced functional groups with positive and negative charges. In contrast to them several other proteins, for example insulin, adrenocorticotropic hormone, myogene, are sorbed with comparable capacities both on the hydrogen and the sodium form of the resin. These proteins do not show a strongly pronounced zwitterion property in the sorption phenomenon. They can have two different electrochemical structures : either the functional groups carrying opposite charges are at a great distance from each other in the protein macromolecule or else they are closely spaced but close to them there is a third group neutralizing the repulsion effect. It is espacially important that the second protein group in contrast to the first one includes only the proteins posessing specific biological properties. They are hormones, enzymes, zymogen. It is quite natural that these proteins in contrast to albumins and globulin have a tertiary electrochemical structure capable of inter-molecular interaction with other electrolytes. If these proteins were deprived of such a property, the possibility to display biological activity consisting of interaction with other electrolytes would disappear. The universal application of the conclusions about the electrochemical structure of proteins can be proved by comparison with the results of sorption experiments carried out for other types of resins.

Table 5 illustrates this. It shows the results of the determinations of the sorption capacity for proteins on the carboxyl cation exchange resin KMT. The study of the screening of the protein molecule charges in the sorption experiment can serve as a proof of the phenomenon of the electrostatic repulsion observed on the sodium forms of the resins.

table 5 The capacity of sorption of proteins on carboxylic cation-exchanger KMT in H- and Na-form

Protein	Capacity mg/gr H	Capacity mg/gr Na
serum albumin	120	0
insulin	156	16
γ-globulin	140	40
AKTU	160	77
Chymotrypsinogen	284	136

table 6 Capacity of sorption of native and reduction proteins on cation-exchangers (mg/gr)

Substance		Dowex 50x1	KMT	form
RN-ase	nat.	750	490	Na
	red.	960	260	
trypsin	nat.	1320	0	Na
	red.	260	6	
chymotrypsin	nat.	1640	960	Na
	red.	0	0	
serum albumin	nat.	1450	780	H
	red.	100	40	

The curves in fig. 2 show that an increase in the ionic strength of the solution causes a more intensive sorption on the SBS sulforesin not only of the serum albumin and γ-globulin but also of insulin. The sorption capacity for proteins can also be increased by passing from the aqueous to the aqueous-acetonic solution because it reduces the degree of the ionization of the carboxyl groups.

The analysis of the topography of the charges in the tertiary structure of proteins is very important for understanding the mechanism of the intermolecular interaction of proteins with other electrolytes. Thus, the interaction between enzyme and substrate for the case that the substrate is an electrolyte, is largely determined by the position of the charged functional groups since the appearance of local dipoles may result in a disturbance of the intermolecular interaction.

Difference in the degree of the swelling of ion exchange resins in water (and in aqueous solutions) depends on the degree of the cross-linking of the resin and on the properties of functional groups. This difference causes variations in the penetration of large ions through the grains of the sorbent. When the degree of the swelling of the resin rises, it results in an increase in the capacity of large ions to penetrate through the grains and to be sorbed by the sorbents. Thus, the ions of organic substances having molecular weight from 500 to 1000 can be sorbed at great capacities exceeding the weight of dry sorbent. The resin swelling coefficient (ratio of the volumes of the swollen resin and of the dry one) in this case is equal to 3-4.

Differences in the penetration capacity of the protein molecules through ion exchange resins have been used for studying the size variations of the molecules when the tertiary structure of the proteins is disturbed in the process of disulfide bond breaking due to the reduction by tioglycolic acid.

The sulfhydryl groups of the reduced proteins were acetamidated to avoid oxidation by the oxygen of the air. The sorption of proteins was carried out finaly on ground sorbents in static conditions. The results of the experiments are shown in table 6.

The above presented data show that the molecules of trypsin and especially those of the serum albumin and chymotrypsin lose their ability to penetrate through the grains of the sorbent. Naturally, this is caused by an increase of the size of macromolecules. This takes place even in chymotrypsin through three fragments are formed out of each macromolecule. Ribonuclease behaves quite differently. The breaking of disulfide bridges and further acetamidation bring about an increase in the penetration of these macromolecules into the resin, that is, a decrease of their size.

The method of using the protein sorption by ion exchange resins for the analysis of the size of macromolecules can be compared to the chromatographic fractionation of the proteins by gel filtration on Sefadex. As we know the gel filtration method makes possible the separation of substances precisely according to the size of the molecules. Figures 3 and 4 show the chromatograms for both native and reduced ribonuclease and for trypsin. (For each substance the experiment was carried out separately.) The acetamidated reduced ribonuclease moves more slowly than the native one in complete accordance with the sorption measurements.

These last results are interesting because a denser packing appears after the tertiary structure is disturbed and the disulfide bonds holding it are broken. The coiling of protein with the formation of a dense packing takes place not only at the expense of tyrosine-carboxyl hydrogen bonds but also as a result of the participation of the amide groups. These experimental data explain a complete reactivation of ribonuclease in reoxidation observed by ANFINSET and others. At the same time this phenomenon should not be regarded as universal since in many proteins the breaking of the S-S-bonds causes such a considerable uncoiling of the macromolecule and such an increase of its size that the reproduction of the native structure in reoxidation is considered as impossible because of statistical considerations since the number of the S-S groups is too great. From this point of view low yield of active trypsin in reoxidation is quite natural.

The interaction of polypeptides and proteins with cation exchange resins in the acid or hydrogen form can be used not only for sorptional fractionation of the mixtures of such substances but also for studying many structural characteristics of the zwitterions and for studying the process of proteolysis, that is the breaking of the peptide bond. The latter process is a catalytical one and usually takes place at a temperature close to 100°C. The hydrolysis of proteins and peptides by sulforesins is in many respects similar to the ordinary homogeneous acid hydrolysis and is investigated in detail. Nevertheless it gives a new possibility for studying the processes of the transformation of polypeptides by combining catalysis and sorption. Both the initial polypeptides and the transformation products can exist in a sorbed state on a sufficient amount of ion exchange resin. Evidently the process in solution and in sorbed state will proceed in a different way. The free energy of the polypeptide transformation on the resin may be calculated from the free energy of the polypeptide transformation in solution and from that of the sorption of the initial and the transformation products.

The free energy of sorption for certain standard states can be calculated from sorption isoterms. The values of the free energy of sorption on sulfo cation exchange resins SDW are shown in table 7 for amino acids and polypeptides.

The data in this table show that the zwitterion sorption by the H-cation exchange resins is accompanied by a considerable decrease in free energy equal to 2000-4000 cal/mole, while in the cation sorption the change of free energy is very low. The free energy of the zwitterion sorption is determined by the change in the entropy of the system.

Taking into consideration low zwitterion sorption properties of tripeptides (as well as tetrapeptides) and high zwitterion properties of amino acids and higher polypeptides and proteins one may expect peptides to undergo disproportionation during catalytical transformation and to form both amino acids and higher peptides.

It should be pointed out once more that the compensation of the free energy of the sorption is sufficient for suppressing the free energy of the peptide bond hydrolysis in solution. In order to intensify the synthesis and to reduce the hydrolysis of polypeptides it is necessary to use instead of sulforesins a carboxyl resin KMT on which the zwitterion sorption is increased in the series amino acids-decapeptides.

In order to realize the synthesis of the peptide bond between the catalyst and the sorbent we have carried out experiments for studying the catalytic transformation of protein hydrolysates on carboxyl resin KMT at 70°C. The experiments have shown that the hydrolysates on the catalyst undergo certain changes. On the one hand side the changes of the amine nitrogen are not observed and on the other hand side the amount of dialyzed products increases as well as the amount of the substances precipitated by trichloroacetic acid. Thus we observe here the disproportionation phenomenon. It is possible to intensify the synthetic effects by working with a certain fraction of protein hydrolysates. For obtaining that we precipitate high molecular weight polypeptides by trichloro-acetic acid and separate low molecular components by dialyzis. The transformation of these polypeptide fractions on the KMT resin results in a total decrease of the end amine groups (table 8).

The change in the dialyzability of the products of transformation and of the degree of their precipitation by trichloro-acetic acid (table 9) is also evidence of the synthetic process. The results of a direct determination of the average molecular weight and a higher hydrolysability of the products under the action of enzymes than that of the initial polypeptides (fig.5) - all this warrants the same conclusions.

table 7 The free energy of sorption of aminoacids and polypeptides on-sulforesins

Substance	Resin		F^0
Alanine	SDW-3	6,2	-3250
Glycil-leucine	-	6,0	-4220
Alanine	WF	6,6	-2100

table 8 The transformation of fractionated protein's hydrolysates with the carboxylic cation-exchanger KMT

Initial protein	Conditions of hydrolysis	Outcome $\frac{N_{NH_2}}{N_{tot}}$
Horse serum albumin	ferment (pepsin) 20 hour	0,65
	3N HCl - 20 min	0,41
	3N HCl - 40 min	0,79
Casein	3N HCl - 2 hour	0,79
Horse serum globulin	1N HCl - 4 hour	0,75

table 9 Dialysis precipitation with T.Cl.A.A. of initial hydrolysates transformation's products on the KMT-resin

	% of precipitat.	% of dialysis
Initial	0	63
Synthesis	22	30

The change of the molecular weight and size of polypeptides in the process of catalytic transformation can be demonstrated also by chromatographic method on sefadex. As we can see in fig.6, the products of transformation in some cases move on a column rapidly than the initial polypeptides. This indicates an increase in the molecular weight. In other cases the chromatography demonstrates the splitting of the peptide bond. These experiments deal with fine fractions obtained after acid hydrolysis of serum albumin and preparative chromatographic separation of the peptide mixture on amphoteric resin or sefadex. It is interesting that the synthetic phenomena take only in a mixture of several peptides. Pure peptide can only be splited on resins. This indicates that the process of synthesis of peptide bonds on ion exchange depends not only on thermodynamic energetic characteristics but also on the correlation between the structure of peptids and resins.

D I S K U S S I O N S B E M E R K U N G E N

BERG : In Tabelle 6 ist die Zunahme der Kapazität für reduzierte Ribonuclease im Gegensatz zu den anderen Beispielen sehr bemerkenswert. Ist durch andere Methoden auch eine Kôntraktion des RNAse-Moleküls nachgewiesen worden ? Wie verhält sich Lysozym ?

SAMSONOV : The size variability of ribonuclease and some other proteins in disulfide bond breaking has been studied not only by the sorption method but also by ordinary hydrodynamic methods: sedimentation and diffusion. The results obtained by these methods correlate with each other.

The absorption method is of interest because of its simplicity as well as because of the short time of the experiment.

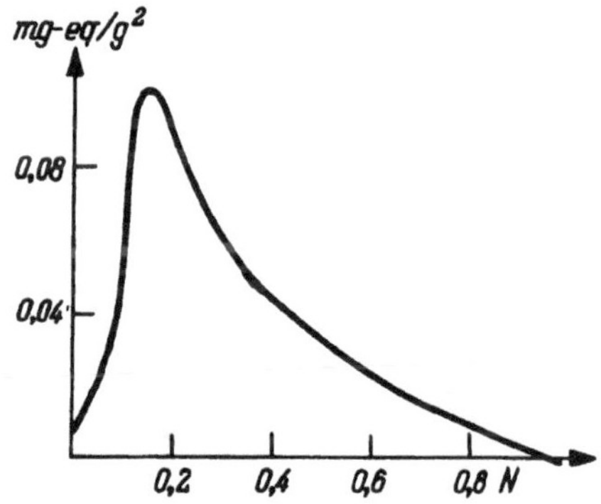

Fig.1 (II) Dependence of sorption capacity of alanin on sulforesin SDB-3 (Na) on NaCl concentration. Initial concentration of alanin - 0,01 N.

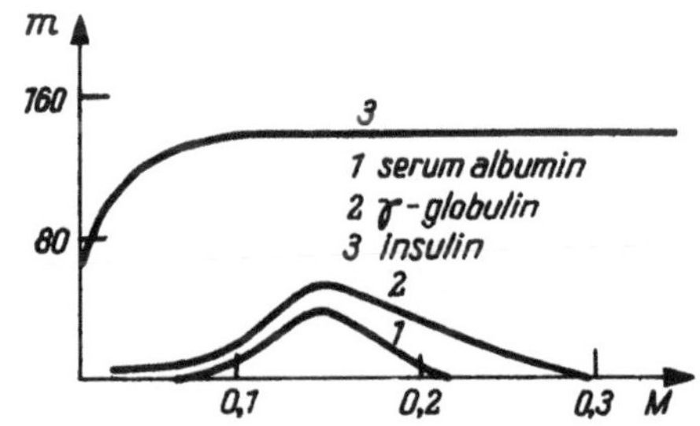

Fig.2 (YII) Dependence of sorption capacity of proteins on sulforesin SBS-3 (Na) on concentration of Na^+ in phosphate buffer. 1-serum albumin (horse), 2-γ-globulin (horse), 3-insulin.

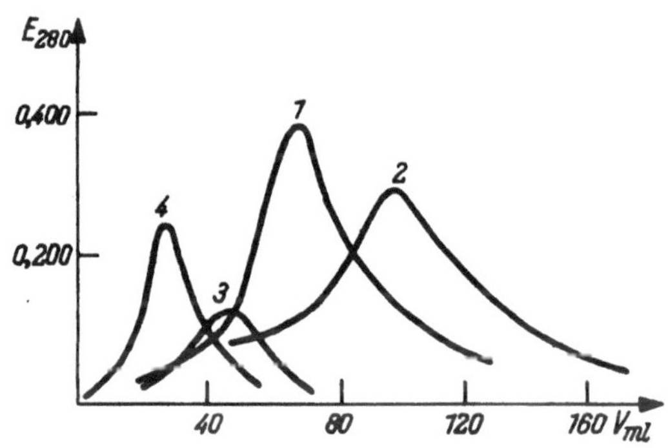

Fig.3 (IX) Chromatography of native and reduced ribonuclease on sefadex G-50. 1-native, 2-reduced (acetamidated), 3-reduced, 4-reduced (carboxymethylated).

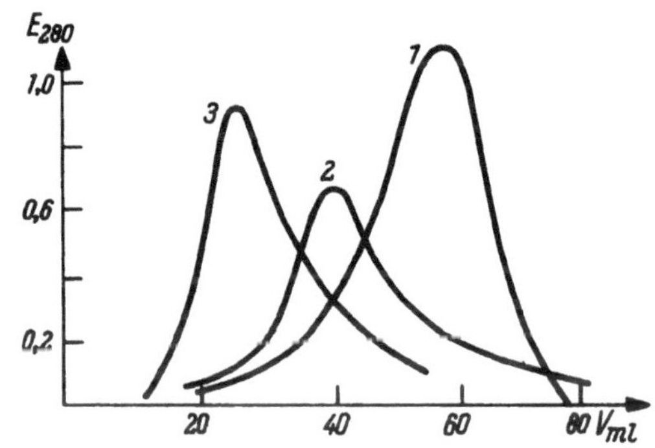

Fig.4 (X) Chromatography of native and reduced trypsin on sefadex G-50. 1-native, 2-reduced, 3-reduced (carboxymethylated).

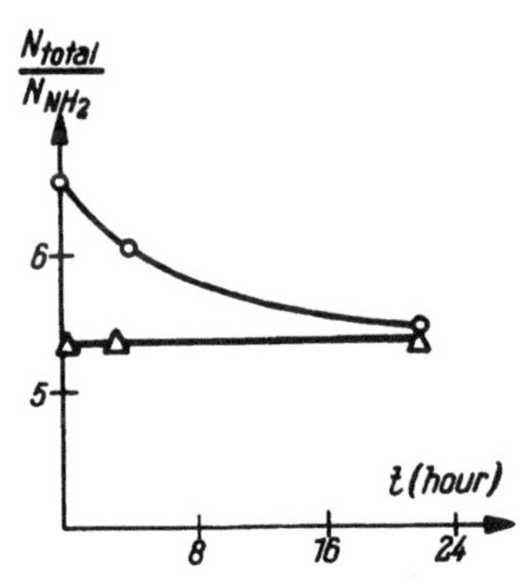

Fig.5 (XIY) Fermentative hydrolysis of polypeptide mixtures (Δ-initial mixture, O-mixture after catalytic transformation on carboxylic resin).

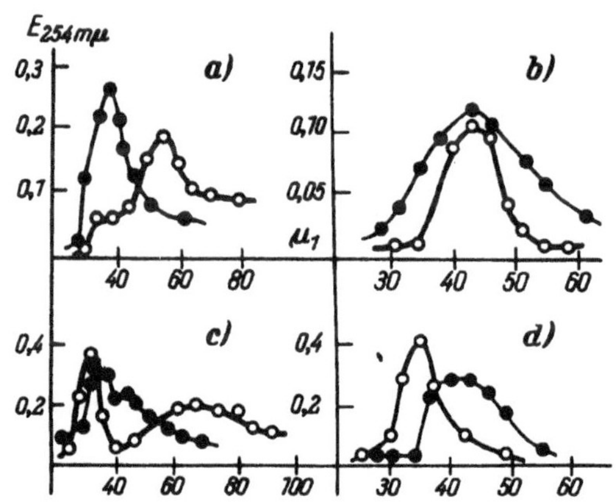

Fig.6 (XX) Chromatographic analysis of polypeptide mixtures on sefadex G-25 (○-○-○ initial mixtures, ●-●-● products of transformation). Molecular weight (average) a-227, b-370, c-2200, d-1500.

C Proteine

Formation of Free Radicals under the Influence of Mechanical Forces on Proteins and Migration of Free Valence

P. BUTJAGIN and G. ABAGJAN

Institute of the Chemical Physics of the Academy, Moscow

S U M M A R Y

If high molecular compounds are influenced mechanically their chemical bonds are broken. Thereby formation of free radicals arises followed by further reactions : migration of free valences, recombinations, oxydations etc. We have stated by investigations on synthetic polymers that the yields of mechanochemical reactions are of the same order of magnitude as the yields of radiochemical processes and that the influence of very different energy kinds acts on the same position of the molecule in many cases.

From native high molecular compounds (in most cases proteins of different composition) we have derived high concentrations of free radicals with characteristical ESR-spectra. At low temperatures (77°K) the primary unstable intermediate products were fixed and during freezing-in we have observed the migration processes of the free valence which leads to formation of stable free radicals. Velocity and direction of the migration are dependent of the humidity of the presence of oxygen and of the temperature.

We investigated the nature of the primary act and early secondary processes occuring by mechano-chemical reactions in proteins, or more exactly, by mechanical destruction of the protein molecules. Now we have obtained the first results.

The method of mechanical action on the macromolecules, the conditions of experiments and the methods of analyses were chosen in such a way that we could catch and observe the unstable intermediate products of the destruction - free radicals.

Free radicals have the longest life time in the solid state at low temperature. For experiments in such conditions a glass vibrating micromill was used.

It could hold from three hundred to six hundred milligrammes of substance. The mill connected with the vacuumplant is shown in fig.1. The glass vessel with glass balls and the substance is placed under the vibrator and is connected with the vacuumplant by means of a glass spring. The frequency of the vibrator is fifty periods per second, the amplitude from five to seven millimetres.

The glass vessel with glass balls for the milling and the measurements of the e.s.r. spectra is shown in fig.2. The experiments were made in high vacuum, in oxygen, in inerte gases and in solid solutions at temperatures from -196°C to +200°C.

Free radicals were investigated by electron spin resonance method. The wave length of our radiospectrometer type ИХФ-2 was three centimeters. The surface area and the molecular weight were also measured.

Basic experiments were made with gelatine, zein, ovalbumine, serum albumine, hemoglobine, kazein and keratin of horn and hoof. The destruction of polymethylmethacrylate was investigated for the purpose of comparison.

At first the kinetics of the destruction were investigated. In this part of the work we wanted to obtain some preliminary information about the nature of mechano-chemical reactions in macromolecules. Some measurements were made with polymethylmethacrylate, otherones with gelatine. The results are shown in fig.3.

The logarithm of destruction rate constant is plotted against the reverse intensity of mechanical action, (mechanical energy per second per gramm polymer). The experimental data are in good agreement with law ARRHENIUS, but instead of the temperature the intensity of the mechanical force arises.

The destruction can be considered to be a spontaneous dissociation of macromolecules catalyzed by mechanical action. In a mechanically strained molecule the equilibrium distances between the main chain atomes undergo a change. Therefore both the bond energy and activation energy of the destruction process decrease: $E = E_0 - f(I)$. This function $f(I)$ is determined by the shape of the interacting main chain atoms. It should be supposed that MAXWELL-BOLTZMAN equilibrium is preserved in the system. This method of analysing experimental data was used by SCHURKOW, BUCHE and others.

On the other hand the deformation rate may turn out to be greater than the rate with which the system returns to equilibrium. Then the mechanical action can be considered as a source of unequilibrated states in the system. From this point of view the increased intensity of action not only decreased the activation energy, but increased the total energy level in the system as compared with the equilibrium energy level at a given temperature.

Than : $K = K_0 \exp\left(-\frac{E - f(I)}{kT + \alpha I}\right)$, if kT is much smaller than αI and $f(I) \ll E$ we can write approximately : $K = K_0 \exp\left(-\frac{E}{\alpha I}\right)$. The experiment (fig.3) confirms that this mechanism can exist.

Therefore the first stage of the mechanical destruction is the appearance of unequilibrated states and the second stage is the localization of the excess energy at given bond. The two stages depends both on the method of action and on the structure of macromolecule.

Thus, mechano-chemical processes must have much in common with radiation-, photochemical and other processes. In all cases the point of damage is greatly determined by the secondary processes of the migration of energy and after the rupture of bond by the migration of free valence.

Now let us analyse the places of bond rupture in protein.

The electron spin resonance spectra of seven proteins are shown in fig.4. Before measurements were carried out all samples were dried in vacuum and milled in vacuum at temperature $-196^{\circ}C$ in liquid nitrogen.

The concentration of free radicals in some samples reached from 10^{19} to 10^{20} radicals per gram. All the spectra are assymetrical complex and consist of several groups of lines belonging to different types of free radicals. The width of the spectra is about 50 oersted and the "g"-factor is close to this of free electrons.

Most of the spectra are like to another. In the ceratine spectrum we could see additional lines near the "g"-factor 2,024.

In further experiments we made attempts to separate the different types of radicals and to obtain their individual spectra. For this purpose we measured the spectra of milled solid diluted solutions of proteins at the temperature of liquid nitrogen. The aim of this experiments was to diminish the intramolecular transfer of free valence and to check the secondary processes. In solution the mechanical forces discrupt only the macromolecules, the solvent molecules remain intact.

Water, heavy water, acetic acid, sulphuric acid and sodium alkali were used as solvents.

Similar spectra were found in all solutions (see fig. 5a). It is a triplet with the distance of 19 oersted between the lines. The nature of the protein and solvents did not influence the shape of the spectrum. In insufficient vacuum in presence of a small concentration of oxygen an additional line can be seen in the spectrum at low temperature. After heating in oxygen this additional line becomes the spectrum of peroxide radicals.

After heating in vacuum the triplet becomes a doublet (fig.5-b, c). The overall intensity of the spectrum remains the same. At the melting point of the solvent all the radicals disappear.

In the deuterium proteins the double line did not appear. The triplet disappears, the single line remains.

In the dried proteins the secondary radicals with a double line spectrum are much more stable than in solutions. Changes in the spectrum of gelatine when it was heated from -196° to $+150^{\circ}C$ are shown in fig.6. At room temperature the spectrum becomes a double line, when heated further it becomes a single line. The time of heating is 20 min at the given temperature. When cystines containing proteins were heated above the room temperature they showed the typical spectrum of cystine. This phenomena was investigated in cooperation with CAREL ULBERT.

All the spectra which we observed in proteins consist mainly of four signals : triplet, doublet, the peroxide assimetric line and the cystine line.

Pure triplet can be observed only when free radicals are stabilized in crystals at low temperatures. We have too little data to interprete this spectrum. By bond rupture two types of free radicals should be found but we observe only one. Therefore we probably witness the results of so far unknown processes of decay or transition of some type of radicals. Most likely the free valence is localized on the carbon α-atom of the main chain.

These radicals are very active in the chemical reactions. They combine with oxygen at $-130^{\circ}C$. About -100° to -80° they become the most stable form.

Free radicals with doublet spectrum are secondary radicals. They appear as a result of the transfer of free valences. Such radicals were discovered in radiationchemical investigations of proteins. The results of experiments with deuterium gelatine in our work and continued by PULATOV and KAUSHIN show that the free valence is most likely localized on the oxygen atom participating in the hydrogen bond. This is the point of view of BLUMENFELD and KALMANSON.

Thus the low temperature transition triplet-doublet found in proteins is the transfer of an free valence of a very short distance. It is probably a transformation inside the molecule which leads to a stabilization of free valence.

When cystin is present in protein we can obcerve the second transition. It takes place at higher temperatures about 30...50°C.

This transition occurs over longer distances than the transition triplet-doublet. It was observed in molecular mixture of cystin and gelatine which indicated that there is a possibility of intermolecular transfer of free valence. Migration to cystine was also observed in the radiationchemistry investigation of proteins.

The results of investigation of e.s.r. spectra can be represented in a simple way :

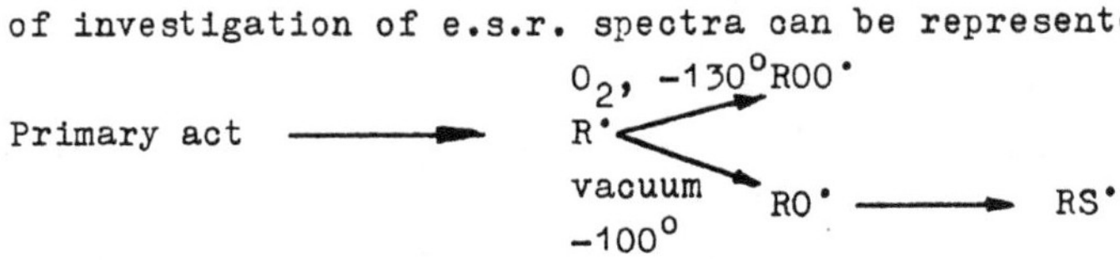

R˙ - free radicals with triplet line spectrums, localisation of free valence on the α-carbon atom (?), RO˙ - secondary radicals, ROO˙ - peroxide radicals, RS˙ - cystine radicals.

Up to date it was unknown that the types of damages by mechanical forces are the same as this by radiation action on proteins.

D I S C U S S I O N - Q U E S T I O N S

BERG : Verlauf der Radikalbildung und Messung; Aggregatzustand ?

BUTJAGIN : The investigated solution of polymer was pumped in glass mill. After sealing the vessel with solutions and balls (fig.2) was put on the vibrator and become cool with liquid nitrogen. It was the matter of experimentator to prevent the balls and solution from freasing together in monolithic "pulp". In order to achieve it the freezing must be done gradually. After milling the vessel was turned down and the powder of frozen solution was filled in the tube of vessel. The e.s.r. measurements were made in DEWAR plastic vessel placed in the cavity of the e.s.r. spectrometer. These methods were described in the work : P.BUTGAGIN, Dokl.A.N.SSSR 148, 129 (1963).

BERG : Keratin und Albumin enthalten beide Cystin. Weshalb sind die Spektren so verschieden ?

BUTJAGIN : The contain of cystine in albumine is about 10 time lower than in ceratine. Therefore we could observe the cystine-spectrum in albumine only after migration of free valence to the S-atom at 30-50°C.

BERG : Die Spektren sind für ein Makromolekül in Eis excellent scharf und charakteristisch.

BUTJAGIN : Danke sehr.

DOSE : Ihre Ergebnisse zeigen eine große Ähnlichkeit mit den ESR-Signalen, die man bei Bestrahlung dieser Proteine beobachtet. Gestatten Sie mir daher die Frage, ob beim Erwärmen der bei tieferen Temperaturen geschädigten Proteinen auch das Schwefel-Signal auf Kosten des Glycin-Signals verstärkt wird. Weiterhin möchte ich Sie fragen, ob man auch hier beim Anwärmen ähnlich wie bei den bestrahlten Proteinen eine Thermolumineszenz beobachtet.

BUTJAGIN : In the cystine containing proteins as well as in other proteins with added cystine, in all cases we observed the migration of free valence to a S-atom. These phenomena was investigated in special work : K.ULBERT, P.BUTJAGIN, Dokl.A.N.SSSR 149, 1194 (1963).

The grinding always was accompanied by luminescence. However during the heating of the samples of polymers from -196°C to room temperature we could not discover thermoluminescence. The thermoluminescence is usually connected with charge recombination. By mechanical destruction there is no ionisation of the substance.

DAMERAU : Wurden bei der mechanischen Radikalerzeugung wie bei der Bestrahlung Sättigungdeffekte bezüglich der Radikalausbeute beobachtet ?

BUTJAGIN : We have observed the effect of saturation in free radical concentrations only after destruction at room temperature or by experiments with solid solutions. At the low temperatures (-196°C) concentration of free radicals increases unceasing to $10^{19} - 10^{20} g^{-1}$. We have milled the samples in this conditions 10-20 h. The longer grindings are senseless because the glass balls weared out.

LASURKIN : Numbers of radicals 10^{19}-10^{20} per gramm are not unusual. In our experiments with A.USSATY we obtained such high values if the irradiation was carried out at high vacuum, were also the gaseous products of the radiochemical reactions were removed.

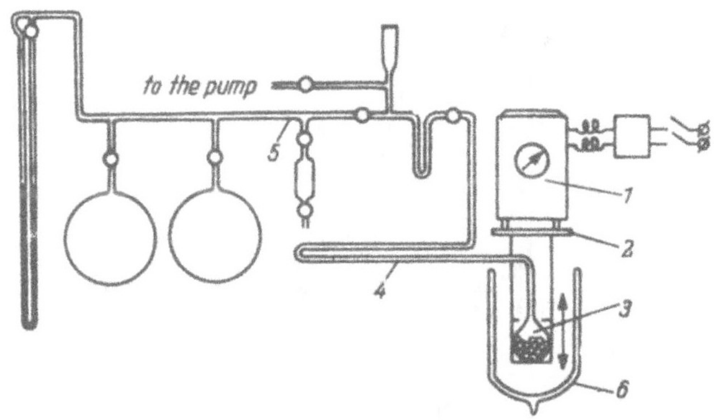

Fig.1 Vibration mill connected with vacuumplant. 1,2-vibrator, 3-glass vessel with glass balls, 4-glass spring, 5-vacuumtube, 6-DEWAR vessel

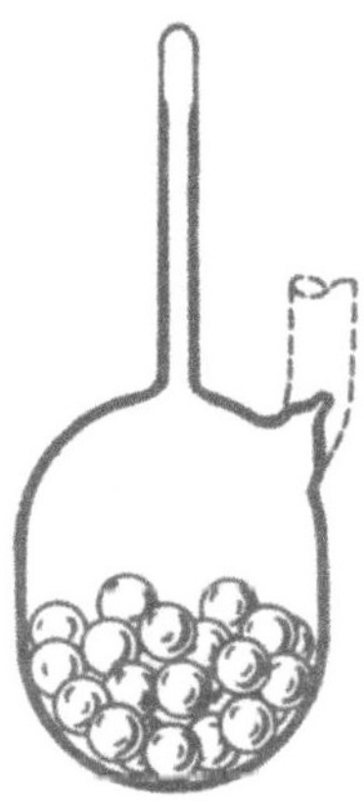

Fig.2 Glass vessel for milling and measurements of e.s.r. spectrum

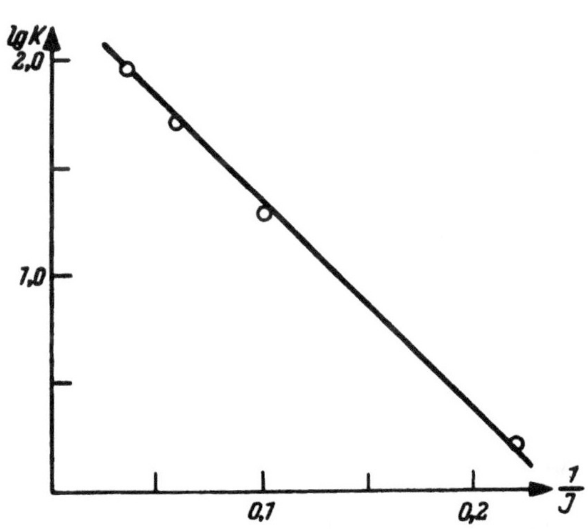

Fig.3 The kinetic of destruction of polymethylmethacrylate at 20°C. K-rate constante by molecular weight decrease, I-the intensity of mechanical force (energy per unit time).

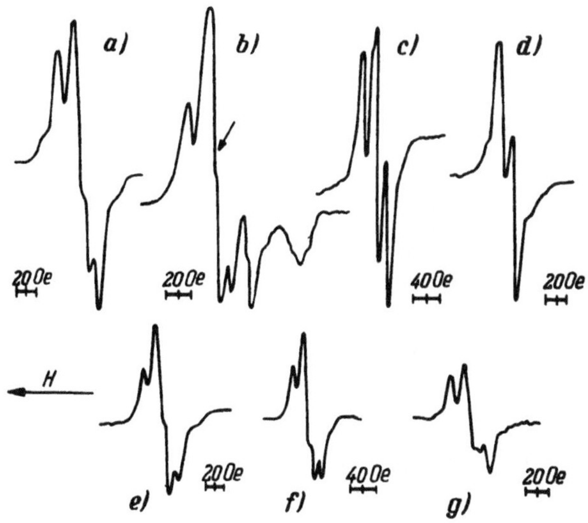

Fig.4 The e.s.r. spectra of proteins at $-196^{\circ}C$; a-gelatine, b-ceratine, c-casein, d-serum-albumin, e-egg albumin, f-zein, g-hemoglobin

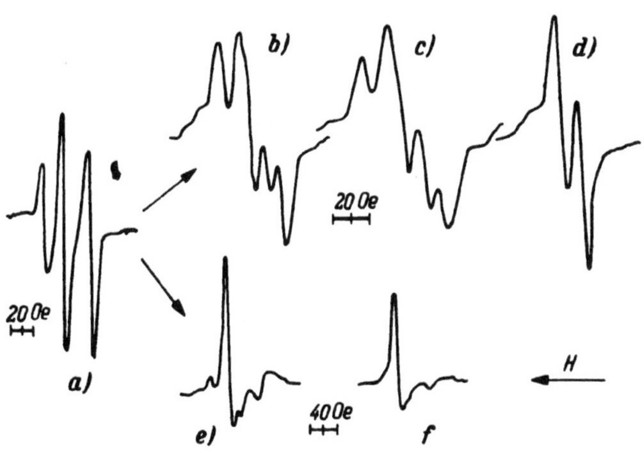

Fig.5 The e.s.r. spectra of diluted protein solution. a, in vacuum at $-196^{\circ}C$; b, c, d - after heating in vacuum at $-78^{\circ}C$; e-in presence small concentration of oxigen (10^{-2}mm hg); f-after heating at $-98^{\circ}C$

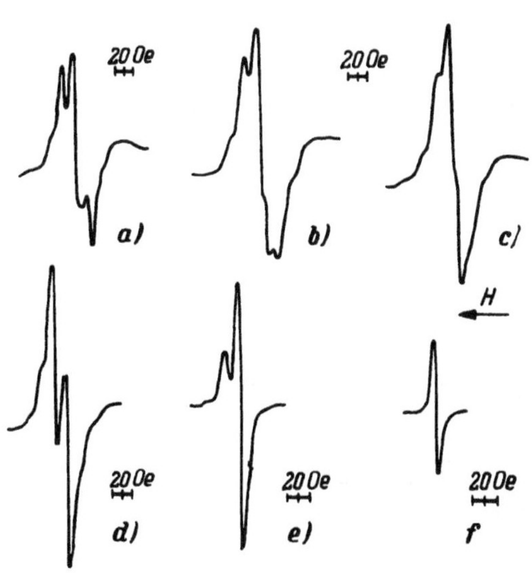

Fig.6 The e.s.r. spectra of protein after heating in vacuum. a- -196°, b- -95°, c- -78°, d- -20°, e- 100°, f- $150^{\circ}C$

C Proteine

Kinetik der thermischen Denaturierung der Glucoseoxydase Notatin

O. BODMANN

Institut für physikalische Chemie der Universität, Mainz

S U M M A R Y

The irreversible thermal denaturation of the enzyme Notatin at temperatures of about 60°C is found to be a first order reaction with respect to concentration dependence of the inactivation velocity. On the other hand the course of the inactivation with time appears to be due to a reaction of higher oder (nearly of second order).

LAIDLER [1] in a more comprehensive survey of kinetic phenomena in denaturation reactions represents four possibilities to cause such behaviour :

1. Denaturation equilibrium is slowly approached,
2. The denaturation process is inhibited by the product of reaction,
3. The protein is not homogeneous,
4. The denaturation process consists of a series of consecutive steps.

The first, second and fourth possibilities could be excluded experimentally by denaturation kinetics. With respect to the third possibility it is shown that these heterogeneities can be explained at least partly as having originated from preparative manipulations.

I. E I N L E I T U N G

Bei Denaturierungsvorgängen findet man häufig, daß der zeitliche Ablauf der Prozesse nicht mit der Reaktionsordnung übereinstimmt, gleichgültig ob die Denaturierung durch chemische oder physikalische Einflüsse hervorgerufen wird. Formal reaktionskinetisch läßt sich ein derartiges Verhalten einfach dadurch ausdrücken, daß man zwischen der "konzentrationsbezogenen" und der "zeitbezogenen" Reaktionsordnung unterscheidet [1].

Die konzentrationsbezogene, wahre Reaktionsordnung erhält man aus der Konzentrationsabhängigkeit der Anfangsgeschwindigkeiten, die zeitbezogene, bei komplizierteren Prozessen oft nur scheinbare Ordnung, wird aus der Zeitfunktion des Umsatzes bestimmt.

Zur Messung des Denaturierungsgrades gibt es neben der klassischen Methode, bei welcher der koagulierte Anteil bestimmt wird, eine Reihe anderer chemischer und physikalischer Eigenschaften, durch deren Veränderung Denaturierungsprozesse äußerlich in Erscheinung treten und messend verfolgt werden können. Bei Enzymen ist z.B. die als biochemische Aktivität bezeichnete katalytische Wirksamkeit eine Eigenschaft der nativen Moleküle. Sie geht bei der Denaturierung meist völlig verloren. Unter geeigneten und normierten Bedingungen gemessen, ist die Aktivität eines Enzymes proportional der Konzentration der nativen Moleküle. Denaturierungsgrad und Inaktivierungsgrad sind vielfach einander gleich [2].

Ursachen für scheinbare Diskrepanzen der Reaktionsordnung bei Denaturierungsprozessen können sehr verschiedenartig sein. Einmal können sie an komplexen Vorgängen im Reaktionsmechanismus liegen, ebenso gut aber auch stofflicher Natur sein. Der Reaktionsmechanismus der Denaturierung besteht aus dem Nebeneinander und Hintereinander vieler Elementarschritte, von denen nicht einer allein geschwindigkeitsbestimmend für den Gesamtprozess zu sein braucht. Ausgehend von seinen nativen Konformationszuständen erleidet das Protein im Laufe des thermischen Denaturierungsprozesses eine Reihe inner- und zwischenmolekularer Konformationsänderungen, bis es schließlich als Endprodukt irreversibel zwischenmolekular gebunden, meist vollständig koaguliert als amorpher Niederschlag vorliegt. Native und nichtnative Bindungen werden teils reversibel, teils irreversibel gelöst und andere neugebildet.

Die thermische Inaktivierung des Pepsins ist unter den Denaturierungsprozessen sicherlich als Sonderfall zu betrachten. Nach CASEY und LAIDLER [3] erfolgt die Inaktivierung dieses Enzyms über einen aus fünf Einzelmolekülen bestehenden Reaktionskomplex. Bei kleinen Anfangskonzentrationen fanden sie, daß die Reaktion fünfter Ordnung ist, während größere Konzentrationen eine Reaktion erster Ordnung zeigen. Vom Beginn bis zum Ende eines Denaturierungsversuches kann bei Pepsin die Reaktion von der ersten Ordnung zur fünften überwechseln, wenn die Konzentration des aktiven Pepsins entsprechend abgenommen hat. Bei den meisten anderen Proteinen findet man dagegen als wahre Reaktionsordnung ausschließlich die erste. Wenn daneben der zeitliche Ablauf der Reaktion gleichzeitig eine höhere Ordnung als die erste vortäuscht, wie das z.B. in dem hier behandelten Fall bei der thermischen Inaktivierung des Notatins der Fall ist, dann sind die Ursachen dafür nicht in einer höheren Molekularität der Reaktion zu suchen, sondern in einem zwar unimolekularen bzw. quasi-unimolekularen aber komplexen Reaktionsmechanismus, oder aber in einer gewissen Uneinheitlichkeit des Enzyms.

LAIDLER [1] behandelt zusammenfassend alle Kombinationen von zeit- und konzentrationsbezogenen Reaktionsordnungen. Durch geeignete experimentelle Ansätze kann man unter verschiedenen Typen geschwindigkeitsbestimmender Schritte unterscheiden, oder auch die Frage stofflicher Einheitlichkeit entscheiden.

Identitätsfragen bei Enzymen mit Hilfe der Denaturierungsgeschwindigkeit zu klären, wurde in anderem Zusammenhang bereits 1895 von TAMMANN [4] vorgeschlagen. Wegen unkontrollierbarer Einflüsse fremder Begleitstoffe, wie z.B. anderer Proteine, Kohlenhydrate, Ionen usw. auf die Denaturierungsgeschwindigkeit, ist der Nachweis der Identität sogenannter Heteroenzyme [5], gleich wirkender Enzyme verschiedener Herkunft, sehr erschwert. TAMMANNS Vorschlag kam deswegen keine praktische Bedeutung zu.

Umgekehrt läßt sich jedoch der Nachweis nicht identischer Formen eines Enzyms im gleichen Versuchsmaterial unter Umständen verhältnismäßig einfach denaturierungskinetisch führen. Hierbei kann es sich einmal um die mit dem Begriff "Isoenzyme" bezeichnete genetische Multiplizität eines Enzyms handeln, wie sie u.a. bei Milchsäuredehydrogenasen gefunden wurde [5] [6], oder aber um nicht native Abarten eines Enzyms, die noch biochemische Aktivität besitzen. Heterogenitäten nicht enzymatischer Plasmaproteine lassen sich mit Hilfe anderer Denaturierungskriterien mit der gleichen Methode nachweisen. FRIČ [7] konnte z.B. durch fraktionierte Denaturierung zeigen, daß menschliches Serumalbumin als heterogenes, aus Komponenten unterschiedlicher Wärmebeständigkeit zusammengesetztes Protein vorliegt.

Unsere folgenden kinetischen Messungen der thermischen Inaktivierung der Glucoseoxydase "Notatin" [8] machen es wahrscheinlich, daß nicht mehr alle aktiven Enzymmoleküle unserer Präparate in nativem Zustand vorlagen.

II. Präparate und Methoden

Enzympräparate

Notatin [8] wurde aus dem Kulturfiltrat von Submerskulturen eines Penicillium notatum Stammes gewonnen, der uns freundlicherweise durch Herrn Dr. LUMB von der Boots Pure Drug Co.Ltd. Nottingham als lyophilisierte Sporensuspension unter der Bezeichnung FD 446 zur Verfügung gestellt wurde.

Aus dem Kulturfiltrat wurde die Glucoseoxydase nach VAN BRUGGEN u.a. [9] durch Uranylacetatlösung ausgefällt, mit 0,2 m Phosphatpuffer vom pH 6,8 extrahiert und mit Ammonsulfat fraktioniert ausgefällt. Nach Dialyse gegen destilliertes Wasser und Gefriertrocknung wurden die Enzympräparate bei etwa 2°C aufbewahrt.

Bei den Denaturierungsversuchen wurden verschiedene Fraktionen des gleichen Ansatzes 106 verwendet, welche sich im Reinheitsgrad erheblich unterschieden (Notatingehalt etwa zwischen 10 und 50 %), und das Präparat A aus den gesammelten Hauptfraktionen der Ammonsulfatfällung mehrerer Ansätze. Qualitativ zeigten alle Präparate das gleiche typische Bild der im folgenden beschriebenen Denaturierungskinetik. Quantitativ waren von Präparat zu Präparat Unterschiede vorhanden, doch traten sie kaum in Erscheinung.

Denaturierungsversuche

Eine Standardlösung der Notatinpräparate in Phosphatpuffer vom pH 5,6 war, mit oder ohne Zusätzen von Äthylendiamintetraacetat, bei 2°C aufbewahrt, bis zu zwei Wochen haltbar. Mit Ausnahme der Versuche der Tabelle 1 wurden alle anderen Denaturierungsversuche so durchgeführt, daß in einem 20 ml-Kölbchen aus Jenaer Glas 9 ml Pufferlösung im Thermostaten auf die Denaturierungstemperatur gebracht wurden, und nach etwa 10 Minuten 1 ml der Standardlösung oder einer anderen näher bezeichneten Enzymlösung von Zimmertemperatur zugegeben wurde. Die Denaturierungsdauer bezieht sich stets auf diesen Zeitpunkt als den Nullpunkt der Reaktion. Während des Versuches wurde das Kölbchen ständig durch einen Vibromischer geschüttelt. Nach angemessenen Versuchszeiten wurden jeweils etwa 0,5 bis 1 ml von der Lösung entnommen, im Eisbad abgekühlt, und alsbald auf die restliche Aktivität untersucht.

Aktivitätsmessung

Die Glucoseoxydaseaktivität wurde manometrisch teils bei 25°C unter reinem Sauerstoff, teils bei 20°C unter Luft atmosphärischer Zusammensetzung in Warburgapparaten der Modelle V oder S der Herstellerfirma Braun in Melsungen gemessen. Die Reaktionsgefäße enthielten als Substratlösung jeweils 3 ml einer 0,1 m Glucoselösung in 0,3 m Phosphatpuffer mit Katalasezusatz, 0,1 oder 0,2 ml der Enzymlösung aus den Denaturierungsversuchen kamen in die birnenförmigen Gefäßansätze.

Die Druckabnahme wurde jeweils 90 Minuten lang in Abständen von 10 Minuten abgelesen und der Mittelwert der Geschwindigkeit auf die Aktivitätseinheit µMol Sauerstoffverbrauch pro Minute bei 0,1 ml Enzymlösung im Warburggefäß umgerechnet. Da bei 20° unter atmosphärischer Luft nur Geschwindigkeiten mit einem Sauerstoffverbrauch von höchstens 0,08 $\mu\mathrm{Mol.min}^{-1}$ der Konzentration an aktivem Enzym proportional waren, wurden Enzymlösungen höherer Aktivitäten zur Aktivitätsmessung entsprechend verdünnt.

III. Reaktionsordnung der Inaktivierung

Variiert man bei den thermischen Denaturierungsexperimenten unter sonst gleichen Bedingungen die Anfangskonzentration an Notatin, dann zeigt sich strenge Proportionalität zwischen der Inaktivierungsgeschwindigkeit $\frac{\Delta A}{\Delta t}$ und der Anfangsaktivität A_o.

$$- (\frac{dA}{dt})_{t=0} = k^{I} \cdot A_{o} \quad (1)$$

Da die katalytische Wirksamkeit A als molekulare Eigenschaft der Enzyme proportional der Konzentration c_A an aktiven Molekülen zu setzen ist,

$$A = a \cdot c_A , \quad (2)$$

ist k^I in Gl.(1) mit der Geschwindigkeitskonstanten eines Prozesses 1. Ordnung identisch wie er durch Gl.(3) definiert ist.

$$- (\frac{dc_A}{dt})_{t-0} = k^{I} \cdot c_{Ao} \quad (3)$$

In Gl.(2) ist a als katalytische Wirksamkeit der Konzentrationseinheit ein konstanter, nur vom Maßsystem abhängiger Umrechnungsfaktor. Anstelle der nicht messbaren differentiellen Anfangsgeschwindigkeit nach Gl.(1) werden in Tab.1 die zwischen der 5. und 15. Minute der Denaturierungsexperimente gemessenen Aktivitätsverluste $\Delta A = A_5 - A_{15}$ von Notatinlösungen wiedergegeben. Wie aus der letzten Spalte der Tabelle ersichtlich ist, findet man selbst bei 200 facher Verdünnung der Ausgangslösung innerhalb der Meßgenauigkeit den unveränderten Wert einer Geschwindigkeitskonstanten erster Ordnung.

Tabelle 1 Thermische Inaktivierung von Notatin (Präparat 106 Fr.4) Denaturierung bei 60,0°C in 0,3 m Phosphatpuffer, pH 5,6

A_5	A_{15}	$\bar{A}$	$\frac{\Delta A}{\Delta t}$	$\frac{1}{\bar{A}} \cdot \frac{\Delta A}{\Delta t} \cdot 10^2$
Sauerstoffverbrauch in [µMol . min^{-1}]				[min^{-1}]
8,8	7,4	8,1	0,14	1,7
4,6	3,6	4,1	0,10	2,4
$2,2_5$	1,8	2,0	0,045	2,3
0,86	0,74	0,80	0,012	1,5
0,45	0,35	0,40	0,010	2,5
$0,23_5$	$0,19_5$	$0,21_5$	0,004	1,9
0,22	$0,17_5$	0,20	0,0045	2,3
0,089	0,074	0,081	0,0015	1,8
0,044	0,036	0,040	0,0008	2,0
				2,0 ± 0,1

Dementsprechend könnte man erwarten, daß auch das Zeitgesetz der 1. Ordnung erfüllt ist,

$$\log A/A_o = - k^{I}t \quad (1a)$$

Logarithmisch gegen die Zeit aufgetragen, sollte die Aktivität eine Gerade mit der Neigung $-k^I$ ergeben. Das ist nach Abb.1 jedoch keineswegs der Fall. Zwar ist die experimentelle nichtlineare Zeitfunktior

$$\log A/A_o = f(t) \tag{4}$$

unabhängig von der Anfangsaktivität A_o, was in der Tab.2 noch deutlicher zum Ausdruck kommt. Dort werden unmittelbar die Zahlenwerte der Abb.1 verglichen.

Tabelle 2 Zeitverlauf der thermischen Inaktivierung von Notatin (Präparat A). Denaturierung bei 60,0°C in 0,3 m Phosphatpuffer, pH 5,6

Enzymkonzentration	0,02 mg/ml	0,2 mg/ml	2,0 mg/ml	10,0 mg/ml
Zeit min	A/A_o			
0	(1,00)	(1,00)	(1,00)	(1,00)
2	0,83	0,83	0,82	—
5	—	—	—	0,75
10	0,71	0,79	0,72	—
15	—	—	—	0,66
30	0,57	0,64	0,59	0,58
60	0,46	0,51	0,45	0,46
120	0,32	0,37	0,33	0,34
180	—	—	—	0,27
200	0,22	0,26	0,22	—
300	0,16	0,17	0,15	0,17
400	$0{,}11_4$	$0{,}11_8$	—	—
410	—	—	—	$0{,}11_3$
420	—	—	$0{,}08_8$	—

Die Anfangskonzentrationen wurden in einem Bereich von 1:500 variiert. Daß der gesamte Kurvenverlauf von der Anfangskonzentration so völlig unabhängig ist, zeigt, daß nicht nur am Beginn der Reaktion, sondern überall bei gleichen Denaturierungszeiten t konzentrationsbezogen die 1.Ordnung in Erscheinung tritt. Die Neigung der Tangenten der Kurve in Abb.1 gibt in jedem Punkt ein wirkliches k^I, eine echte Geschwindigkeitskonstante 1.Ordnung wieder. Allerdings nimmt der Wert von k^I mit fortschreitendem Prozess ab.

$$k^I = \frac{-d \log A/Ao}{dt} = f'(t) = \varphi(A/Ao) \tag{5}$$

Wasserstoffionen-Konzentration und Temperatur scheinen in gewissen Bereichen kooperierende Faktoren der Denaturierung zu sein. Verändert man nämlich den pH-Wert der Lösung, so läßt sich eine entsprechende Temperatur finden, bei der wieder die unveränderte Inaktivierungskurve auftritt.

Wie Abb.2 zeigt, stimmen die Inaktivierungskurven von Notatin, gemessen bei pH 5,6 und 60°C mit denen bei pH 5,0 und 61°C sowie bei pH 4,5 und 59,5°C überein. Das Stabilitätsmaximum liegt danach in der Nähe von pH 5. Nach der Inaktivierungskurve der Abb.2 fällt die Geschwindigkeitskonstante k^I im Laufe des Prozesses bis auf etwa ein Drittel des anfänglichen Wertes ab, wie aus der Tab.3 ersichtlich ist.

Tabelle 3 Abhängigkeit der Geschwindigkeitskonstanten der Inaktivierung vom Inaktivierungsgrad (nach Abb.2)

A/Ao	0,85	0,7	0,5	0,4	0,3	0,2
$k^I \cdot 10^2 \left[min^{-1}\right]$	1,0	0,9	0,8	0,6	0,4	0,3

Wäre die wahre Reaktionsordnung nicht bekannt, könnte man aus dem zeitlichen Ablauf leicht irrtümlich auf eine höhere als die erste Ordnung schließen.

Abbildung 3 enthält z.B. den Reaktionsablauf in einer Auftragung entsprechend der 2. Ordnung, welche danach den Reaktionsablauf wenigstens bis zum Umsatz von etwa 75 % gut wiederzugeben scheint.

Dieser scheinbare Wert 2 einer scheinbaren Reaktionsordnung n_t, der sich aus dem zeitlichen Ablauf der Denaturierung hier ergibt, ist natürlich in strengem Sinne nicht definiert. Die wahre Reaktionsordnung n_c hat den Wert 1, das ergab sich eindeutig aus der Konzentrationsabhängigkeit der Geschwindigkeiten.

IV. Reaktionsmechanismen und ihre Unterscheidung

Für die scheinbare Diskrepanz

$$n_c = 1 \; ; \; n_t > 1 \tag{6}$$

gibt es nach LAIDLER [1)] vier Grundtypen möglicher Reaktionsmechanismen. Am einfachsten kann man sie voneinander unterscheiden, wenn man den Denaturierungsversuch unterbricht, nachdem ein großer Teil des Enzyms inaktiviert ist, und nach genügend langer Unterbrechungszeit (bei thermischer Denaturierung durch Abkühlung) wieder unter den ursprünglichen Bedingungen beginnend mit der Zeit t' fortsetzt.

Nach den verschiedenen Reaktionstypen erhält man dann die in Abb.4 schematisch dargestellten Aktivitätskurven. Der erste Mechanismus, der ein Verhalten nach (6) zur Folge hat, ist der eines sich sehr langsam einstellenden Denaturierungsgleichgewichtes.

$$X \overset{K}{\rightleftharpoons} D \tag{7}$$

Die aktive Komponente X lagert sich reversibel in das inaktive D um. K ist die Gleichgewichtskonstante des Prozesses. Die Aktivität

$$A = a_X \cdot c_X \tag{2a}$$

nähert sich asymptotisch einem endlichen Grenzwert A_∞.

$$A_\infty = \frac{A_o}{K + 1} \tag{7a}$$

Wird die Denaturierung genügend lange unterbrochen, dann bildet sich der Ausgangszustand mit der Anfangsaktivität A_o vollständig zurück. Der gleiche Vorgang läßt sich beliebig oft wiederholen. In Kurve 1 der Abb.4 ist das schematisch dargestellt.

Logarithmische Aktivitätskurven mit gleicher Anfangsneigung bei t_o und t_o', jedoch abweichender Anfangsaktivitäten $A_0' < A_0$ erhält man nach Fortsetzung der unterbrochenen Denaturierung bei dem vierten Typ des LAIDLERschen Schemas. Hier ist in der Gesamtreaktion einem irreversiblen Schritt k_ω ein Konformationsgleichgewicht, oder eine Folge solcher Gleichgewichte vorgelagert.

$$X \underset{}{\overset{K_1}{\rightleftharpoons}} Y \overset{K_2}{\rightleftharpoons} \ldots\ldots\ldots\ldots \overset{K_n}{\rightleftharpoons} Z \overset{k}{\longrightarrow} D \qquad (8)$$

Wenn die Gesamtaktivität

$$A = a_X c_X + a_Y c_Y + \ldots\ldots + a_Z c_Z \qquad (2b)$$

sich in der Weise zusammensetzt, daß alle reversiblen Konformationen X, Y .... Z des Enzyms die gleichen spezifischen Aktivitäten besitzen

$$a_X = a_Y = \ldots\ldots\ldots = a_Z \, , \qquad (8a)$$

dann beginnt, wie Kurve 4a in Abb.4 zeigt, die Fortsetzung der Denaturierung mit der gleichen Aktivität, wie sie zur Zeit t_u unmittelbar vor der Unterbrechung der Reaktion vorlag. Für den Fall, daß

$$a_X > a_Y > \ldots\ldots\ldots > a_Z \, , \qquad (8b)$$

erhalten wir die Kurve 4b mit partieller Reaktivierung des Enzyms. Umgekehrt liefert die Fortsetzung der Denaturierung bei

$$a_X < a_Y < \ldots\ldots\ldots < a_Z \qquad (8c)$$

eine während der Unterbrechung weiter abgesunkene Aktivität. Alle diese Kurven 4a, 4b, 4c und 1 laufen parallel zueinander. WRIGHT und SCHOMAKER [10] fanden im Falle der Harnstoffdenaturierung des Diphtherie-Antitoxins einen Mechanismus nach dem Schema (8) mit zwei gleich aktiven Formen X und Y.

In Abb.5 ist der experimentelle Verlauf der thermischen Inaktivierung des Notatins wiedergegeben. Wir finden weder Regeneration der Aktivität, noch Änderung der Geschwindigkeit als Folge einer Unterbrechung der Denaturierung. Die Mechanismen mit reversiblen geschwindigkeitsbestimmenden Reaktionsschritten kommen daher nicht in Betracht. Der Prozess läuft nach Unterbrechung entsprechend der gestrichelten Kurve in Abb.4 weiter, so daß nach LAIDLERs Schema noch der zweite und dritte Reaktionstyp zu diskutieren sind.

Nach dem Ergebnis der Tab.2 bzw. der Abb.1, wo der gesamte Kurvenverlauf der Inaktivierung von der Gesamtkonzentration völlig unabhängig ist, ist der zweite Reaktionstyp, ein durch Reaktionsprodukte gehemmter Prozess nicht zu erwarten. Diese Frage läßt sich jedoch eindeutig klären, wenn man Inaktivierungskurven mit und ohne Zusatz von denaturiertem Enzym vergleicht [1].

Das ist in Abb.6 geschehen. Die Zugabe einer größeren Menge des denaturierten Enzyms hat auf den Kurvenverlauf keinerlei Einfluß im Sinne einer produktgehemmten Reaktion. Als restliche Möglichkeit bleibt nunmehr lediglich der dritte Fall übrig, der auf Uneinheitlichkeit des Enzyms zurückzuführen ist.

V. Uneinheitlichkeit von Enzymen

Gleichartige enzymatische Wirkung erfordert nicht notwendigerweise vollkommen identische Gesamtstrukturen der aktiven Moleküle. Auch unter den Glucoseoxydasen existieren mindestens zwei unterscheidbare Heterozyme. Neben der einen hier behandelten Glucoseoxydase Notatin aus Penicillium notatum gibt es im Aspergillus niger eine Glucoseoxydase, die sich im Spektrum der prosthetischen Gruppe [11], in der Aminosäurenanalyse und in den kinetischen Konstanten des MICHAELIS-MENTEN-Schemas [11] charakteristisch vom Notatin unterscheidet. Lange Zeit wurden beide Enzyme für identisch gehalten, weil ihre Unterschiede teilweise nur so geringfügig sind, daß sie erst in unmittelbaren Vergleichsmessungen deutlich werden. Die von uns früher beschriebene größere Abweichung der Molekulargewichte [12] ist offenbar nicht reel. Bei Wiederholung der früheren Messungen unter besseren experimentellen Bedingungen und ebenfalls auch bei den reinen Enzymen zeigen beide Glucoseoxydasen fast genau die gleichen Sedimentations- und Diffusionskonstanten [11].

Bei Mikroorganismen, wie z.B. bei den Schimmelpilzen, die sich durch ungeheure Mannigfaltigkeiten von Mutanten auszeichnen, ist die Entscheidung oft schwierig, ob genetische Multiplizitäten von Enzymen als Heterozyme oder als Isozyme [5] zu klassifizieren sind. Nur eine biologische Differenzierung der Kulturen kann das entscheiden. Das denaturierungskinetische Bild von Isozymen und Heterozymen ist jedenfalls gleich. Es liegt in beiden Fällen ein Gemisch genetisch verschiedener aktiver Komponenten X, Y ..... Z vor, und im Normalfall wird jede dieser Komponenten unabhängig von den übrigen mit der ihr arteigenen Geschwindigkeitskonstanten k_X, k_Y ....., inaktiviert werden, nämlich durch Umlagerung in das entsprechende denaturierte Produkt D_X, D_Y usw.

$$X \xrightarrow{k_X} D_X , \qquad (9)$$
$$Y \xrightarrow{k_Y} D_Y ,$$
$$Z \xrightarrow{k_Z} D_Z ,$$

Die Gesamtaktivität A fällt nach Gl. (9a) ab

$$A = A_{X_o} \cdot e^{-k_X t} + A_{Y_o} \cdot e^{-k_Y t} + \ldots\ldots + A_{Z_o} \cdot e^{-k_Z t} , \qquad (9a)$$

wobei die A_{X_o}, A_{Y_o} usw. die Anfangszusammensetzung des Enzyms in Aktivitätseinheiten angeben. Eine Unterbrechung der Denaturierung hat bei diesem Mechanismus 3a keinen Einfluß auf den weiteren Verlauf des Prozesses (gestrichelte Kurve in Abb.4).

Neben diesen beiden Arten genetischer Uneinheitlichkeit gleichwirkender Enzyme gibt es auch Uneinheitlichkeiten nicht genetischer Herkunft, wo eine gemeinsame native Form des Enzyms Ursprung anderer nichtnativer Enzymformen ist.

Wenn die native Form X des Enzyms direkt oder indirekt über eine irreversible Folge von aktiven Zwischenstufen Y,......Z in die inaktiven Formen D übergeht, wie im Reaktionsschema (10) formuliert ist,

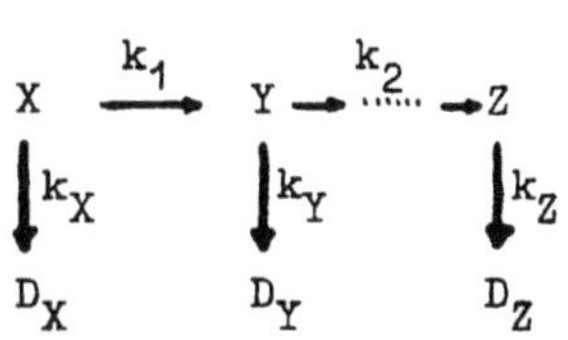

(10)

fällt die Aktivität entsprechend Gl.(10a) ab

$$A = B_1 \, e^{-\lambda_1 t} + B_2 e^{-\lambda_2 t} + \ldots\ldots \qquad (10a)$$

Hier besitzen die B_1 und λ_1 nicht mehr die einfache Bedeutung, wie die entsprechenden Parameter in Gl. (9a). Es sind komplexe Funktionen der Geschwindigkeitskonstanten des

Schemas (10) und der Anfangszusammensetzung, wie es in ähnlicher Weise auch bei der formal gleichen Aktivitätsgleichung des Schemas (8) der Fall ist.

Im Zusammenhang mit der Kinetik der Harnstoffdenaturierung von Diphtherie-Antitoxin diskutierten WRIGHT und SCHOMAKER [10] dieses Schema mit speziell zwei aktiven Stufen X und Y unter der Voraussetzung, daß zum Beginn der Denaturierung alles Enzym noch in X-Form vorliegt. COLEMAN und ELEY [13] fanden das gleiche Schema bei den gleichen Voraussetzungen zur Erklärung der Inaktivierungskinetik der Acetylcholinesterase bei thermischer Denaturierung gleich gut brauchbar, wie das Schema (9) für genetische Duplizität. Den Mechanismus (8) der Inaktivierung mit vorgelagertem Gleichgewicht hatten sie aus Gründen des Wertes der Aktivierungsentropie außer Betracht gelassen. Eine unterbrochene Denaturierung wäre ein eindeutigeres Kriterium dafür gewesen.

Falls ein Mechanismus wie im Schema (10) mit irreversiblen, nicht deaktivierenden Schritten in Frage kommt, dann ist es nicht unwahrscheinlich, daß eine solche künstliche Multiplizität der Formen auch bereits vor dem Denaturierungsversuch, bereits durch die Präparation auftreten kann. Selbst bei nur zwei aktiven Stufen X und Y, entsprechend einer aus zwei Gliedern bestehenden Reihe (10a), wird die quantitative Behandlung physikalisch uninteressant, da die unabhängigen Parameter λ_1, λ_2 und B_1 wegen der nicht definierten Anfangsbedingungen unendlich viele Kombinationen der drei Geschwindigkeitskonstanten zulassen. Um zu wissen, ob ein Enzym vollständig in seiner nativen Form vorliegt, ist es natürlich von größtem Interesse, schon qualitativ über künstliche Multiplizität entscheiden zu können.

Bei Notatin fanden wir die Inaktivierungskurven von Präparationsmethoden abhängig. Die Kurve a in Abb.7 entstand nach der üblichen Weise. Nach 300 Minuten, bei etwa 20 % Restaktivität wurde die Denaturierung abgebrochen und die Lösung nach verschiedener Weise aufgearbeitet. Lediglich vom unlöslichen denaturierten Proteinteil durch Zentrifugieren befreit mit anschließender kurzer Dialyse erhält man für den Rest des aktiven Enzyms die Inaktivierungskurve b. Diese Kurve unterscheidet sich nicht von der Kurve bei Fortsetzung der Denaturierung ohne präparative Behandlung. Außer der Zentrifugation und Dialyse war bei der Kurve c noch eine Ammonsulfatfällung vorausgegangen, bei Kurve d zusätzlich eine Gefriertrocknung. Das Ergebnis, daß die Kurve c von b abweicht und c und d gleich sind, war insofern überraschend, als nach der Erfahrung die Ammonsulfatfällung keinen Aktivitätsverlust ergibt, dagegen die Gefriertrocknung erhebliche Veränderungen des Notatins hervorruft. Meistens sinkt die Gesamtaktivität bei dieser Operation um 50 bis 70 % ab [14]. Fraktioniereffekte bei der Ammonsulfatfällung bis zu 90 % iger Sättigung dürften ausgeschlossen sein. Dagegen spielt hier vielleicht ein spezifischer Ionen-effekt eine Rolle. WALTER [14] beobachtete, daß z.B. Zusätze von Sulfationen eine sehr langsame Aktivitätssteigerung um etwa das 3 fache zur Folge hat.

Denaturierende und die Aktivität verändernde Einflüsse bieten ein recht kompliziertes Bild des Enzyms Notatin. Zur Zeit sind wir noch nicht in der Lage, die genetischen und künstlichen Multiplizitäten auseinanderzuhalten, da wir weder den nativen Anteil der Präparate kennen, noch das Verhalten der reinen nativen Komponente. Möglicherweise ist sogar der Begriff "nativ" exakt nur in der biologisch natürlichen Umgebung definiert.

Herrn Prof.Dr.G.V.SCHULZ danke ich für Anregung und Diskussionen zu dieser Arbeit. Ferner danke ich Frau H.BARBASCH und Frau M.v.BALUGYANSZKY für die Durchführung der Messungen und der Deutschen Forschungsgemeinschaft für Sachbeihilfen.

L I T E R A T U R

1) LAIDLER,K.J., The Chemical Kinetics of Enzyme Action. Oxford (1958) S.358 ff., vgl. auch WRIGHT,G.G., V.SCHOMAKER, J.Am.Chem.Soc. 70, 356 (1948)

2) OKUNUKI,K., Adv. Enzymol. 23, 29 (1961);
OKUNUKI,K., HAGIHARA,B., H.MATSUBARA, T.NAKAYAMA, J.Biochem. (Tokyo) 43, 453 (1956)

3) CASEY,E.J., K.J.LAIDLER, J.Am.Chem.Soc. 73, 1455 (1951)

4) TAMMANN,G., Z.phys.Chem. 18, 426 (1895)

5) WIELAND,Th., G.PFLEIDERER, Adv. Enzymol. 25, 329 (1963)

6) CAHN,R.D., N.O.KAPLAN, L.SEVINE, E.ZWILLING, Science 136, 962 (1962)

7) FRIC,I., II.Jena.Symp. (1963) über Physikalische Chemie biogener Makromoleküle

8) COULTHARD,C.E., R.MICHAELIS, W.F.SHORT, G.SYKES, G.E.H.SKRIMSHIRE, A.F.B.STANDFAST, J.H.BIRKINSHAW, H.RAISTRICK, Biochem.J. 39, 24 (1945)

9) VAN BRUGGEN,J.T., F.J.REITHEL, C.K.CAIN, P.A.KATZMANN, E.A.DOISY, J.Biol.Chem. 148, 365 (1943)

10) WRIGHT,G.G., V.SCHOMAKER, loc.cit.

11) WALTER,M., O.BODMANN, G.V.SCHULZ, noch nicht veröffentlicht.

12) BODMANN,O., D.KRANZ, G.V.SCHULZ, Makromol.Chem. 41, 225 (1960)
SCHULZ,G.V., D.LAUE, O.BODMANN, Makromol.Chem. 31, 75 (1959)

13) COLEMAN,M.H., D.D.ELEY, Biochim.Biophys.Acta 67, 646 (1963)

14) WALTER,M., Diplomarbeit Mainz 1961.

D I S K U S S I O N S B E M E R K U N G E N

SAUERBIER : Sie haben angenommen, daß es sich um die Inaktivierung zweier Fraktionen mit unterschiedlicher Empfindlichkeit handelt. Beide sollen nach einer Reaktion 1. Ordnung inaktiviert werden. Sie müßten demzufolge bei genügend starker Inaktivierung in einen Bereich kommen, in dem Sie in der halblogarithmischen Darstellung eine Gerade erhalten. Haben Sie bis in diesen Bereich messen können ?

BODMANN : Das ist richtig, wir haben es mit einer fraktionierten Denaturierung zu tun, und nach genügend langer Denaturierungszeit sollten wir eine Gerade mit der Neigung der Geschwindigkeitskonstanten für die Inaktivierung der stabileren Komponente erhalten. Unsere experimentelle Kurve, wie z.B. die in Abb.2, konnten wir durch die Funktion

$$A/Ao = 0{,}6 \cdot e^{-0{,}018\ t} + 0{,}4 \cdot e^{-0{,}0028\ t}$$

gut anpassen und deshalb auf die Annahme einer dritten Komponente verzichten. Gemessen haben wir bis 400 Minuten, wo die angenommene Grenzneigung bis auf etwa 2 % erreicht wird. Ob die Kurve tatsächlich mit dieser Neigung weiterläuft, oder ob noch weitere Komponenten vorhanden sind, das läßt sich aus unserem Experiment nicht entscheiden.

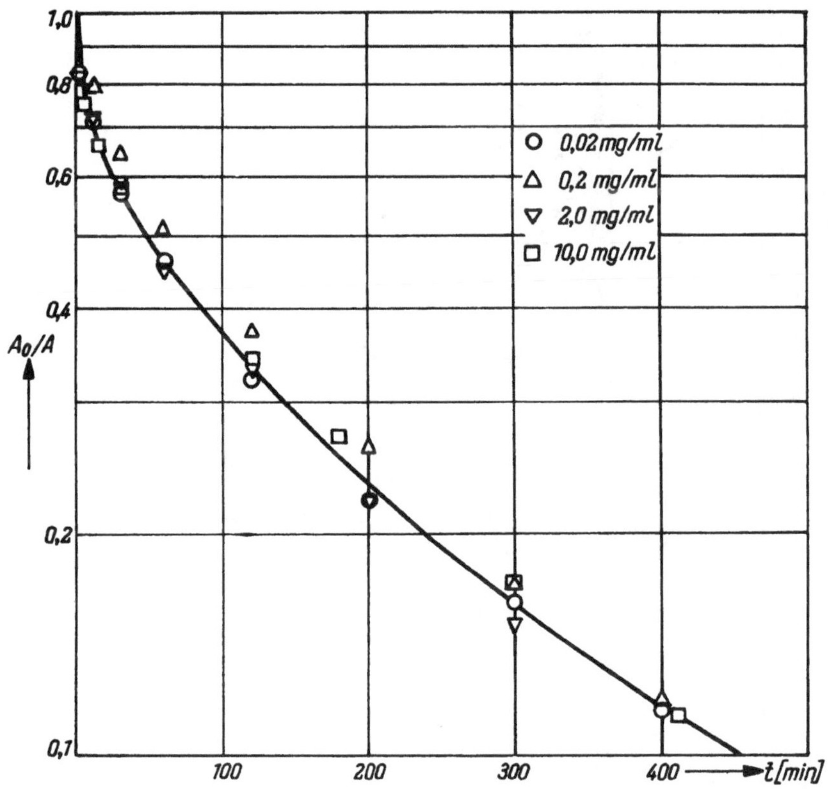

Abb.1 Zeitverlauf der thermischen Inaktivierung von Notatin (Präparat A). Denaturierung bei 60°C in 0,3m Phosphatpuffer, pH 5,6

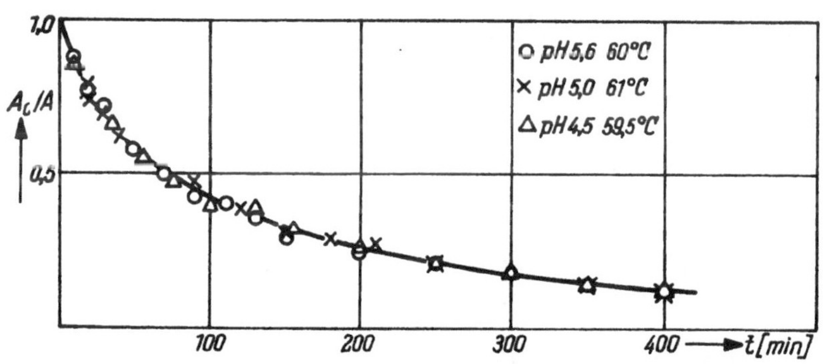

Abb.2 Übereinstimmende Inaktivierungskurven unter veränderten Bedingungen der Temperatur und des pH (Präparat 106 Fr.3)

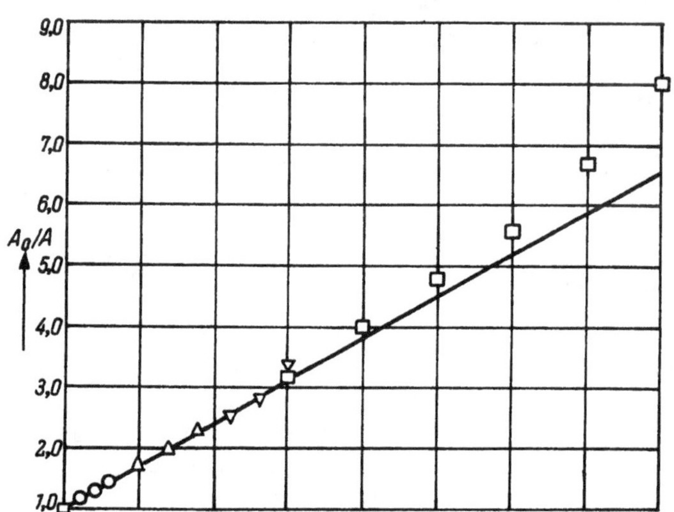

Abb.3 Inaktivierung von Notatin (Präparat 106 Fr.3) in der Auftragung einer Reaktion 2.Ordnung.

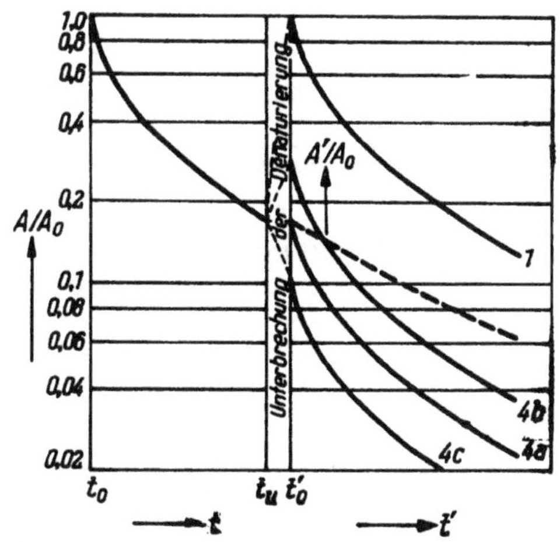

Abb.4 Inaktivierung als Gleichgewichtsreaktion oder bei vorgelagertem Gleichgewicht. Schema des Reaktionsverlaufes vor und nach Unterbrechung der Denaturierung.

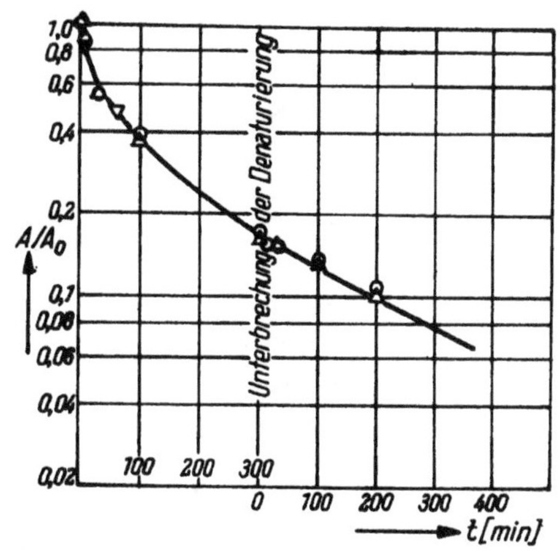

Abb.5 Experimenteller Reaktionsverlauf (Notatin Präp.A) vor und nach Unterbrechung der Denaturierung.

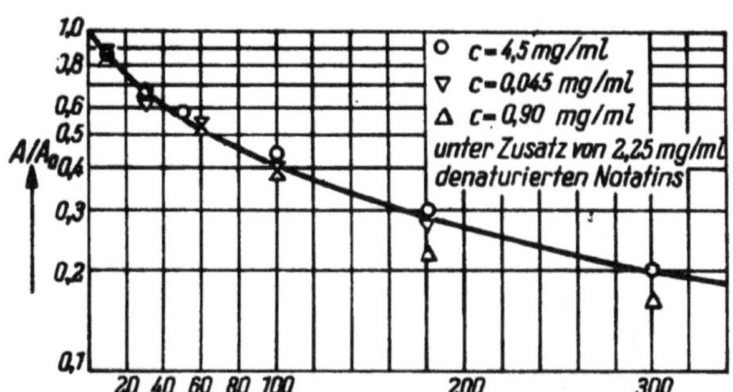

Abb.6 Inaktivierung mit und ohne Zusatz von denaturiertem Notatin (Präparat 106 Fr.3).

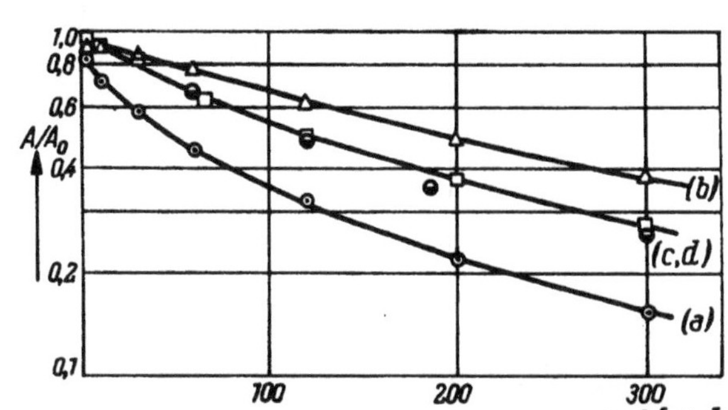

Abb.7 Inaktivierungskurven (Notatin Präparat A). Normaler Versuch (a) und Fortsetzung eines nach 300 Minuten abgebrochenen Versuches (b und c). b) nach Dialyse gegen Wasser, c) außerdem Ammonsulfatfällung (◘) und d) noch zusätzlich Gefriertrocknung (◉).

C Proteine

Ultrazentrifugenuntersuchungen von wärmedenaturierten Rinderserumalbuminen

R. MÜNCHMEYER, L. SEGAL und J. SEGAL

Institut für Tierernährungslehre und Institut für allgemeine Biologie der Humboldt-Universität, Berlin

S U M M A R Y

The sedimentation behaviour of two samples of crystalline bovine serum albumin (BSA) of different industrial provenance has been investigated by means of an analytical ultracentrifuge.

The BSA's were dissolved in MICHAELIS buffer of pH 8,6 and Heated to 65 - 90°C. The sedimentation coefficients determined at various BSA concentrations were used to estimate graphically the sedimentation constants, s_0, of the individual albumin components at c = 0.

In the solutions heated at 70°C, besides the native BSA component there is another one having a 4-fold faster migrating sedimentation maximum; after heating at 90°C, only one broad maximum being 6 to 7-fold faster remains.

The distribution functions of the sedimentation constants of native and denatured BSA have been calculated.

Die Wärmedenaturierung von gelöstem Rinderserumalbumin (RSA) wurde von L.SEGAL [2] bei verschiedenen Temperaturen mit Hilfe der Papierelektrophorese, durch Trübungsmessungen und durch Messungen der UV-Absorption untersucht. Bei über 60° erhitzten Lösungen zeigt sich im Elektropherogramm neben unverändertem Albumin eine zweite, langsamer laufende Fraktion, d.h. das Albumin ist positiver geworden, während die UV-Absorption derjenigen leicht angesäuerter Albuminlösungen gleicht. Durch Trübungsmessungen wurde festgestellt, daß bei Lösungen, die auf 61 - 65° erhitzt worden waren, eine Erhöhung des Dipolmomentes und eine Verschiebung des IEP nach der alkalischen Seite eintritt. Lösungen, die auf 80 - 85° erhitzt worden waren, zeigen im Elektropherogramm nur eine einzige im Globulinbereich laufende Fraktion. Die Fraktion des nativen Albumins ist verschwunden. Die Absorption ist im kurzwelligen UV (240 nm) erhöht, ebenso das Dipolmoment. Noch höhere Temperaturen bewirken eine Homogenisierung des Albumins, wobei das elektrophoretische Bild verwischt wird.

Um neben der Ladungsveränderung auch die Aggregation der Teilchen, also die Veränderung ihres Molekulargewichtes, zu verfolgen, wurden die einzelnen erhitzten RSA-Lösungen und das native Rinderserumalbumin in der Ultrazentrifuge untersucht.

E X P E R I M E N T E L L E S

Die Experimente wurden in einer luftgetriebenen PHYWE-Ultrazentrifuge bei 48 000 Upm und einer Temperatur von 20,0 - 22,5°C durchgeführt. Die 3 %igen Rinderserumalbuminlösungen in MICHAELIS-Puffer, pH 8,6, 0,015 m (2 kristallisierte RSA-Proben der Firmen Behringwerke und Armour Company) wurden 20' auf Temperaturen zwischen 65°C und 90°C erwärmt. Zur Untersuchung in der Ultrazentrifuge wurden die Lösungen auf Konzentrationen zwischen 0,4 % und 1,2 % Eiweiß mit MICHAELIS-Puffer verdünnt. Die Bereitung der Lösungen und die Untersuchungen in der Ultrazentrifuge erfolgten an demselben Tage. Die Sedimentation des nativen RSA, der auf 70°C und 90°C erwärmten Lösungen wurde wiederholt bei den angegebenen Konzentrationen durchgeführt. Die reziproken Werte der Sedimentationskoeffizienten wurden als Funktion der Konzentration auf unendliche Verdünnung extrapoliert.

E R G E B N I S S E

Abbildung 1 zeigt die schematische Darstellung der Gradientenkurve einer Lösung von nativem RSA.

Die Abbildung 2 stellt die Gradientenkurve der auf 70°C erhitzten RSA-Lösung dar. Daraus ist zu ersehen, daß neben nativem Albumin eine mit ihrem Maximum ca. 4 mal schneller laufende Fraktion enthalten ist. Die relativen Prozentgehalte der Komponenten betragen 30 % und 70 %.

Die auf 90°C erhitzten Proben ergaben eine sehr flache Gradientenkurve, deren Maximum eine Sedimentationskonstante von s_0 = 35 S besitzt.

In der Tab.1 sind die für die Maxima der Gradientenkurve ermittelten Sedimentationskonstanten aufgeführt.

Während sich in der Elektrophorese auf 60 - 65°C erhitzte Proben bereits in zwei Fraktionen aufteilen, bleibt die Molekelgröße unverändert, da die Ultrazentrifugenuntersuchung nur eine Komponente mit der Sedimentationskonstante des nativen Albumins zeigt. Es handelt sich aber bei dieser Denaturierungsstufe nur um eine Ladungsveränderung einen Teiles der Moleküle. Erst bei den auf 70°C erhitzten Lösungen treten sowohl in der Elektrophorese als auch in der Ultrazentrifuge der größte Teil (ca. 70 %) des RSA in einer Komponente auf, die sich von der des nativen RSA unterscheidet. Es ist eine eindeutige Vergrößerung der Teilchen eingetreten. Ähnliche Ergebnisse haben STOKROVA, BARTL und SPONAR [4)] bei ihren Untersuchungen erhalten.

Bei den auf 90°C erhitzten Lösungen sind sowohl in der Ultrazentrifuge als auch in der Elektrophorese die native Albuminkomponente und die bei 70°C auftretende Komponente nicht mehr zu finden. Die Gradientenkurve ist sehr breit, was auf eine hohe Polydispersität des denaturierten Albumins hinweist. Ein Einfluß der Diffusion kann bei den vorliegenden Sedimentationskonstanten und der Kürze der Untersuchung ausgeschlossen werden.

Die starke Verbreiterung der Gradientenkurven bei den erhitzten Lösungen ließ die Berechnung einer Sedimentationskonstantenverteilung wünschenswert erscheinen. SCHOLTAN [3)] fand bei Serumalbumin, das unter Umgehung der kristallinen Phase gewonnen worden war, eine Molekulargewichtsverteilung. Wir bedienten uns bei der Aufstellung der Sedimentationskonstantenverteilung der von SCHOLTAN [3)] modifizierten Methode. Zu diesem Zweck werden die Gradientenkurven unter Berücksichtigung des Verdünnungseffektes in Integralkurven umgerechnet, durch die der Konzentrationsverlauf in der Ultrazentrifugenzelle beschrieben wird (Abb.4).

Aus diesen Kurven kann die zeitliche Abhängigkeit der einzelnen Konzentrationsstufen vom Zellenort bestimmt werden.

Tabelle 1 Sedimentationskonstanten von nativen und wärmedenaturierten Rinderserumalbuminen
(gelöst in MICHAELIS-Puffer, pH-Wert 8,6 0,015 m).
Die Verteilungskurven der Sedimentationskonstanten wurden aus den Ergebnissen der Versuchsreihen 9, 12 und 15 berechnet.

Hersteller	Versuchsreihe Nr.	Erhitzungstemperatur	Erhitzungsdauer	$s_{20,H_2O} \cdot 10^{13}$ d.Kurvenmax. Komponente 1	Komponente 2	Komponente 3
Behring-Werke	1	-	nicht erhitzt (nativ)	4,9		
	2	-	"	5,4		
	3	75°	20'	5,2	23,9	
	4	"	"	4,6	17,2	
	5	"	"	5,2	22,1	
	6	90°	20'			31,5
Armour Co.	7	-	nicht	5,3		
	8	-	erhitzt	4,8		
	9	-	(nativ)	4,9		
	10	70°	20'	5,4	20,7	
	11	"	"	4,9	18,5	
	12	"	"	5,2	25,7	
	13	90°	20'			29,5
	14	"	"			30,5
	15	"	"			32,7

Aus der Neigung dieser Geraden (Abb.5) können ausreichend genau die Sedimentationskoeffizienten berechnet werden. Wenn die Werte der Sedimentationskoeffizienten der einzelnen Konzentrationsstufen für verschiedene Konzentrationen bestimmmt wurden, können die Sedimentationskoeffizienten oder ihre Kehrwerte als Funktion der Konzentration dargestellt und auf unendliche Verdünnung extrapoliert werden. Diese s_0-Werte geben die Verteilung der Sedimentationskonstanten wieder.

Aus der Abb.6 ist zu ersehen, daß schon das native Albumin eine gewisse Verteilung besitzt, also nicht einheitlich ist. Die dieser Verteilungskurve zugrunde liegenden Werte der Sedimentationskonstanten s_0 sind aus s_c-Werten für eine Konzentration c = 0,4 % nach der Gleichung $s_0 = s_0/1-k_s c$ mit einem Wert von 0,102 (100 ml/g) für k_s nach CHARLWOOD[1]) berechnet worden.

Die auf 70°C erhitzte RSA-Lösung zeigt neben einer geringen Menge unveränderten Albumins schon eine relativ breite Verteilung von s_0 = 2 S bis s_0 = 29 S.

In der auf 90°C erhitzten RSA-Lösung finden sich Teilchen mit einer Sedimentationskonstanten von 25 S bis hinauf zu 54 S. Die Wärmedenaturierung des Albumins bei 90°C führt somit zur Polydispersion.

Die Ergebnisse unserer Versuche machen es deutlich, daß die Sedimentationskonstante der Gradientenkurvenmaxima bei Untersuchungen über Denaturierungsvorgänge nur recht unzureichend die tatsächlichen Verhältnisse wiedergeben.

Erst die Bestimmung der Sedimentationskonstantenverteilungskurve läßt eine eindeutigere Aussage über die Massenverteilung der verschiedenen Makromoleküle zu.

Zusammenfassend möchten wir feststellen, daß bei der Erwärmung von Albuminlösungen bis zu einer Temperatur von 60° nur Ladungsveränderungen durch Aufbrechen eines Teiles der H-Brücken auftreten. Die Aggregation beginnt erst bei 70°, die durch die Erhöhung des Dipolmomentes der Albuminteilchen gefördert wird. Bei den auf 90° erhitzten Lösungen werden Seitenkettenbrücken aufgebrochen, die Tendenz zur regellosen Aggregation ist dadurch erhöht.

L I T E R A T U R

1) CHARLWOOD,R.A., Biochem.J. 56, 259 (1954)
2) SEGAL,L., Acta biol.med.germ. 10, 488 (1963)
3) SCHOLTAN,W., Makromolekulare Chem. 36, 162 (1960)
4) STOKROVA,S., P.BARTL, J.SPONAR, Collect. czechoslov.chem.Commun. 25, 1258 (1961)

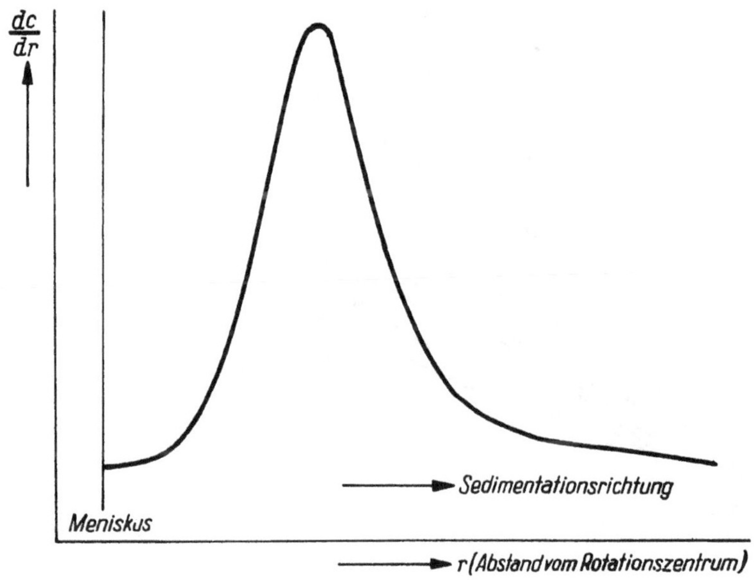

Abb.1 Gradientenkurve von nativem RSA; konz.0,9% gelöst in MICHAELIS-Puffer, pH-Wert 8,6, PHILPOT-Winkel 30°.

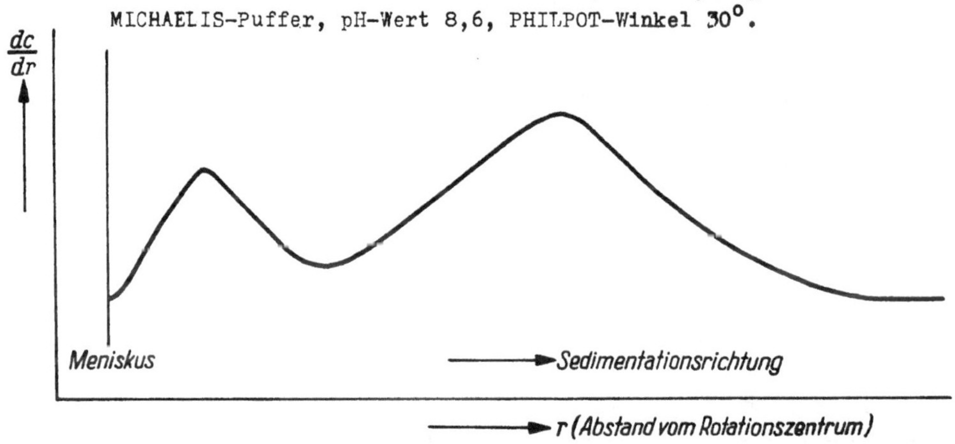

Abb.2 Gradientenkurve von RSA; konz. 0,9% gelöst in MICHAELIS-Puffer, pH-Wert 8,6, auf 70° erhitzt (20'),PHILPOT-Winkel 30°.

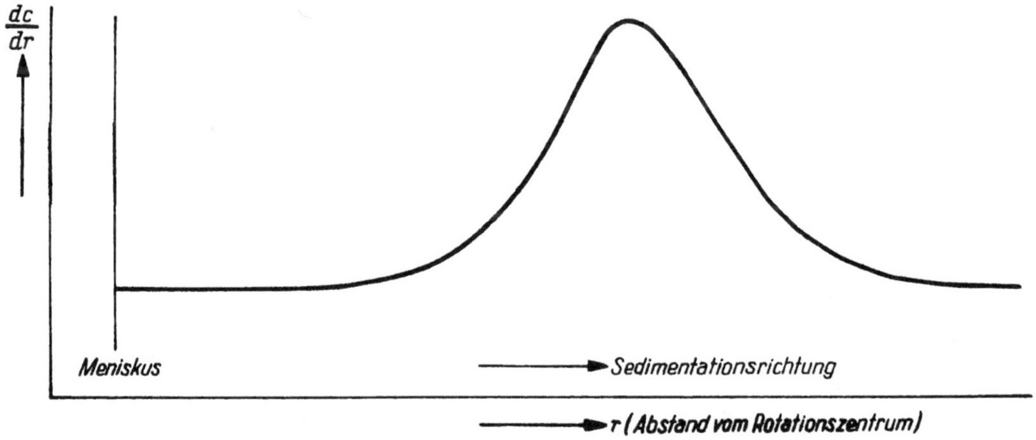

Abb.3 Gradientenkurve von RSA; konz. 0,9% gelöst in MICHAELIS-Puffer, pH-Wert 8,6, auf 90° erhitzt (20'), PHILPOT-Winkel 30°.

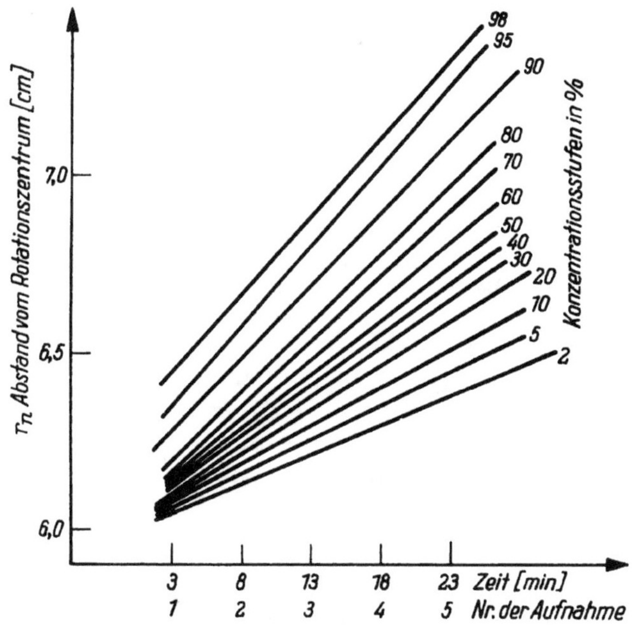

Abb.4 Zeitliche Abhängigkeit des Sedimentationsweges der verschiedenen Konzentrationsstufen von RSA, 20' auf 90°C erhitzt (Konzentration 0,6 %).

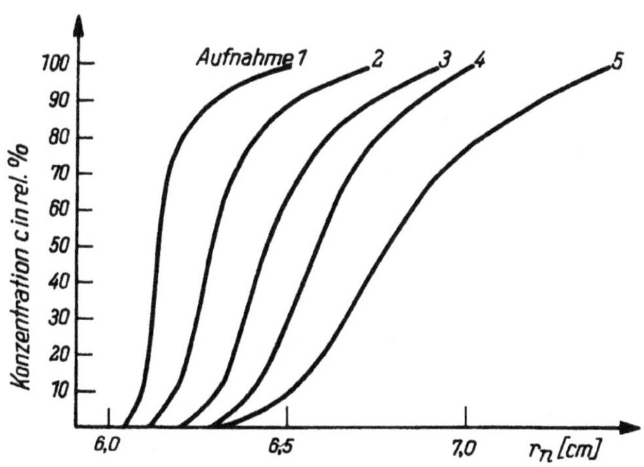

Abb.5 Zeitliche Abhängigkeit des Konzentrationsverlaufes in der Ultrazentrifugenzelle bei der Sedimentation von RSA (20' auf 90°C erhitzt, konz. 0,6%). (Zeitlicher Abstand der Aufnahmen: je 5 Min.) r_n: Abstand vom Rotationszentrum.

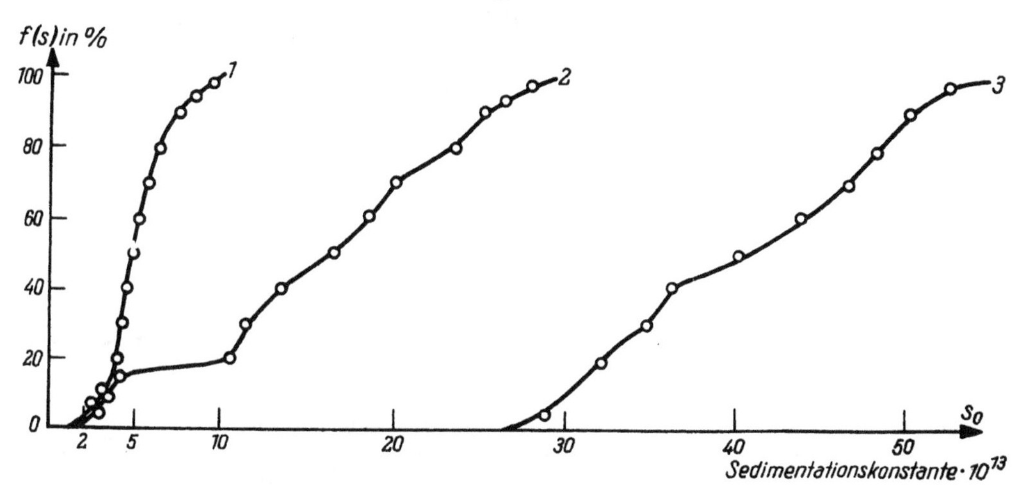

Abb.6 Verteilung der Sedimentationskonstanten für natives (1) für die bei 70°C (2) und bei 90°C (3) denaturierten RSA.

C Proteine

Zum Mechanismus der Inaktivierung von Proteinen durch Strahlungen

K. DOSE

Max-Planck-Institut für Biophysik, Frankfurt/M.

S U M M A R Y

In many cases disulfides are involved in stabilizing tertiary protein structures. UV-light of 2537 Å splits these linkages reductively and with high selection yielding thiols. Comparisons of dose-effect relationships between this photoreduction and the decrease in biological activity of various proteins have been presented. The close connection between disulfide rupture and inactivation of trypsin and ribonuclease has been demonstrated very recently. This paper deals with the inactivation of lysozyme and insulin. From the data presented here it is evident that at least 3 of 4 disulfides in lysozyme must be ruptured untill an inactivation can be observed. No disulfide seems to be crucial here. In the case of insulin, however, rupturing a single of 3 disulfides leads to an inactivation of the hormone. The UV-light action is compared with the X-ray action. The experimental data are discussed.

E I N L E I T U N G

Durch Bestrahlung von Proteinen wird eine Vielzahl chemischer und physikalischer Prozesse ausgelöst. Jedoch wissen wir noch wenig über den tatsächlichen Verlauf dieser Prozesse und den Mechanismus der durch sie direkt oder indirekt bewirkten Proteininaktivierung. Über einige Hypothesen zu dieser Frage wurde kürzlich zusammenfassend berichtet [1]. Die hier zu betrachtenden Prozesse lassen sich etwa wie folgt zusammenfassen : 1. Spezifische strahlenchemische Veränderungen von Aminosäureresten unmittelbar am Wirkungsort (aktives Zentrum, active site) des Proteins [2)-5)]. 2. Spezifische physikalische Veränderungen des Wirkungsortes selbst oder benachbarter Teile des Moleküls, die für den Aufbau des Wirkungszentrums mitbestimmend sind [6)-9)]. 3. Unspezifische Veränderungen (chemische und physikalische) der gesamten Proteinstruktur, durch die schließlich auch Struktur und Funktion des aktiven Zentrums beeinflußt werden [10)-14)]. Innerhalb eines Dosisbereiches von biologischem Interesse (vielleicht bis etwa 1 Mrad oder 100 Quanten UV-Licht pro Molekül) beobachten wir hauptsächlich Veränderungen der sekundären bzw. der tertiären Proteinstruktur und wesentlich weniger der primären Struktur, wenn man die Proteine unter einigermaßen physiologischen Konzentrationsverhältnissen bestrahlt [9].
Im Sinne der Nomenklatur von K.LINDERSTRÖM-LANG soll unter einer s e k u n d ä r e n S t r u k t u r eines Proteins hier die besondere Konformation der Polypeptidketten z.B. als α-Helices im Sinne von PAULING und COREY [15] verstanden werden. In Lösung

entspricht das Verhältnis zwischen den in regelmäßiger Konformation befindlichen Polypeptidketten und dem weitgehend zufälligungeordnetem Anteil einem echten thermischen Gleichgewicht. Die Gleichgewichtslage hängt dabei sehr stark vom Milieu und von der Art des betrachteten Eiweißmoleküls ab. Durch die t e r t i ä r e S t r u k t u r eines Proteins wird die räumliche Anordnung der zu α-Helices oder anderen konformeren Formen orientierten Peptidketten im Sinne der Arbeiten von KENDREW et al. [16] und PERUTZ et al.[17] festgelegt. Eine strenge Unterscheidung zwischen sekundärer und tertiärer Struktur wird jedoch nicht immer möglich sein, da auch die tertiäre Struktur durch die Konformation der Peptidketten bestimmt wird. Unter der p r i m ä r e n S t r u k t u r schließlich versteht man die Aminosäure-Sequenz wie sie erstmals für das Insulin-Molekül durch SANGER et al. [18],[19] gezeigt werden konnte.

Der Nachweis von Veränderungen der primären Struktur durch Bestrahlung von Proteinen in höherer Konzentration (1 % und mehr) oder in vivo mit Strahlendosen bis 1 Mrad oder bis 100 Quanten pro Mol UV-Licht ist äußerst schwierig, sofern man nach definierten molekularen Veränderungen sucht. Daher konzentriert sich das Interesse des Strahlenbiochemikers heute mehr auf Veränderungen der sekundären und tertiären Proteinstruktur. Die spezifische Konformation der Polypeptidketten in einem Proteinmolekül (sekundäre bzw. tertiäre Struktur) wird durch relativ schwache Bindungen stabilisiert. Erwähnt seien hier die Wasserstoffbrücken, die verschiedenen elektrostatischen Bindungen, die VAN-DER-VAALS-Bindungen und in vielen Fällen auch Disulfidbindungen. Die letztere Gruppe ist von besonderem Interesse, weil sie in vielen Fällen eine Schlüsselrolle bei der Stabilisierung einer spezifischen Struktur zu spielen scheint. Einige besonders charakteristische Eigenschaften seien hier erwähnt : Unter den uns heute als gesichert bekannten Bindungen, welche die tertiäre Struktur stabilisieren, sind nur die Disulfide covalent. Da diese nur zwischen zwei Cystein-Resten bestehen können, ist auch die Spezifität dieser Querverbindung zwischen zwei verschiedenen Polypeptidketten oder innerhalb einer einzigen Kette denkbar groß. Aber wie die übrigen vorher erwähnten Bindungen zählen auch die Disulfidbindungen zu den relativ schwachen Bindungen [20]-[22]. In Proteinen können sie bereits durch relativ kleine Energien von 20-30 kcal/Mol gebrochen werden. Es ist bemerkenswert, daß die Aktivierungsenergien für die thermische Proteindenaturierung die gleiche Größenordnung haben. Weiterhin sind die Disulfidgruppen chromophore Gruppen für UV-Licht um 2.500 Å. Ihre Extinktionskoeffizienten sind hier den aromatischen Aminosäuren vergleichbar.

AUGENSTEIN und GHIRON [24] untersuchten die Beziehung zwischen der UV-Inaktivierung von Trypsin und dem Bruch der Disulfidbrücken durch UV-Licht (2.537 Å). Ihre quantitativen Überlegungen zeigten, daß beim Trypsin etwa zwei der sechs Disulfidbrücken gebrochen werden müssen, bis eine irreversible Inaktivierung des Enzyms manifest wird. Von weiteren Daten, die von AUGENSTEIN et al. über die Beteiligung des Bruchs von Wasserstoffbrücken bei der Inaktivierung des Trypsins angegeben werden [25], läßt sich herleiten, daß die zur Inaktivierung führende Reaktionsfolge offenbar durch die Anregung einer oder zwei "essentieller" Disulfidbrücken eingeleitet wird. Ähnliche Versuche mit Ribonuclease zeigen, daß ein oder zwei Disulfidgruppen offenbar auch hier eine Schlüsselrolle spielen.

Von anderen Versuchen [27] über die Photochemie von Aminosäuren und Peptiden in wäßrigen Lösungen wissen wir, daß Cystein und Cystin stärker strahlenempfindlich sind als die anderen normalerweise in Proteinen auffindbaren Aminosäuren. In der Regel verhält sich dabei peptidgebundenes Cystein oder Cystin ganz ähnlich wie in freier Form. Jedoch ist bei peptidgebundenem Cystin die photochemische Reduktion zum Cystein, beim Cystein dagegen die reduktive Entschwefelung zum Alanin,die klar dominierende Photoreaktion. Im Gegensatz hierzu sind mehr Hauptreaktionen am photochemischen Abbau dieser Aminosäuren beteiligt, wenn sie in freier Form vorliegen. Wir konnten allerdings zeigen, daß z.B. die SH-Verbindung praktisch das ausschließliche Photoprodukt des Cystins ist, wenn man die UV-Bestrahlung unter reinem Stickstoff bei pH-Werten unter 5 durchführt. Auch ein denkbarer Disulfidaustausch dürfte unter diesen Bedingungen klein sein [28]. Die Quelle des Wasserstoffs bei dieser Photoreduktion ist noch unbekannt. Die Ergebnisse deuten jedoch daraufhin, daß der Wasserstoff schließlich aus dem Wasser der Lösung stammt [27]. Der Mechanismus dieser Reaktion könnte einem ähnlichen Prozess gleichen, der während der Photosynthese eine wesentliche Rolle zu spielen scheint (Reduktion der Thioctsäure [29]). Die Ergebnisse über die Disulfidspaltung sowohl in Trypsin und Ribonuclease als auch die hier beschriebenen Ergebnisse über die Bestrahlung von Lysozym und Insulin stehen in recht guter Übereinstimmung mit den photochemischen Versuchen an peptidgebundenem Cystin z.B. in Glutathion. Die initialen Quantenausbeuten für die Reduktion sind jedoch bei den Proteinen stets wesentlich höher.

Lysozym und Insulin wurden für diese Versuche gewählt, weil die primäre Struktur des Insulins vollständig [18],[19] und die primäre Struktur des Lysozyms fast vollständig bekannt sind [30]. Weiterhin vermeidet man Komplikationen wie Auto-Reaktivierungserscheinungen von geschädigten Enzymen, die man häufig beim Arbeiten mit Proteasen insbesondere Trypsin beobachtet.

M A T E R I A L U N D M E T H O D E N

M a t e r i a l :

Das Lysozym aus Hühnereiweiß war dreifach kristallisiert. Es wurde von der Sigma Chemical Company, St. Louis, Missouri, bezogen. Das kristallisierte Insulin erhielten wir von den Farbwerken Hoechst. Die Proteine wurden in dreifach destilliertem Wasser (0,8 mg/ml) gelöst. Die pH-Werte wurden durch Zufügen von einigen Tropfen 0,1 n HCl mit einer Glaselektrode eingestellt. Die pH-Werte nahmen in diesen schwach gepufferten Lösungen während der Bestrahlungen bis zu 0,5 pH-Einheiten ab.

B e s t i m m u n g d e r S H - u n d S-S - G r u p p e n :

Die SH-Gruppen wurden routinemäßig meist mit p-Chlormercuribenzoesäure (pCMB) nach BOYER et al. [31] bestimmt. Diese Ergebnisse wurden mit Jodessigsäure nach MOORE et al. [32] und mit N-Äthylmaleinimid nach ALEXANDER [33] kontrolliert. Die Zahl der S-S-Gruppen (in Röntgen-bestrahlten Proteinen) wurde in Anlehnung an MOORE et al. [32] bestimmt.

U V - B e s t r a h l u n g :

Die UV-Bestrahlungen wurden ausgeführt wie kürzlich beschrieben wurde [27]. Die Temperatur war 20°C. Die Quanten-Intensität an der Stelle der Quarz-Zelle (Oberfläche) war 0,4 µMole Quanten/min/cm^2. Dieser Wert beruht auf den Ergebnissen der Monochloressigsäure-Aktinometrie nach SMITH et al. [37] und MORING-CLAESSON [35].

R ö n t g e n b e s t r a h l u n g :

Die Quelle für die Röntgenbestrahlung war eine SIEMENS-REINIGER Dermopanröhre, die bei 45 kV, 25 m Amp und bei einer Filterung mit 200 µAl in Luft auf der Höhe der Oberfläche der bestrahlten Lösungen eine Dosisleistung von 2.10^5 rad/min hatte. Die Temperatur war 20°C. Die Lösungen wurden bei einer Schichtdicke von 2 mm in kleinen Al-Schalen bestrahlt. Die Al-Schalen befanden sich in besonderen Behältern, die eine Bestrahlung unter Stickstoff oder einem anderen Gas gestatteten [36].

T e s t d e r L y s o z y m - A k t i v i t ä t :

Im allgemeinen wurden die Teste in Anlehnung an CASELLI [37] durchgeführt. Der Abbau der Zellwände von Micrococcus lysodeicticus wurde bei pH 7,0 und 20°C mit Hilfe eines direkt anzeigenden Photometers (EPPENDORF - MEDEOR) gemessen. Die Strahlung dieses mit einer Hg-Hochdrucklampe ausgerüsteten Photometers wurde mit einem Spezialfilter für Nephelometrie (Transmissionsbereich 400 - 600 mμ) gefiltert.

E r g e b n i s s e u n d D i s k u s s i o n :

Abbildung 1 zeigt die Ergebnisse, welche nach UV-Bestrahlung von Lysozym bei pH 5 in sauerstoffreier, wäßriger Lösung erhalten wurden. Die Form der zwei Dosiseffektkurven für die Inaktivierung und Disulfidreduktion ist im Dosisbereich unter 100 Quanten pro Molekül sehr ähnlich. Eine genaue Angabe einer Strahlendosis ist jedoch sehr kritisch hier, weil die Extinktion der Lösung während der Bestrahlung ansteigt. Für die Berechnung der hier angegebenen Dosen wurde eine "Halbwertsextinktion" zugrunde gelegt. Der Unterschied der Dosiseffektkurven für die Enzym-Inaktivierung und die Disulfidspaltung für Strahlendosen oberhalb von 100 Quanten pro Molekül überrascht nicht; denn in einer vorangegangenen Arbeit konnte schon gezeigt werden, daß die photochemische Spaltung des peptidgebundenen Cystins nach Spaltung eines bestimmten Anteils des Disulfids zu einem Gleichgewichtszustand kommt [27].

Offenbar werden hier etwa 80-90 % des proteingebundenen Cystins, das an einer photochemischen Umsetzung teilneimmt, reduktiv gespalten, wobei pro Disulfid fast zwei SH-Gruppen gebildet werden. Man kann errechnen, daß wenigstens drei der vier Disulfide des Lysozyms gespalten werden müssen, ehe eine Inaktivierung des Enzyms deutlich wird. Die initialen Quantenausbeuten für die reduktive Spaltung des Cystins in Lysozym und auch in Insulin haben die gleiche Größenordnung. Ähnliche quantitative Ergebnisse liegen bekanntlich für die photochemische Spaltung der Disulfide in Trypsin [24] und Ribonuclease [26] vor. Die Dosiseffektkurven für die photochemische Spaltung der Disulfidbrücken in Lysozym und Insulin werden in Abb.2 einander gegenübergestellt. Man erkennt aber auch, daß die Verhältnisse komplizierter werden, wenn man die Dosis erhöht. Die Unterschiede zwischen den beiden Kurven sind offensichtlich durch die verschiedenen Strukturen dieser beiden Proteine gegeben. Obwohl die Aminosäurezusammensetzung und primäre Struktur dieser beiden Proteine klar verschieden sind (z.B. Insulin enthält kein Tryptophan), dürften für eine eingehende Diskussion noch mehr Kenntnisse über die sekundäre und tertiäre Struktur dieser Proteine erforderlich sein. Doch ist es schon jetzt klar, daß das Lysozym keine besonders ausgezeichnete Disulfidbrücke enthält, während im Insulin ein Cystinrest offensichtlich mit größerer Leichtigkeit gespalten werden kann als die übrigen zwei. Weiterhin scheint schon die Spaltung dieses Disulfid-Gliedes auszureichen, um eine Inaktivierung dieses Hormons einzuleiten [38].

Die Tabelle in der Abb.2 zeigt einige initiale Quantenausbeuten. Berechnet man diese Ausbeuten für die insgesamt pro Proteinmolekül absorbierte Energie, so erhält man Werte in der Größenordnung von einigen 10^{-2}. Auch der Wert für oxydiertes Glutathion hat diese Größenordnung (der Wert für freies Cystin liegt um etwa 20 % niedriger). Berücksichtigt man für die Quantenausbeuten jedoch nur das Licht, das von den Disulfid-Gliedern selbst absorbiert wird, so erhält man für die Proteine Werte von einigen 10^{-1}. Allerdings ist eine genaue Ermittlung der molekularen Extinktionskoeffizienten für proteingebundenes Cystin sehr schwierig. Bei pH 1,0 fanden wir einen Wert von 276 für freies Cystin und bei pH 3,2 281 für oxydiertes Glutathion (bei 2537 Å). Diese Werte sind etwa 2,5 mal so groß wie die von LYONS [39] gefundenen Werte und etwa 0,55 mal so groß wie die von OTEY und GREENSTEIN [40] für peptidgebundenes Cystin (meist Glycin-Cystin-Peptide) berechneten Werte. Innerhalb dieses noch relativ großen Bereichs der Unsicherheit stimmen unsere für Insulin und Lysozym gefundenen Werte jedoch mit den Angaben von AUGENSTEIN und GHIRON [24] überein.

Für die Vergrößerung der Quantenausbeuten bei der Spaltung von proteingebundenem Cystin gegenüber freiem oder in kleinen Peptiden gebundenem Cystin um das rund zwanzigfache gibt es mehrere Erklärungen :

a) Der tatsächliche molare Extinktionskoeffizient liegt wesentlich höher als hier angenommen wurde. Dagegen spricht aber auch, daß Aminosäuremischungen gleicher Zusammensetzung wie im Lysozym oder Insulin annähernd die gleiche Absorption bei 2537 Å haben wie die betreffenden Proteine.

b) Disulfidbrücken in Proteinen könnten teilweise besonders gespannt sein oder bei ihrer Anregung oder bei der Anregung benachbarter Moleküle gespannt werden, so daß ihre photochemische Spaltung auf diese Weise erleichtert wird [41].

c) Die Wahrscheinlichkeit eines Energie-Transfers von aromatischen Chromophoren zu den Cystinresten oder umgekehrt (Energie-Dissipation) ist nur hinreichend groß, wenn die betreffenden Gruppen bis auf ganz wenige Å einander nahe gebracht werden können (siehe auch [42] !). Dies könnte geschehen, wenn die beiden Reste einander in der Peptidkette benachbart sind oder wenn die beiden Reste sich z.B. in zwei übereinanderliegenden Windungen der α-Helix berühren können. Im Insulin grenzt ein Tyrosinrest unmittelbar an einen Cystinrest. Im Lysozym ist offenbar keine aromatische Aminosäure einem Cystin unmittelbar benachbart. Unglücklicherweise ist aber die sekundäre Struktur beider Proteine noch unbekannt, so daß die Möglichkeit eines Transfers von Anregungsenergie hier nicht ausgeschlossen werden kann.

d) Möglich ist auch, daß die absorbierte Energie von anderen Orten der Proteinmolekel durch eine Photoleitung zu den Disulfiden gelangen oder umgekehrt überschüssige Energie auf diese Weise von den Disulfiden abgegeben werden kann. Diese Möglichkeit wurde grundsätzlich durch verschiedene Autoren diskutiert ([25],[43],[44]). Auch die ESR-Daten von GORDY et al. [45] über den Übergang von C-Radikal-Signalen zu S-Radikal-Signalen beim "Altern" von in festem Zustand bestrahlten Proteinen müssen hier erwähnt werden. Spezielle Daten für die hier konkret gestellten Fragen sind aber leider noch nicht verfügbar.

e) Weiterhin muß die Möglichkeit eines Wasserstoff- bzw. Elektronentransfers von einer sterisch besonders günstig liegenden Gruppe zum Disulfid geprüft werden. Arbeiten, besonders über die Beteiligung des Tyrosins bei diesen Prozessen, sind im Fortschreiten [46].

Unter wohl definierten Bedingungen während der Bestrahlung scheint UV-Licht von 2537 Å ein recht brauchbares Werkzeug bei der Untersuchung einer Beziehung zwischen Proteininaktivierung und Disulfidbrücken-Spaltung zu sein. Nur die gleichzeitige Wirkung des Lichtes auf andere schwache Bindungen (Wasserstoffbrücken, elektrostatische und VAN-DER-VAALS-Bindungen) bringt natürlich einige Komplikationen, die sich aber durch geeignete Experimente und theoretische Betrachtungen [25] weitgehend eliminieren lassen. Auch bei der Einwirkung anderer reduzierender Agentien auf die Disulfidbrücken von Proteinen läßt sich eine Beeinflussung anderer Bindungen des Proteins zweifellos nie ganz vermeiden.

Natürlich entsteht auch jetzt die Frage, ob energiereiche Strahlen, wie z.B. Röntgenstrahlen, mit vergleichbarer oder vielleicht sogar größerer Selektivität Disulfidbrücken in Proteinen spalten können. Von radiationschemischen Versuchen mit

Aminosäuren und einfachen Peptiden wissen wir aber, daß Röntgenstrahlen bei weitem nicht so spezifisch wirken wie UV-Licht [47]. Auch bei der Bestrahlung von Proteinen gibt es dann natürlich einen Wettbewerb unter einer sehr großen Zahl von Reaktionen, an denen alle Aminosäuren in einem verhältnismäßig ähnlichem Umfang beteiligt sind. Auf die Frage des Angriffs auf die Disulfidbrücken selbst findet sich folgende Antwort [48] : Durch Bestrahlung unter Stickstoff und bei pH-Werten unterhalb 5,0 kann man für die Bestrahlung von reinem Cystin und oxydiertem Glutathion erreichen, daß die reduktive Spaltung der Disulfidbrücken die dominierende radiationschemische Reaktion ist. Ihr Anteil an dem gesamten Spektrum der radiationschemischen Reaktionen der Disulfide beträgt hier etwas mehr als 50 %. In Konkurrenz laufen jedoch andere Reaktionen ab, die z.B. bei Beteiligung der OH-Radikale durch Bildung einer neuen S-O-Bindung eingeleitet werden und sehr häufig schließlich mit einem oxydativen Abbau des Disulfids zu Cysteinsäure oder freiem Sulfat enden. Bezieht man sämtliche Bindungen ein, welche die Struktur einer Proteinmolekel stabilisieren, dann kann man jedoch grundsätzlich keine definitive Voraussage mehr darüber machen, ob es bei Bestrahlung eines Proteins irgendeine vorherrschende Reaktion geben wird, welche die Inaktivierung dieses Proteins bedingt. Die Gruppe von AUGENSTEIN [20)-22] und auch die Gruppe von HUTCHINSON [49] postulierte früher, daß eine bestimmte direkte Beziehung zwischen der Proteininaktivierung und einem Bruch von Disulfidbrücken auch bei Bestrahlung mit energiereichen Strahlungen bestehen sollte.

Abbildung 3 zeigt nun die vergleichbaren Ergebnisse für die Bestrahlung von Lysozym in wäßriger Lösung. Die Verläufe der beiden Dosiseffektkurven differieren zu stark, als daß irgendeine einfachere Beziehung zwischen beiden Kurvenverläufen augenscheinlich würde, selbst wenn wir versuchen, einen wesentlichen Teil der indirekten Wirkung durch Bestrahlung des kristallisierten Enzyms auszuschließen (siehe Abb.4 !). Im letzteren Fall wurde kristallisiertes Lysozym unter einer N_2-Atmosphäre bestrahlt. Aber kurze Zeit danach wurde das bestrahlte Protein in dreifach destilliertem Wasser, das auch Luftsauerstoff enthielt, gelöst. Daher sind diese Versuche nicht ganz mit den anderen Experimenten vergleichbar, bei denen Sauerstoff von allen intermediär gebildeten Radikalen oder angeregten Zuständen bis zu deren Absättigung bzw. Übergang in stabilere Systeme praktisch völlig ferngehalten werden konnte. Eine weitere hier nicht vergleichbare Eigenschaft ist die Änderung der Proteinstruktur (sekundäre bzw. tertiäre Struktur), die beim Übergang vom festen Zustand zu einer relativ verdünnten Lösung (ungefähr 1 mg/ml) sicherlich regelmäßig auftritt. Auch aus diesem Grund dürfte die direkte Wirkung der Strahlung auf das im Kristallverband befindliche Proteinmolekül anders sein als auf dasselbe, aber gelöste Molekül.

Die Inaktivierung des gelösten Lysozyms durch Röntgenstrahlen kann wie folgt diskutiert werden.

Verschiedene andere Autoren ([51)-53]), aber auch wir fanden, daß die Tyrosin-Komponente des Lysozyms - wie in vielen anderen Proteinen auch - relativ schnell während einer Bestrahlung verändert wird. Qualitativ gleicht dieser Effekt ganz der Bestrahlung von gelöstem Tyrosin selbst ([54)-58]). Andererseits wissen wir, daß Lysozym einen oder mehrere Tyrosinreste enthält, die auch sonst relativ leicht oxydierbar sind, z.B. durch Tyrosinase [50]. Gleichzeitig wird das Enzym inaktiviert. Daher hat man vermutet [50], daß mindestens einer dieser Tyrosinreste zum aktiven Zentrum des Proteins gehört. Aus diesem Grund scheint auch die Inaktivierung des Lysozyms durch Röntgenstrahlen ein gutes Beispiel für den Wettbewerb von mindestens zwei unabhängigen Mechanismen darzustellen, die beide zur Inaktivierung führen.

a) Der Angriff auf die allgemeine Struktur des Moleküls mit möglicher Beteiligung der Disulfid-Brücken.

b) Der unmittelbare Angriff auf das aktive Zentrum unter Oxydation von Tyrosinresten.

Die Inaktivierungsdosis (D = 37 % - Dosis) ist etwa 0,3 Mrad. Die entsprechende Dosis für die Spaltung der Disulfidbrücken ist jedoch hier nicht direkt bestimmbar, weil es in diesem Dosisbereich zur Ausbildung eines strahlenchemischen Gleichgewichts für die (teilweise reversible) Spaltung der Disulfide kommt. Da der Kurvenverlauf in Abb.3 die Annäherung an ein Gleichgewicht für Dosen oberhalb 0,3 Mrad deutlich erkennen läßt, erscheint es möglich, daß auch ein erheblicher (strahlenchemischer) Disulfidaustausch an der Inaktivierung des Lysozyms beteiligt sein kann. Weitere Versuche zur Klärung dieser Frage wurden bereits begonnen.

Der Autor möchte auch an dieser Stelle Fräulein Hildegard STAPELFELDT (Frankfurt/M.) und Herrn Cesare FIORE (Universität von Neapel) für ihre Mitarbeit sehr herzlich danken. Weiterhin dankt er den Herren Dr.L.AUGENSTEIN (Michigan State University), Dr.R.MASON (Imperial College, London) und Dr. F.HUTCHINSON (Yale University, New Haven, Vonn.) für sehr anregende Diskussionen der Ergebnisse.

Für die Überlassung des Insulins sei den Farbwerken HOECHST auch an dieser Stelle herzlich gedankt.

L I T E R A T U R

1) AUGENSTEIN,L., J.CARTER, D.NELSON, H.YOCKEY, Radiat.Res.Suppl. 2, 19 (1960)
2) GALSTON,A.W., R.S.BAKER, Science 109, 485 (1949)
3) WEIL,L., S.JAMES, A.BUCHERT, Arch.Biochem.Biophys. 46, 266 (1953)
4) KOSHLAND,D.E., W.RAY, M.J.ERWIN, Federat.Proc. 17, 1145 (1958)
5) RAY,W., D.E.KOSHLAND, Brockhaven Symp. in Biology, 13, 135 (1960)
6) AUGENSTEIN,L., R.RAY, J.Phys.Chem. 61, 1385 (1957)
7) AUGENSTEIN,L., Symp. on Information Theory in Biology, ed.H.YOCKEY, New York, Pergamon Press, 1956, p. 262
8) AUGENSTEIN,L., Science 129, 718 (1959)
9) DOSE,K., B.RAJEWSKY, Strahlentherap., Suppl. 49, 215 (1962)
DOSE,K., Habilitationsschrift, Frankfurt/M., 1962
10) McLAREN,A., P.FINKELSTEIN, J.Am.Chem.Soc. 72, 5423 (1950)
11) McLAREN,A., P.FINKELSTEIN, J.Am.Chem.Soc. 72, 5423 (1950)
12) PLATZMAN,R., J.FRANK, Symp. on Information Theory in Biology, ed.H.YOCKEY, New York, Pergamon Press, 1956, p. 262
13) ARNOW,L.E., Physiol.Rev. 16, 671 (1936)
14) FRICKE,H., Cold Spring Harbor Symp. Quant.Biol. 6, 164 (1938)
15) PAULING,L., B.COREY, Proc.Nat.Acad.Sci. (USA) 39, 255 (1953)
16) KENDREW,J.C., H.C.WATSON, B.E.STRANDBERG, R.E.DICKERSON,D.C.PHILLIPS, V.C.SHORE, Nature [London] 190, 663 (1961)
17) PERUTZ,M.F., M.G.ROSSMANN, A.F.CULLIS, H.MUIRHEAD, G.WILL, A.C.T.NORTH, Nature [London] 185, 416 (1960)
18) SANGER,F., E.O.P.THOMPSON, Biochem.J. 53, 353, 306 (1953)
19) RYLE,A.P., F.SANGER, L.F.SMITH, R.KITAI, Biochem. J. 60,541 (1955)
20) AUGENSTEIN,L., R.RAY, J.Phys. Chem. 61, 1385 (1957)
21) AUGENSTEIN,L., Symp. on Information Theory in Biology, H.YOCKEY, ed., Pergamon Press, New York, p. 287 (1956)
22) AUGENSTEIN,L., Information Theory in Biology, H.QUASTLER, ed., University of Illinois Press, Urbana, p. 119 (1953)

23) STEARN,A., Advances in Enzymol., 9, 25 (1949)
24) AUGENSTEIN,L., C.A.GHIRON, Proc.Nat.Acad.Sci. (USA), 47, 1530 (1961)
25) AUGENSTEIN,L., C.A.GHIRON, K.L.GRIST, R.MASON, Proc.Nat.Acad.Sci. (USA), 47, 1733 (1961)
26) AUGENSTEIN,L., persönliche Mitteilung
27) DOSE,K., B.RAJEWSKY, Photochemistry and Photobiology, 2, 181 (1962)
28) KLOTZ,I., J.AYERS, J.HO, M.HOROWITZ, R.HEINEY, J.Am.Chem.Soc. 80, 2132 (1958)
29) CALVIN,M., Angew.Chem. 68, 253 (1956)
30) JOLLES,J., M.P.JOLLES, C.R.hebd. Seances Acad.Sci., t. 253, 2773 (1961)
31) BOYER,P.D., H.A.LARDY, P.H.PHILLIPS, J.Biol.Chem. 146, 673 (1942)
32) MOORE,S., R.D.COLE, H.G.GUNDLACH, W.H.STEIN, 8.Symp., Internat.Kongr.Biochem., Wien (1958)
33) ALEXANDER,N.M., Anal.Chem. 30, 1292 (1958)
34) SMITH,R.N., P.A.LEIGHTON, W.G.LEIGHTON, J.Am.Chem.Soc. 61, 2299 (1939)
35) MORING-CLAESSON,I., Ark.Kemi 10, 21 (1956)
36) PETER,G., Doktordissertation, Frankfurt/M., 1960
37) CASELLI,P., Boll.Soc. ital.Biol. sperim. 22, 57 (1946)
38) DOSE,K., in Vorbereitung
39) LYONS,W., Nature [London] 162, 1004 (1948)
40) OTEY,M., J.GREENSTEIN, Arch.Biochem.Biophys. 53, 501 (1954)
41) LUMRY,R., H.EYRING, J.Phys.Chem. 58, 110 (1954)
42) FOERSTER,T., Radiat.Res.,Suppl. 2, 326 (1960)
43) CARDEW,M., D.ELEY, Disc.Faraday Soc. 27, 115 (1959)
44) EVANS,M., J.GERGELEY, Biochim. biophysica Acta 3, 188 (1949)
45) GORDY,W., Radiat.Res.,Suppl. 1, 491 (1959)
46) AUGENSTEIN,L., persönliche Mitteilung
47) SWALLOW,A.J., Radiationchemistry of Organic Compounds, Pergamon Press, 1960,p.200
48) DOSE,K., in Vorbereitung
49) RAY,D.K., F.HUTCHINSON, H.J.MOROWITZ, Nature [London] 186, 312 (1960)
50) KERTESZ,D., P.CASELLI, Bull.Soc.Chim.biol. 32, 583 (1950)
51) BARRON,E.S.G., Ann.New York Acad.Sci. 59, 574 (1955)
52) BARRON,E.S.G., J.AMBROSE, P.JOHNSON, Radiat.Res. 2, 180 (1958)
53) SHALEK,R.J., T.L.GILEPSIE, Radiat.Res. 9, 180 (1958)
54) BARRON,E.S.G., P.FINKELSTEIN, Arch.Biochem.Biophys. 41, 212 (1952)
55) NOSWORTH,J., C.B.ALLSOPP, J.Colloid Sci. 11, 565 (1956)
56) HIRSCH,H.M., Radiat.Res. 2, 9 (1956)
57) STENSTRÖM,W., Radiology 13, 437 (1929)
58) STENSTRÖM,W., A.LOHMANN, J.Biol.Chem. 79, 673 (1928)

D I S K U S S I O N S B E M E R K U N G E N

BERG : Weshalb wird bei Insulin mit 254 nm nur 1 -S-S-Brücke gebrochen ? (Energiedissipation unterschiedlich !) Als Lösungsreaktion eines Chromophors ist das ungewöhnlich.

DOSE : Es ist vorläufig noch nicht klar, ob die höheren Quantenausbeuten für die photochemische Spaltung der Disulfide in Proteinen auf eine hier günstigere Energiedissipation (Bindungsenergie des Disulfids wesentlich kleiner als die Quantenenergie der Strahlung) oder auf günstigere Möglichkeiten für eine Energieleitung oder Energieübertragung von anderen Chromophoren zu den Disulfidbrücken zurückgeführt werden müssen.

Zu dieser Frage laufen bereits weitere Untersuchungen. Ebenso beschäftigt uns die Frage, ob eine bestimmte Disulfidgruppe des Insulins gegenüber den anderen in irgendeiner Weise bevorzugt ist.

LASURKIN : Vielleicht liegt die Ursache des Effektes in verschiedenen Spannungen bei verschiedenen -S-S-Brücken ?

DOSE : Ich glaube, daß dies die einfachste und für einen Chemiker die sympatischste Erklärung ist. Aber wir müssen noch ausscheiden, daß nicht noch andere Prozesse (Energiedissipation, Energieleitung, Energieübertragung, Käfigeffekt usw.) hier wesentlich mitbeteiligt sind.

BEHLKE : Sie erwähnten die leichte Oxydierbarkeit und die benachbarte Stellung von Tyrosin- und Cysteinresten. Ist es möglich, daß sich bei der Oxydation eine R-S-O-R Bindung ausbildet oder muß eine Chinonstruktur bei der Tyrosinoxydation angenommen werden ?

DOSE : Ich halte dies für eine wichtige Frage und darf dazu bemerken, daß sich z.Z. L.AUGENSTEIN's Gruppe (East Lansing, Mich.) hiermit beschäftigt. Gesicherte Ergebnisse sind mir jedoch leider noch nicht bekannt.

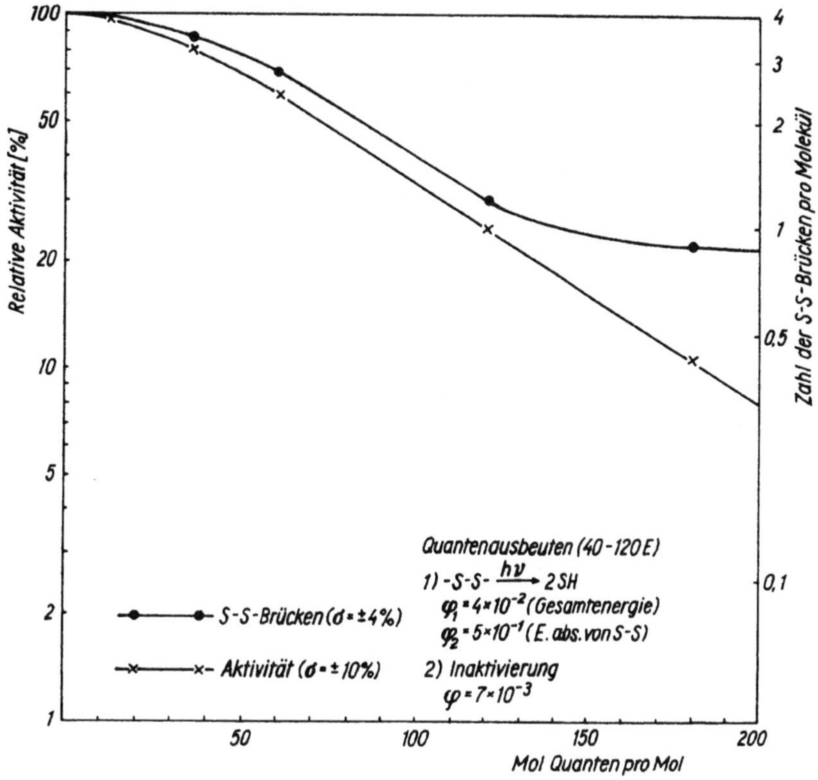

Abb.1 Abhängigkeit der Lysozym-Inaktivierung und der Disulfid-spaltung von der Strahlendosis: UV-Licht; λ = 2537 Å; Protein-konzentration 0,8 mg/ml; pH 5,0.

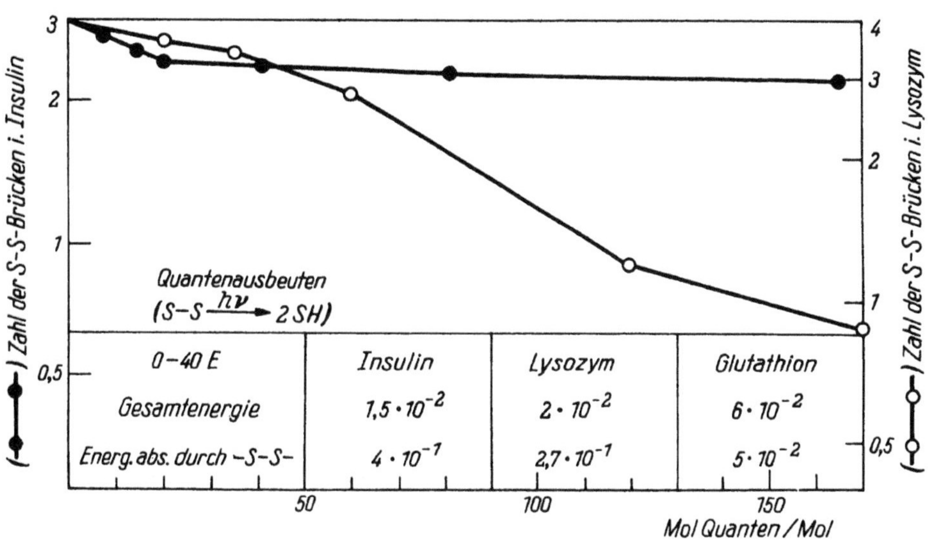

Abb.2 Vergleich der Wirkung des UV-Lichtes (2537 Å) auf den Bruch der Disulfid-Brücken in Insulin (schwarze Kreise) und Lysozym (weiße Kreise): Proteinkonzentration 0,8 mg/ML; pH 2,0 (Insulin und pH 5,0 Lysozym).

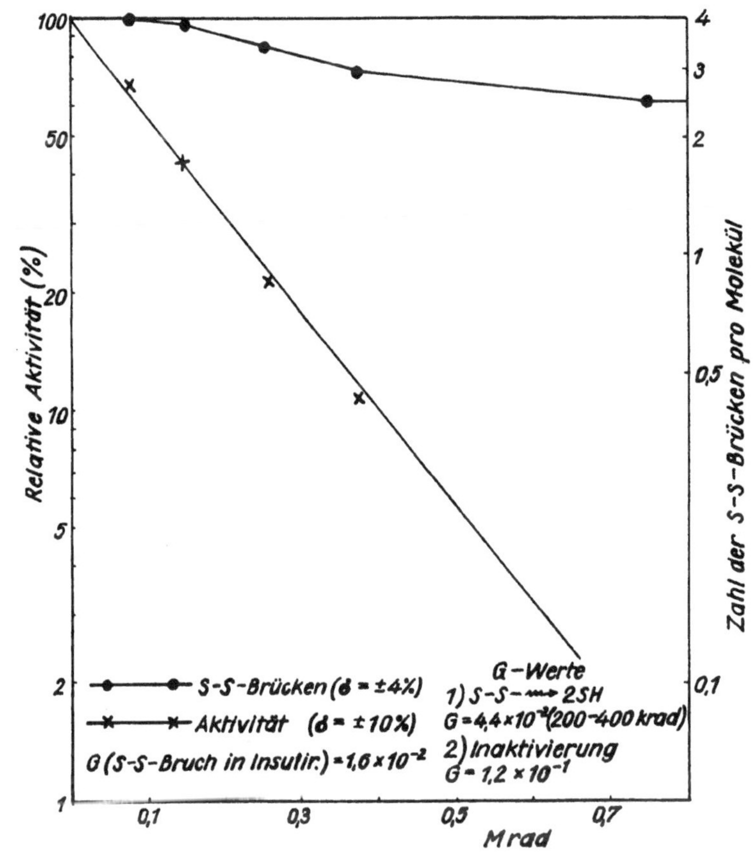

Abb.3 Abhängigkeit der Lysozym-Inaktivierung (Kreuze) und des Disulfid-Bruchs (schwarze Kreise) von der Röntgenstrahlendosis: 45 kV; 25 m Amp. 200 µAl-Filter; Proteinkonzentration 0,8 mg/ml; pH 5,0.

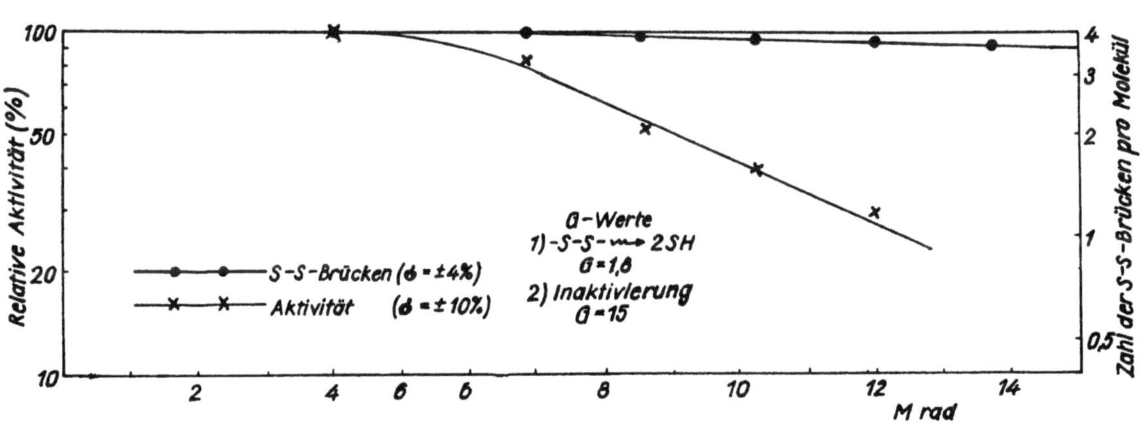

Abb.4 Abhängigkeit der Inaktivierung (Kreuze) und des Disulfidbruchs (schwarze Kreise) von der Röntgenstrahlendosis. Bestrahlung von kristallisiertem Lysozym.

C Proteine

Polarographic Studies of Proteins and their Degradation Products. VI. The Catalytic Effect of Cysteine on the Acid Denaturation of Albumin

O. H. MÜLLER

Department of Physiology, State University, Syracuse, New York

S U M M A R Y

Crystalline bovine albumin was exposed to dilute solutions of hydrochloric acid in the presence of different amounts of cysteine and then subjected to several polarographic tests. It was found that cysteine catalyzed the denaturation of the protein in hydrochloric acid which resulted in a marked increase in the second protein wave in TriCo buffer. This effect could also be demonstrated in DiCo buffer by comparison of the alkaline denaturation of untreated and "acid-cysteine exposed" protein after extensive dialysis of the latter which removed cysteine and hydrochloric acid. These results were compared to the behavior of a less well purified sample of bovine albumin "powder" which, upon the addition of cysteine, produced a marked increase in the second protein wave in a TriCo buffer (in which cysteine itself is quite inactive) even when the cysteine was added in neutral solution.

I N T R O D U C T I O N

It is known that dilute solutions of cysteine give rise to a catalytic polarographic wave with a single peak in divalent cobalt (DiCo) buffer, while no such effect is observed in trivalent cobalt (TriCo) buffer [1]. It is also known that proteins produce the so-called protein double-wave (two-peaked catalytic wave) in either di- or trivalent cobalt buffer. The fact that the second protein peak occurs at about the same potential as the cysteine peak in DiCo buffer and that its height correlates well with the cysteine content of the protein has led to the assumption that sulfhydryl groups which are still part of the protein molecule are responsible for this peak [1]. The height of this peak undergoes considerable changes when the protein is subjected to alkaline denaturation, which suggests that more sulfhydryl groups of the protein molecule are made reactive [2]. In such denaturation studies it has been customary to neglect the effect of any free cysteine that may have been split off from the protein because similarly large changes in the protein double wave were found in either DiCo or TriCo buffers. However, the possibility exists that occasionally sufficient cysteine would be present to affect the d i g e s t t e s t [3] which is carried out in DiCo buffer. Hence cysteine was deliberately added to different protein solutions and the results studied in divalent and trivalent cobalt buffers. The results were quite unexpected and first led us to suggest that a complex was formed between the cysteine and the protein or some split product [4].

Further investigation revealed, however, that the results were caused by the denaturation of proteins calatyzed by cysteine in the presence of acid or of some yet unknown material in an impure serum albumin preparation. The present report is concerned with these findings and with some additional observations on the effect of cysteine additions on protein double-waves in DiCo buffer.

R E S U L T S

E f f e c t   o f   a d d e d   c y s t e i n e   o n   t h e   p r o t e i n   d o u b l e - w a v e   i n   T r i C o   b u f f e r s

Since the cysteine wave is superimposed on the second protein wave in DiCo buffers which thus becomes unusually large and difficult to measure, we set out to analyze protein-cysteine mixtures in TriCo buffer. We knew that free cysteine in the concentrations used does not produce a catalytic wave in such a buffer and we expected therefore no changes in the polarograms resulting from the addition of cysteine. To our surprise, however, there was a change in the second protein wave and this change was much more marked with a sample of relatively impure bovine albumin powder (PBA) than with some crystalline bovine albumin (CBA).(The samples had been obtained from the Armour Laboratories, Chicago, Ill., and were labelled "Crystallized Bovine Plasma Albumin" Lot 43, and "Bovine Albumin Powder (Fraction V from Bovine Plasma), Control No. C1405.) Two typical polarograms which illustrate these findings are represented in Fig.1. It will become clear from the further discussion that the composition of the solutions used was fortuitous for several reasons.

In this experiment, 1.0 ml. aliquots of a 5 g./100 ml. solution of albumin in 0,9 g./100 ml. NaCl were mixed with x ml. of 0.01 M acid cysteine solution and (1-x) ml. 0.01 N HCl, and with 1.0 ml. 0.01 N NaOH solution. Within 30 minutes after mixing, 0.05 ml. samples of the mixtures were added to 10.0 ml. of a s p e c i a l trivalent cobalt buffer (16.0 ml. of 0.01 M hexamminecobaltic chloride + 10.0 ml. 1 N NH_4Cl + 64.0 ml. distilled water + 10.0 ml. 1 N NH_4OH) and polarographed immediately. This composition of the buffer was chosen in order to make possible a comparison with standard divalent cobalt buffer in which the cobalt concentration is 1.6×10^{-3} M. The cysteine solution had been prepared in 0.1 N HCl to ensure its stability. By accident, however, only 1/10 of the acid had been neutralized in the above mixtures, so that increased cysteine concentration was accompanied by increasing acidity.

Notice that the cysteine effect is very small in the lower polarogram of Fig.1 which was obtained with CBA, while it is very marked when PBA was used (upper polarogram). In the latter case the second protein wave grows larger than the first protein wave which remains essentially unaltered. The minimum following the second protein wave is also altered. This means that the third protein wave (so-called "pre-natrium" wave) is also increased with additions of cysteine. Since this wave depends on the presence of nitrogenous groups rather than sulfhydryl groups [5] it will not be considered further in this report.

The first two curves on each polarogram reestablish the fact that even with the greatest amount of cysteine added in this experiment, no measurable wave is found in this TriCo buffer in the absence of albumin. This led us to the conclusion that the change effected by the external cysteine must have been in the intramolecular cystine or cysteine of the protein; the more vigorous reaction of the PBA we ascribed to impurities.

A more systematic research was then started in which the effect of different cysteine concentrations and different hydrochloric acid concentrations and their combination was investigated as a function of time. We also varied the ammonium hydroxide concentration in the 0.001 M TriCo buffer from the standard value [1] of 0.8 M to 0.4 M and 0.1 M and found the intermediate concentration most suited for showing the effects. It soon became clear that the observation of the cysteine effect was made possible by the fortuitous admixture of HCl which came about by the accidental failure to neutralize all the acid of the cysteine solutions.

Apparently the effect is caused largely by acid denaturation that is catalyzed further by the presence of cysteine. This is illustrated by means of the four polarograms reproduced in Figures 2 and 3. The solutions were prepared as follows : To 1.0 ml. of 5 g./100 ml. CBA in distilled water was added 1.0 ml. 10^{-2} M cysteine solution (cysteine hydrochloride made up in distilled water) plus x ml. 0.1 N HCl and (1-x) ml. of distilled water. As control served a solution containing no cysteine, viz., 1.0 ml CBA + 1.0 ml. HCl + 1.0 ml. H_2O. The pH of the mixtures was measured with a BECKMAN pH meter. It was 4.1 for the solution containing the least acid (0.01 ml.) and decreased to 3.9, 3.7, 3.0, and 1.9 as more acid was added to the cysteine-containing solutions. The pH of the cysteine-free solution was 2.0. For the polarographic analyses 0.05 ml. samples of these mixtures were added to 10.0 ml. of the i n t e r m e d i a t e TriCo buffer (1×10^{-3} M hexamminecobaltic chloride in 0.1 N NH_4Cl and 0.4 N NH_4OH). The four polarograms of Fig.2 and 3 were obtained at different times after mixing. The results of the first day are shown in Fig. 2; the upper polarogram was obtained within 30 minutes after mixing, the lower polarogram after 2-3 hours' standing at 23°C. Figure 3 shows the results of the second day, the top polarogram after additional storage for 18 hours in a refrigerator, or about 21 hours after mixing, and the bottom polarogram after 5 more hours' standing at room temperature (23°C.).

Notice that the most acid cysteine-containing solution (pH 1.9) shows an almost immediate effect on the second protein wave which seems to have reached a limit after standing overnight in the refrigerator. The next most acid cysteine-containing solution (pH 3.0) has not yet reached this limit after 5 more hours at room temperature, but it, too, shows a definite, although small effect soon after mixing. In contrast to these two solutions are the two last acid solutions with pH 3.9 and 3.7 in which there is no apparent effect although they contained the same amount of cysteine as the others. Similarly, no effect is found in the control solution which contained no cysteine but was quite acid (pH 2.0).

Results such as these at first led us to the belief that the cysteine was actually incorporated or bound into the protein molecule so that it became polarographically active in TriCo buffer [4]. The polarographic differences between cysteine-free and cysteine-exposed acid protein solutions persisted even after extensive dialysis of the solutions against water. Standing for several hours in the alkaline cobalt buffer did not significantly alter the situation. We therefore anticipated differences in the cysteine content of the cysteine-free and cysteine-exposed dialyzed material. However, hydrolysis of these proteins in 5 N HCl for 8 hours and subsequent polarographic analysis of the hydrolysates did not reveal any significant difference in their cysteine content. This indicated that the observed polarographic changes reflected intramolecular rearrangements brought about by the presence of cysteine and acid.

Further studies demonstrated that, given sufficient time, the acid CBA solution will show increases in the second protein wave even in the absence of cysteine, but the presence of cysteine will always produce a more rapid increase. As an example, some results are shown graphically in Fig.4. Here the wave height (in microamps./mm.^{2}) at the peak of the second portion of the double-wave is plotted as a function of time. The two experiments of Fig.4 were carried out with different concentrations of NaCl as added electrolyte, since YANG and FOSTER [6] had found marked changes in the intrinsic viscosity and optical rotation of bovine plasma albumin associated with acid binding when different amounts of sodium chloride had been added to the solutions. Both these parameters reached a maximum at pH 2.2 - 2.7; it was highest in the absence of added electrolyte, was markedly reduced in the presence of 0.02 N NaCl, and was almost abolished in the presence of 0.10 N NaCl. The effects were essentially instantaneous and were explained as caused by a swelling or expansion of the molecule, a reversible phenomenon which in the case of a denaturation is supposed to be followed by irreversible steps to an unfolded protein and then to an aggregated protein [6)].

Our experiment was not an exact parallel of YANG and FOSTER's [6)] but should reflect this expansion of the molecule if it had any bearing on our reaction. Solutions of 1.67 g./100 ml. CBA in 0.033 N HCl and containing from 0 to 0.0033 M cysteine were made up in either 0.0007 N or 0.0333 N NaCl. These solutions were kept at 25°C. in a water bath. After known intervals of time, 0.05 ml. samples of the mixtures were added to 10.0 ml. of intermediate TriCo buffer (see above). In Fig. 4 are plotted corresponding data for zero, 0.00067, 0.00133, and 0.00266 M cysteine solutions. It is obvious that the polarographic effects take a long time to develop and that the influence of NaCl is negligible when the cysteine concentration is relatively high. On the other hand, when there is little or no cysteine present, the addition of NaCl seems to speed up the alteration in the protein molecule brought about by the acid. One may thus conclude that the swelling or expansion of the albumin molecule, which undoubtedly still exists under our experimental conditions is not directly related to the rate of denaturation that is measured by the polarographic reaction. On the contrary, there may be an inverse relationship since there is more denaturation at higher ionic strength, where YANG and FOSTER [6)] found the swelling to be minimal.

A control value has also been plotted in part B of Fig.4 to show the effect of HCl without the addition of either NaCl or cysteine under otherwise identical conditions. The points plotted are the mean values of from 3 to 5 different experiments, with ranges as indicated. It becomes obvious that the presence of as small a concentration of NaCl as 0.0007 N has a tremendous influence on the progress of the denaturation and that it was a fortuitous circumstance that led us to dissolve our protein samples in physiological saline solution in our early experiments (see Fig.1).

On the basis of these findings, additional tests were performed with the impure PBA preparation. Solutions of it were exposed for several days to either cysteine in water or to hydrochloric acid, or simultaneously to HCl and cysteine. These experiments clearly demonstrated that there was some reactivity between PBA and cysteine in the absence of acid, but they brought out the far more striking fact that the marked increase in the second protein wave in the intermediate TriCo buffer (see Fig.1) was the same in an acid medium whether cysteine is present or not. This suggests that something similar to cysteine, perhaps a polypeptide with sulfhydryl groups, is present and catalyzes the reaction so that any added cysteine is ineffective.

The last conclusion finds additional support in the analysis of sulfosalicylic acid filtrates made from the various protein-containing solutions that were carried out in standard TriCo buffer. The crystalline product, CBA, gives a negative filtrate test before or after KOH digestion, i.e., it yields no filterable material that causes a catalytic polarographic wave in TriCo buffer [7)]. In contrast to this, we found a small amount of filtrate material when we subjected a solution of the impure PBA to this test. Perhaps it is this material which acts as a catalyst.

It should be added here that when CBA or PBA is exposed to acid cysteine solutions for short periods of time (e.g., 30 minutes), no additional filterable material is found. However, when the exposure to HCl or to HCl plus cysteine is continued for several days at room temperature, good-sized polarographic filtrate waves are obtained even for CBA. There is no difference in the wave heights between "acid-cysteine-" and "acid only" exposed material under these circumstances, but the PBA waves are about 5 times as large as those of CBA.

Digest tests on dialysates of cysteine- and acid-cysteine-exposed solutions of CBA

In order to see how the acid-exposure and acid-cysteine-exposure would affect the alkaline denaturation of the proteins, digest tests were made of such pretreated CBA solutions after extensive dialysis against distilled water. Since it was known that pepsin treatment of CBA in acid produced a marked increase in the protein double wave [1], it was entirely possible that pretreatment of CBA solutions with acid alone would produce a more marked reaction in the alkaline digest test than untreated CBA. We found that comparable solutions produced protein double waves of about the same height at the peak of the digest reaction, and that it did not matter whether the pretreatment with HCl was in the presence of cysteine or not. However, there was one major distinguishing feature, namely, that the pretreated solutions appeared to have become partially denatured and that the alkaline denaturation merely completed the process. This becomes clear from an examination of the two polarograms reproduced in Fig.5. The solutions were prepared as follows: To 10.0 ml. of 5 g./100 ml. CBA (in distilled water) was added 10.0 ml. of 0.1 N HCl and 10.0 ml. of either distilled water or of 0.01 M cysteine (in distilled water). After being kept for 6 days in a water bath at 25°C., the resulting mixtures were dialyzed against distilled water in the refrigerator for 5 days during which time the water was frequently changed. Then, to 1.0 ml. of each dialyzed solution was added 0.25 ml. of 1 N KOH, and, after known intervals of time, 0.05 ml. samples of the digest mixture were added to 10.0 ml. of a standard DiCo buffer [1]. The two polarograms of Fig.5 are typical of the results obtained. All indicate that the limiting height of the second protein peak is of the same magnitude whether the protein was exposed only to HCl or to a mixture of HCl and cysteine. This suggests that the same maximum number of cysteine groups has been exposed in both solutions during alkaline digestion. However, note that in both instances the control curves (over the blank), which were obtained with undigested material, showed a second wave that was higher than the first. Since it is known that fresh CBA under similar conditions produces only a very small second wave and that this is always smaller than the first wave [1,2] it must be concluded that the two solutions here were partially denatured before the alkaline digestion, and that exposure to cysteine (in the presence of HCl) had caused a greater denaturation in the same time than simple acid exposure. Concerning the half-life of the decay reaction [2] of the CBA after exposure to either acid or cysteine plus acid we have only preliminary observations which indicate that in either case this half-life is prolonged but more so in the CBA that has been exposed to acid only.

Effect of added cysteine on protein double-wave in DiCo buffer

When increasing amounts of freshly prepared cysteine solutions (in water) are added to aliquots of a standard DiCo buffer containing some beef serum, the protein double-wave is altered as seen in Fig.6. As might have been anticipated on the basis of the earlier comments, the second protein wave increases markedly, while there is hardly any effect (outside of a slight depression) on the first wave. Certainly no distinction is possible between the added cysteine and the original second protein wave, yet a displacement of the potentials at the current peaks towards more negative values with increasing cysteine concentrations is very noticeable. This same displacement is also found when pure cysteine solutions of increasing concentrations are analyzed in DiCo buffer; it is far greater than can be accounted for by IR corrections, and must therefore be a characteristic of the catalytic wave. A theoretical explanation of this potential shift must await a better understanding of the nature of the catalytic wave and of the factors which determine its magnitude. However, recognition of this shift becomes of practical importance from the standpoint of quantitative analysis as will become apparent from the data of Fig.6.

The solutions for this polarogram were prepared as follows: To 10.0 ml. standard DiCo buffer was added 0.05 ml. of beef plasma (diluted five-fold with distilled water) plus x ml. 10^{-4} M cysteine (in distilled water) and (1.0 - x) ml. distilled water. (The solution marked 2.0 ml. cysteine actually was prepared by adding 0.2 ml. of 10^{-3} M cysteine (in H_2O) and 0.8 ml. distilled water.)

It has been customary to measure the height of the second protein wave as the difference between the peak height and the current observed at the same potential with a DiCo buffer blank (e.g. the horizontal line drawn into Fig.6). This will ordinarily give correct values, but not so when the second peak is quite small and therefore very close to the first peak, as for instance in the cysteine-free solution, of Fig.6. As indicated by the dotted line, the first peak is symmetrical and extends into the region of the second peak. Hence the net current of the second peak is the difference indicated by the length of the arrow, rather than the difference between the peak and the base line of the cobalt diffusion current. As soon as sufficient cysteine has been added (e.g. 0.2 ml.) the peak of the second wave is sufficiently negative to the first wave to no longer be affected by it (see dotted line). Evaluation of the curves of Fig.6 has been done by both

Table 1 Evaluation of Curves of Figure 6
(All concentrations are in units of 10^{-6} M)

Figure 6 "ml. cysteine added"	Actual concentration of added cysteine in final solution	Concentration of "cysteine" determined from wave height	Calculated concentration of added cysteine method (a)+	method (b)+
0	0	3.6 (a)+	0	
0	0	2.8 (b)+		0
0.2	1.8	4.6	1.0	1.8
0.5	4.6	7.6	4.0	4.8
1.0	9.1	12.3	8.7	9.5
2.0	18.2	21.0	17.4	18.2

+(a): Measured as difference between peak of second wave and cobalt diffusion current.

(b): Measured as difference between peak of second wave and extrapolation of first wave.

methods, as shown in Table 1. Here the second peak of the protein double-wave has been evaluated i n t e r m s o f "c y s t e i n e," even in the cysteine-free solution, on the basis of a cysteine calibration curve. Note that the results obtained by the extrapolation of the first wave, method (b), show better agreement with the known cysteine concentrations than method (a). It should be pointed out, however, that the evaluation of the curves becomes rather difficult and the experimental error becomes large when actual peaks are no longer observed on the waves.

Table 2 Cysteine additions to solutions of crystalline bovine albumin (All concentrations are in units of 10^{-6} M)

	Before KOH digestion		After KOH digestion	
Actual concentration in final solution	Concentration of "cysteine" calculated from wave height	Calculated concentration of added cysteine	Concentration of "cysteine" calculated from wave height	Calculated concentration of added cysteine
0	1.6	0	15.5	0
0.8	2.5	0.9	16.0	0.5
2.0	2.7	1.1	17.0	1.5
4.0	6.2	4.6	20.0	4.5
8.0	11.0	9.4	23.0	7.5

Similar evaluations have been carried out on dilute solutions of crystalline bovine albumin (CBA) containing different amounts of cysteine already before aliquots of the mixture were added to a fixed quantity of DiCo buffer. The results were entirely analogous. Table 2 represents the data of such an experiment along with a parallel experiment in which the same protein-cysteine mixtures had been subjected to alkaline denaturation for 30 minutes (as in the d i g e s t t e s t) before being added to the DiCo buffer. Again the height of the second protein wave was evaluated in terms of "cysteine" on the basis of a calibration curve. In view of the aforementioned difficulties of measurement, the results demonstrate satisfactorily that the presence of cysteine during the alkaline denaturation has no significant effect on the digestion and that no cysteine is liberated from the protein during this digestion. (In a similar experiment it could be ascertained that solutions of pure cysteine remained essentially unchanged during alkaline digestions lasting as log as 120 minutes.)

D I S C U S S I O N

The polarographic findings reported here should be compared with numerous other data on the behavior of bovine albumin in acid media that have appeared in the literature during the past decade. An excellent review of this subject is given by FOSTER [8]. The anomalous steepness of the titration curve [9], changes in the intrinsic viscosity and optical rotation and the effect of ionic strength on these parameters [6], increase in the depolarization of fluorescence of dye conjugates [10][11], interaction with cationic detergent ions [12], decrease in the sedimentation coefficient and the effect of ionic strength thereon [13] and the anomalous uptake of heat [14] of bovine albumin in acid solution have all been explained in terms of an e x p a n s i o n of the protein molecule which is rapid and reversible. In addition to this, a further slow molecular change occurs on standing, wherefore TANFORD et al. have suggested the existence of an "expandable" form [15], which undergoes continuous expansion, increasing with charge and decreasing with ionic strength, which is followed by a slow third stage, possibly an aggregation. The entire process, including the final slow change, was found to be reversible. AOKI and FOSTER [16] subsequently made a carful investigation of the electrophoretic behavior of bovine albumin over the region of pH 2.0 to 4.7. They found an equilibrium between two forms of the protein which they called N (for normal) and F (for fast-moving boundary) which could be expressed by the equation $N + 3\ H^+ \rightleftharpoons F$. At an ionic strength of 0.02 (chloride), a temperature of $0^oC.$, and a pH 4.0 (where LUETCHER first observed two boundaries in electrophoretic analysis [17]) the two forms were present in about equal concentrations; at pH 3.5 there was essentially no N-form left. This isomerization was reversible and had a half-life of about two hours. In this connection it is worthy of note that SCHMID [18] was able to isolate from COHN's Fraction VI of pooled human plasma an albumin that was homogeneous during electrophoresis at pH 4.0. Since this homogeneity was lost after treatment with cysteine and deionizing with ion-exchange resins, SCHMID postulated that the "electrophoretically homogeneous" albumin contains a "blocking agent" which is removable with cysteine.

Reactions between thiol-containing compounds and albumin have been studied in connection with d e n a t u r a t i o n . For instance, HUGGINS et al. [19] explained the need for SH-groups in the production (by urea) of gels from albumin by means of a sulfhydryl-disulfide chain reaction. Gel formation is prevented if the one free sulfhydryl group per albumin molecule is made unreactive (e.g. with iodoacetate), but the presence of very small amounts of cysteine enhances the rate of gelation of albumin by urea. It is postulated that the free sulfhydryl group of one albumin molecule reacts with a disulfide group in a neighboring molecule to form an intermolecular disulfide while generating a new sulfhydryl group, which is then available for a repetition of the process, either with another albumin molecule or with another disulfide group in the same molecule. This "mercaptan interchange" reaction thus replaces intramolecular disulfide linkages by intermolecular linkages.

Besides sulfhydryl blocking reagents such as iodoacetate or silver ion, there have been found other protective agents, for example caprylate [20] and acetyltryptophane [21] which protect albumin against denaturation by heat [22] or by urea [23], but no real understanding of stabilization has so far arisen [8].

There is a definite effect of pH on the denaturation of albumin which shows an optimal rate near pH 4 and a minimal rate near pH 5.5. This fact has been incorporated into a theory [24] which presumes a single labile form of the protein in protonic equilibrium with forms that are less susceptible toward denaturation. In another theory [8], the increase in rate of denaturation, as the pH is decreased from 5.5 to 4, is supposed to be produced by the shift of the N to the F form of the albumin. This is based on the fact that the pH dependence of the rate corresponds to an uptake of 3 hydrogen ions for the conversion of the isoelectric into the labile form [24] which is the same number of hydrogen ions postulated for the isomerization [16].

The pH optimum of 4 for the denaturation implies that at lower pH some other reaction competes. HENCE, FOSTER and YANG [25] suggested that the pH expansion at low pH is such a competing process and that the expanded form may serve as a "reversibly protected" form of the albumin. Taking into account that the pH dependence of expansion corresponds to an uptake of 2 hydrogen ions [24], these authors formulated the reaction mechanism as follows:

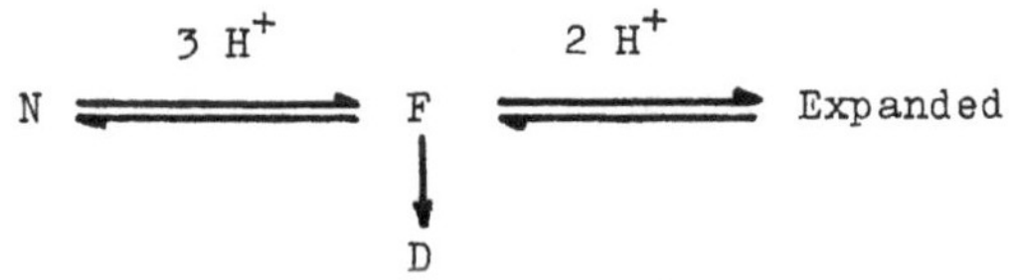

where D is the denatured form as defined by isoelectric precipitation [8)24)].

In a different application of the "mercaptan interchange" reaction KLOTZ et al. [26)] investigated the reaction between the lone sulfhydryl group of albumin and an externally-added small-molecule disulfide. This compound, the disodium salt of 2,2'-(2-hydroxy-6-sulfonaphthyl-1-azo)-diphenyl disulfide (abbreviated as DSSD), was colored and had a characteristic spectrum when cleaved and bound to protein through a disulfide linkage. Their studies showed clearly that there was no chain reaction but a stoichiometry of two protein thiols reacting with one molecule of DSSD. This led the authors to the conclusion that the DSSD must have reacted with the available sulfhydryl group of albumin as well as with one of its disulfide groups according to the following two equations :

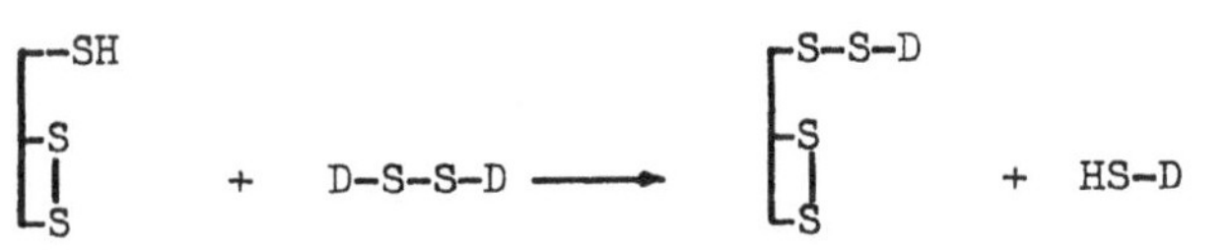
(1)

and

(2)

The final product contained as many sulfhydryl groups as the starting material but these sulfhydryl groups are now "masked" since they are not accessible to further reaction with more DSSD. In confirmation of this explanation, the dye 4-(p-dimethylaminobenzeneazo)-phenylmercuric acetate, which is known to bind to hindered as well as to accessible sulfhydryl groups, was found to react readily with the "masked" mercaptan groups. Also, an amperometric silver titration demonstrated that these newly formed mercaptan groups are markedly different from those normally present in albumin. KLOTZ et al. point out that equation [2)] cannot be a simple direct reduction of the disulfide group of the protein must be involved in this reduction reaction. This implies that the HS-D produced in equation [1)] "in some way seeks out the protein molecules which still have their original thiol groups and reacts only with t h e i r disulfide groups and in such a way that the thiol group is simultaneously removed" [26)]. This conclusion then led to the postulate of hydride ion transfer between groups of different oxidation states separated in space but connected by a bridge of oriented water molecules [26)].

In comparison with these findings, the polarographic effects reported in this paper lead us to the following conclusions :

1) The polarographically recognizable change must be similar to the slow molecular change reported by TANFORD et al. [15)] which follows the change of albumin into the expandable form (form F of AOKI and FOSTER [16)]) and the expanded form. This follows from the fact that the change in albumin is negligible at pH 4 (Figures 2 and 3) and does take a long time in the absence of cysteine (Fig.4).
2) The increase in the rate of transformation of the albumin in a sufficiently acid solution produced by an increase in ionic strength (Fig.4) can be explained by

assuming that the expanded form is a "reversibly protected" form of the albumin as suggested by FOSTER and YANG [25]. Since there is less tendency to form this expanded form at higher ionic strength [6], more of the albumin remains in the "expandable form" (form F) which is the labile form of albumin. Interesting in this connection is the observation of MARHA [27] that higher fatty acids, down to capric acid depress the second protein wave of albumin while the lower fatty acids up to valeric acid have no effect. This seems to represent a direct polarographic demonstration of the protective action of these compounds [20][21][22][23].

3) The catalytic effect of cysteine on the rate of transformation of albumin is more difficult to explain. One might postulate a mechanism similar to that proposed by SCHMID [18] in which a blocking agent, which hinders the transformation, is removed. We prefer, however, to postulate the following extension of the mechanism suggested by KLOTZ et al. [26] : The reactions studied by these authors were carried out mainly in an acetate buffer of pH 6 with only one significant exception below pH 4. It is of interest that this reaction, carried out at pH 2.5, was the only one, free of additional reagents, in which an abnormally higher stoichiometric ratio of albumin to DSSD was found. Since KLOTZ et al. [26] state that their reactions in buffers were completed in every case within 2 minutes it is entirely possible that prolonged exposure of albumin to acid b e f o r e reaction with DSSD would have given still higher stoichiometric ratios.

We feel that prolonged acid exposure produces additional changes in the molecule besides the expansion which is supposed to be instantaneous (see our conclusion [1]). This "prolonged expansion" makes disulfide groups of the albumin more accessible. Cysteine and other thiols could hasten this process by combining with some of the disulfide groups by a mutual oxidation-reduction reaction in a manner schematically represented as follows:

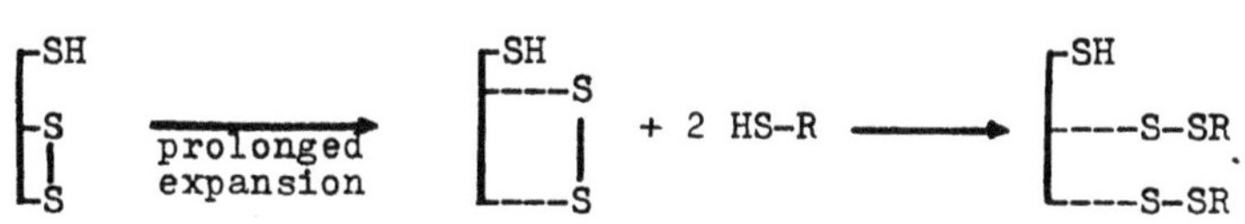

The longer the exposure to acid, the more disulfide groups of each albumin molecule could thus react until either all or a certain maximum number of the 16 available groups have reacted.

The accessible disulfide groups as well as the disulfide linkages formed with cysteine are polarographically indistinguishable. Both would be reduced to the corresponding thiol structure at a relatively positive potential and could then react to give the catalytic wave (cf. the behavior of cystine and its cobaltcatalyzed reaction [1]).

In addition to the foregoing conclusions we feel that the following speculations may also be appropriate at this time :

The fact that the second portion of the catalytic double-wave of protein and the cysteine wave are indistinguishable and that we were able to calculate "cysteine" concentrations from the height of the second protein wave caused us to attempt similar calculations based on polarograms of cysteine-free albumin solutions under different conditions. For instances in the case under consideration in Table 2, the concentration of CBA in the final solution was 0.010 g./100 ml. or about 1.5×10^{-6} M. Evaluation of the second protein wave in terms of "cysteine" for the undigested solution indicated a concentration of 1.6×10^{-6} M, i.e., about one reactive sulfhydryl group per molecule of protein. This is in agreement with the generally accepted value. However, alkaline digestion for 30 minutes must have resulted in the activation of 9 more of such active groups since the "cysteine" concentration calculated on the basis of a calibration curve was now about

$15.5x10^{-6}$ M. If all of the 16 disulfide groups in the albumin molecule had been converted into sulfhydryl groups that are catalytically active at the dropping mercury electrode, the currents would have been about three times as large. We speculate therefore (a) that only certain disulfide groups of the albumin can be activated in this fashion, and (b) that probably only one of any two groups formed by electrochemical reduction of the accessible disulfide groups in a protein will become polarographically active because of steric hindrance. This should explain why alkaline denaturation does not produce more groups with catalytic activity at the dropping mercury electrode. Without knowing the exact protein concentration of the dialysates in Fig.5, we can only guess that perhaps one-half and two-thirds as many disulfide groups (or perhaps 5 and 6 disulfide groups per molecule) were made accessible by the acid and by the acid-plus-cysteine exposure of 6 days at 25°C. as by alkaline digestion. These speculations may seem somewhat premature at present; however they may well serve as a working hypothesis for future experimentation.

Acknowledgement

This work was supported in part by grants No. C-716 and C-2358 from the Division of Research Grants and Fellowships of the National Institute of Health, U.S. Public Health Service, which is gratefully acknowledged. I also wish to thank Miss Patricia FOLSE and Mrs. Marilynn RICHTARICK for valuable assistance.

LITERATURE

1) MÜLLER,O.H., "Polarographic Analysis of Proteins, Amino Acids, and Other Compounds by Means of the BRDIČKA Reaction," in D. GLICK, ed., *Methods of Biochemical Analysis*, Vol. 11, Interscience, New York-London, 1962 p. 329
2) MÜLLER,O.H., "Polarographic Studies of Proteins and Their Degradation Products, IV. The Activation and Decay Reactions of Proteins During the Digest Test." Paper presented at the Symposium "*Die Polarographie in der Chemotherapie, Biochemie und Biologie*, Jena 1962, (in print:). Akad.V.1964
3) MÜLLER,O.H., and J.S.Jr.DAVIS, Arch Biochem., 15, 39 (1947)
4) MÜLLER,O.H., Federation Proc., 14, 105 (1955)
5) MILLAR,G.J., Biochem. J., 53, 385 (1953)
6) YANG, J.T., and J.F.FOSTER, J.Am.Chem.Soc., 76, 1588 (1954)
7) MÜLLER,O.H., "Polarographic Behavior of Various Plasma Protein Fractions" in T.SHEDLOVKY, ed., *Electrochemistry in Biology and Medicine*, Wiley, New York-London, 1955, p. 301
8) FOSTER,J.F., "Plasma Albumin", in PUTNAM,F.W., *The Plasma Proteins*, Vol.1, Academic Press, New York-London, 1960, p. 179
9) TANFORD,C., J.Am.Chem.Soc., 72 441 (1950)
10) WEBER,G., Biochem.J., 51, 155 (1952)
11) HARRINGTON,W,P.JOHNSON, R.OTTEWILL, Biochem.J., 62, 569 (1956)
12) FOSTER,J.F., J.T.YANG, J.Am.Chem.Soc., 76, 1015 (1954)
13) REICHMANN,M., P.CHARLWOOD, Can.J.Chem., 32, 1092 (1954)
14) GUTFREUND,H., J.F.STURTEVANT, J.Am.Chem.Soc., 76, 1595 (1954)
15) TANFORD,C., J.G.BUZZELL, D.G.RANDS, S.A.SWANSON, J.Am.Chem.Soc., 77, 6421 (1955)
16) AOKI,K., J.F.FOSTER, J.Am.Chem.Soc., 79 3885 (1957)

17) LUETSCHER,J.A., J.Am.Chem.Soc., 61 2888 (1939)
18) SCHMID,K., J.Am.Chem.Soc., 79 4679 (1957)
19) HUGGINS,C., D.F.TAPLEY, E.V.JENSEN, Nature, 167, 592 (1951)
20) BOYER,P.D., F.G.LUM, G.A.BALLOU, J.M.LUCK, R.G.RICE, J.Biol.Chem., 162, 181 (1946)
21) SCATCHARD,G., L.E.STRONG, W.L.HUGHES, J.N.ASHWORTH, A.H.SPARROW, J.Clin.Invest., 24, 671 (1945)
22) HOCH,H., A.CHANUTIN, Arch.Biochem.Biophys., 51, 271 (1954)
23) BOYER,P.D., G.A.BALLOU, J.M.LUCK, J.Biol.Chem., 162, 199 (1946)
24) TIMASHEFF,S.N., R.J.GIBBS, Arch.Biochem.Biophys., 70 547 (1957)
25) FOSTER,J.F., J.T.YANG, J.Am.Chem.Soc., 77, 3895 (1955)
26) KLOTZ,I.M., J.AYERS, J.Y.C.HO, M.H.HOROWITZ, R.E.HEINEY, J.Am.Chem.Soc., 80 2132 (1958)
27) MARHA,K., Collection Czechoslov.Chem.Commun., 22, 153 (1957).

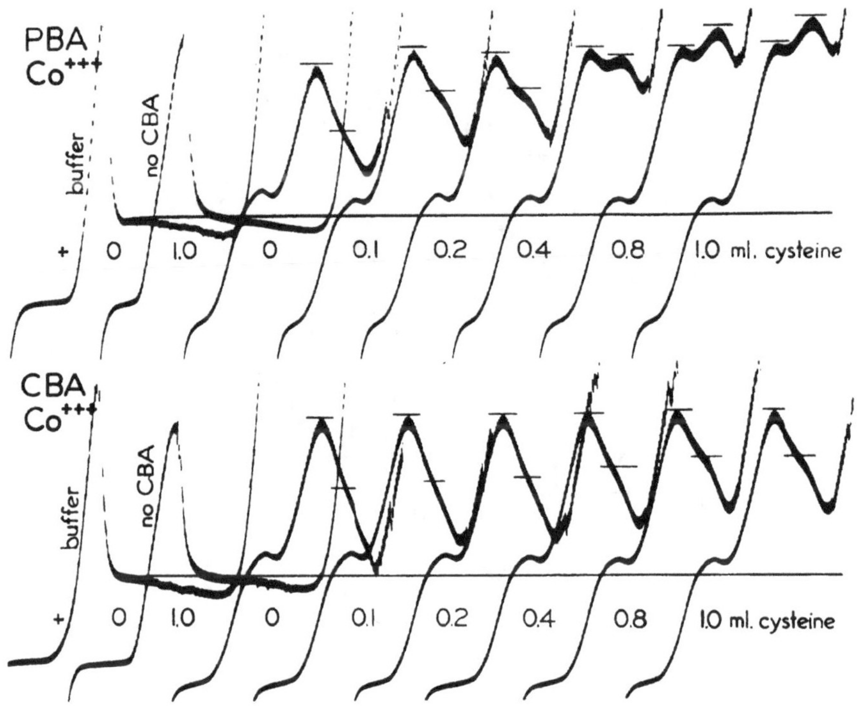

Fig.1 Effect of cysteine addition on the protein double-waves of bovine albumin powder (PBA) and crystalline bovine albumin (CBA) in a special trivalent cobalt buffer (see text).

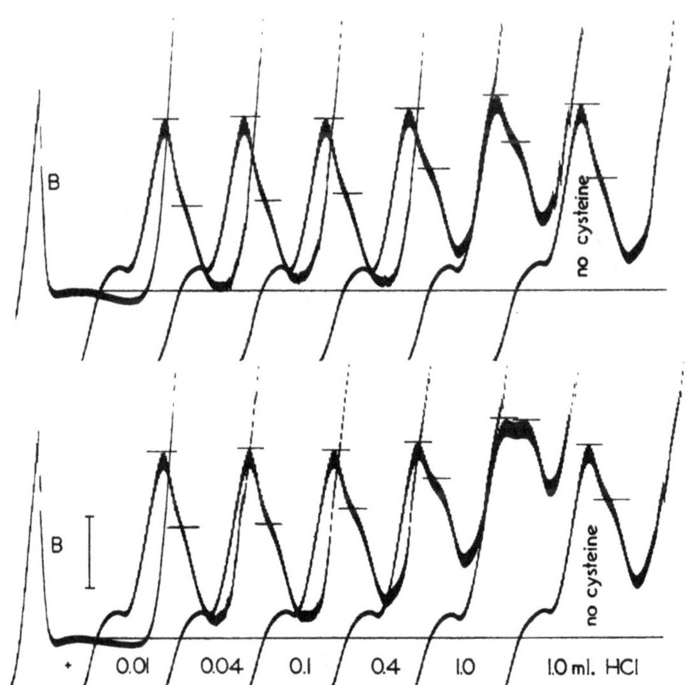

Fig.2 Effect of acid on the protein waves of CBA in an intermediate trivalent cobalt buffer (see text). B=buffer blank. Upper polarogram taken within 30 minutes of mixing; lower polarogram taken after 2-3 hours' standing at 23°C.

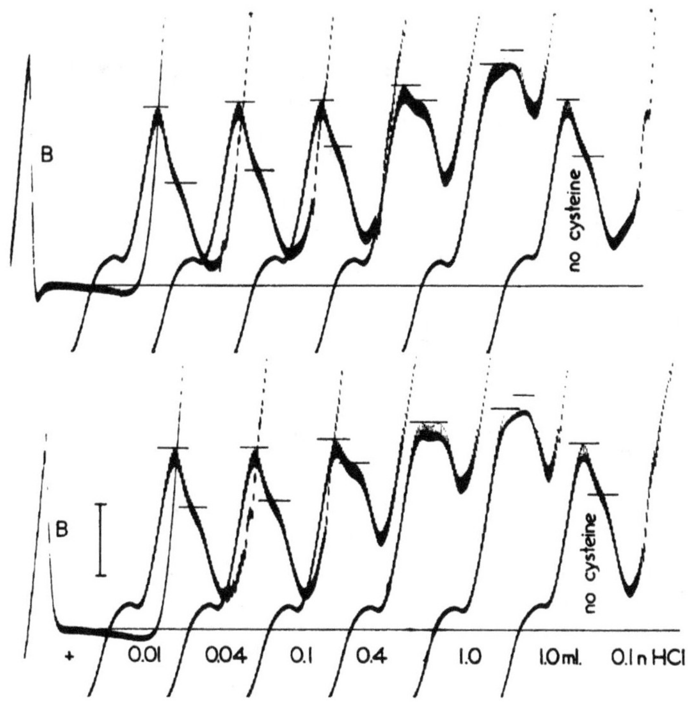

Fig.3 Continuation of Fig.2. Upper polarogram taken 21 hours after mixing (see text); lower polarogram after 5 more hours' standing at 23°C.

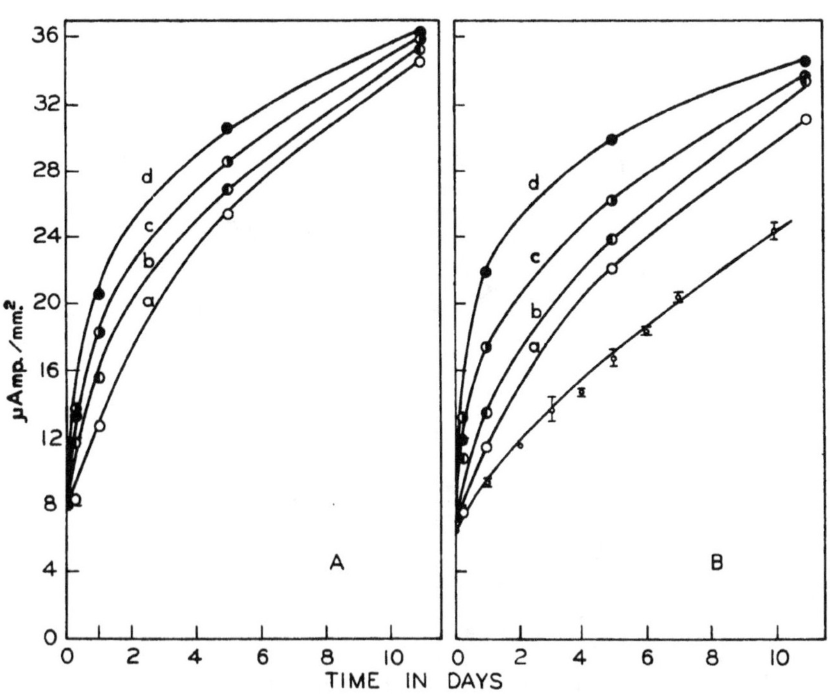

Fig.4 Effect of standing at 25°C. on second protein wave in intermediate trivalent cobalt buffer. Solutions: 1.67 g./100 ml. CBA in 0.0333 N HCl in presence of (a) 0, (b) 0.00067, (c) 0.00133, and (d) 0.00266 M cysteine and in either (A) 0.0007 N or (B) 0.0333 N NaCl. Lowest curve on the right side in absence of NaCl (see text).

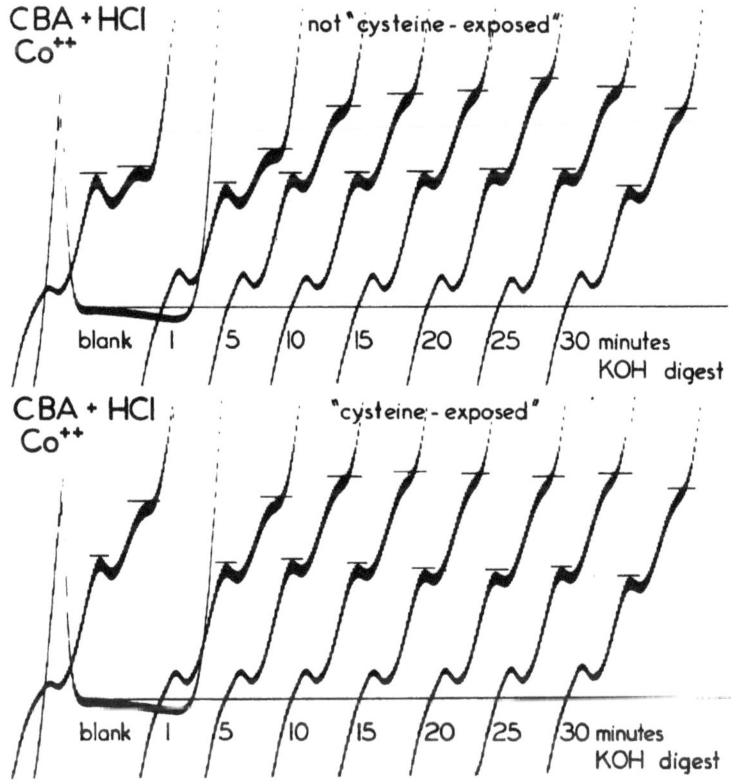

Fig.5 Alkaline digestion of dialysates of acid - CBA solutions, analyzed in standard DiCo buffer. Upper polarogram: not "cysteine-exposed"; lower polarogram: "cysteine-exposed" (see text).

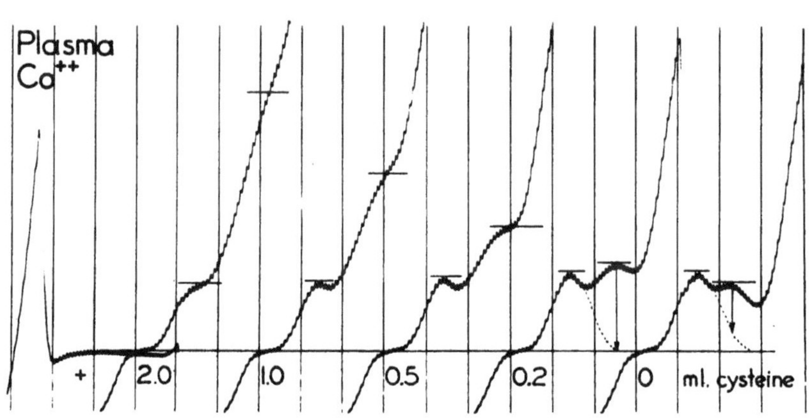

Fig.6 Effect of added cysteine on protein double-wave of plasma in standard DiCo buffer (see text).

C/D Proteine – Nucleinsäuren

Polarographische Beiträge zur Photolyse von Proteinen und Nucleinsäuren

H. BERG

Institut für Mikrobiologie und experimentelle Therapie der Akademie, Jena

S U M M A R Y

The possibilities of polarography to characterize biogenic macromolecules and their component bases on electron-exchange-reactions and adsorption-phenomena. By means of this, we intend to investigate the photolytic changes of the secondary- and tertiary-structure and also of the stability of bases. The following results treat of

1. Proteins : Their tertiary-structure mainly depends on disulfide bridges, which are polarographically active after reduction and complex formation with cobalt ions. Some well known proteins such as ribonuclease, lysozyme (and serumalbumine in comparison with it) we have studied during photochemical denaturation in presence and absence of oxygen or cobalt salts. The course of height of different catalytic- and complex-waves was measured under several conditions. An interpretation of the results is given.
2. Nucleic acids : A dependence between molecular weight and adsorption at the mercury electrode (that means the damping of polarographic maxima) was found.

In this manner the photolytic chopping of the double helix was recorded in presence and absence of oxygen and a distinct difference between nucleic acids of various kinds was discovered.

By means of the reduction waves of pyrimidines the photolytic splitting and some dark reactions were studied.

Some advantages of the polarographic techniques are shown :

a) the macromolecules may be measured in extremely low concentrations (10^{-3} to 10^{-5}%),
b) the macromolecules only diffuse and are adsorbed. Therefore shear-denaturation is prevented.

E I N L E I T U N G

Für Struktur- und Reaktionsprobleme biogener Makromoleküle ist die Polarographie bisher relativ selten zur Anwendung gelangt.

Bei Dextranfraktionen konnte eine Beziehung zwischen Molekulargewicht und Maximadämpfungsfähigkeit aufgefunden werden[1)2)]. Danach werden kleinere Moleküle kompakter an der Tropfelektrode adsorbiert als große und bringen dadurch die den Depolarisator zusätzlich herantransportierenden Strömungen in der angrenzenden Flüssigkeit zum Stillstand.

Bei den Proteinen haben sich insbesondere O.H.MÜLLER [3)4)] und I.IWANOW [5)6)] um die

Zuordnung der katalytischen Kobaltkomplexwellen zur sekundären und tertiären Struktur bemüht. Danach sollen mindestens zwei Komplexe mit zwei- oder dreiwertigen Kobaltionen existieren, die auch nach der Reduktion des Zentralatoms die Abscheidung von Wasserstoff erleichtern. Nach IWANOW entsteht der leichter adsorbierbare und reduzierbare Komplex mit dem globulären Teil des Moleküles, während die stärker adsorbierbaren aglobulären Anteile die 2.Welle verursachen.

Die katalytische Wasserstoffentwicklung [7] stellt man sich, nachdem Disulfidbrücken des adsorbierten Proteins reduziert worden sind, als dreistufigen Mechanismus vor :

Reduktion :

$$\left[Co^{III}(SCH_2\ CH\ NH_2\ COO)_2(H_2O)_2\right]^- + e^- \rightleftharpoons \left[Co^{II}.....\right]^{2-} + 2e^- \rightarrow \left[Co^{0}.....\right]^{4-} \quad (1)$$

Protonisierung :

$$\left[Co^{0}\ .....\right]^{4-} \quad + \quad H^+ \quad \rightleftharpoons H\left[Co^{0}\ .....\right]^{3-} \quad (2)$$

Wasserstoffkatalyse :

$$H\left[Co^{0}\ .....\right]^{3-} \quad + \quad e^- \quad \longrightarrow \left[Co^{0}\ .....\right]^{4-} + H \quad (3)$$

Eine vollständige Klärung ist damit aber noch keineswegs erreicht.

Die Komplexbildung hängt von mehreren funktionellen Gruppen [8] ab, was bei schwefelhaltigen Niedermolekularen sich in der Reihenfolge :

$$SH,\ NH_2,\ COOH > SH,\ NH_2 > SH,\ COOH > SH,\ COOR$$

auswirkt. Davon gehen zahlreiche analytische Untersuchungen aus, die hier nur erwähnt werden sollen.

Auf dem Gebiet der Nucleinsäuren liegen außer den oszillopolarographischen Bestimmungen auch des T_M-Wertes von PALEČEK [9] eine Reihe Messungen an Basen [10][11] vor, sodaß ihre Elektronenaufnahmen, der Mechanismus und einige Protonisierungsgleichgewichte bekannt sind.

In Tabelle 1 sind die polarographischen Kriterien für spezielle Eigenschaften biogener Makromoleküle zusammengefaßt.

Tabelle 1 Polarographische Möglichkeiten zur Ermittlung von Moleküleigenschaften

Polarographisches Kriterium	Moleküleigenschaft	Beispiele
1. Diffusionsstufe	Reduktions- oder Oxydationsgruppe	Adenin, Cytosin 4-Amino-2,6 Dimethyl-pyrimidin u.a. Pyrimidine, sowie ihre N-Oxyde.Cystin-Cystein
2. Katalytische Kobalt-Welle nach BRDIČKA	-SH oder -S-S- Gruppen	Albumine, RNase Lysozym
3. Katalytische Welle	Peptidbindung	Lysozym, RNase
4. Maximadämpfung (Kapazitätseffekte)	Adsorbierbarkeit (mäßige Löslichkeit)	Proteine Nucleinsäuren Polysaccharide

Diese Möglichkeiten werden in vorliegender Arbeit auf folgende Fragestellungen [12] angesetzt :

1. Proteinstrukturen
   1.1 Sprengung der -S-S-Brücken unter Änderung der Tertiärstruktur und Weiterreaktion der bei 1.1 entstandenen -SH Gruppen.
   1.2 Öffnung von Peptidbindungen unter Zerstörung der Primärstruktur.
2. Nucleinsäurestrukturen
   2.1 Photochemische Primärvorgänge.
   2.2 Kettenbrüche bei G-C Grenztypen.
   2.3 Sauerstoffeinfluß.
3. Nucleinsäurebasen
   3.1 Reduktionsmechanismen.
   3.2 Photolyse.
   3.3 Nachreaktionen.

Da noch zu wenig Zusammenhänge zwischen Struktur und polarographischer Meßgröße aufgeklärt sind, wurden möglichst bekannte Substanzen ausgewählt (Tabelle 2) und 3). Unter ihnen ist jedoch, abgesehen von den niedermolekularen Bausteinen, nur die Ribonuclease als weitgehend erforscht anzusehen.

Tabelle 2 Proteine. (Die Abteilung Biochemie des Instituts stellte Präparate von Ribonuclease, Trypsin und Chymotrypsin zur Verfügung.)

Protein	$\bar{M}_W$	Anzahl -S-S-	Isoel. Punkt	$\phi_{253,7}$ (Quantenausbeute [13] der Enzyminaktivierung)
Ribonuclease	13 600	4	9,5	0,027
Lysozym	14 000	4	10,5	0,024
Trypsin	15 000	6	10,5	0,015
Chymotrypsin	22 000	5 (?)	8,6	0,005
Serumalbumin	70 000	16	4,7	

Tabelle 3 Nucleinsäuren und Basen. (Die Präparate wurden von der Abteilung Biochemie des Instituts zur Verfügung gestellt.)

Nucleinsäuren, Basen	$\bar{M}_W$	G + C Gehalt	T_M-Wert	M.P. °C
Proteus mirabilis-DNA	25.10^6	42 %	87°C	
Strep.chrysomallus-DNA	$1,5.10^6$	71 %	98°C	
2,6-Dimethyl-4-amino-pyrimidin	123	-	-	180

E X P E R I M E N T E L L E S

Die Gleichspannungspolarogramme der Konzentrationsabhängigkeiten wurden mit Tintenschreiberpolarographen (ungarische Fabrikate) registriert, wobei die Meßlösung sich in einer temperierbaren Zelle unter den üblichen Bedingungen befand.

Meßlösungen für die Proteine:

1. Phosphatpuffer pH 4,5 - 7 - 10,5 - 0,1n NaOH
2. 0,1m NH_4OH, 0,1m NH_4Cl, 10^{-3}m Co^{3+} oder Co^{2+} (Kobalt Solution nach BRDIČKA[16]), jeweils mit 0,01 % Protein

Meßlösungen für die DNA :

2.10^{-4}m Citratpuffer, 2.10^{-3}m NaCl

Meßlösungen für die Basen :

Phosphatpuffer pH 8, 20 % Äthanol

Die Bestrahlung erfolgte in temperierten Quarzzellen mit oder ohne Essigsäurefilter (5 cm Schichtdicke, 10 %) mittels :

	Bestrahlungsleistung	in Entfern.	Filter
Quecksilberhochdrucklampe U 500 (BGW) :	$6,7.10^5$erg $s^{-1}cm^{-2}$	10 cm	ohne
Quecksilberniederdrucklampe (wassergekühlt, Eigenbau) :	$0,4.10^5$erg $s^{-1}cm^{-2}$	10 cm	+Essigsäure

Dabei waren die Lösungen luftgesättigt ($\vartheta = 24^{\circ}C$) oder nach Durchleiten von Argon weitgehend vom Sauerstoff befreit.

E R G E B N I S S E

In der Reihenfolge der aufgeführten Problemstellungen soll ein Überblick unserer vorläufigen Meßergebnisse gebracht werden.

1. Z u r P h o t o l y s e d e r P r o t e i n s t r u k t u r

Besondere Beachtung wurde den Unterschieden zwischen Ribonuclease und Serumalbumin geschenkt. Dabei sollen nun die polarographischen Kriterien von Disulfidbrücke und von Peptidbindung während der Bestrahlung gegenübergestellt werden.

1.1 Die -S-S-Brücken und ihre Photolyse

Die Konzentrationsabhängigkeit der katalytischen Protein-Kobaltwellen ergibt einen Grenzwert der Gesamthöhen von erster und zweiter Welle (Abb.1), wobei die negativere Teilwelle anfangs die niedrigere ist. Werden derartige Proteinlösungen, die im Bereich λ = 240 bis 310 nm Absorptionsmaxima aufweisen, mit Wellenlängen $>$ 240 nm bestrahlt, so treten charakteristische Veränderungen im Stufenbild auf, und zwar ohne wesentlichen Einfluß des pH-Wertes und der Temperatur. Bisher lassen sich zwei Typen von Dosis-Wirkungskurven unterscheiden. Der bereits bekannte Typ [14)12)] wird durch Serumalbumin unterhalb seiner Sättigungskonzentration repräsentiert :

Die vordere Wellenhöhe bleibt nahezu konstant und sinkt später gegen Null ab.
Die Höhe der Doppelwelle geht durch ein Maximum (Abb.2).

Der zweite Typ wurde bei der Ribonuclease [15)] aufgefunden :

Beide Wellen nehmen ab, wobei sie in zunehmendem Maße miteinander verschmelzen (Abb.2).

Daran ändert sich qualitativ nichts, ob Kobaltionen während der Bestrahlung zugegen sind oder erst nachträglich hinzugefügt werden. Auch der pH-Wert übt keinen entscheidenden Einfluß auf den Verlauf der Wellenhöhen aus.

1.2 Die Peptidbindungen und ihre Photolyse

Die am Endanstieg liegende katalytische "Peptidwelle" (frühere Bezeichnung "Praenatriumwelle"), als deren Ursache die katalytische Wasserstoffabscheidung aus protonisierten Gruppen nach

$$-\underset{\underset{O}{\|}}{C}-\overset{\overset{H}{|}}{\underset{\underset{H}{|}}{N^{+}}}- \; + \; e^{-} \longrightarrow -\underset{\underset{O}{\|}}{C}-\underset{\underset{H}{|}}{N}- \; + \; H \qquad (4)$$

anzusehen ist, eignet sich ebenfalls zum Nachweis der Denaturierung. Für die Höhe dieser Peptidwelle ist wiederum die Zahl der an der Elektrodenoberfläche adsorbierten Peptidbindungen maßgebend. Daher sollte auch der Photolyseverlauf sich ähnlich auswirken, wie es bei den katalytischen Protein-Kobaltwellen (1.1.) gefunden wurde. In Abb.3 ist die Abhängigkeit der Wellenhöhe von der Bestrahlungszeit für RNase wiedergegeben. Der Bestrahlungseffekt wird nach einer Latenzzeit im Absinken der Wellen deutlich. Entsprechendes zeigen die Dämpfungsisothermen der Sauerstoffmaxima (Abb.3).

Beim Serumalbumin durchläuft die Peptid-Wellenhöhe ein Maximum und ähnelt darin den katalytischen Protein-Kobaltwellen. Allerdings ist eine zeitliche Verschiebung festzustellen, was in Abb.4 schematisiert dargestellt ist.

1.3. Diskussion zur Protein-Photolyse

Den Verlauf der Meßgrößen in Abhängigkeit von der Bestrahlungsdosis bringt für die beiden Typen die Abb.4. Übereinstimmend ist den ausgezogenen Kurven für Serumalbumin zu entnehmen, daß durch Öffnung der -S-S- und Wasserstoffbrücken eine Auflockerung des Knäules, ähnlich wie im Falle der alkalischen Denaturierung[16], herbeigeführt wird, wodurch

a) auf Kosten der nativen -S-S-Gruppen mehr freie -SH und -S-S-Gruppen als Kobaltkomplexe adsorbiert werden (Kurve A-P_{1+2} und A-P_1)

b) weitere Peptidgruppen adsorbiert werden (Kurve A-K)

c) sich aufgrund von a) und b), sowie neuer Bruchstücke, welche an der Elektrodenoberfläche haften, die Maximadämpfung verstärkt.

Die gestrichelten Linien für RNase bieten demgegenüber das Bild eines Moleküls, dessen volle elektrochemische Aktivität bereits im nativen Zustand vorhanden ist und mit der Bestrahlungsdauer absinkt, während die Maximadämpfungsfähigkeit, d.h. die Adsorbierbarkeit, sich wie beim Serumalbumin erhöht. Anstelle der Auflockerung soll nach SAMSONOW (vgl. Symposiumsvortrag) eine Kontraktion des denaturierten Moleküls während der Disulfidbrücken-Öffnung erfolgen.

Zieht man außerdem die entsprechenden Ergebnisse für Lysozym, Trypsin und Chymotrypsin in Betracht, so sind als Ursachen der beiden polarographisch unterscheidbaren Typen das MOLEKULARGEWICHT und die TERTIÄRSTRUKTUR anzusehen.

Zum Einfluß des Molekulargewichtes kann vorerst festgestellt werden, daß oberhalb $M_w > 20\,000$ der Typ I (Vertreter: Serumalbumin, Chymotrypsin, Chymotrypsinogen) gefunden wird, während unterhalb der Typ II (Vertreter: RNase Lysozym, Trypsin) vorherrscht. Insulin dürfte aufgrund enzymatischer Abbaustudien [16] zu Typ II gehören.

Hinsichtlich der Tertiärstruktur läßt sich nur soviel sagen, daß eine starke Vernetzung im nativen Molekül durch -S-S-Brücken den Typ I erwarten läßt. Beiden Typen gemeinsam ist eine beträchtliche zeitliche Verzögerung des Absinkens der "Peptidwelle" gegenüber den beiden "Kobaltkomplexwellen".

Wir führen diese Tatsache darauf zurück, daß die aus den -S-S-Brücken entstandenen -SH-Gruppen photochemisch weiterreagieren, z.B. zu katalytisch inaktiven Resten wie Sulfonen oder Schwefelwasserstoff [17], noch bevor eine Addition [18] und Koagulation [19][20] der radikalischen Kettenbruchstücke einsetzt. Letzteres vermindert gleichzeitig die katalytische Aktivität der Peptidgruppen (Gl.(4)). Die zur polarographischen Inaktivierung führende Weiterreaktion der -SH-Gruppen konnte experimentell durch Bestrahlung von Cystin im Gemisch mit anderen Aminosäuren gezeigt werden, indem die katalytische Welle mit Co^{2+} Lösung nach einer Latenzzeit abnimmt.

Als Modellbeispiel für Makromoleküle ist die Ribonuclease mit vier -S-S-Brücken am ehesten diskutierbar (Abb.5). (Vom Insulin mit nur zwei Brücken fehlen z.Z. noch die entsprechenden Daten.)

Wir nehmen an, daß infolge der Anordnung des sechs Helixes die Disulfidbrücken a, a' gegenüber den b, b' stabilisiert und stärker abgeschirmt werden.
Das hätte zur Folge :

a) Nachdem an der Elektrodenoberfläche alle vier Disulfidbrücken zu -SH-Gruppen reduziert worden sind, bilden sich die Komplexe $Co^{III}a$, $Co^{III}a'$, $Co^{III}b$, $Co^{III}b'$. Die beiden ersteren wären aus sterischen Gründen schwächer, die beiden letzteren stärker an der Elektrodenoberfläche adsorbiert, wodurch der positivere und negativere Teil der Kobaltwelle gedeutet werden könnte.
b) Auf Lichteinwirkung (hν) dürften die Brücken b, b' zuerst gespalten werden, worauf die freien Helixenden benachbarter Moleküle die Vernetzung k_1 (Abb.5) einleiten. Schließlich bilden sich Koagulate (K), die weniger Kontakt mit der Elektrodenoberfläche haben, und photochemisch veränderte Bruchstücke, die ebenfalls keine katalytische Aktivität besitzen, jedoch noch das Sauerstoffmaximum zu dämpfen vermögen.

Auf weitere Folgereaktionen, wodurch neue polarographisch aktive Gruppen wie Aldehyde, Peroxyde und Hydroxylamine gebildet werden können, soll an anderer Stelle eingegangen werden.

2. Zur Photolyse der DNA-Struktur

Vorläufig sind Rückschlüsse nur aus Adsorptionsmessungen zu ziehen :

a) Dämpfung polarographischer Maxima 1. und 2.Art
b) Erniedrigung der Elektrokapillarkurve
c) Änderung der Differentialkapazität
d) Beeinflussung der Strom-Spannungs- oder Strom-Zeit-Kurven von Testdepolarisatoren.

In vorliegender Meßserie bedienten wir uns der Maximadämpfung als Anzeige für photochemische Veränderungen an der DNA-Doppelhelix (Abb.6).

2.1. Photochemische Primärvorgänge

Mit spezifischen Methoden wurden die bekannten photochemischen Reaktionen aufgefunden : Hydratisierung, Thymindimerisierung, Cytosindesaminierung.

Allerdings werden Hydratation und Steifheit des Moleküls dabei nicht ausreichend verändert, daß die Maximadämpfungsfähigkeit dadurch merklich beeinflußt wird. Es muß noch gezeigt werden, ob die anderen Adsorptionsverfahren b) c) d) hierauf empfindlicher ansprechen oder ob Adduktgleichgewichte (DNA + Farbstoff) sich besser dazu eignen.

2.2. Kettenbrüche bei G-C-Grenztypen

Einfache Kettenbrüche, welche die Knäuelbildung erleichtern, und Doppelbrüche, die das mittlere Molekulargewicht vermindern, verstärken die Maximadämpfung. Diese Feststellung wurde schon im Verlauf der thermischen Denaturierung von Proteus-DNA getroffen. In

Abb.7 ist die Dämpfungsisotherme (Dämpfung $D = 1 - \frac{h}{h_0}$ über der prozentualen Konzentration c aufgetragen, wobei h_0 die Ausgangshöhe des Sauerstoffmaximums h die Höhe nach DNA-Zusatz bedeuten) für die unbehandelte und für thermisch denaturierte Proben dargestellt. Durch Kettenbruch und Flexibilitätserhöhung sind die Dämpfungshalbwertskonzentrationen bei 1 und 2 beträchtlich erniedrigt worden.

Entsprechendes resultiert aus der Bestrahlung für die beiden Grenztypen : Proteus mirabilis-DNA und Streptomyces chrysomallus-DNA in entlüfteten Pufferlösungen. Die ungefilterte Strahlung des Quarz-Quecksilberhochdruckbrenners bewirkt von Anbeginn eine Zunahme der Maximadämpfung bei gleichbleibender DNA-Konzentration von 6.10^{-4}% (Abb.8).

2.3. Sauerstoffeinflüsse

In Gegenwart von Luftsauerstoff ist eine Latenzzeit in der relativen Verminderung des Sauerstoffmaximums (h/h_0) zu beobachten (Abb.6), dennoch bleibt der deutliche Stabilitätsunterschied zwischen den beiden DNA erhalten. Schließlich zeigt das Polarogramm der entlüfteten Photolyse-Lösung eine erhebliche Konzentration an Wasserstoffperoxyd (Abb.9), welches auf Zusatz von Katalase zerstört wird.

Unsere Niederdrucklampe mit Essigsäurefilter bringt demgegenüber entsprechend schwächere Effekte hervor.

2.4 Diskussion zur DNA-Photolyse

Trotz häufiger Untersuchung der Bestrahlungseffekte besteht in vielen Punkten noch keine Übereinstimmung [21][22]. Auch unsere vorläufigen Ergebnisse sind nicht ohne besondere Annahmen zu interpretieren, insbesondere die anscheinende Schutzwirkung des Sauerstoffes gegen Kettenbrüche. Allgemein sind 4 Bestrahlungskombinationen zu unterscheiden:

I	Anaerob,	$\lambda > 250$ nm
II	Anaerob,	$\lambda \leq 250$ nm
III	Aerob,	$\lambda > 250$ nm
IV	Aerob,	$\lambda \leq 250$ nm .

Die Versuche von ALEXANDER mit der filterfreien Quarz-Niederdrucklampe sind am ehesten den unseren (filterfreie Quarz-Hochdrucklampe) vergleichbar.
Er findet aus Lichtstreuungsmessungen für Kombination II :
eine starke Zunahme von Kettenbrüchen und Knäuelungen; für Kombination IV :
eine starke Zunahme von Kettenbrüchen, jedoch eine geringe Knäuelung.

Unsere Ergebnisse nach Schema der Abb.8 wären auf dieser Basis so zu deuten, daß der Maximadämpfungseffekt in erster Linie durch zunehmende Flexibilität (coiling, Knäuelung) vergrößert wird, während die verhältnismäßig seltenen Kettenbrüche erst in zweiter Linie die Adsorptionsschicht begünstigen. Eine kurzfristige Erwärmung über T_M nach der Bestrahlung verstärkt beides und sorgt für die Ablösung der Bruchstücke aus der Doppelhelix.

3. Zur Photolyse von Nucleinsäurebasen

Die kinetische Untersuchung der Basen wird häufig dadurch erleichtert, daß ihre polarographischen Reduktionsstufen und die der entstehenden Produkte registrierbar sind.

3.1. Reduktionsmechanismen

Ob eine Base an der Tropfelektrode reduzierbar ist, hängt vielfach vom pH-Wert der Lösung ab, wobei die protonisierten Formen begünstigt sind. Eingehendere Messungen stammen von CAVALIERI [10] und ELVING [11], wonach Reduktionsstufen bei Cytosin, 2-Aminopyrimidin, 2-Amino-4-methylpyrimidin, 4-Amino-2,6-dimethylpyrimidin, Adenin, Purin, Hypoxanthin u.a. gefunden worden sind.

Bisher konnten drei Hauptmechanismen wahrscheinlich gemacht werden :

Radikalkombination :

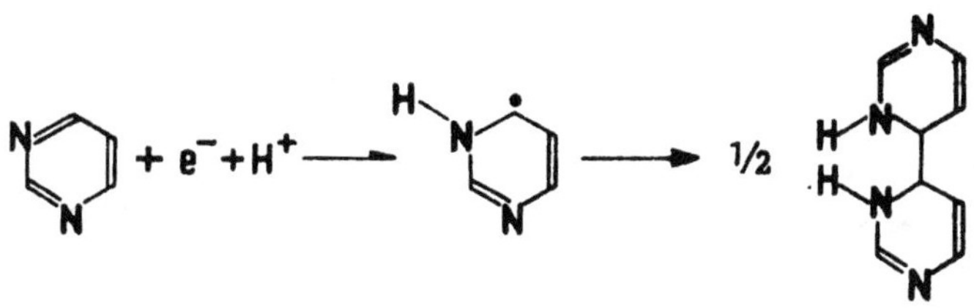

(5)

Aufhebung einer Doppelbindung :

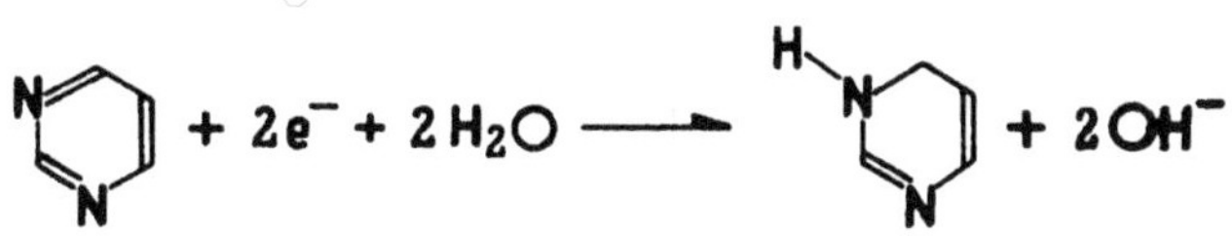

(6)

mit Abspaltungsreaktion :

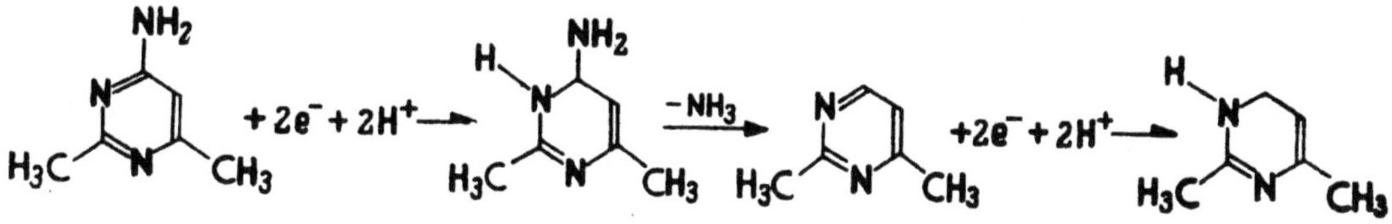

(7)

Sofern aus den Elektrodenprozessen (5 - 7) ausmeßbare Grenzströme resultieren, ist die Verfolgung einer chemischen oder photochemischen Reaktion in der Gesamtlösung möglich.

Leider verschmelzen manche der Reduktionsstufen mit dem Endanstieg des Polarogramms, sodaß ihre Höhen nicht ausgewertet werden können. Dann bleibt nur noch die Möglichkeit, daß ein Reaktionsprodukt positivere Stufen verursacht. In jedem Fall ist es ratsam, die pH-Abhängigkeit der Stufenhöhe zu kennen, um die Elektronenaufnahme unabhängig von Protonisierungsgleichgewichten ermitteln zu können.

3.2. Die Photolyse von 4-Amino-2,6-dimethylpyrimidin

Unter den bisher bestrahlten Pyrimidinderivaten nimmt das 4-Amino-2,6-dimethylpyrimidin wegen seiner Folgereaktionen eine Sonderstellung ein [23]. Die Reduktionsstufe des Ausgangsproduktes (A), die nach Gl.(7) entsteht und bei -1,75 V ihr Halbstufenpotential (pH 8, gegen NCE) hat, nimmt während der Bestrahlung ab (die Absorptionsmaxima liegen bei 236 nm und 270 nm). Die Quantenausbeute ist pH-abhängig und verringert sich in saurer Lösung. Das Photolyseprodukt B verursacht eine gut ausgebildete positivere Stufe mit $\pi_{1/2}$ = -1,4 V (Abb.10). Gegen weitere Bestrahlung ist es verhältnismäßig stabil, jedoch ergeben sich aus kinetischen Daten Hinweise auf eine merkliche Photo-Nebenreaktion.

3.3. Nachreaktionen und Mechanismen

In der einsetzenden Dunkelreaktion von B nimmt seine positivere Stufe wieder langsam ab, ohne daß in Abwesenheit von A ein weiteres Produkt im Polarogramm angezeigt wird. Dennoch könnte das Folgeprodukt C in der Tetraalkylammoniumsalz-Grundlösung eine Stufe negativer als -1,9 Volt hervorrufen. Abbildung 11 zeigt in relativen Konzentrationseinheiten den Verlauf der Photolyse von A, das daraus gebildete B und die Dunkelreaktion 1.Ordnung des Photolysenproduktes als Kurve B_1.

Unsere polarographische Registrierung steht im Einklang mit dem Reaktionsschema nach WIERZCHOWSKI und SHUGAR [23]

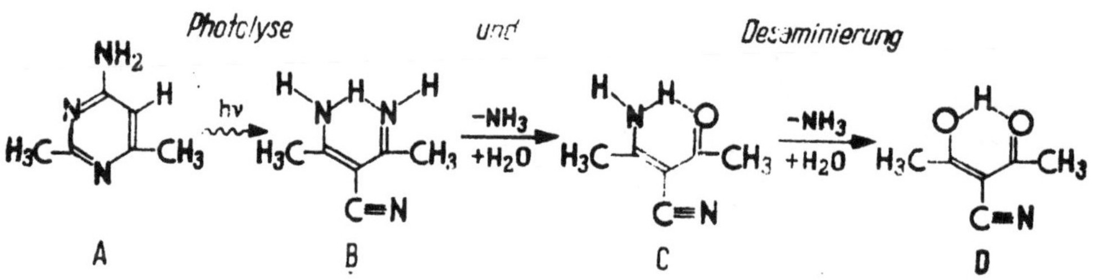

8)

Die Reduktionsstufe von B könnte von der Elektronenaufnahme in die N=C-C=C Gruppierung herrühren, während die C≡N-Bindung erhalten bleibt :

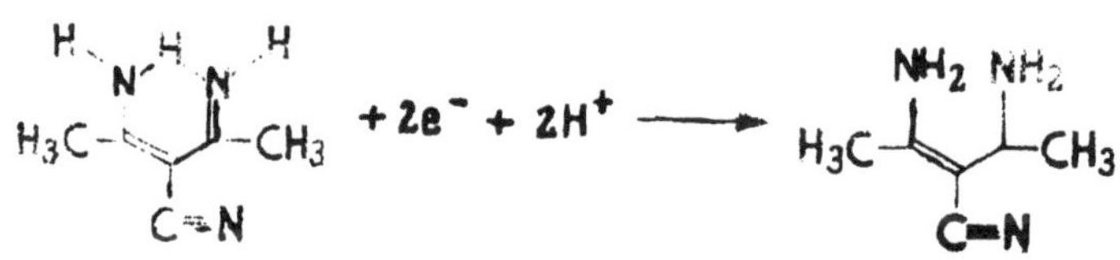

(9)

Zur Reduktion der Desaminierungsprodukte C und D mit der Gruppierung O=C-C=C bedarf es erheblich negativerer Potentiale.

In entsprechender Weise kann die Polarographie für kinetische Messungen anderer Basen oder Nucleoside herangezogen werden.

Zusammenfassend muß festgestellt werden, daß abgesehen von analytischen Erfolgen insbesondere auf dem Gebiet der Krebsdiagnose, die ausgiebige Nutzung der polarographischen Methoden bei Strukturproblemen von Makromolekülen und ihren Bausteinen sich noch im Anfangsstadium befindet. Eine Forcierung dieser Entwicklung besteht vor allem in der weitergehenden Spezifizierung quantitativer Aussagemöglichkeiten.

Für die Mitarbeit an dieser Aufgabe danke ich Fräulein LORBEER, Frau FIEDLER, Herrn BÄR und Herrn KITTLER. Herrn Dr.H.VENNER und Herrn Dr.R.NÜSKE (Abt.Biochemie) gebührt unser Dank für die bereitwillige Unterstützung mit wertvollen Substanzen.

L I T E R A T U R

1) BERG,H., Pharmazie 11, 239 (1956)
2) LÜCKE,H., H.BERG, 1.Jena.Symp. (1962) Polarographie in der Chemoth., Biochem. u. Biol., Akad.V., Berlin (1964)
3) MÜLLER,O.H., 1.Jena.Symp. (1962) Polarographie in der Chemoth., Biochem. u.Biol., Akad.V., Berlin (1964)
4) MÜLLER,O.H., 2.Jena.Symp. (1963) Physik.Chem.biogener Makromol., Akad.V., Berlin (1964)
5) IWANOW,I., E.RACHLEJEWA, V.KRYLOVA, Ukrain.biochem.J. 34, 678 (1962)

6) IWANOW,I., E.RACHLEJEWA, 1.Jena.Symp. (1962) Polarographie in der Chemoth.,Biochem. u.Biol., Akad.V., Berlin (1964)
7) NEZU,H., Rev.of Polarogr. Japan 10, 154 (1962)
8) BŘEZINA,M., V.GULTJAJ, Coll.czechoslov.chem.Commun. 28, 181 (1963)
9) PALEČEK,E., 1.Jena.Symp. (1962) Polarographie in der Chemoth.,Biochem.u.Biol., Akad.V., Berlin (1964)
10) CAVALIERI,L., B.LOWY, Arch.Biochem.Biophysik 35, 83 (1952)
11) SMITH,D., P.ELVING, J.Amer.chem.Soc. 84, 2741 (1962)
12) BERG,H., 1.Jena.Symp. (1962) Polarographie in der Chemoth., Biochem.u.Biol., Akad.V., Berlin (1964)
13) McLAREN,A., R.LUSE, Science 22, 836 (1961)
14) WENIG,K., O.JIROVEC, Biochem.Z. 295, 405 (1938)
15) BERG,H., Ricerca Sci. 27 (1957) Suppl.3, 184
16) BRDIČKA,R., Research 1, 25 (1947)
17) DOSE,K., B.RAJEWSKY, Progreß in Photobiology S.612, Elsev.Publ.Comp. (1961)
18) RIDEAL,E., R.ROBERTS, Proc.Roy.Soc. A 205, 391 (1951)
19) ENGELHARD,H., K.-R.EIKENBERG, Z.Naturfschg. 11b, 626 (1956)
20) REINERT,K.-E., CH.EILHAUER, unveröff.Versuche
21) MOROSON,H., P.ALEXANDER, Radiat.Res. 14, 29 (1961)
22) MARMUR,J., W.ANDERSON, L.MATTHEWS, K.BERUS, E.GAJEWSKA, D.LANE, P.DOTY, J.cellular. comparat. Physiol. 58, 33 (1961)
23) WIERZCHOWSKI,K., D.SHUGAR, Photochem.Photobiol. 2, 377 (1963)

D I S K U S S I O N S B E M E R K U N G E N

CLAESSON : Welche Möglichkeiten bietet die Polarographie zur Registrierung schneller Reaktionen bei Makromolekülen ?

BERG : Die erläuterten Methoden lassen folgende Modifikationen im Hinblick auf die Verfolgung schneller Vorgänge zu.

a) Bei Austausch von Elektronen : Messung der Austauschgeschwindigkeit, der Dissoziationsgeschwindigkeit u.a. Umlagerungen bei NA-Basen oder bestimmten Aminosäuren.
b) Katalytische Wellen : Protonisierungsgeschwindigkeit.
c) Adsorption : Geschwindigkeit der Adsorption und Desorption von Makromolekülen an der Elektrode. Geschwindigkeit der Desorption bestimmter Farbstoffe aus dem Makromoleküladdukt.

WARSCHAWSKI : Welche Kationen können ebenso wie Kobalt mit Proteinen Komplexe bilden ?

BERG : Katalytische Wellen wurden vor allem auch mit Zn^{2+} und Ni^{2+} beobachtet.

SAUERBIER : Mit welcher UV-Quelle haben Sie die DNA bestrahlt ? Ist es möglich, daß die Kettenbrüche durch UV-Quanten der Wellenlängen unterhalb 260 nm hervorgerufen worden sind ?

BERG : Es kamen mit qualitativ gleichem Ergebnis die Höchstdrucklampe HBO 500 und eine wassergekühlte Niederdrucklampe jeweils mit 5 cm Essigsäurefilter (1:1) zur Anwendung, sodaß Kettenbrüche schon auf Wellenlängen $>$ 240 nm zurückzuführen sind.

LÖBER : Wie erklären Sie die Verschiebung des Maximums mit zunehmender Dämpfung ?

BERG : Die Dämpfung des Maximums ist zumeist gekoppelt mit einer Positivierung des Abbruchpotentials.

LÖBER : Die Maximumdämpfung sollte außer auf das Molekulargewicht auch auf Änderungen des Adsorptionsvermögens ansprechen. Könnte man damit vielleicht Aussagen über den Bindungsmechanismus der F-DNA-Komplexe erhalten ?

BERG : Das ist nicht so einfach meßbar, da der freie Farbstoff seinerseits auch adsorbiert wird und Maxima dämpfen kann. Andererseits besteht die Möglichkeit, den freien Farbstoff F aufgrund seiner Reduktionsstufe zu messen und damit das Gleichgewicht :

$$DNA + F \rightleftharpoons [DNA - F]$$ zu bestimmen.

LAPANJE : How do you interpret the lowering of maxima in the case of nucleic acids ? Do you ascribe this effect to an increased total number of negative charges and polar groups, respectively, in degradation products ?

BERG : The reason is a change in molecular weight and the coiling during photolysis.

Ergänzung zu Abschnitt 1.3 über die

Kobaltkomplex-Doppelwelle

Neben der hier gegebenen Hypothese der Doppelwelle infolge sterischer Effekte auf die Komplexbildung der SH-Gruppen tritt nach unseren neueren Wechselstrommessungen mit Insulin, Ribonuclease u.a. im Konzentrationsbereich 0,002 bis 0,01 % als zweite allgemeinere Ursache das Adsorptions-Desorptionsverhalten der Proteine an der Elektrodenoberfläche hervor. Auf den BREYER-Wechselstrompolarogrammen erscheinen Desorptionspeaks, die im Potentialbereich der Kobaltkomplex-Doppelwelle liegen. Daraus ist zu entnehmen, daß die 1. Welle von der katalytischen Wasserstoffabscheidung des adsorbierten Kobaltproteinkomplexes herrührt, während die 2. Welle dem sich desorbierenden sowie dem herandiffundierenden Kobaltproteinkomplex zuzuschreiben ist. Diese zweite Deutung scheint mit den bisher bekannten experimentellen Abhängigkeiten der Doppelwelle nicht im Widerspruch zu stehen.

Die beobachteten Bestrahlungseffekte auf beide Wellenhöhen lassen sich aus der potentialabhängigen Adsorptionsisotherme des Kobaltproteinkomplexes erklären, wobei die Konzentrationsänderung der aktiven SH-Gruppen durch Photoreduktion der Disulfidbrücken und Umwandlung der Sulfhydryl-Gruppen bewirkt wird.

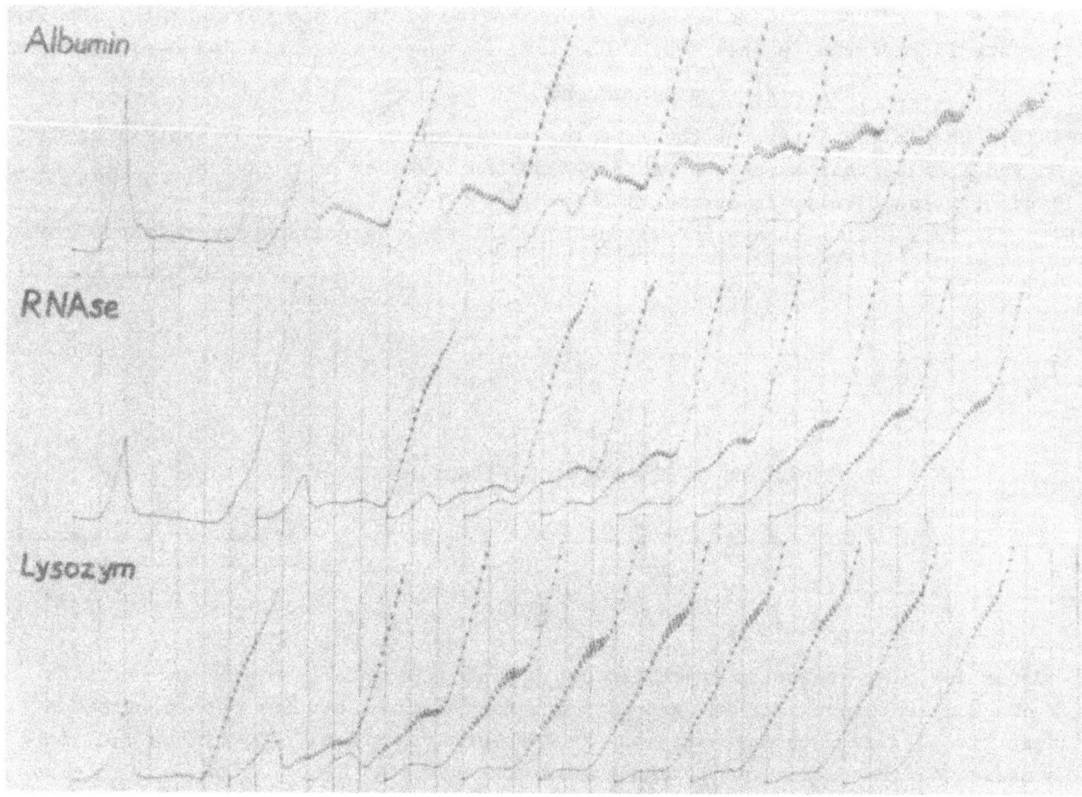

Abb.1 Konzentrationsabhängigkeit der katalytischen Protein-Kobaltwellen nach BRDIČKA (Co^{3+}) für Serumalbumin, Ribonuclease, Lysozym :
v.l.n.r.: 1 - 2 - 3,5 - 5 - 7,5 - 10 - 12,5 - 15 - 17,5 (20) . 10^{-3} %. Die ersten Polarogramme weisen außerdem die "Peptidwelle" auf.

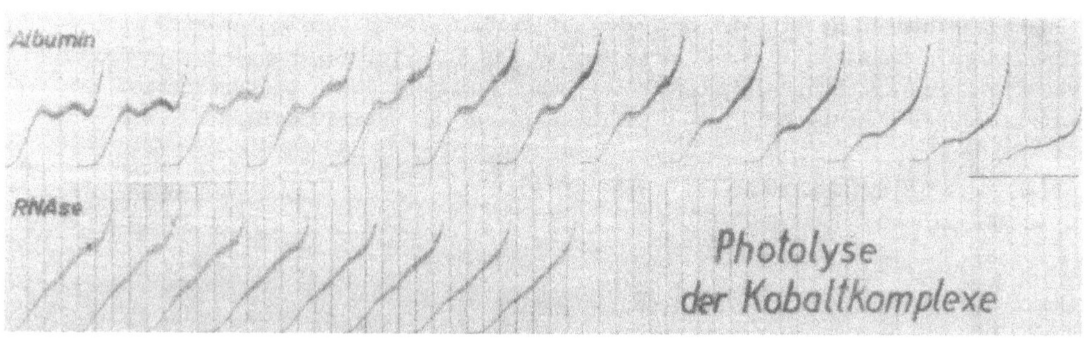

Abb.2 Photolyseverlauf der Protein-Kobaltkomplexe für Serumalbumin und Ribonuclease (0,01 %). (10^{-3}m Co^{3+}-Lösung, PRK-4-Hochdruckbrenner),
v.l.n.r.: 0 - 0,5 - 1,5 - 3,5 - 6,5 - 10,5 - 15,5 - 20,5 - (25,5 -30,5 - 35,5 - 40,5 - 45,5) min.

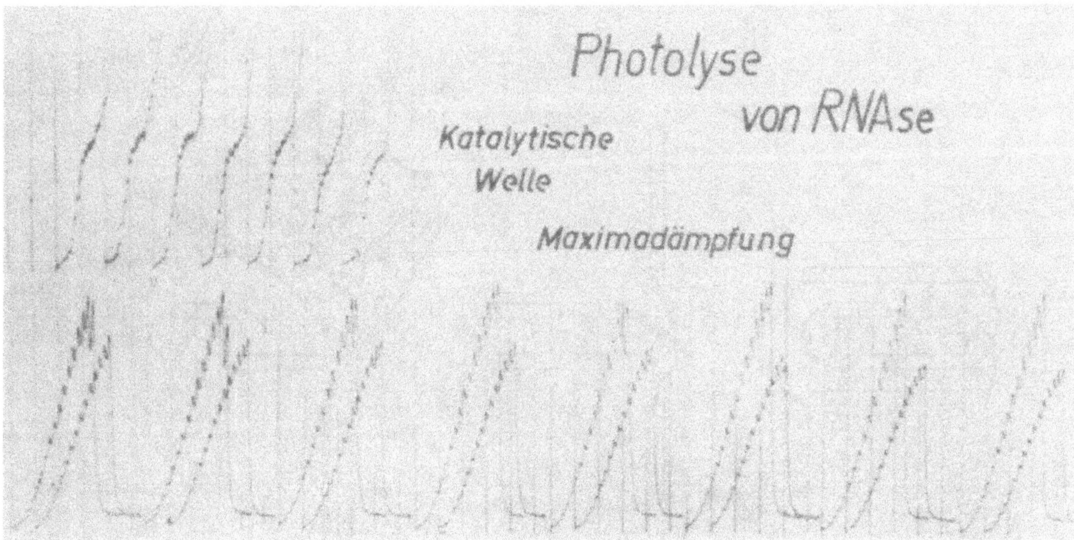

Abb.3 Photolyseverlauf von Ribonuclease ($2,5 \cdot 10^{-4}$%, pH 7, ohne Sauerstoff)
obere Reihe: katalytische Peptidwelle nach 0 - 0,5 - 1 - 5 - 10 - 20 - 30 min
untere Reihe: Dämpfung des Sauerstoffmaximums. Gleiche Bestrahlungsproben wurden nach 0 - 0,5 - 1 - 5 - 10 - 20 - 30 - 40 min entnommen und einer sauerstoffgesättigten Grundlösung zugefügt (daneben ungedämpftes Maximum).

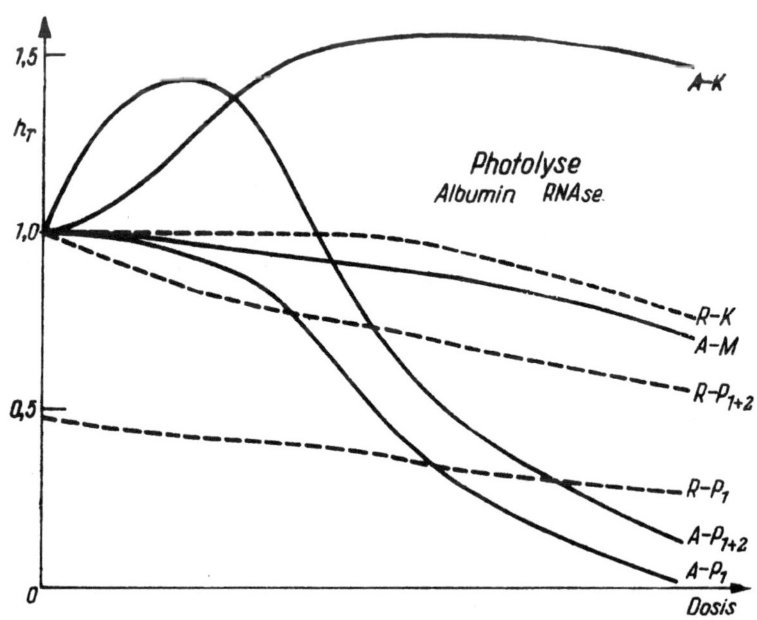

Abb.4 Schema der Kurvenverläufe für Typ 1 (Serumalbumin: A) und Typ 2 (RNAse: R) : A-K - katalytische Peptidwelle, A-M - Maximadämpfung, A-P_{1+2} - beide Protein-Kobaltwellen, A-P_1 - erste Protein-Kobaltwelle, R-K - katalytische Peptidwelle, R-P_{1+2} - beide Protein-Kobaltwellen, R-P_1- erste Protein-Kobaltwelle. h_r - relative Höhen der Wellen und Maxima.

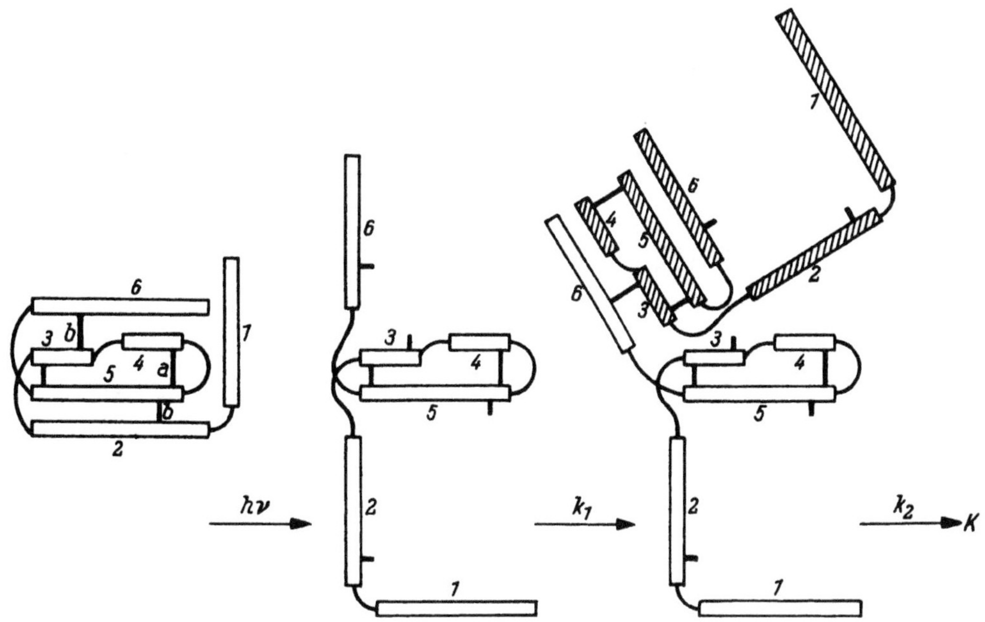

Abb.5 Modellvorstellung für Photolyse der Disulfidbrücken, Vernetzung (k_1) und Koagulation (k_2) bei Ribonuclease.

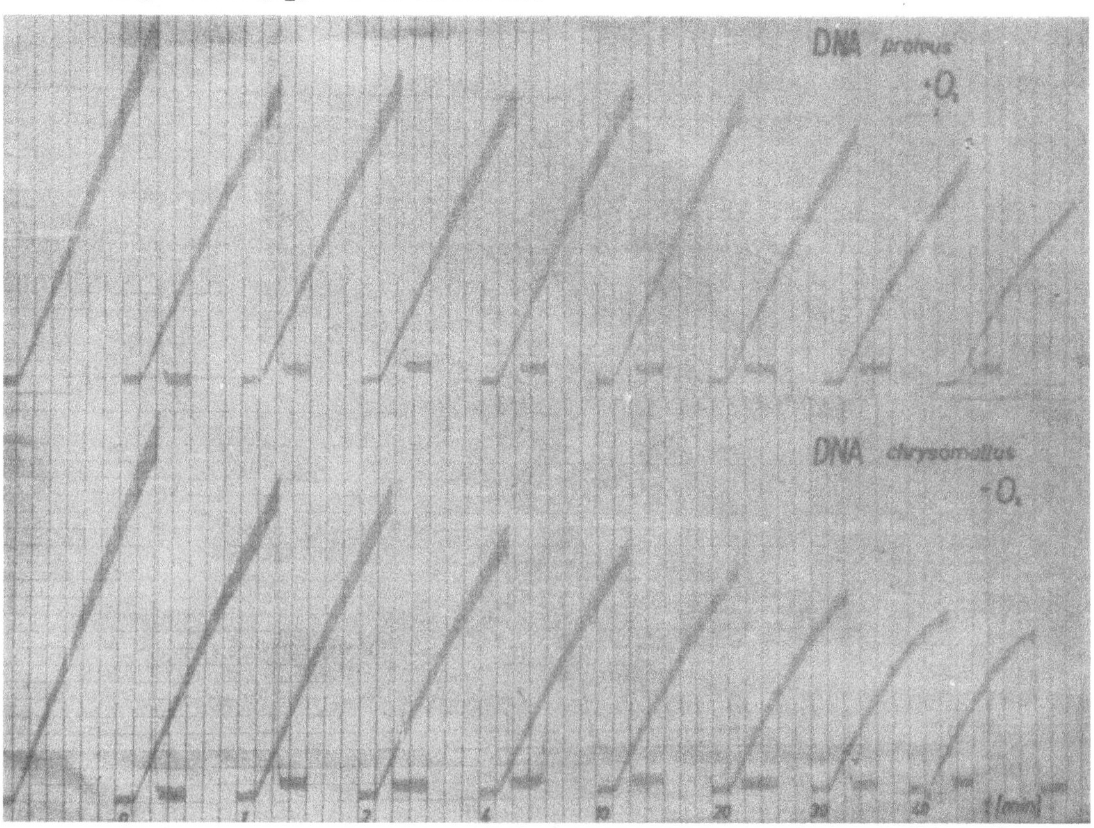

Abb.6 Maximadämpfung durch Photolyse von DNA (links Maxima in der Grundlösung 2.10^{-3}m NaCl, 2.10^{-4}m Na-Citrat). Obere Reihe : Proteus-DNA (6.10^{-4}%), luftgesättigte Lösung; untere Reihe:Chrysomallus-DNA (6.10^{-4}%),ohne Sauerstoff.

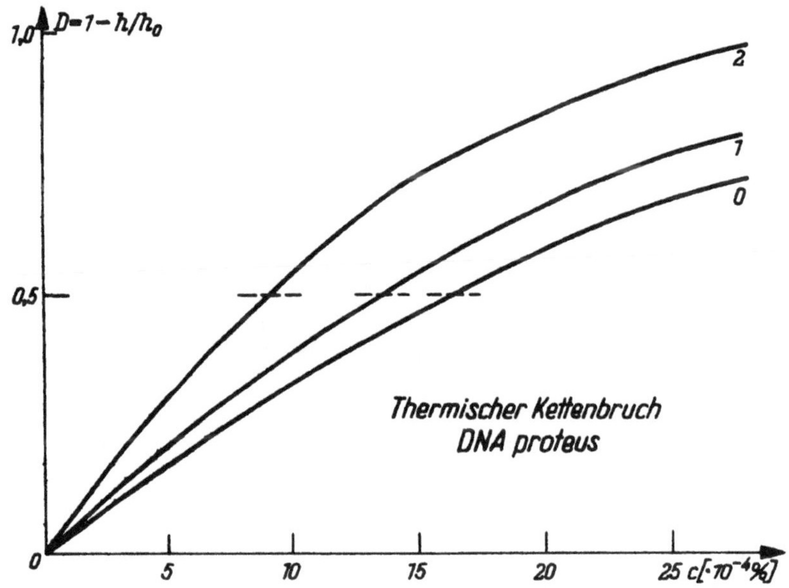

Abb.7 Maximadämpfungsisothermen (24°C) für Proteus-DNA bei thermischer Denaturierung (nach Messungen von H.J.BEER)

0 - Ausgangsprobe, Dämpfungshalbwertskonzentration 16,5.10^{-4}%

1 - 10 Minuten bei 85°C Dämpfungshalbwertskonz. 13,5.10^{-4}%

2 wie 1 und 20 Stunden bei 45°C Dämpfungshalbwertskonz. 9 .10^{-4}%

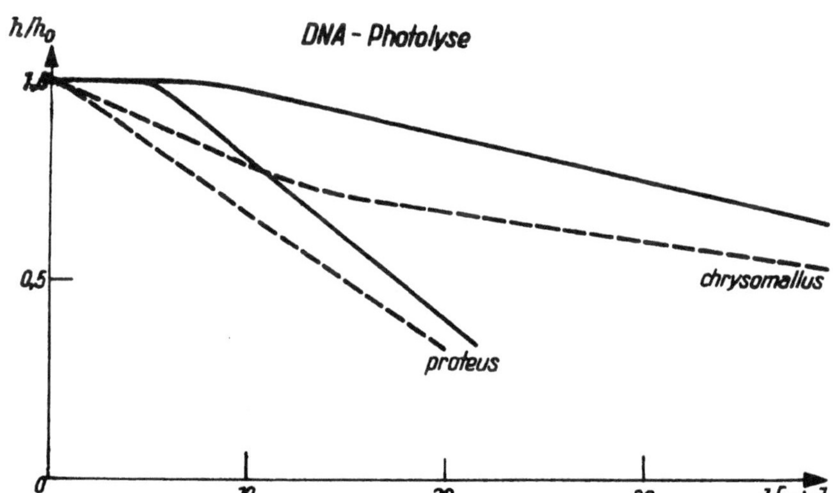

Abb.8 Relative Maximahöhen in Abhängigkeit von der Bestrahlungszeit als Maß für die Photolyse von Proteus-DNA und Chrysomallus-DNA :

———— luftgesättigte Lösungen

- - - - - ohne Sauerstoff.

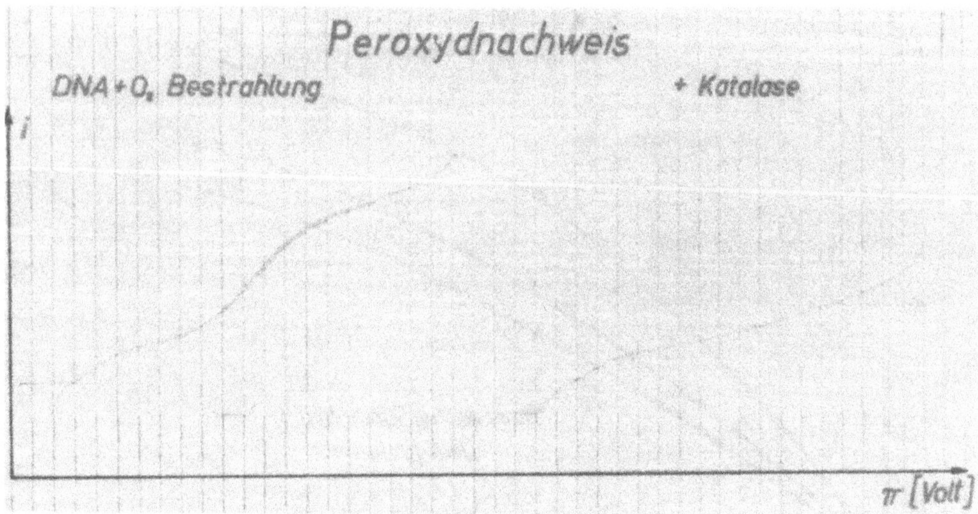

Abb.9 Peroxydbildung während der Photolyse von Proteus-DNA in Gegenwart von Luftsauerstoff. l.: Polarogramm mit Peroxydstufe nach der Bestrahlung
r.: dasselbe nach Einwirkung von Katalase.

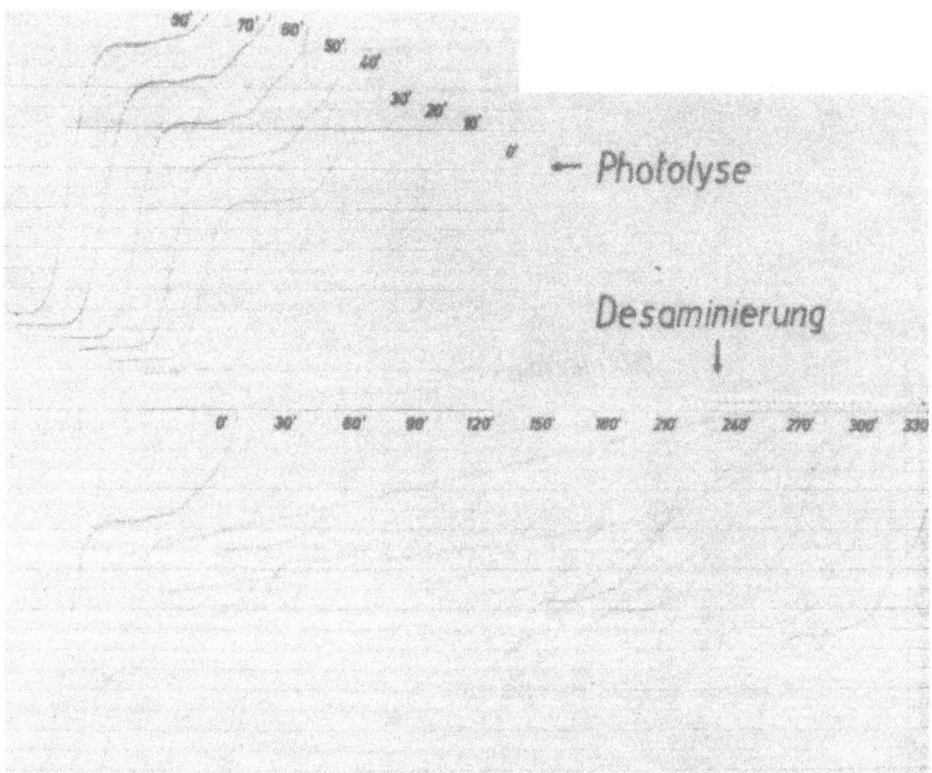

Abb.10 Photolyse (oben) und Desaminierung (unten) von 4-Amino-2,6-dimethylpyrimidin ($4 \cdot 10^{-4}$m, 20 % Äthanol, pH 8)

obere Reihe v.r.n.l.: Bestrahlungszeiten 0 - 10 - 20 - 30 - 40 - 50 - 60 - 70 - 80 min

untere Reihe v.l.n.r.: Abnahme der Stufe des Bestrahlungsproduktes B (negative Reststufe von A sichtbar).

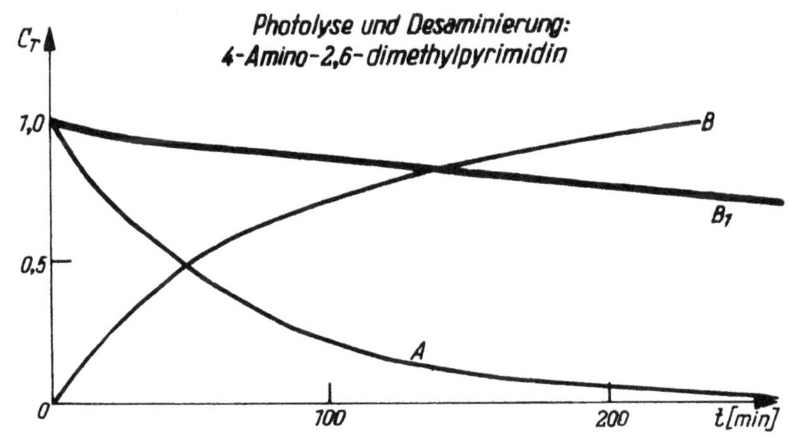

Abb.11 Schematischer Verlauf von Photolyse und Desaminierung nach Abb.10 :

A - relative Abnahme von 4-Amino-2,6-dimethylpyrimidin

B - relative Bildung des Photolyseproduktes

B_1 - relative Desaminierung des Photolyseproduktes.

Generaldiskussion II

Zur Struktur der Proteine und Polysaccharide

Chairman: K. DOSE (Frankfurt/M.)

DOSE : Einleitende Worte zur Eröffnung der Diskussion. Kurze Zusammenfassung der wesentlichsten Problemstellungen auf dem Gebiet der Struktur der Proteine und Polysaccharide. Diskussion wird sich auf Fragen aus dem Proteingebiet beschränken. Vorschlag zu folgender Aufteilung der aktuellsten Fragenkomplexe :

I. a. primäre, b. sekundäre, c. tertiäre Proteinstruktur

II. Homogenität der zu untersuchenden Proteine (Methodische Fragen, Isoenzyme, multiple Enzyme u.ä.)

III. Bioenergetik der Proteine (Denaturierung, Energieleitung u.ä.).

Über Fragen der primären Struktur soll nicht diskutiert werden, da hier physikalisch-chemische Fragen nicht im Vordergrund stehen. Zum Problem der sekundären und tertiären Proteinstruktur wird die Diskussion auf das Falten-Trommel-Modell nach J.SEGAL gelenkt. Die Möglichkeit einer so regelmäßigen Faltung ist von einer besonderen, regelmäßigen Sequenz von Aminosäuren der Peptidketten abhängig. Bisherige Berichte über eine gesetzmäßige Periodizität in der Primärstruktur konnten aber nicht bestätigt werden. Dagegen haben die Arbeiten von KENDREW und PERUTZ inzwischen weltweite Anerkennung gefunden.

MÜNCHMEYER konnte leider hierzu keine Einzelheiten im Zusammenhang mit seinem Vortrag bieten, während

BEHLKE auf den zu erwartenden Tagungsband über Molekulare Zellphysiologie (Berlin, 13.-17.8.63) hinwies, wo J.SEGAL elektronenmikroskopische Aufnahmen und darauf sichtbare "Löcher" für sein Modell gedeutet hat. Allerdings meldeten sich Gegner, die entschieden für die Struktur nach KENDREW und PERUTZ sprachen, zu Wort.

BERG : Mit Hinweis auf die elektronenmikroskopische Aufnahme aus dem Vortrag von St. CLAESSON wurde davor gewarnt ohne Berücksichtigung der möglichen Denaturierungen durch die Präparationstechnik Hypothesen aufzustellen. Welche Umstände dabei zu beachten sind, wird auch aus dem Vortrag von J.RICHTER, Basel, deutlich. Andererseits steht die streifenartige Struktur von DNS-Fäden in guter Übereinstimmung mit dem Modell nach WATSON und CRICK, wie durch jüngste Elmi-Aufnahmen gezeigt wurde.

DOSE : Wenn auch mehr oder weniger regelmäßige Faltungen in Proteinen außer Zweifel sind, so erfordert das SEGALsche Modell doch einige bisher nicht bekannte Regelmäßigkeiten der Primärstrukturen, so daß dieses Modell eine Ausnahme, aber nie die Regel sein dürfte.

BODMANN : Im Falle des Fibrinogens brachten elektronenmikroskopische Aufnahmen entgegen der bisherigen Vorstellung über ein Stäbchenmodell nach neueren amerikanischen Arbeiten eine Aufklärung. Dabei handelt es sich um Anordnungen von Knäulen, wie schon früher von Dr.ENGEL in unserem Institut aufgefunden, aber nicht publiziert wurde.

LUMRY : Es gibt inzwischen viele Röntgenstrukturanalysen von globulären Proteinen, bei denen das Vorhandensein von gewendelten Strukturen, die andere als die α-Helix sind, einen Widerspruch herausfordert. Aber auch die α-Helix ist hier nur von sekundärer Bedeutung. Anders liegen die Verhältnisse natürlich bei den Faserproteinen. Bei der Denaturierung treten,unter Änderung der Wasserstoffbrückenbindung, Knäuel-Strukturen auf, jedoch wissen wir noch zu wenig über den denaturierten Zustand.

DOSE : Da Röntgenfeinstruktur-Untersuchungen nicht auf der Tagesordnung standen, wollen wir die Tertiärstruktur nicht weiter diskutieren, sondern uns dem Problem Nr.2, der Heterogenität, zuwenden. Diese beeinflußt entscheidend die Reproduzierbarkeit der Messungen aus verschiedenen Laboratorien, wobei wir Verunreinigungen außer Betracht lassen wollen. Es geht hierbei vielmehr um

1. schwer trennbare und engverwandte Proteine wie z.B. die sog. Isoenzyme
2. Aufarbeitungseffekte (das Material wird durch die Aufarbeitung infolge teilweiser Denaturierung heterogen).

BODMANN : Im Hinblick auf die Enzymeinheitlichkeit sind elektrophoretische Analysen aus dem Laboratorium von KAPLAN der Milchsäure-Dehydrogenase von Hühnerembryonen zu nennen. Er fand eine Abhängigkeit in der Weise, daß er zunächst zwei extrem unterschiedliche Enzymformen feststellte. Mit steigendem Alter des Objektes schien eine Hybridisierung einzutreten, die sich durch das Auftreten neuer multipler Formen äußerte. Diese Formen sind offenbar durch Kombination zweier verschiedener Untereinheiten im wechselndem Mengenverhältnis entstanden.

LUMRY : Man muß annehmen, daß die in Bezug auf ihre Aminosäurezusammensetzung definierbare Heterogenität der Proteine durch verschiedene Gene bedingt ist. Zum Beispiel gibt es 3 Arten von Enolasen mit verschiedener Aminosäurezusammensetzung. Eine weitere Art von Heterogenität von großer Bedeutung haben wir in der tertiären Struktur hinsichtlich der Faltung und der Lage der Disulfidbrücken.

DOSE : WIELAND und PFLEIDERER haben schon vor vielen Jahren die Heterogenität der Enzyme untersucht. Beispielsweise konnte die aus Rattenleber isolierte und kristallisierte Lactatdehydrogenase (LDH) in 5 verschiedene LDH-Fraktionen aufgespalten werden (pH 8,6 Veronalpuffer). Es scheint so, daß alle Organe desselben Tieres diese 5 verschiedenen LDH's enthalten. Der Unterschied besteht nur darin, daß die Anteile der einzelnen LDH's von Organ zu Organ verschieden sind. Man hat diese einzelnen Fraktionen mit Trypsin partiell hydrolysiert und die so erhaltenen Polypeptide elektrophoretisch und chromatographisch analysiert, so wie es INGRAM bei der Analyse des Sichelzellenhämoglobins tat. Weiterhin hat man quantitative Aminosäureanalysen der Totalhydrolysate durchgeführt. Es zeigte sich, daß die Primärstruktur aller dieser LDH's sehr ähnlich sein muß; dagegen müßten sie sich in Bezug auf ihre sekundäre und tertiäre Struktur sehr unterscheiden, da ihre isoelektrischen Punkte, Wirkungsoptima, Michaeliskonstanten sehr verschieden sind. Hinweis auf BODMAN's Bemerkung bezüglich des Aufbaus dieser verschiedenen Enzymformen aus 2 Typen von Untereinheiten.

CLAESSON : Machte grundsätzliche Bemerkungen zur Definition der Heterogenität. Er möchte z.B. Dissoziationsformen oder Dimere bzw. Polymere nicht als heterogene Substanzen bezeichnet wissen. Er stellt die Frage nach den Gründen, die zur Einführung des Begriffs Iso-Enzym geführt haben.

DOSE : weist darauf hin, daß die Einführung der Bezeichnung Iso-Enzym nur phänomenologisch zu verstehen ist : Man kristallisierte ein bestimmtes Enzym und unterwarf es danach noch einer weiteren Trennprozedur (meist Elektrophorese oder Chromatographie) und stellte oft eine Auftrennung in mehrere Proteinfraktionen identischer Spezifität fest. Diese Formen bezeichnete man als Iso-Enzyme, zunächst ohne Rücksicht auf die bis dahin noch unbekannte molekulare Struktur des Proteins. Stellt man aber nun strukturchemische Betrachtungen in den Vordergrund, dann wäre die Bezeichnung Iso-Enzym hier mißverständlich.

BODMANN : Weist auf Enzym-Beispiele von KAPLAN hin, wonach diese trotz verschiedener Aminosäurefolge gleiches Molekulargewicht und gleiche Aktivität besitzen sollen. Man müßte also von Heterogenität in der Primärstruktur und Homogenität in der Sekundär- und Tertiärstruktur sprechen.

BERG : Bei dieser Bezeichnung geraten die Definitionen durcheinander. Man muß erwarten, daß bei heterogener Primärstruktur auch die Sekundär- und Tertiärstruktur unterschiedlich sind und sich trotz großer Ähnlichkeit etwas anders auswirken. Trotzdem wäre es angebracht, sich über eventuelle neue Definitionen einig zu werden, wenn der ursprüngliche Begriff der Heterogenität nicht mehr ausreicht.

BODMANN : Erwähnte, daß eine teilweise Änderung des Moleküls ohne merkliche Einbuße an Enzymaktivität möglich ist.

CLAESSON : Selbst bei größter Ähnlichkeit sollte man streng genommen nicht von Homogenität sprechen.

LUMRY : Ich unterscheide 4 Möglichkeiten der Heterogenität :

1. Die Aminosäurezusammensetzung.
2. Die Lage der Disulfidbrücken bei den Proteinen und
3. die sterische Anordnung in der Sekundär- und Tertiärstruktur (Faltung der Helix-Bereiche).
4. Die unterschiedliche Kombination von Subproteinen zu einem zusammengesetzten Protein wie z.B. Hämoglobin.

DOSE : Vertritt auch diese Auffassung, aber weist auf die experimentellen Schwierigkeiten hin, die bei der Unterscheidung zwischen diesen Möglichkeiten auftreten. Man sollte aber statt von Iso-Enzymen besser von multiplen Enzym- bzw. Proteinformen sprechen, um weniger mißverständlich zu sein. Versucht die Diskussion auf Fragen der Protein-Bioenergetik, Energieleitung, Energieübertragung u.ä. zu lenken. Jedoch keine Diskussionsbemerkung. Stellt die Frage nach weiteren Diskussionsbemerkungen zu irgendeinem Thema des Fachgebietes.

BERG : Von wesentlicher Bedeutung erscheinen mir zwei Phänomene aus dem Vortrag von LUMRY :

1. Die mehrfache Änderung der Molekülform von Hämoglobin während der Sauerstoffaufnahme. Hierbei ist der seltene Fall zu verzeichnen, daß die Umwandlung von chemischer in mechanische Energie in zwei Schritten erfolgt und das noch reversibel.
2. Die zweistufige Extinktionstemperatur-Kurve mit 2 T_M-Werten beim α-Chymotrypsin. Dieser Fall ist bei Nucleinsäure nur in Gemischen nachweisbar, wie in dem Vortrag von VENNER gezeigt wird.

LUMRY : Ursache für die beiden T_M-Werte sind stark unterschiedliche Helix-Bereiche im Molekül.

DOSE : Schlußworte zur Beendigung der Diskussion.
Dankesworte an alle Diskussionspartner und an die Veranstalter der Tagung.

D Nucleinsäuren

The Denaturation of Deoxyribonucleic Acid in Dilute Solution

D. O. JORDAN

Department of Physical and Inorganic Chemistry of the University, Adelaide

S U M M A R Y

The structure of deoxyribonucleic acid (DNA) proposed by WATSON and CRICK was based on X-ray diffraction studies of fibres of DNA. The evidence for this structure being retained in solution is less direct, but nevertheless the results of investigations using a variety of physico-chemical techniques support the WATSON-CRICK structure; these investigations will be discussed. The relationship of the structure of ribonucleic acid to that of DNA will be considered.

The nature of the forces, both specific (hydrogen bonds) and nonspecific (polar and hydrophobic bonds) which bind the two polynucleotide chains together will be considered. The denaturation of DNA, which is caused by the breaking of the bonds and the consequent separation of the polynucleotide chains will be discussed with reference to the various ways whereby denaturation can be caused, viz. heat, treatment with acid and alkali, solution at low ionic strength, and change of solvent. Methods for studying the kinetics of denaturation will be considered together with the mechanism of the reaction.

The thermal denaturation of DNA occurs with reasonably homogeneous samples over a fairly narrow temperature range. This process represents the breakdown of the ordered, WATSON-CRICK structure of DNA to a more random structure. It has been established by studies using N^{15} and deuterium labelled DNA in admixture with unlabelled DNA from the same organism [1] and also by denaturation in the presence of formaldehyde [2], that on denaturation the single polynucleotide strands of the original double helix separate into single strands. The final product obtained on cooling the heated DNA to room temperature depends very much on the nature of the cooling process. On cooling slowly, the chains recombine to form a material which in many of its properties (e.g. T_m value, sharpness of the denaturation temperature range, and biological activity) closely resembles that of the original DNA; such material is generally referred to as renatured DNA. Rapid cooling, however, produces a material which is very different in its properties from those of the original DNA and may best be regarded as a heterogeneous aggregate of the DNA polynucleotide chains with random hydrogen bonds between such groups as happen to become adjacent. The renaturing process does not produce a material identical with the original DNA and we have shown [3] from chromatographic studies that heat denaturation produces a marked change in the chromatographic profile from that of the native DNA and this we interpret as being indicative of some degradation of the DNA.

An alternative method of denaturing DNA, originally observed by THOMAS [4] and more recently fully investigated by INMAN and JORDAN [5], may also be employed to study the reversibility and mechanism of the denaturation process and has the advantage that chromatographic studies suggest that little degradation of the DNA accompanies the denaturation. This method involves the solution of the DNA in electrolyte solutions at concentrations below critical values for both the DNA and electrolyte. The critical concentration may be determined by following the variation of some physical property which changes when the ordered DNA structure breaks down, e.g. electrical conductivity, optical density, solution density, etc. (Fig. 1). The critical concentration for DNA denaturation is found to vary with the electrolyte concentration (Fig. 2). It is thus possible to denature DNA by solution below the critical concentration and, by adding electrolyte, produce an environment in which the native DNA would be stable and which would thus favour the renaturation of denatured DNA.

We will now compare heat denaturation and dilution denaturation for the DNA of B. megatherium [6]. The native DNA has a T_m of 88° C (Fig. 3, curve a). Heating to 100° C and slow cooling yields a sample with the thermal denaturation curve c (Fig. 3), indicating that the material possesses substantial amounts of reformed complementary regions. This material is generally referred to as 'renatured DNA'. The thermal denaturation curve of a sample heated to 100° C and cooled rapidly (Fig. 3, curve b) shows evidence of almost a complete lack of renaturation. A similar cycle of operations has been performed with calf thymus DNA (Fig. 4). The interesting feature of these results is that in contrast to the DNA from B. megatherium, slow cooling of heated calf thymus DNA does not produce renaturation. This result is generally ascribed to the greater genetic heterogeneity of this DNA compared with the bacterial DNA and therefore the lower probability of complementary strands coming together during the cooling process.

That changes in ionic strength can produce denaturation and renaturation of DNA in a similar manner to that produced by heating [6] is demonstrated for B. megatherium DNA in Fig. 5. Curve a is identical with curve a in Fig. 3 and represents the thermal denaturation curve of native DNA in 0.2 M NaCl. Curve b shows the variation of the relative absorbance with temperature for an 0.001 % solution of B. megatherium DNA in pure water. This concentration is below the critical DNA concentration for denaturation. From the low hyperchromic shift on heating and the high value of $\varepsilon(P)$, [8330], it is evident that the DNA has been denatured.

The addition of sodium chloride to produce a concentration of 0.2 M produces an environment in which native DNA is stable at temperatures below the onset of heat denaturation. Curve c (Fig. 5) shows the variation of the relative absorbance with temperature for a solution of B. megatherium DNA in pure water which has been brought to 0.2 M NaCl by the slow addition of solid sodium chloride over a 24 hour period. The curve is very similar to that obtained for the slow cooled, heat denatured DNA (Fig. 3, curve c) which was interpreted as indicating that the DNA had been very largely renatured. Rapid addition of solid NaCl to the solution of denatured DNA produces a material showing a thermal denaturation curve (curve d) very similar to that obtained by the slow addition of sodium chloride; the rate of addition of sodium chloride thus appears, for this DNA, to have little effect on the extent of renaturation. The result with calf thymus DNA is surprisingly different, after solution denaturation, the slow addition of sodium chloride produces a material with a thermal denaturation curve very close to that of native DNA (Fig. 6, curve c). This curve thus shows a higher degree of renaturation than that shown, under similar conditions, by the DNA of B. megatherium. Rapid addition of sodium chloride leads to a material with a thermal denaturation curve, d, which indicates a lower degree of renaturation and is similar to that obtained for the DNA of B. megatherium (Fig. 5, curves c and d).

Confirmation of the results of the thermal denaturation comes from density gradient sedimentation and electron microscopy. Denaturation of DNA in the presence of formaldehyde has been shown by THOMAS and BERNS [2] to lead to combination of the formaldehyde with the liberated amino groups, thus preventing the formation of either specific or random hydrogen bonds on the transfer of the DNA to an environment favouring renaturation. This enables a distinction to be made between double helical and single strand DNA by density gradient centrifugation. Fig. 7 shows the photometric trace of the ultraviolet photographs of the equilibrium distribution of B. megatherium DNA. Heat denaturation produces a density increase of 0.022 g./c.c. similar to that observed by other workers. Dilution denaturation, however, yields a material having a density higher than that of heat denatured DNA, the increase of density being 0.032 g./c.c. (Fig. 7). This difference is not at present understood, but must reflect a difference in the structure of the hydrodynamic unit of these two forms of DNA in combination with formaldehyde. Renaturation of the dilution denatured DNA by addition of sodium chloride yields a material having the same density as the original native DNA. This behaviour is in contrast to that of renatured, heat denatured DNA, which has a density slightly higher than that of native DNA. DOTY, MARMUR, EIGNER and SCHILDKRAUT [1] give the value for the density of the renatured DNA from D. pneumoniae as 1.704 g./c.c. compared with the value of 1.700 g./c.c. for the native material. This difference is attributed to the presence of regions of denatured DNA at the ends of the renatured molecules arising from incomplete recombination of the single strands. Treatment of such renatured, heat denatured DNA with E. coli phosphodiesterase, which preferentially hydrolyses single strand DNA, yields a material having a density exactly equivalent to that of the original DNA.

Electron micrographs of native B. megatherium DNA show the characteristic rod-like molecules of the WATSON-CRICK structure. The dilution denatured material shows the presence of irregularly coiled molecules similar to that observed by INMAN and JORDAN [5] for dilution denatured calf thymus DNA and to that observed by MACLEAN and HALL [7] for the single strand DNA from bacteriophage ØX 174, for RNA, and for fast cooled heat denatured DNA from D. pneumoniae. Renatured dilution denatured DNA shows under the electron microscope a rod-like structure very similar to the native DNA, although the diameter is somewhat greater.

From these results it would appear that dilution denaturation is a more completely reversible process than heat denaturation. However, the evidence that strand separation has occurred is by no means conclusive and at the moment we consider that our results are best explained in terms of incomplete rupture of the bonds leading to perfect renaturation. This interpretation would explain the unexpected behaviour of calf thymus DNA, in that renaturation occurs after dilution denaturation, but not heat denaturation. This point is being further studied.

The mechanism of the denaturation process produced by solution in low ionic strength solvents is not fully understood. The most probable explanation is that the repulsion between the charged groups on the two polynucleotide chains overcomes the forces of attraction between the chains, thus causing the breakdown of the structure. From the experimental values of the charge fraction determined by INMAN and JORDAN [5] and the Coulombic repulsion force between the polynucleotide chains calculated as a function of the number of charges on the polyion [8], the force of repulsion between the two strands at the critical concentration may be estimated. The value obtained, 3 k.cals. per phosphorus atom, is sufficient to cause the breakdown of the molecule.

L I T E R A T U R

1) DOTY,P., J.MARMUR, J.EIGNER, C.L.SCHILDKRAUT, Proc. nat. Acad. Sci. USA 46, 461 (1960)
SCHILDKRAUT,C.L., J.MARMUR, P.DOTY, J. molecular Biol. 3, 595 (1961)
2) THOMAS,C.A., K.I.BERNS, J. molecular Biol. 3, 289 (1961)
THOMAS,C.A., K.I.BERNS, J. molecular Biol. 4, 309 (1962)
3) COLVILL,A.J.E., D.O.JORDAN, unpublished results
4) THOMAS,R., Biochim. biophysica Acta 14, 231 (1954)
5) INMAN,R.B., D.O.JORDAN, Biochim. biophysica Acta 42, 421, 427, 530 (1960); 43, 9, 206 (1960)
6) COLVILL,A.J.E., D.O.JORDAN, J. molecular Biol., in press
7) MACLEAN,E.C., C.E.HALL, J. molecular Biol. 4, 173 (1962)
8) SCHILDKRAUT,C.L., S.LIFSON, unpublished results.

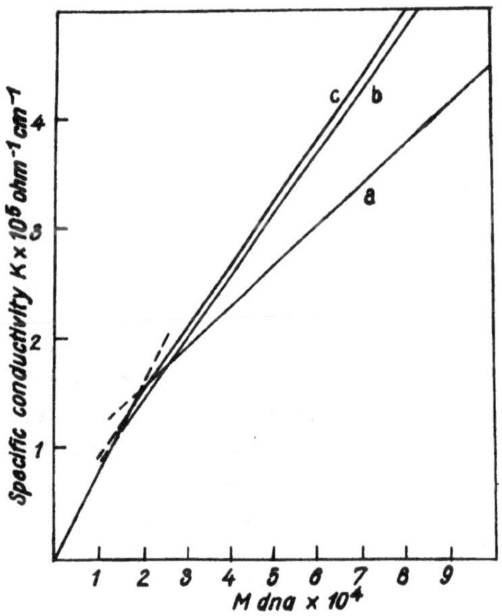

Fig.1 Variation of specific conductivity with concentration for solutions of calf thymus DNA (a) Native DNA; (b) dilution denatured DNA; (c)heat denatured DNA.

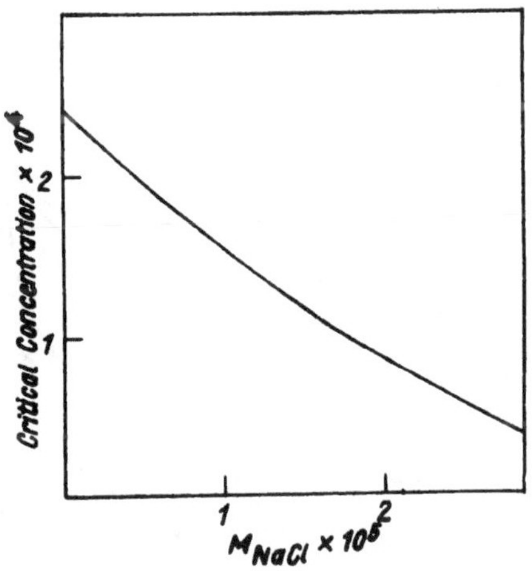

Fig.2 Variation of critical concentration for denaturation with sodium chloride concentration (calf thymus DNA).

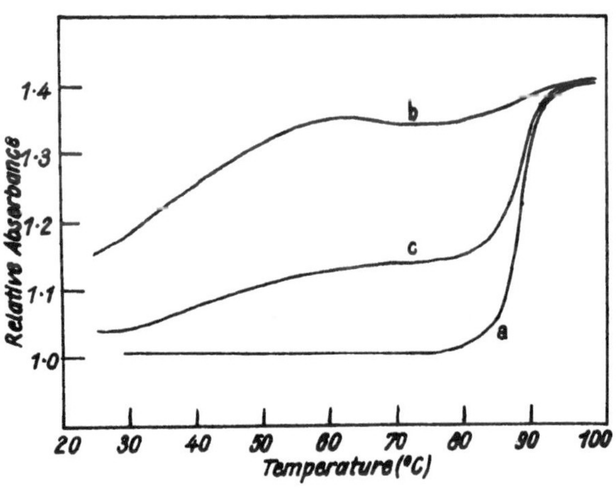

Fig.3 Variation of relative absorbance with temperatur for B. megatherium DNA. Curve a, native; curve b, quick cooled; curve c, slow cooled.

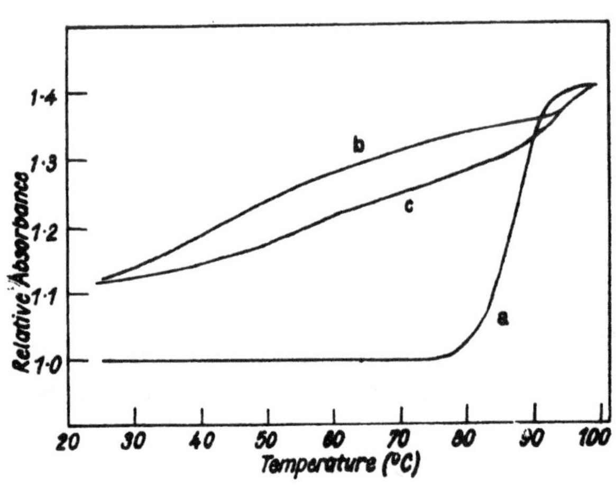

Fig.4 Variation of relative absorbance with temperature for calf thymus DNA in 0.2 M NaCl. Curve a, native; curve b, quick cooled; curve c, slow cooled.

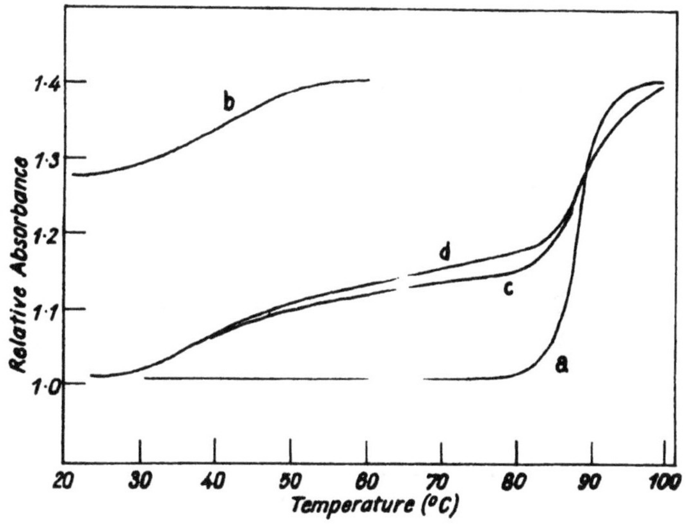

Fig.5 Variation of relative absorbance with temperature for B.megatherium DNA. Curve a, native; curve b, dilution denatured; curve c, dilution denatured,renatured by slow addition of NaCl; curve d,dilution denatured,renatured by rapid addition of NaCl.

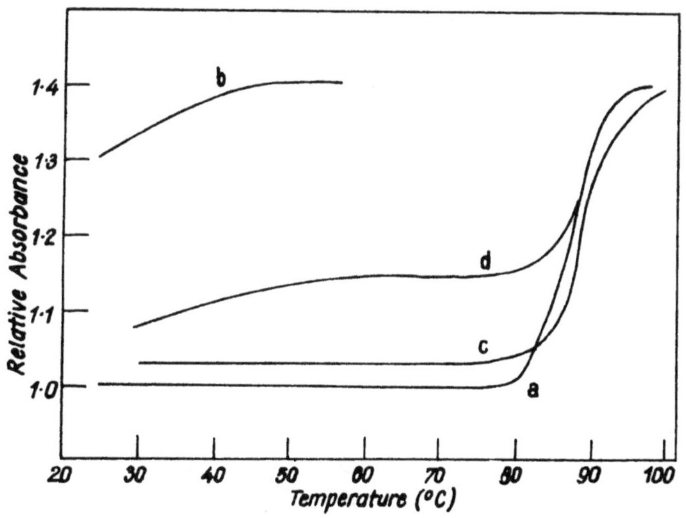

Fig.6 Variation of relative absorbance with temperature for calf thymus DNA. Curve a, native; curve b, dilution denatured; curve c, dilution denatured, renatured by slow addition of NaCl; curve d, dilution denatured, renatured by rapid addition of NaCl.

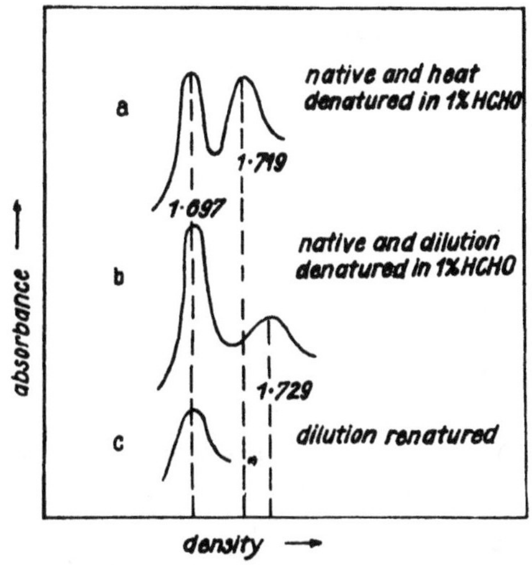

Fig.7 Equilibrium concentration distributions for B.megatherium DNA banded in CsCl gradient.

D Nucleinsäuren

Viscosity and Sedimentation Characteristics of DNA Solutions

J. B. T. ATEN

Medical Biological Laboratory of the National Defence Research Organization TNO, Rijswijk

SUMMARY

DNA was isolated from T4 bacteriophage by means of HERSHEY's phenol extraction method [1]. For some experiments the high molecular weight DNA samples obtained were degraded by mechanical shear [2].

Viscosity measurements were made with an UBBELOHDE viscometer as modified by VAN DUIN and HERMANS [10][11], permitting measurements at very low shear-rates. Determination of the limiting viscosity number gave no special difficulties.

Sedimentation measurements at various rotor speeds and DNA concentrations were performed with a Phywe analytical ultracentrifuge, equipped with a photoelectric recording ultra-violet absorption optical system [12]. Anomalies encountered during the centrifugation of DNA solutions at high rotor speeds are reported.

The value of viscosity measurements and sedimentation analysis for molecular weight determinations is discussed.

INTRODUCTION

In the absence of molecular weight determinations by any absolute method it has been common practice to use light scattering, sedimentation and viscosity measurements to assess the molecular weight of deoxyribonucleic acid (DNA). These methods appeared to give consistent results for molecular weights up to about 10^7 daltons. When it became possible to isolate samples of bacteriophage DNA with a molecular weight of about 10^8 daltons [1] the light scattering method became useless. On the other hand the homogeneity of this DNA as compared with e.g. calf thymus DNA facilitates the study of the difficulties involved in molecular weight determination. Moreover it can easily be degraded in a reproducible way by means of mechanical shear forces [2].

We carried out sedimentation and viscosity measurements on broken or unbroken samples with the intention to contribute to the solution of the problem of molecular weight determination.

One of the points we wanted to study was the dependence of the sedimentation coefficient on the angular velocity of the ultracentrifuge rotor. Some authors [3)-5] reported such a dependence, whereas others [6)-9] did not find any influence of the angular velocity.

During our experiments with different rotor speeds we observed an anomaly especially at higher speeds, which may explain the contradictions.

I should like to ask your attention for this anomaly and for our results, obtained with samples containing broken or unbroken DNA molecules.

MATERIALS AND METHODS

DNA solutions. DNA was isolated from T4 bacteriophage by means of the phenol extraction method as described by MANDELL and HERSHEY [1]. The DNA obtained was dissolved in a solution containing 0.14 M NaCl and 0.02 M sodium citrate, pH 7.

The sedimentation coefficients and limiting viscosity numbers were the same for the different samples prepared.

The concentration of DNA was determined from extinction measurements at 260 mμ, assuming $E_{260}^{1\%} = 200$.

It proved necessary to store DNA solutions for at least 2 days at 4° C before use in order to assure a homogeneous distribution of DNA molecules free from lumps of higher concentration. Stirring was avoided which could produce uncontrolled breakage by shear. Moreover extreme care was taken to prevent contamination of the solution with uv absorbing non-sedimenting material. It was found that rubber stoppers sometimes introduced such contaminants.

Mechanical shear. Degradation of high molecular weight DNA was carried out by forcing solutions at different speeds through a hypodermic needle (No. 20) connected to a syringe with a motor driven piston. In some cases an atomizer was used.

As might be expected the extinction of the solutions is not influenced by this treatment.

Viscometer. Viscosity measurements were made by means of an UBBELOHDE type of viscometer as modified by VAN DUIN [10] and HERMANS [11]. In this instrument the bulbs have been replaced by a calibrated tube of uniform diameter. Measurements at very low shear rates (about 10 sec^{-1}) are possible.

The time intervals were measured, both for solution and solvent, between the moments at which the meniscus passes fixed but rather arbitrarily chosen points in the calibrated tube. From these intervals and the distance between the fixed points the value of the specific viscosity was calculated. These values were plotted vs. the height in the calibrated tube above the equilibrium position. Extrapolation to zero height gives the value of the specific viscosity at zero shear rate (Fig. 1). The extrapolated specific viscosity was plotted vs. DNA concentration (Fig. 2). The slope of the line through these points represents the limiting viscosity number $[\eta]$.

Ultracentrifuges. Some years ago we constructed an ultra-violet absorption optical system with photoelectric recording for the air driven Phywe ultracentrifuge in our laboratory [12]. Using this equipment, which was used for the greater part of the experiments the extinction as a function of distance to the axis of rotation can be read from the recorder strips.

Some experiments were also done with a Spinco ultracentrifuge model E, as commercially available, equipped with an ultra-violet absorption optical system with photographic recording.

Sedimentation coefficients were calculated from a plot of the logarithm of the distance of the boundary to the axis of rotation (log x) vs. time (t). The values were corrected in the usual way for the effect of solvent viscosity and density in order to obtain the value for standard conditions (20° C in water).

As the value of the sedimentation coefficient, especially of unbroken DNA, is strongly dependent on the concentration it is of utmost importance to determine the concentration as accurately as possible. To our experience the reproducibility of the sedimentation coefficient is very good when two samples of the same, homogeneous solution are taken, whereas the coefficients, measured on two different solutions with the same apparent concentration, may differ one or two units in the s-value.

RESULTS AND DISCUSSION

Viscosity measurements. The viscosity measurements gave no difficulties. The limiting viscosity numbers, obtained on 11 samples, containing broken or unbroken DNA molecules, are given in the third column of Table I.

Sedimentation experiments. Studying the dependence of the sedimentation coefficients on the angular velocity we were confronted with an effect which is probably the same as the "pronounced sloping of the plateau" and the "loss of DNA from the plateau", described by THOMAS and PINKERTON [13], and "the upper curvature of the densitometer trace", mentioned as "an unexplained anomaly" by DAVISON and FREIFELDER [14].

We should like to describe the effect as a decrease of the concentration in the plateau region, far exceeding the radial dilution effect and increasing with increasing angular velocity.

This effect is illustrated by Fig. 3. The initial DNA concentrations were 40 μg/ml and 9 μg/ml respectively. It follows from the figure that the effect is already observed during acceleration of the rotor. The final rotor speed was 45,000 r.p.m. The curves shown were recorded with intervals of 5 minutes.

Fig. 4 gives the results obtained on the same solution at 15,000 r.p.m. Under these circumstances the concentration decrease can be fully accounted for by the radial dilution effect.

The same solution was also analysed in a Spinco ultracentrifuge model E, using a 12 mm Kel-F cell at 44,700 r.p.m. The intervals between the pictures were 2 minutes. Fig. 5 gives the results.

Experiments with a Spinco 12 mm Kel-F cell, mounted in the Phywe ultracentrifuge, gave similar results.

These experiments show that the effect is not due to a special type of disturbance, inherent to a particular type of ultracentrifuge.

Attempts to find the origin of the effect revealed the following experimental facts:

The effect is not due to the special form of DNA molecules in solution. It is also observed with purified bacteriophage ØX 174 particles with a sedimentation coefficient of 114 S. These particles are nearly spherical.

Experiments with a synthetic boundary cell, filled with two DNA solutions containing 20 and 45 μg/ml respectively, gave two normally sedimenting boundaries at low rotor speeds. At higher speeds, however, the synthetic boundary was disturbed and, despite a gradual loss of material, the concentration became the same on both sides of the original boundary (Fig. 6). These experiments indicate that convections occur throughout the cell.

The effect is also observed when the DNA has been dissolved in 1 M NaCl. According to DAVISON and FREIFELDER [14] this solvent should have a stabilizing effect on the boundary.

The lower the molecular weight of the samples the higher is the speed at which the ultracentrifuge can be run before the effect is observed. For unbroken DNA it is not possible to go much higher than about 18,000 r.p.m., whereas for samples degraded to an average molecular weight of about 10^6 daltons a rotor speed of about 40,000 r.p.m.

does not cause the effect. This might suggest that there is a correlation between the rotor speed at which the effect occurs and the molecular weight or perhaps the number of particles per unit volume.

We may conclude that the effect

a) is independent of the type of ultracentrifuge,
b) is dependent on the rotor speed,
c) is dependent on the molecular weight or perhaps on the number of particles per milliliter,
d) is independent of the shape of the particles,
e) is caused by convections in the ultracentrifuge cell.

The origin of these convective disturbances is not yet clear. It seems that a decrease of the density of the solution near the cell bottom initiates the convections. This might be caused by the influence of the hydrostatic pressure on the hydratation condition, size and shape of the DNA molecules.

An effect, associated with the concentration decrease in the plateau region is the curvature of the log x vs. t plots: the curvature is upwards with unbroken DNA, downwards with DNA of low molecular weight, whereas both may occur with material that has been subjected to rather mild mechanical shear (Fig. 7). These curvatures may also be influenced by the heterogeneity of the samples and can be explained as follows:

With unbroken molecules the upward curvature is caused by the concentration decrease during the experiment, resulting in an increased apparent s-value.

When the samples have been degraded to a large extent, accompanied by an increase of the heterogeneity, a small number of larger molecules will sediment without boundary formation [15], causing a concentration decrease in the plateau region. The s-value of the remaining molecules is lower than the original average value, giving a downward curvature of the log x vs. t curve.

When the DNA molecules have been degraded to a smaller extent both curvatures may occur, depending on rotor speed and size-distribution of the samples.

As to the effect of the angular velocity on the value of the sedimentation coefficient, we found no effect at different speeds below the value at which the concentration decrease occurs. That means no effect was found for unbroken DNA between 9,000 and 15,000 or 13,000 r.p.m., a small concentration decrease taking place at 18,000 r.p.m. With smaller DNA molecules the speed could be increased to about 40,000 r.p.m. and no angular velocity effect was observed.

It is therefore assumed that all s-values, obtained at 15,000 r.p.m., are reliable.

Fig. 8 shows some sedimentation results obtained at 15,000 r.p.m. on samples that had been subjected to different shear rates. From this figure it appears that the sedimentation coefficient of unbroken DNA is strongly dependent on the concentration. When the molecular weight decreases this influence of concentration becomes less.

S-values at infinite dilution (S_0) are given in the second column of Table I.

The contradictions in the literature mentioned in the introduction may be caused by the effect described above, and can now be explained as follows: At low concentrations the log x vs. t curve, and therefore the apparent s-value, is influenced by the abnormal concentration decrease in the plateau region. At high DNA concentrations, as required when a Philpot-Svensson optical system is used, the s-values become less concentration dependent, the boundary stability increases and an abnormal concentration decrease in the plateau region probably does not take place, therefore no speed effect is observed.

Molecular weight assignment. Several semi-empirical relationships have been proposed for the limiting viscosity number $[\eta]$, the sedimentation coefficient at infinite dilution S_0 and the weight average molecular weight M_w [3)16-20)]. Our values of S_0 and $[\eta]$ can be substituted into the proposed equations. The results are given in Table I, columns 4-10.

Table I

1	2	3	4	5	6	7	8	9	10
Sample No.	S_0	$[\eta]$	$M_w \times 10^{-6}$, calculated with equation						
			a)	b)	c)	d)	e)	f)	g)
1	65	327	117	62	402	140	29	206	138
2	63	325	111	60	363	129	29	188	133
3	61	307	103	56	327	117	27	172	126
4	59	257	90	49	294	108	23	156	118
5	53	243	74	42	208	81	22	115	97
6	48	188	56	33	151	62	18	87	81
7	43	167	45	27	106	46	16	63	66
8	31	110	22.4	14.4	37	19	11	25	36
9	26	77	14.4	9.7	21	12	8	15	26
10	23	–	–	–	14	8.4	–	10.6	21
11	18	31	5.3	4.0	6.3	4.4	3.5	5.3	13.3
12	14.3	16.4	2.7	2.2	3.0	2.3	2.0	2.7	8.7

a) MANDELKERN-FLORY [16)]: $M_w = \left[\frac{S_0 [\eta]^{1/3} \eta_0 \; N.10^{-13}}{\beta\,(1 - V)} \right]^{3/2}$ with: $\beta = 2.5 \times 10^6$, $V = 0.55$, $\eta_0 = 0.01002$, $\rho = 0.9982$

b) EIGNER [3)]: $\log M_w = 4.269 + 1.33 \log (S_0 [\eta]^{1/3})$

c) CAVALIERI-ROSENBERG [17)]: $S = 0.141\, M_w^{0.37}$

d) DOTY-MCGILL-RICE [18)]: $S = 0.063\, M_w^{0.37}$

e) DOTY-MCGILL-RICE [18)]: $[\eta] = 1.45 \times 10^{-6} M_w^{1.12}$

f) HERSHEY [19)]: $S = 0.080\, M^{0.35}$

g) RUBENSTEIN-THOMAS-HERSHEY [20)]: $S = 0.00244\, M^{0.543}$

Some remarks should be made:

a) Most empirical relations have been based on values of M_w from light-scattering experiments with samples of low molecular weight material. Therefore the rather good agreement in the low molecular weight region (columns 5, 6, 7 and 8) is not surprising.

b) Equation g) (column 10) has been calibrated in the high molecular weight region by means of autoradiography [20)]. However, this method yields a number average molecular weight M_n, rather than a weight average one. The results in column 10 are in principle not comparable with the other columns. A difference between column 10 and the other columns in the low molecular weight region is to be expected.

c) It appears that the MANDELKERN-FLORY equation [16], column 4, using $ß = 2.5 \times 10^6$ and the DOTY-MCGILL-RICE equation [18], column 7, give a rather good approximation of the true molecular weight, both in the high and low molecular weight region.

Conclusion. Some authors [13][20] have questioned the value of physical-chemical methods for the determination of molecular weights. However, making allowance for some anomalous phenomena during sedimentation, it seems possible to assess the molecular weight of native DNA by means of sedimentation and viscosity measurements, using the MANDELKERN-FLORY equation [16] or the DOTY-MCGILL-RICE equation [18].

LITERATURE

1) MANDELL, J.D., A.D.HERSHEY, Analyt. Biochem. 1, 66 (1960)
2) a) DAVISON,P.F., Proc. nat. Acad. Sci. USA 45, 1560 (1959)
   b) LEVINTHAL,C., P.F.DAVISON, J. molecular Biol. 3, 674 (1961)
3) EIGNER,J., Thesis Harvard University, 1960
4) EIGNER,J., C.SCHILDKRAUT, P.DOTY, Biochim. biophysica Acta 55, 13 (1962)
5) HEARST,J.E., J.VINOGRAD, Arch. Biochem. Biophysics 92, 206 (1961)
6) KOENIG,V.L., J.D.PERRINGS, J. Colloid Sci. 8, 452 (1953)
7) GORING,D.A.I., C.CHEPESWICK, J. physic. Chem. 60, 506 (1956)
8) KAHLER,H., J.SHACK, J. physic. Chem. 61, 649 (1957)
9) SCHUMAKER,V.N., H.K.SCHACHMAN, Biochim. biophysica Acta 23, 628 (1957)
10) VAN DUIN,P.J., Thesis Leiden 1957
11) a) HERMANS,J., Thesis Leiden 1958
    b) HERMANS,J., J.J.HERMANS, Proc., Kon. nederl. Akad. Wetensch. 59, 324 (1958)
12) ATEN,J.B.T., A.SCHOUTEN, J. Sci. Instruments 38, 325 (1961)
13) THOMAS,C.A. Jr., T.C.PINKERTON, J. molecular Biol. 5, 356 (1962)
14) DAVISON,P.F., D.FREIFELDER, J. molecular Biol. 5, 643 (1962)
15) MARKHAM, R. in: Advances in Virus Research, Vol. 9, 1962
16) MANDELKERN,L., P.J.FLORY, J. chem. Physics 20, 212 (1952)
17) CAVALIERI,L.F., B.H.ROSENBERG, J. Amer. Chem. Soc. 81, 5136 (1959)
18) DOTY,P., B.B.MCGILL, S.A.RICE, Proc. nat. Acad. Sci. USA 44, 432 (1958)
19) a) HERSHEY,A.D., private communication from Dr. EIGNER,J.
    b) BURGI,E., A.D.HERSHEY, Biophys. J. 3, 309 (1963)
20) RUBENSTEIN,I., C.A.THOMAS, Jr., A.D.HERSHEY, Proc. nat. Acad. Sci. USA 47, 1113 (1961)

DISCUSSION-QUESTIONS

JORDAN. - Could you tell me what is the temperature control on the Phywe ultracentrifuge. I agree with Dr. CROTHERS comments that the equations used are semi-empirical equations derived at low molecular weights and extrapolations to high molecular weights are therefore open to doubt and the value of the 'constant' ß is not certain.

Broad boundaries can be obtained by using a synthetic boundary cell and forming boundaries between 2 concentrations of DNA. We have done this for polyelectrolytes and for DNA showing that the sharp boundary is due to photoelectrolytic effects but convection is not necessarily eliminated as an additional cause.

ATEN. - There is only a rather poor temperature control on this instrument. With some experience it is however possible to keep the temperature increase within 1° C per hour.

BERG. - Ihre Tabelle 1 mit den nach verschiedenen Gleichungen berechneten Molekulargewichten ist sehr aufschlußreich. Bei hohen M-Werten sind die Differenzen nicht mehr zu vernachlässigen. Man sollte daher auf jeden Fall Diffusionsmessungen durchführen.

ATEN. - One should use as many methods as possible to assess the true molecular weight by intercomparison.

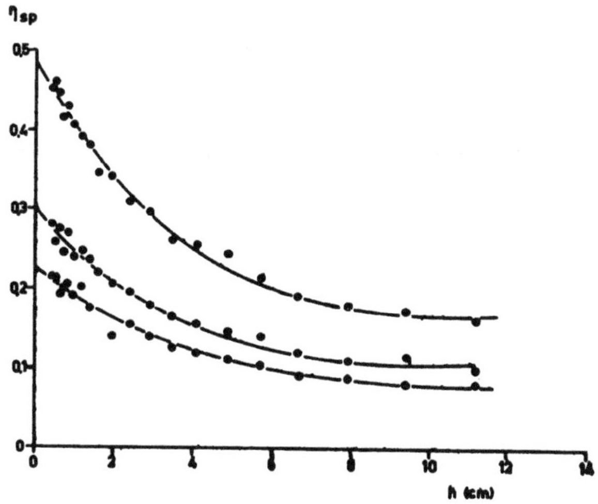

Fig.1 η_{sp} vs. height of meniscus above equilibrium position. DNA concentrations: 8.13, 9.53 and 14.69 µg/ml.

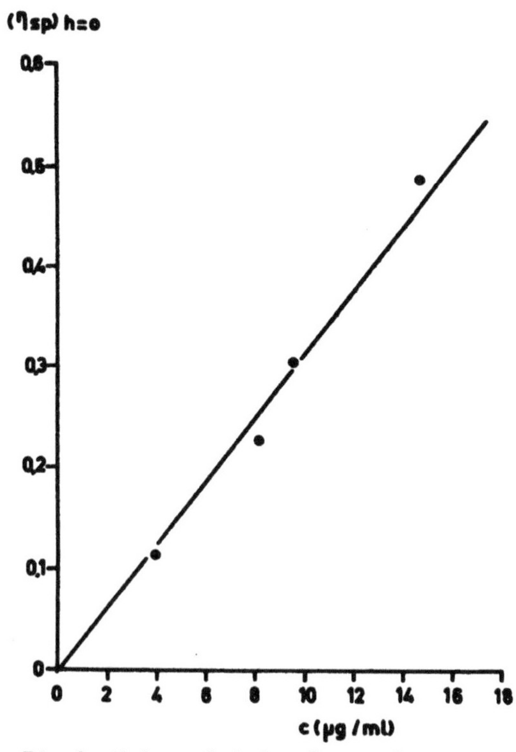

Fig.2 Extrapolated values of η_{sp} plotted vs.concentration. DNA concentrations: 3.98, 8.13, 9.53 and 14.69 µg/ml.

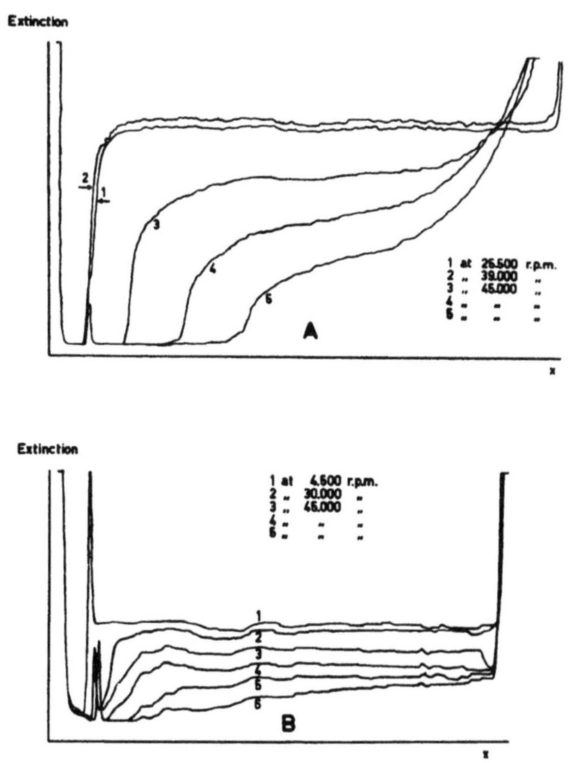

Fig.3 Decrease of concentration in plateau region.Initial DNA concentrations:A. 40 µg/ml B. 9 µg/ml.Time interval: 5minutes. Phywe ultracentrifuge.

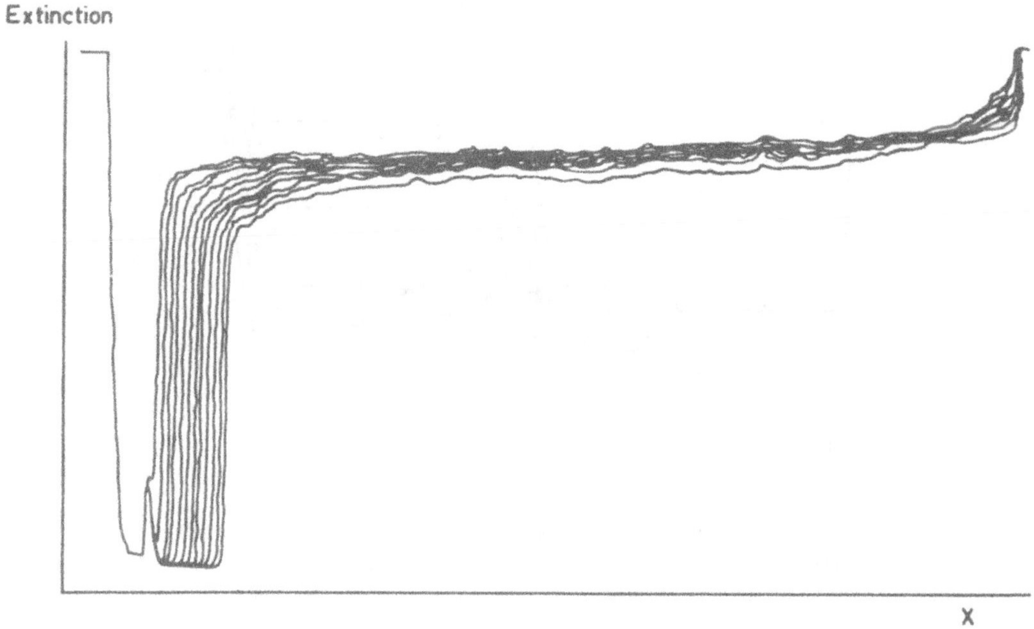

Fig.4 Sedimentation of DNA at 15,000 r.p.m. Initial DNA concentration: 40 µg/ml. Time interval: 5 minutes. Phywe ultracentrifuge.

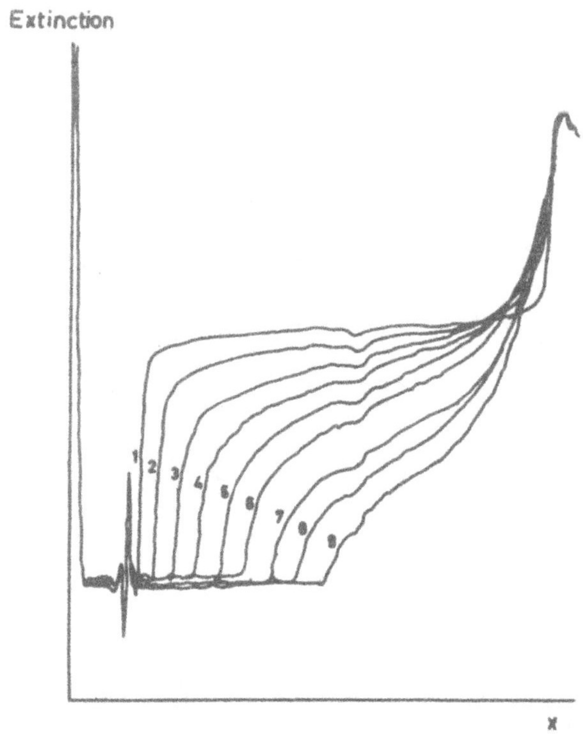

Fig.5 Decrease of concentration in plateau region. Initial DNA concentration: 40 µg/ml. Time interval: 2 minutes. Spinco ultracentrifuge. Rotor speed: 44,700 r.p.m.

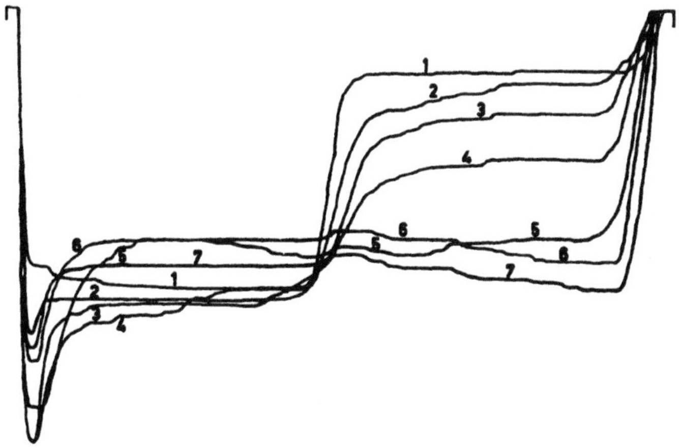

Fig.6 Experiment with synthetic boundary cell. Initial concentrations: bottom 45 µg/ml. top 20 µg/ml. Time interval: 2 minutes. Rotor speeds: 21,000 r.p.m. (1), 27,000 r.p.m. (2), 31,000 r.p.m.(3), 36,000 r.p.m. (4-7)

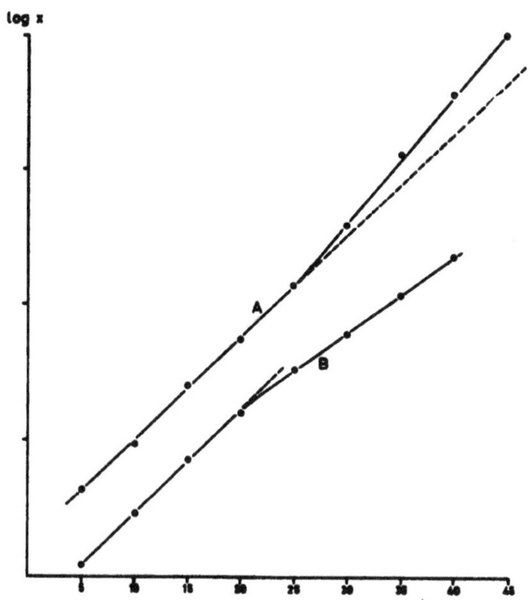

Fig.7 Examples of log x vs. t curves. A. DNA before shear. Concentration 31.4 µg/ml., 21,000 r.p.m.B. DNA after shear. Concentration 17.1 µg/ml., 36,000 r.p.m.

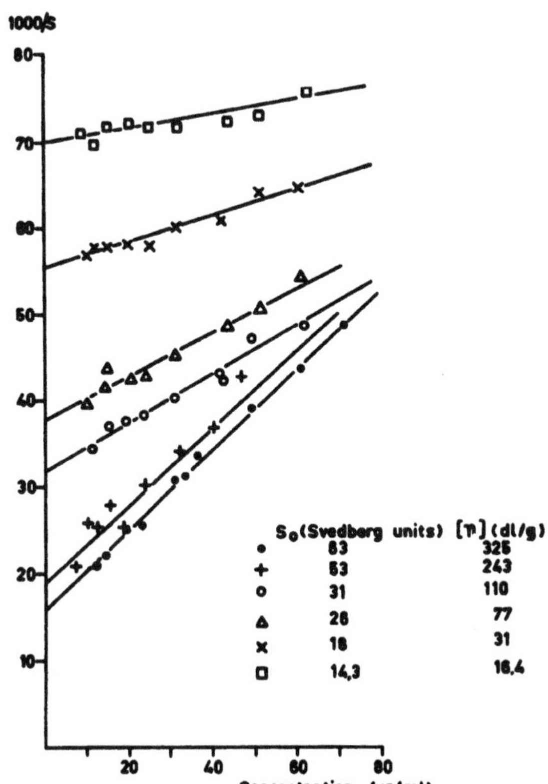

Fig.8 Sedimentation coefficients of broken and unbroken DNA molecules at various concentrations. Angular velocity: 15,000 r.p.m.

D Nucleinsäuren

Untersuchung der Sedimentationskonstantenverteilung verschiedenartig isolierter Nucleinsäuren

H. TRIEBEL

Institut für Mikrobiologie und experimentelle Therapie der Akademie, Jena

S U M M A R Y

By ultracentrifugation we investigated the influence of the preparation conditions (low and high speed mixing applicated during the procedures of isolation and deproteinisation) on the mean molecular weight and on the s_{20}-distribution curves of some DNA samples which have been isolated from different organisms.

In contrast to the results of SCHUMAKER und SCHACHMAN we found that the concentration of the DNA solution strongly influences the shape of the s_{20}-distribution curves even when the concentrations are below 0,004 g/100 ml.

1. E I N F Ü H R U N G

Bestimmungen des Molekulargewichts und der molekularen Uneinheitlichkeit sind in der Molekularbiologie der DNA von besonderer Bedeutung, umsomehr, als diese Kenngrößen stark von der Präparationstechnik abhängen. Einerseits besteht die Gefahr, daß die an DNA gemessenen Molekulargewichte durch den Gehalt an Restproteinen, der zu einer Aggregation der DNA-Moleküle führt [1)2)], beeinflußt werden, andererseits aber sind die DNA-Moleküle extrem empfindlich gegen mechanische Beanspruchung und können leicht einem Scherabbau unterliegen [3)4)5)]. Seit den Arbeiten von THOMAS und BERNS [6)] sowie BURGI und HERSHEY [7)] ist bekannt, daß DNA-Präparate mit gegenüber früheren Präparationen 5-10 mal höheren Molekulargewichten erhalten werden können, wenn Scherkräfte bei der Isolierung so weit wie möglich ausgeschaltet werden.

Wir haben Sedimentationsmessungen an 8 aus verschiedenen Organismen isolierten DNA-Proben durchgeführt und den Einfluß der Präparationsbedingungen auf Molekulargewicht und Massenverteilung der DNA untersucht.

2. E X P E R I M E N T E L L E S

2.1 D e s o x y r i b o n u c l e i n s ä u r e n

Tabelle 1 gibt eine Übersicht über die Herkunft der Präparate sowie über die zugehörigen Analysendaten. In Tab.1 bedeuten : $\varepsilon_P(max)$ die unmittelbar nach der Isolierung gemessene, auf P bezogene maximale Extinktion; T_m die in SSC-Puffer (0,15m NaCl + 0,015m Na-citrat, pH 7,0) gemessene "Schmelztemperatur" und Δ T das zwischen 17 % und 83 % des

Extinktionsanstiegs (bei λ = 260 nm) gemessene Temperaturintervall, das ein Maß für die Heterogenität in der Basenzusammensetzung der DNA darstellt.

Tabelle 1

Organismus	Probe	N-Geh. [%]	P-Geh. [%]	N:P Atom-verh.	ε_P(max)
Kalbsthymus	KT 32	14,01	8,28	3,73	7260
Kalbsthymus	KT 35	15,01	7,63	4,35	7600
Streptomyces chrysomallus	Str. 7	11,15	6,42	3,84	6620
Streptomyces chrysomallus	Str. 8	12,06	6,89	3,87	6580
Escherichia coli	E.coli 4	6,47	5,34	2,68	4580
Bacillus subtilis	Sub. 1	10,62	6,54	3,59	6450
Proteus mirabilis	Prot.67	6,59	3,26	4,46	7900
Proteus mirabilis (L-Form)	Prot.80	12,63	7,15	3,88	7800

Probe	G+C [Mol-%]	T_m [°C]	ΔT [°C]	Hyperchromie λ=260 nm [%]	Prot.-Geh. nach LOWRY [%]
KT 32	44,3	87,0	6,5	34,0	0,1
KT 35		87,2	7,0	32,7	0,5
Str. 7	70,8	96,8	4,0	27,1	0,2
Str. 8		97,2	5,0	24,4	0,6
E.coli 4	49,7	91,2	4,3	32,4	0,2
Sub. 1		88,6	4,1	25,3	
Prot. 67	42,1	87,1	4,4	32,7	1,4
Prot. 80	40,3	87,2	4,4	31,0	1,0

Die DNA wurde isoliert : aus Bacillus subtilis nach der Methode von MARMUR [8], aus Streptomyces chrysomallus nach ZIMMER und VENNER [9], aus E.coli und Proteus mirabilis nach SARFERT und VENNER [10], aus Kalbsthymus nach KAY, SIMMONS und DOUNCE [11]. Bei der Präparation wurde besonderer Wert auf eine möglichst gute Deproteinisierung [12] gelegt, um eine Aggregatbildung der DNA-Moleküle weitgehend zu verhindern. Der nach dem LOWRY-Test bestimmte Eiweißgehalt liegt (mit Ausnahme der DNA aus Proteus mirabilis) in der Größenordnung °/oo (Tab.1).

Während der Aufarbeitung der Zellen und der anschließenden Feinreinigung der DNA wurden mehrfach ein hochtouriger Mixer (Drehzahl : 14 000 UpM) bzw. ein Laborrührwerk (ca. 250 UpM) eingesetzt (s. 3.23).

2.2 Ultrazentrifugation

Die Sedimentationsmessungen wurden teils an einer analytischen Ultrazentrifuge vom Typ Spinco E mit UV-Optik durchgeführt, teils an einer älteren Phywe-Ultrazentrifuge, die wir mit einer UV-Absorptionsoptik ausgerüstet haben. Bei sämtlichen Untersuchungen wurde die DNA in SSC-Puffer gelöst.

3. ERGEBNISSE UND DISKUSSION

3.1 Mittlere Sedimentationskonstanten und Molekulargewichte

In Abb.1 und 2 sind die Konzentrationsabhängigkeiten der mittleren Sedimentationskonstanten verschiedener Präparate dargestellt. Als Konzentrationsmaß dient die mit einer 0,5 cm-Küvette gemessene Extinktion bei λ = 260 nm. Die Darstellungen enthalten zugleich die mittleren Molekulargewichte, die nach der Gleichung von DOTY, McGILL und RICE [13]

$$s^{o}_{20,w} = 0{,}063.M^{0,37} \tag{1}$$

berechnet wurden. Für die DNA Prot. 67 wurde bei $c \approx 0{,}001$ g/100 ml ein s-Wert von 12,0 Sv gemessen, entsprechend einem mittleren Molekulargewicht von $M \approx 2.10^6$. Die Molekulargewichte der verschiedenen Mikroorganismen-DNA liegen gegenüber den bei Kalbsthymus-DNA gemessenen Werten unerwartet niedrig, was - ebenso wie die Diskrepanzen zwischen den Molekulargewichtswerten der beiden DNA-Präparate aus Kalbsthymus - auf einen während der Isolierung erfolgten Abbau zurückzuführen ist (s. 3.23).

Höhere Sedimentationskonstanten und eine stärkere Konzentrationsabhängigkeit von s wurden bei dem Präparat Prot. 80 beobachtet (Abb.3). Dem Extrapolationswert s_{20,H_2O}=36,5 Sv entspricht ein mittleres Molekulargewicht von $M \approx 30.10^6$.

3.2 Sedimentationskonstanten-Verteilungskurven

Zur Charakterisierung der molekularen Uneinheitlichkeit wurden integrale Sedimentationskonstanten-Verteilungskurven berechnet. Dazu wurden den aus den Absorptionsaufnahmen gewonnenen Transmissionskurven punktweise Sedimentationskonstanten gemäß [14]

$$s_{20,H_2O} = \frac{2{,}303.\log \frac{r}{r_o}}{\omega^2 t} \cdot \frac{\eta_T^{H_2O}}{\eta_{20}^{H_2O}} \cdot \frac{\eta_T^{SSC}}{\eta_T^{H_2O}} \cdot \frac{1-V_{20}\varrho_{20}}{1-V_T\cdot\varrho_T} \tag{2}$$

zugeordnet und die dazugehörigen relativen Konzentrationen

$$\frac{c}{c_o} = \frac{\left(\frac{r}{r_o}\right)^2 \log \frac{I}{I_1}}{\left(\frac{r_2}{r_o}\right)^2 \log \frac{I_2}{I_1}}$$

bestimmt. I ist dabei die am Photometer gemessene Durchlässigkeit der Photoplatte, die übrigen Größen haben die übliche Bedeutung. Die Indizes von I und r beziehen sich auf den Meniskus (Index 0), auf die Basislinie c = 0 zwischen Meniskus und Grenzschicht (Index 1) und auf den Anfang des Konzentrationsplateaus (Index 2).

3.21 K o n z e n t r a t i o n s e i n f l u ß a u f d i e V e r t e i l u n g s - k u r v e n

Nach Messungen von SCHUMAKER und SCHACHMAN [14] an Kalbsthymus-DNA soll unterhalb c = 0,004 g/100 ml die Konzentration der Lösung keinen Einfluß mehr auf die Gestalt der Verteilungskurve ausüben. Wir haben dieses Ergebnis bei unseren Proben nicht bestätigen können. Abbildung 4 zeigt integrale Verteilungskurven einer Kalbsthymus-DNA für den Konzentrationsbereich c = 0,001-0,0054 g/100 ml ($\hat{=}$ $E_{0,5}$=0,090-0,475). Die Breite der Verteilung nimmt mit fallender Konzentration erheblich zu; bei c = 0,0054 g/100 ml reicht die Verteilung von 3-25 Sv, während die auf unendliche Verdünnung extrapolierte Verteilungskurve einen s-Bereich von 3-57 Sv überstreicht. Demnach lassen sich lediglich die bei c ≈ 0,001 g/100 ml ermittelten Verteilungskurven ohne zu große Fehler ($\frac{\Delta s}{s} \leq 0,10$) zur Charakterisierung einer DNA (mit $M \approx 10^7$) heranziehen. Messungen bei nur einer höheren Konzentration verfälschen das Bild zu stark.

Abbildung 5 zeigt, wie genau die Extrapolation der Verteilungskurven aus Abb.4 auf c = 0 durchführbar ist. Lediglich die oberen Ausläufer der Verteilungskurven bringen eine starke Streuung der s-Werte.

3.22 E i n f l u ß v o n K o n v e k t i o n u n d D i f f u s i o n a u f d i e V e r t e i l u n g s k u r v e n

Um einen möglichen Einfluß thermischer Konvektion auf die Gestalt der Grenzfläche auszuschließen, wurden in einigen Fällen Verteilungskurven für verschiedene Sedimentationszeiten berechnet. Abbildung 6 zeigt am Beispiel einer E.coli-DNA, daß die Grenzfläche stabil bleibt; die zu 4 unterschiedlichen Zeiten eines Experiments ermittelten Verteilungskurven fallen zusammen. Die unsystematischen Abweichungen bei hohen $\frac{c}{c_0}$ - Werten rühren daher, daß die zur Normierung notwendige Festlegung des Plateaubeginns (r_2 in Gl.(3)) bei den Absorptionskurven nicht willkürfrei möglich ist.

Obwohl die niedrigsten Sedimentationskonstanten nur wenige Sv betragen, läßt Abb.6 auch keinen Diffusionseinfluß auf die Gestalt der Verteilungskurven erkennen.

3.23 E i n f l u ß d e r I s o l i e r u n g s b e d i n g u n g e n a u f d i e V e r t e i l u n g s k u r v e n

In Abb.7 sind die (für c < 0,002 g/100 ml berechneten) Verteilungskurven verschiedener Präparate dargestellt. Die Verteilungskurven der DNA aus Streptomyces chrysomallus, E.coli, Proteus mirabilis (Prot. 67) und Bacillus subtilis überstreichen einen viel schmaleren Sedimentationskonstantenbereich und liegen bei niedrigeren Sedimentationswerten als im Falle der Kalbsthymus-DNA (KT 32). Die höchsten vorkommenden Molekulargewichte betragen bei den Mikroorganismen-DNA 5.10^6 - 6.10^6 (bei E.coli-DNA 9.10^6), bei Kalbsthymus-DNA hingegen 63.10^6 (Gültigkeit der Gl.(1) über den gesamten s-Bereich vorausgesetzt).

Die bei den Mikroorganismen-DNA gefundenen niedrigen Molekulargewichte sind für die betreffenden Organismen nicht spezifisch, und die Verteilungskurven der Abb.7 spiegeln ebenfalls nicht den nativen Zustand der DNA-Moleküle wieder. Bei allen 4 Mikroorganismen wurde während der Isolierung und Deproteinisierung der DNA ein hochtouriger Mixer (14 000 UpM) eingesetzt. Die mit der Anwendung des hochtourigen Mixers verbundene Einwirkung starker Scherkräfte auf die DNA hat einen mechanischen Abbau der DNA-Moleküle (Bruch der Doppelhelix) zur Folge. Die gesamte Mixdauer betrug bei den verschiedenen DNA-Proben : 14 min bei Sub.1, 21 min bei Prot. 67, 28 min bei E.coli 4 und 30 min bei Str.8. Die Zellen von Streptomyces chrysomallus wurden mit Quarzmehl aufgearbeitet, wodurch zusätzliche Scherkräfte wirksam wurden. Obwohl die DNA-Präparate aus verschiedenartigen Organismen stammen, ist es sicher kein Zufall, daß die mittleren Molekulargewichte im gleichen Sinne abnehmen, wie die Dauer der Scherwirkung zunimmt (Abb.7; Tabelle 2). Bei der Isolierung und Feinreinigung der Kalbsthymus-DNA KT 32 (Abb.7) wurde nur ein Laborrührwerk in mäßiger Geschwindigkeit verwendet mit dem Resultat : höhere Molekulargewichte.

Die Kalbsthymus-DNA KT 35, deren Verteilungskurve in Abb.8 wiedergegeben ist, entstammt demselben Ansatz und demselben Isolierungsprozeß wie KT 32; die beiden Proben wurden erst nach der Isolierung getrennt und in verschiedenartiger Weise gereinigt. Während im Falle der Probe KT 32 die Deproteinisierung (mit Chloroform und Amylalkohol in 4 Schritten) am Schütteltisch erfolgte, wurde bei der Probe KT 35 mit Phenol am hochtourigen Mixer (7 min) gearbeitet. Die Folge ist wieder ein Scherabbau : die Verteilungskurve ist nach kleineren Molekulargewichten hin verschoben (Abb.8; Tab.2). Zwar sind bei KT 35 auch die höchsten Molekulargewichte vertreten, jedoch in geringeren Konzentrationsanteilen als bei KT 32.

Daß dagegen eine gleichartige Präparation verschiedener Ansätze zu praktisch gleichen Resultaten führt, wird durch die beiden Verteilungskurven für Streptomyces chrysomallus-DNA 7 und 8 illustriert (Abb.8).

Tabelle 2

Präparat	Dauer der Scherbeanspruchung durch hochtourigen Mixer [min]	s^0_{20,H_2O} [Sv]	$M.10^{-6}$ (nach Gl.1)
Prot. 80	0 (nur Schütteltisch)	36,5	30,0
KT 32	0 (nur Rührwerk)	23,6	9,1
KT 35	7	18,7	4,8
Sub.1	14	14,6	2,5
Prot. 67	21	ca. 13,5	ca. 2,0
E.coli 4	28	13,2	1,9
Str. 7	30	10,3 (12,3)	1-(1,5)
Str. 8	30	12,3	1,5

Daß mechanischer Abbau der DNA-Moleküle die Hauptursache für die unterschiedlichen Ergebnisse der Molekulargewichtsbestimmung darstellt, geht auch aus den folgenden Resultaten hervor :

Die DNA Prot. 67 (Abb.7) wurde aus der Normalform des Proteus mirabilis isoliert. Kürzlich wurde in unserem Institut die L-Form [15)16)] von Proteus mirabilis aufgearbeitet [10)] (Prot. 80). Die Bakterien der L-Form besitzen nur die cytoplasmatische Membran; durch das Fehlen der Zellwand ist hier die Gewinnung von DNA mit sehr schonenden Methoden möglich. Die Deproteinisierung erfolgte durch Phenolbehandlung am Schütteltisch. Da auch bei der Isolierung weder ein Laborrührwerk noch ein hochtouriger Mixer eingesetzt wurde, ist ein ebenso starker mechanischer Abbau wie bei Prot. 67 nicht zu erwarten. In der Tat ergaben Sedimentationsmessungen an Prot. 80 viel höhere Sedimentationskonstanten (Abb.3); das mittlere Molekulargewicht liegt um eine Größenordnung höher als bei der unter mechanischer Belastung aus der Normalform isolierten DNA Prot. 67 (Tab.2).

In Abb.9 sind die Verteilungskurven beider Präparate gegenübergestellt. Auffallend ist dabei die breite s-Verteilung der DNA Prot. 80, die von 15-76 Sv ($\hat{=}$ M = $2,5.10^6$ - 210.10^6 nach Gl. (1)) reicht. Diese Distributionskurve wurde aus bei c = 0,004 g/100 ml gewonnenen Absorptionsaufnahmen berechnet; da bei derartig hohen Molekulargewichten mit noch stärkeren Wechselwirkungskräften zu rechnen ist als bei KT 32 (Abb.4), dürfte die dem Grenzfall unendlicher Verdünnung entsprechende Verteilungskurve bis zu Sedimentationskonstanten von mehr als 100 Sv reichen. Wir glauben nicht, daß diese für eine Mikroorganismen-DNA unerwartet starke Heterogenität dem nativen Zustand entspricht; wahrscheinlich spielt selbst hierbei schon ein Abbau der DNA-Moleküle infolge geringfügiger Beanspruchungen (Schütteln, Umgießen der DNA-Lösungen, hochtourige Ultrazentrifugation) eine Rolle.+) Hinzu kommt, daß der obere Ausläufer der Verteilungskurve von Prot. 80 möglicherweise durch den mehrfach beschriebenen [17)18)] Effekt der Drehzahlabhängigkeit der Sedimentationskonstante von hochmolekularer DNA ($M \gtrsim 5.10^7$) nach zu hohen s-Werten hin verfälscht ist.

Die Ergebnisse dieser und entsprechender Untersuchungen können nicht zur Charakterisierung der untersuchten Organismen dienen, vielmehr spiegeln sie nur den Einfluß verschiedenartiger Isolierungsbedingungen auf die resultierende Massenverteilung wieder. Demnach erlauben Messungen des mittleren Molekulargewichts und Heterogenitätsbetrachtungen an DNA nur relative Aussagen, solange nicht durch äußerst schonende und möglichst standardisierte Präparationsbedingungen die DNA in nativer Form isoliert werden kann. Dem Vorhaben standardisierter Isolierungsbedingungen steht andererseits die unterschiedliche Beschaffenheit der Zellwand und der Zellinhaltsstoffe bei verschiedenen Organismen entgegen.

Mein Dank gilt der ČSAV und insbesondere Herrn Dr. SPONAR vom Institut für Organische Chemie und Biochemie in Prag, wo ich den größten Teil der hier mitgeteilten Sedimentationsmessungen durchführen konnte. Herrn Dr. VENNER, Institut für Mikrobiologie und experimentelle Therapie, Jena, danke ich für anregende Diskussion und für die Überlassung der DNA-Präparate, Fräulein SARFERT für die mit großer Sorgfalt ausgeführten Präparationen.

L I T E R A T U R

1) BUTLER,J.A.V., K.V.SHOOTER, Z.physik.Chem. Frankfurt/M. 15, 6 (1958)
2) CAVALIERI,L.F., J.F.DEUTSCH, B.H.ROSENBERG, Biophys.J. 1, 302 (1961)
3) CAVALIERI,L.F., B.H.ROSENBERG, J.Amer.chem.Soc. 81, 5136 (1959)
4) DAVISON,P.F., Proc.nat. Acad.Sci. USA 45, 1560 (1959)
5) HERSHEY,A.D., E.BURGI, J.molecular Biol. 2, 143 (1960)
6) THOMAS,C., K.BERNS, J.molecular Biol. 3, 277 (1961)
7) BURGI,E., A.D.HERSHEY, J.molecular Biol. 3, 458 (1961)
8) MARMUR,J., J.molecular Biol. 3, 208 (1961)
9) ZIMMER,Chr., H.VENNER, Hoppe-Seyler's Z.physiol.Chem. 333, 20 (1963)
10) SARFERT,E., H.VENNER, in Vorbereitung
11) KAY,E.R.M., N.S.SIMMONS, A.L.DOUNCE, J.Amer.chem.Soc. 74, 1724 (1924)
12) SARFERT,E., H.VENNER, Naturwissenschaften 49, 423 (1962)
13) DOTY,P., B.B.McGILL, S.A.RICE, Proc.nat.Acad.Sci. USA 44, 432 (1958)
14) SCHUMAKER,V.N., H.K.SCHACHMAN, Biochim.biophysica Acta Amsterdam 23, 628 (1957)
15) TAUBENECK,U., Z.allg. Mikrobiol. 2, 56 (1962)
16) TAUBENECK,U., Z.allg. Mikrobiol. 2, 132 (1962)
17) HEARST,J.E., J.VINOGRAD, Arch.Biochem.Biophysics 92, 206 (1961)
18) EIGNER,J., C.SCHILDKRAUT, P.DOTY, Biochim.biophysica Acta Amsterdam 55, 13 (1962)

+) Erst kürzlich ist es CAIRNS, J.molecular Biol. 6, 208 (1963), gelungen, aus E.coli unzerstörte DNA-Moleküle von ca. 1 mm Länge ($\hat{=} M \approx 2.10^9$!) zu isolieren.

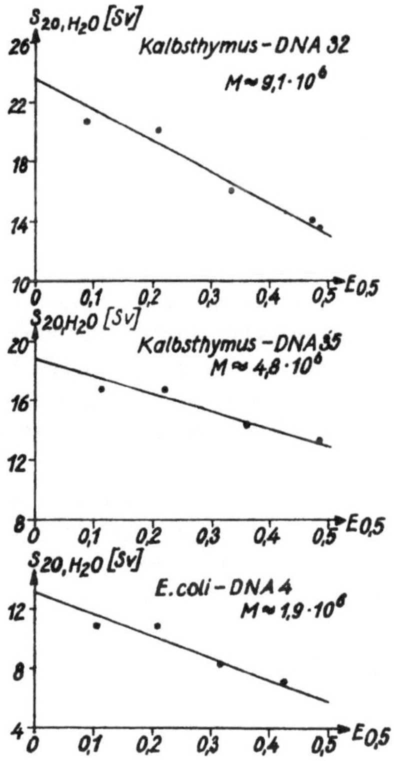

Abb.1 Konz.-Abhängigkeit der mittleren Sedimentationskonstanten verschiedener DNA-Präparate.

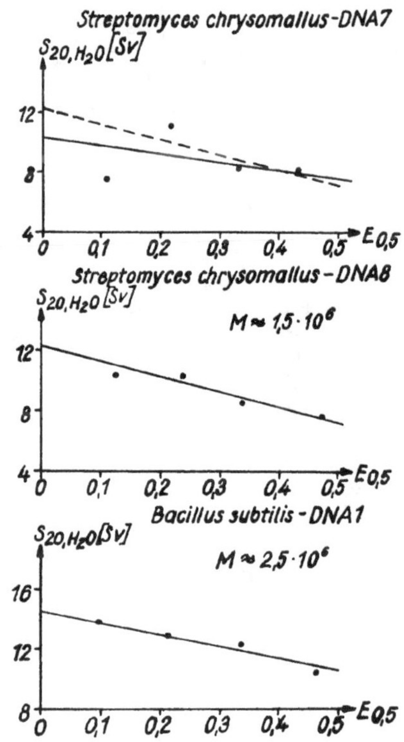

Abb.2 Konz.-Abhängigkeit der mittleren Sedimentationskonstanten verschiedener DNA-Präparate.

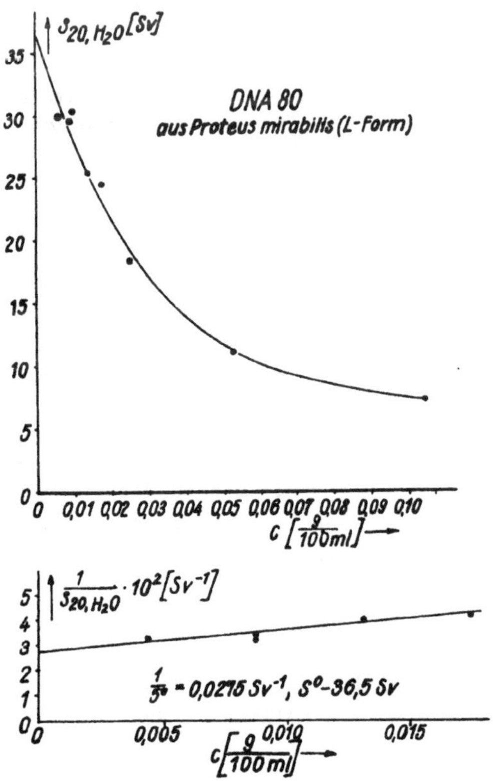

Abb.3 Konz.-Abhängigkeit von s_{20} resp. $1/s_{20}$ einer aus der L-Form von Proteus mirabilis isolierten DNA.

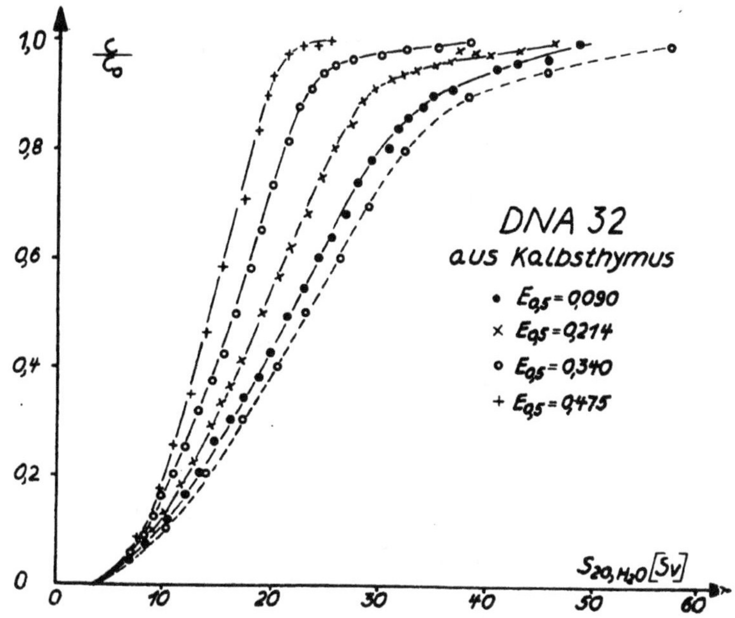

Abb.4 Sedimentationskonstanten-Verteilungskurven einer Kalbsthymus-DNA bei verschiedenen Lösungskonzentrationen ($E_{0,5}$: Extinktion bei λ=260 nm, in 0,5-cm Küvette gemessen), c = 0,001-0,0054 g/100ml. Gestrichelt: auf c=0 extrapolierte Verteilungskurve.

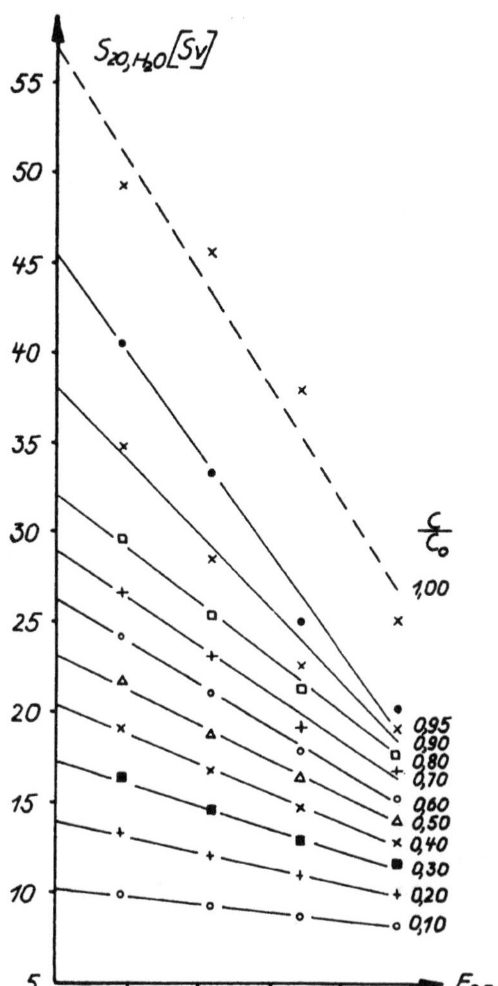

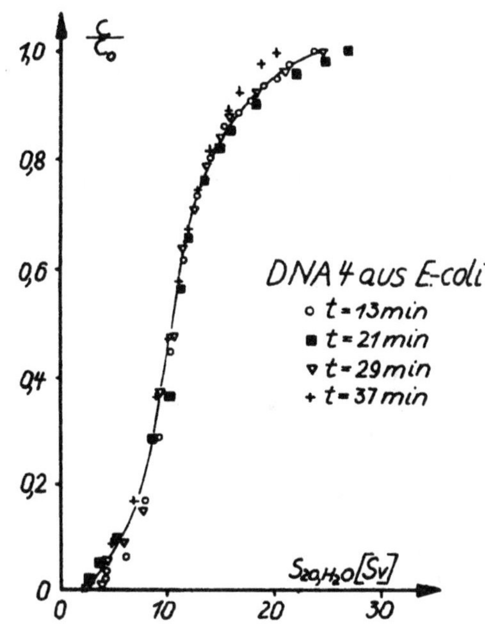

Abb.6 Sedimentationskonstanten-Verteilungskurve einer E.-coli-DNA (c=0,002 g/100ml), berechnet aus zu unterschiedlichen Zeiten eines Experiments exponierten Absorptionsaufnahmen.

Abb.5 Extrapolation der Verteilungskurven aus Abb.4 auf unendliche Verdünnung.

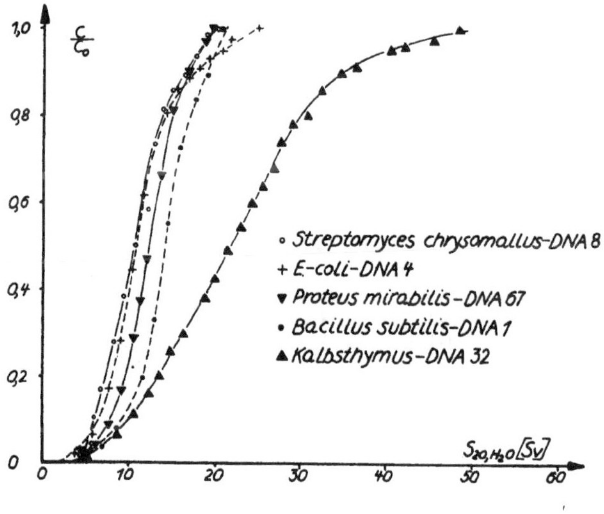

Abb.7 Sedimentationskonstanten-Verteilungskurven einiger DNA-Präparate (Erläuterung s.Text).

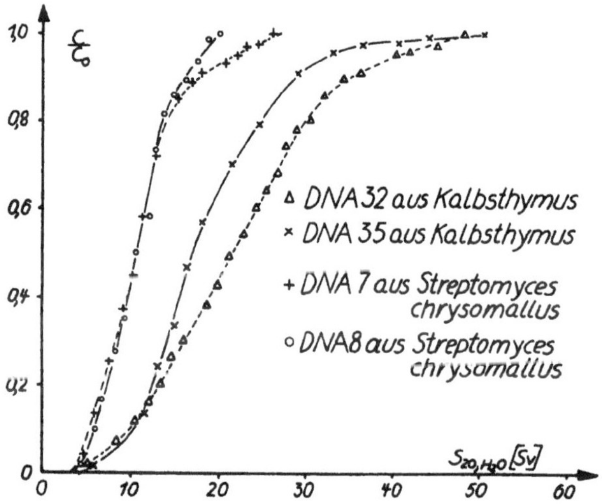

Abb.8 Sedimentationskonstanten-Verteilungskurven einiger DNA-Präparate (Erläuterung s.Text)

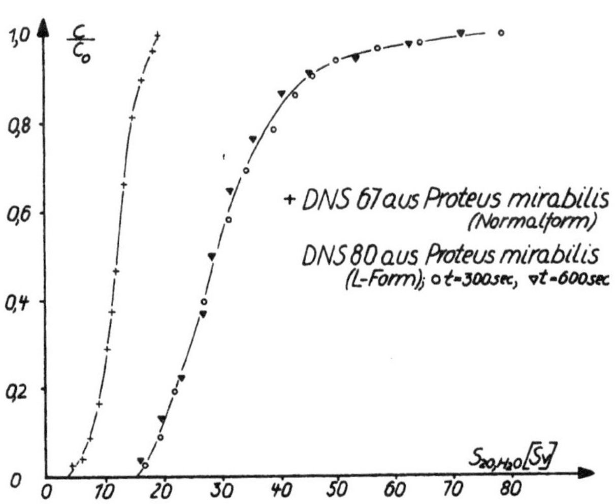

Abb.9 s_{20}-Verteilungskurven von aus der Normalform (unter starkem Scherabbau) und der L-Form (ohne Scherabbau) von Proteus mirabilis isolierter DNA.

D Nucleinsäuren

On the Heterogeneity of Desoxyribonucleic Acids

J. ŠPONAR, L. PIVEČ, Z. ŠORMOVÁ and M. ŠROGL

Institute of Organic Chemistry and Biochemistry and Institute of Microbiology of the Academy, Praha

S U M M A R Y

The heterogeneity of several samples of DNA was studied depending on their origin and on the method of isolation. Special attention was paid to the problem of the heterogeneity of calf thymus DNA which represented a model of the complexity of DNA from higher organisms. The analysis of results enabled us to make conclusions on the subunit structure of the calf thymus DNA. The study of the heterogeneity of B. subtilis DNA revealed significant differences in the transforming activity of fractions.

The heterogeneity of DNA preparations reflects the complexity of the genetic information transferred, but may be also partly due to the method of isolation from the original nucleoprotein. In this work we studied the heterogeneity of several samples of DNA depending on their origin and on the method of isolation. We chose the integral distribution curves of sedimentation coefficients as a measure of heterogeneity in molecular parameters. The distribution curves shown in Fig. 1 were obtained by ultracentrifugation of very dilute DNA solutions (about 10 µg/ml) by means of ultraviolet absorption optics. Strong dependence of the distribution of sedimentation coefficients on the origin of DNA and on the method of isolation appears clearly from the figure.

On the other hand the heterogeneity in chemical composition characterized according to DOTY et al. [1] by the temperature interval in which helix - coil transition occurs, seems to be independent of the method of isolation. Fig. 2 shows the absorption denaturation curves of a sample of bacteriophage DNA and two samples of the calf thymus DNA isolated by phenol and dodecylsulphate. No substantial difference can be found between the temperature intervals of the calf thymus DNA isolated by different methods.

In the light of these results it might seem that the distribution of molecular parameters is the only consequence of some degradation process (mechanical or enzymatic) during the isolation. However, our further results show that this is not the case.

In order to simplify the molecular systems measured, we used the chromatographic fractionation method of HERSHEY, based on selective adsorption of DNA on a column with methylated serum albumin. Fig. 3 shows the elution diagrams of samples used in our previous experiments. Absorbance at 260 mµ (at the left side) and sodium chloride molarity of the elution buffer (at the right side) are plotted against the number of fractions. The separation of DNA into two main fractions occurs. From the absorption spectra and from the analysis of amino acids by paper chromatography it follows that the main fraction eluted first contains a mixture of some peptide material and DNA. We suppose that nucleoprotein particles are present which have not been decomposed on preparation, not even at the high ionic strength of the elution medium.

We have characterized the properties of fractions taken from different parts of the elution curve, especially from the second peak. The absorbing material in the first main fraction usually has quite a low molecular weight and only a slight hyperchromic effect. The denaturation curves of fractions of calf thymus DNA (isolated by the dodecylsulphate method) are plotted in Fig. 4, together with the unfractionated sample. With gradually increasing the molarity of the buffer, the melting points of the fractions decrease. On the other hand, the T_m points of the fractions of the bacteriophage DNA are practically the same. This well agrees with the expected homogeneity in composition of the phage DNA.

In the course of our further work we tried to divide the calf thymus DNA into a number of relatively more homogeneous fractions. Repeated fractionation on methylalbumin columns was used. The results obtained by analysing the fractions are summarized in Table I. The number in the first column determines the order in which the fractions have been eluted. In the second column the average sedimentation coefficients of fractions are given; the third column contains the molecular weights calculated from the sedimentation coefficients providing that the equation given by DOTY et al. [2] is valid. The fourth column shows the values of melting points T_m in standard saline - citrate solution. In the fifth column we compiled the values of the molar fractions of guanine-cytosine obtained from the melting points provided the relation of MARMUR [3] is valid even for the DNA fractions. The assumption that the difference in melting points of fractions is due to that in the guanine - cytosine contents is supported by other measurements on the extreme fractions 1 and 7. The guanine - cytosine values calculated according to SCHILDKRAUT et al. [4] from the buyoant densities of fractions in the CsCl density gradient (column 7 and 8 of Table I) are in accordance with the values obtained from the melting points.

Table I:

Fractions of calf thymus DNA

Number of fr.	$s_{20,w}$	$M.10^{-6}$	T_m °C	p(GC)	ρ	p_ρ (GC)
1	15,3	2,8	92,2	0,56	1,713	0,54
2	18,7	4,8	89,6	0,50	-	-
3	20,5	6,1	88,2	0,48	-	-
4	28,6	15,0	87,1	0,44	-	-
5	20,4	12,1	87,0	0,44	-	-
6	28,8	15,3	86,6	0,43	-	-
7	53,5	81,0	85,1	0,39	1,693	0,35

Comparison of molecular weights and guanine - cytosine contents of fractions reveals an interesting correlation. Fig. 5 shows the plots of the guanine - cytosine molar fraction (p) vs. the molecular weight (M) and vs. the reciprocal value of M. For the construction of these graphs some additional data for calf thymus DNA fractions were used. Within the range of experimental error, all the points (with only one exception) fall on a smooth curve or on a straight line if reciprocal values of M are used. The regularity shown here is unexpected and is evidently at variance with the view that the heterogeneity in molecular weights is due to a random degradation process.

It is not easy to find an adequate explanation for this effect. However, it can be shown that this type of correlation between p and M can be explained as a property of copolymers A(nB)A, the subunits A and B differing in their guanine - cytosine content. Homologous series of this type can result as a final product of a polycondensation reaction of monofunctional (A) and bifunctional (B) monomers. For the resulting polymer it holds that $p = p_B + 2M_A (p_A - p_B) / M$, where p_A and p_B are the guanine-cytosine mole fractions of subunits and M_A is the molecular weight of the subunit A. From the plot of p vs. the reciprocal of M, p_B can easily be determined as intercept on the p axis; neither p_A nor M_A can be obtained explicitly from the slope.

We tried to determine the value of p_A from experiment. Provided that our hypothesis about the existence of DNA in the form of A(nB)A polymer is correct, we must expect discontinuity in the composition for the lowest homologues (AA, ABA). The patterns of sedimentation in CsCl gradient (Fig. 6) accord with this requirement. In the case of unfractionated sample of calf thymus DNA, an additional small peak of higher density indicating the presence of a distinct fraction of a high guanine-cytosine content (see also [4]) was observed. Since the major part of the low molecular weight fraction 1 has the same buyoant density, it can be concluded that fraction 1 consists mainly of molecules of the AA type. Consequently, the value of p_A can be calculated from the buyoant density of fraction 1.

Values of $p_B = 0.39$, $p_A = 0.54$ and $M_A = 1.4 \times 10^6$ were obtained by analysing the experimental data. The molecular weight of the smallest particles in the solution (the dimer AA) then results as 2.8×10^6, which is in a good agreement with measurements (see Table I).

Unlike the DNA of simplest organisms, especially of viruses, calf thymus DNA seems to have a more complicated subunit structure. It seems to be composed of at least two types of subunits arranged in a special fashion. The mechanism of formation of DNA molecules by polymerication of subunits essentially agrees with views of several authors concerning the existence of subunits in DNA and especially with the results of WELSH [5], who isolated calf thymus DNA of a relatively low molecular weight. In view of the strong influence of the method of isolation on the molecular parameters, a polymerisation reaction proceeding during the isolation procedure cannot be excluded.

Finally I should like to mention some recent results dealing with the heterogeneity of the bacterial DNA isolated from Bacillus subtilis, a prototroph, streptomycin resistant strain. A tryptophan dependent, streptomycin sensitive strain was used as receptor for the transformation assays. The B. subtilis DNA was fractionated in the methyl-albumin column. Whereas the absorbing substance present in the first main fraction exhibited no hyperchromic effect and practically no transforming activity, three fractions of the second peak (1, 2, 3 in the order of elution) were analysed for their melting points, sedimentation coefficients and frequencies of transformants of both genetic markers. The results are summarized in Table II. A slight but distinct heterogeneity in composition (as indicated by the melting points of the fractions) is quite apparent. Correlation of the composition with the molecular weight gives a picture which is qualitatively similar to that of the calf thymus DNA. No quantitative conclusions could be drawn because of the small number of fractions measured. The transforming frequencies for both markers have a maximum (in the Fraction 2 taken from the top of the second peak), whose value is distinctly higher than the transforming activity of the unfractionated sample. This is probably due to the fact that substances of low activity were removed in fractionation.

Table II

Fractions of B. subtilis DNA

Number of fr.	$s_{20,w}$	T_m °C	frequency of transformants prototrophy	strept. res.	ratio strept.res./prot.
1	28,8	89,0	0,076	0,0013	(0,017)
2	38,9	88,5	1,68	0,022	0,014
3	37,5	87,0	0,97	0,016	0,017
unfract. sample	-	88,4	1,22	0,017	0,015

The increase of the transforming activity with increasing molecular weight is in accordance with the results of LEDERBERG [6] who fractionated B. subtilis DNA degraded by shear stress. However, in view of the correlation between the composition and the molecular weight of the fractions, specific effects might also be taken into account. The results in the last column of Table II seem to indicate slight difference in the distribution of the two genetic markers (as suggested by ROLFE and EPHRUSSI-TAYLOR [7] and GUILD [8] in the case of Pneumococcus DNA), but further experiments are needed to elucidate this problem.

L I T E R A T U R E

1. DOTY, P., J. MARMUR, N. SUEOKA, Brookhaven Symposia in Biology 12, 1 (1959)
2. DOTY, P., B. B. MCGILL, S. A. RICE, Proc. nat. Acad. Sci. USA 44, 432 (1958)
3. MARMUR, J., P. DOTY, J. molecular Biol. 5, 109 (1962)
4. SCHILDKRAUT, C. L., J. MARMUR, P. DOTY, J. molecular Biol. 4, 430 (1962)
5. WELSH, R. S., Proc. nat. Acad. Sci. USA 48, 887 (1962)
6. NESTER, E. W., A. T. GANESAN, J. LEDERBERG, Proc. nat. Acad. Sci. USA 49, 61 (1963)
7. ROLFE, R., H. EPHRUSSI-TAYLOR, Proc. nat. Acad. Sci. USA 47, 1450 (1961)
8. GUILD, W. R., J. molecular Biol. 6, 214 (1963)

D I S C U S S I O N - Q U E S T I O N S

JORDAN : We have fractionated on ELECTEOLA columns native and sonicated DNA. We found that the fractions obtained are dependent on molecular weight. The T_m is dependent on the molecular weight below about 8.10^6. I would like to suggest that some of Dr. SPONAR's results can be attributed to fractionation into narrower molecular weight bands which have different T_m. Also I would point out that some denaturation may occur on the column.

I would also like to ask Dr. SPONAR about a mechanism for the fractionation. The adsorption of the negatively charged DNA onto basic column would not be expected to be determined by GC composition.

SPONAR : The dependence of T_m on molecular weight found in the case of sonicated samples of bacteriphage DNA (highly homogeneous in composition) is characterized by a slight decrease of T_m with decreasing molecular weight, explainable in terms of stronger influence of the loose ends of the double helix in smaller DNA molecules. (Dr.CROTHERS, pers. commun.). On the other hand, our measurements on fractions of calf thymus DNA revealed a marked increase of T_m with decreasing molecular weight which could not be explained in this way. The accordance of the GC content values calculated from melting points and from buyoant densities in CsCl gradient shows that the fractions probably differ in their composition. For the same reason, denaturation in a greater extent can also be excluded. Moreover, we were not able to find any increase of absorbance of the fractions in the presence of formaldehyde which is characteristic for the reaction of denatured DNA. Therefore we suggest that the fractions really differ in molecular weight and composition as well.

The mechanism of fractionation on methylalbumin columns is rather complex and far from to be clear. However, SUEOKA (J.Mol.Biol. 4, 161 (1962)) pointed out that nucleic acids differing in molecular weight, composition and in the state of the secondary structure could be separated.

ATEN : Can you explain why the sedimentation coefficient distribution, as shown on your first slide, is so broad for T2 DNA and relatively narrow for calf thymus DNA ?

SPONAR : The curves shown are somewhat misleading; the differenes are not so great if you take - for instance - the $s_{20,w}$ interval in question in % of the average $s_{20,w}$ value.

The very big molecules of T2 DNA are much more susceptible to degradation by shear than the much smaller molecules of calf thymus DNA.

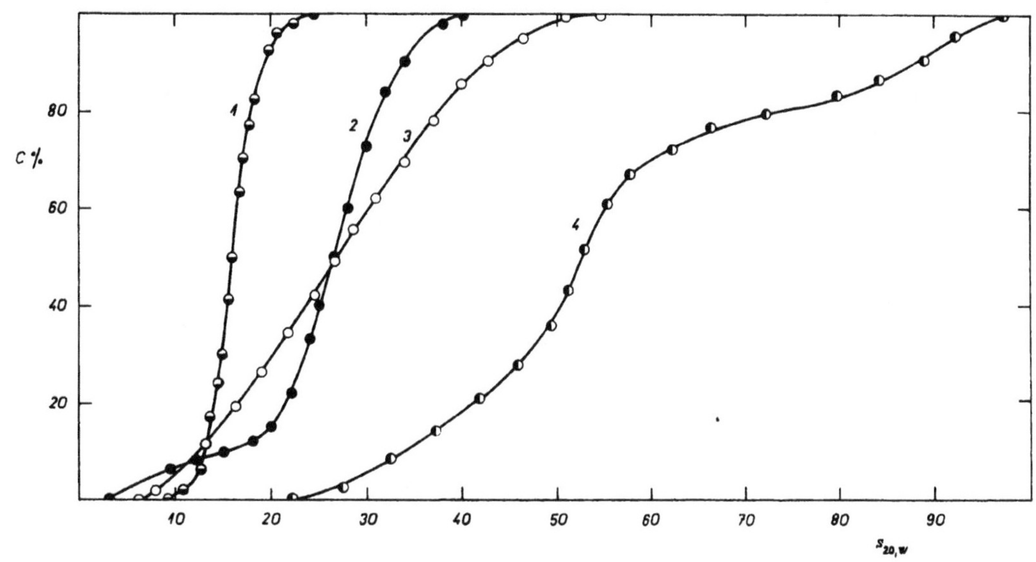

Fig.1 The $s_{20,w}$ distributions of DNA samples
1: Calf thymus DNA isolated by phenol. 2: E.coli DNA.
3: Calf thymus DNA isolated by dodecylsulphate.4:Bacteriophage T 2 DNA

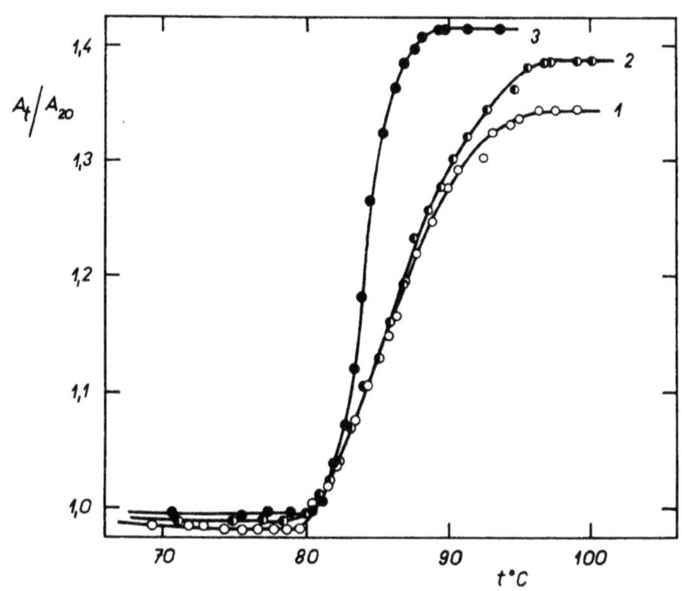

Fig.2 Absorption denaturation curves of DNA samples
1: Calf thymus DNA isolated by phenol. 2: Calf thymus DNA isolated by dodecylsulphate.3:Bacteriophage T2 DNA

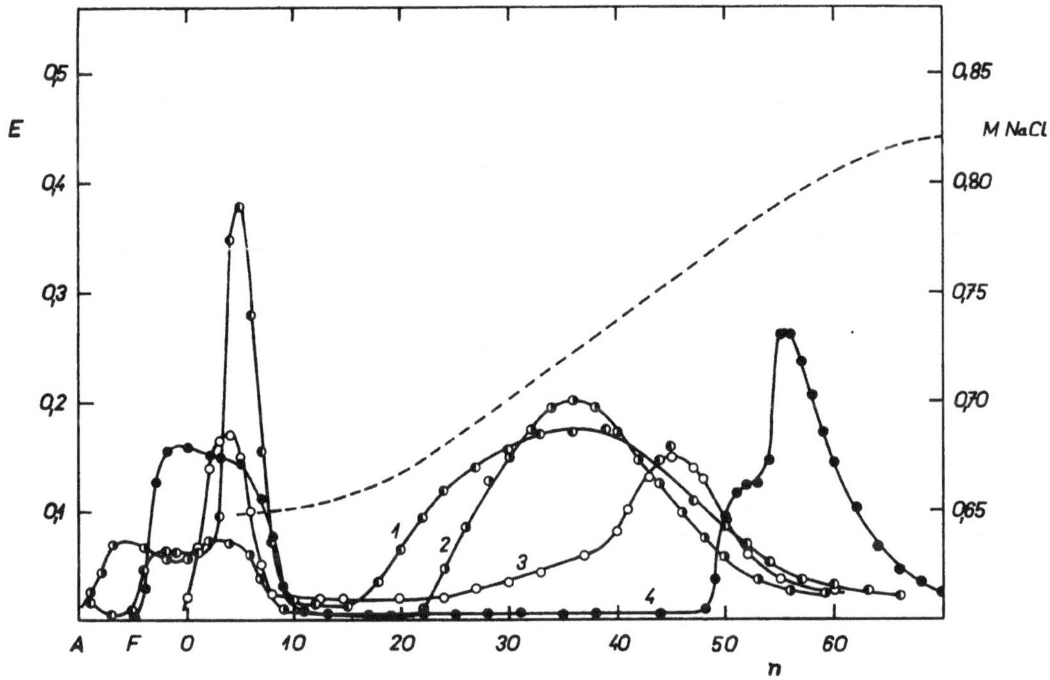

Fig.3 Elution diagrams of DNA samples. 1: Calf thymus DNA isolated by phenol. 2: Calf thymus DNA isolated by dodecylsulphate. 3: E.coli DNA. 4: Bacteriophage T2 DNA.

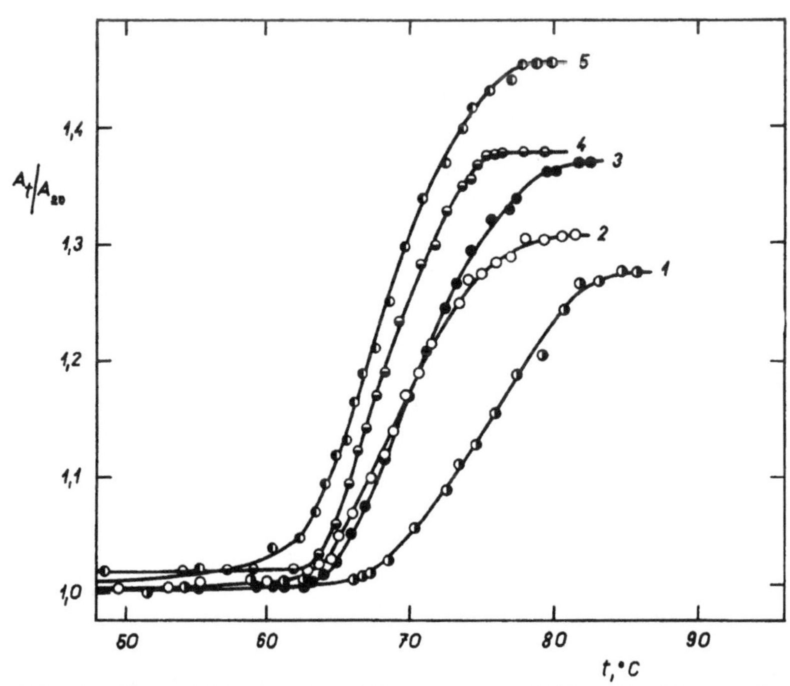

Fig.4 Absorption denaturation curves of the fractions of calf thymus DNA. 1,2,4,5 - DNA fractions in the order of elution. 3 - Unfractionated DNA.

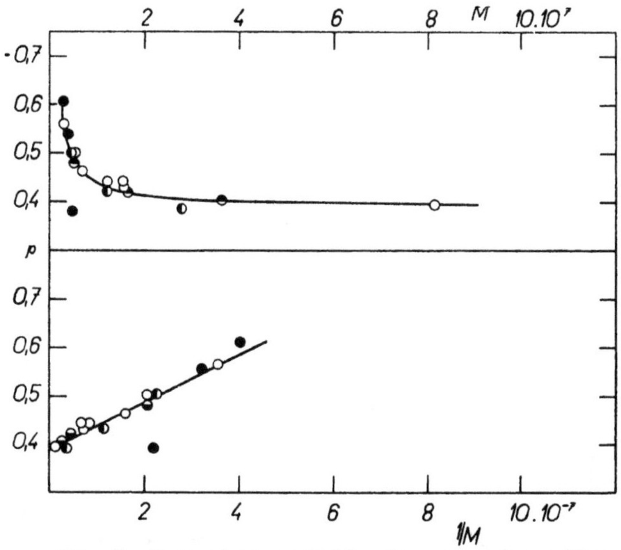

Fig.5 Dependence of the GC content on the molecular weight of DNA fractions

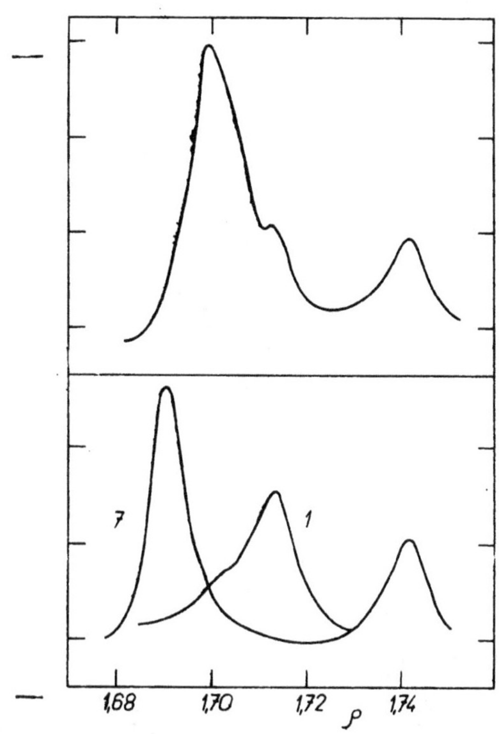

Fig.6 Diagrams of sedimentation in CsCl density gradient. Upper curve: Unfractionated DNA. Lower curves: Fractions 1 and 7. The peak with ρ=1.742 is standard DNA (Pseudomonas aeruginosa N^{15}).

D Nucleinsäuren

Flow Birefringence of Desoxyribonucleic Acids

P. MUNK

Institute of Macromolecular Chemistry of the Academy, Praha

S U M M A R Y

The magnitude of birefringence of DNA solutions at first grows up with increasing velocity gradients faster than linearly, later on the growth is considerably slower. The DNA molecules are deformed in spite of their great rigidity. The deformability cannot obtain a high degree because of the small number of statistical segments in the molecule.

The magnitude of birefringence decreases with time of application the hydrodynamic field. This decrease is fully reversible and has a maximum with varying fieldgradient. We have explained this effect by interaction of the DNA-molecules. In the hydrodynamic field there is perhaps a reversible dissoziation of the bonds of interaction and a deformation of the aggregates.

Flow birefringence measurements can give informations not only about the size and optical properties of macromolecules at rest, but also about the deformation of molecules in the field of hydrodynamic forces. From this deformation we can draw conclusions as to the structure of molecules.

The solutions of deoxyribonucleic acids are strongly birefringent already at very small velocity gradients. In our experiments we have measured several samples of DNA of different origin in the range of velocity gradients from 5 sec^{-1} to 8000 sec^{-1}. The extinction angle χ decreases very quickly with increasing gradient in the range of small gradients; in the range of high gradients it varies only little and approaches a non-zero limit (Fig. 1). This behaviour can be seen well in the plot of the quotient G/cotg 2χ against the velocity gradient (Fig. 2). The limiting slope of this dependence is equal to the tangent of twice the limiting extinction angle. The non-zero extinction angle cannot be explained by existing theories; a new theoretical analysis of the birefringent behaviour of highly oriented macromolecular systems is therefore needed. We have also observed that other macromolecular systems behave similarly.

At first the magnitude of birefringence Δn of nucleic acids grows up with increasing gradient faster than linearly, later on the growth is considerably slower. This course of the dependence can best be seen from the plot of the quotient Δn/Gc against the velocity gradient (Fig. 3); the dependences have a characteristic maximum.

It is evident from the initial increase of this dependence that the molecules are deformed by the hydrodynamic field; this is possible only when forming a coil. This deformation occurs in water - a solvent with low viscosity; accordingly the internal viscosity which acts against the deformation is also relatively low. The conception of the molecule of DNA as a rigid particle - either rigid rod or hardly deformable coil - therefore needs a correction.

The maximum on the above dependence shows that the deformability of the molecular coil is not infinite: the deformation soon saturates. PETERLIN [1] foresees this behaviour for coils with a small number of statistical segments. From the position of the maximum it is possible to estimate that the molecules of DNA with molecular weight approximately ten million have thirty to fifty statistical segments. This value agrees well with estimates made with the aid of other methods.

The relative height of the maximum is smaller and the drop of the dependence past the maximum less than predicted by PETERLIN's theory for soft coils [1]. The molecular coils are therefore not very soft; their internal viscosity is neither infinite nor vanishing. This makes it difficult to interpret the measurements with the aid of present theories, which consider only these limiting cases.

According to our measurements the intrinsic anisotropy of the molecule of DNA has the value $2{,}70 \cdot 10^{-21}$ cm^3 essentially independent of the origin of the sample. This value lies between the values of FRISMAN [2] and LERAY [3] (Table I).

Table I

Anisotropy of the DNA molecule	
This work	$2{,}70 \cdot 10^{-21}$ cm^3
FRISMAN E.V. [2]	$9{,}0 \cdot 10^{-21}$ cm^3
LERAY J. [3]	$1{,}27 \cdot 10^{-21}$ cm^3

From the anisotropy of the statistical segment which equals five thirds of the intrinsic anisotropy of the whole molecule, and from the known anisotropy of the monomer, that is of the pair adenine-thymine or guanine-cytosine nucleotides, it is possible to calculate the number of monomer units forming one statistical segment and accordingly the length of the segment. Using FRISMAN's value $-15 \cdot 10^{-24}$ cm^3 for the anisotropy of the monomer unit, the length of the statistical segment is 1020 Å in excellent agreement with light scattering values of DOTY [4] and FRISMAN [2].

The above values of the magnitude of birefringence and of the extinction angle are the values measured immediately after starting the rotor that is after application of the hydrodynamic field. The magnitude of birefringence as well as the extinction angle at constant gradient, however, change with the time of the application of the hydrodynamic field. The magnitude of birefringence after an induction period of varying length gradually falls to a steady limiting value (Fig. 4). This decrease is fully reversible; if the rotor is stopped for a period long enough (two minutes usually suffice), the repeated course of the time dependence of the magnitude of birefringence is the same as before. The dependence of the magnitude of the relative decrease of birefringence on the velocity gradient has a maximum at gradient values between 300 and 500 sec^{-1}. At gradients low enough as well as high enough the fall of birefringence is absent.

The velocity of the fall increases with increasing velocity gradient; the product of the velocity gradient with the time needed for a half-fall is essentially independent both on the velocity gradient and the concentration. This product is proportional to the number of rotations of single particles around their axis: that means that the number of rotations needed for the process is always approximately the same.

We have observed this behaviour in all our samples of both animal and bacterial origin and prepared by the phenol method as well as the dodecylsulphate one. The magnitude of this effect was different in different samples; it fluctuated also from solution to solution of the same sample. We therefore believe that this phenomenon is caused by the specific interaction of DNA molecules and that it is general for high-molecular samples of DNA, which were not subjected to too strong hydrodynamic fields. The fall effect is namely cancelled by fields with velocity gradients higher than about 5000 sec^{-1}.

The fall effect can be tentatively explained by the following hypothesis: The individual molecules of DNA interact in specific manner; they are bound together with two types of bonds. One type is so weak that the bonds can be disrupted by the action of the hydrodynamic forces; at rest they arise again. The second type of bonds is stronger; it prevents the full dissociation of aggregates and thus causes the reversibility of the whole effect. In the range of intermediate gradients the weak bonds are disrupted and the hydrodynamic forces deform the aggregates in the way schematically shown in Fig. 5. The individual molecules in the deformed aggregate are mutually oriented more or less perpendicularly; the total anisotropy of the aggregate is therefore decreased which causes the fall of birefringence. With the aid of this model we succeeded in interpreting satisfactorily all experimental findings; the conception of the specific interaction is also in line with measurements on DNA with other methods.

Further work is going on in our laboratory on the elucidation of the origin of the specific interaction.

R E F E R E N C E S

1) PETERLIN,A., Polymer 2, 257 (1961)
2) FRISMAN,E.V., VOROBEV,V.J., SHTCHAGINA,L.V., YANOVSKAYA,N.K., Vysokomol. soed. 4, 762 (1962)
3) LERAY,J., J. Chim. Phys. 57, 323 (1960)
4) DOTY,P., MCGILL,B.B., RICE,S.A., Proc. nat. Acad. Sci. USA 44, 432 (1958)

D I S C U S S I O N - Q U E S T I O N S

ZIMMER.- Wie hoch war die Ionenstärke ihrer DNA-Lösungen, die Sie für Ihre Untersuchungen angewendet haben? Die Ionenstärke ist deshalb wichtig, weil dadurch eine Bildung oder Verhinderung von Aggregationen bewirkt werden kann.

MUNK.- Itwas standard saline citrate buffer.

ATEN.- What was the concentration of your DNA solutions?

MUNK.- $c = 10 \cdot 10^{-6} - 45 \cdot 10^{-6}$ g/ml.

BERG.- Wurde Scherabbau beobachtet unter Verringerung des Molekulargewichtes?

MUNK.- Scherabbau beginnt erst bei Geschwindigkeitsgradienten von ca. 5000 sec^{-1}.

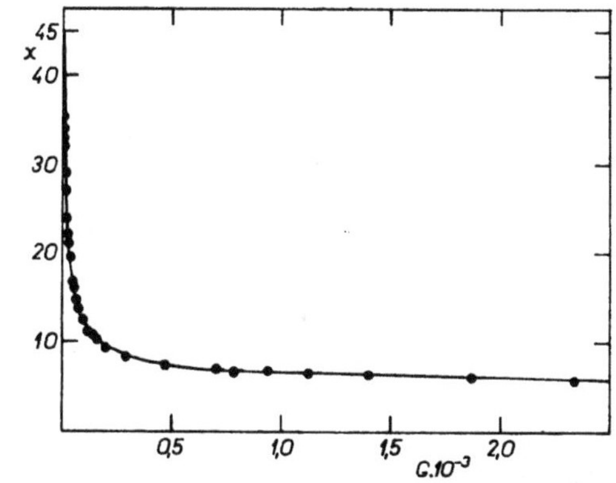

Fig.1 Dependence of the extinction angle on the velocity gradient. DNA from Escherichia coli, $c = 2{,}2.10^{-5}$ g/ml.

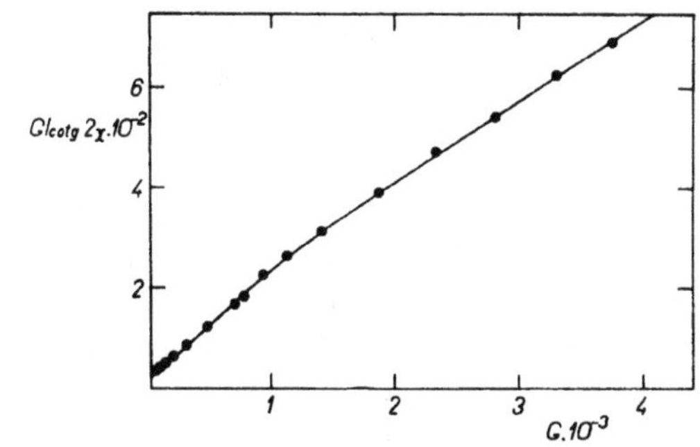

Fig.2 Dependence of the quotient G/cotg 2χ on the velocity gradient. Solution as in Fig.1.

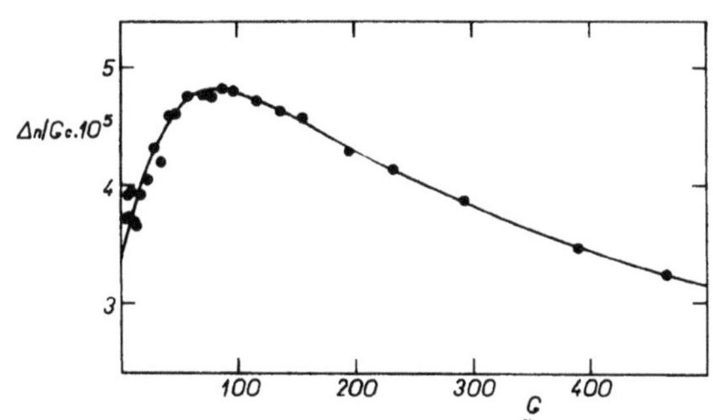

Fig.3 Dependence of the quotient $\Delta n/Gc$ on the velocity gradient. Solution as in Fig.1.

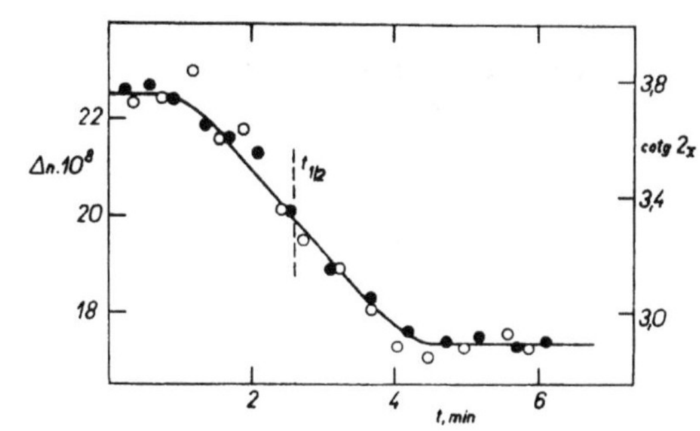

Fig.4 The decrease of the magnitude of birefringence and of the extinction angle with time. Thymus DNA, $c=1{,}8.10^{-5}$g/ml, $G=390\ \sec^{-1}$.
● Magnitude of birefringence
○ Extinction angle. Both scales have common intercept.

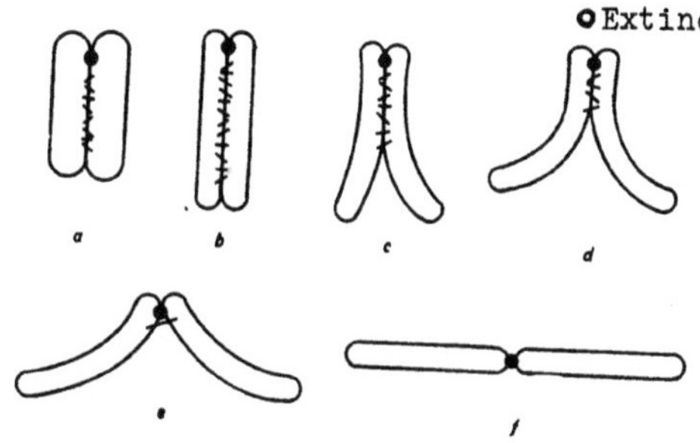

Fig.5 Various phases of the interaction of double molecules with the hydrodynamic field.

D Nucleinsäuren

Physical Chemical Studies on the Secondary Structures of Nucleic Acid Polymers and on the Interactions of Free Bases and Nucleosides

G. K. HELMKAMP, S. I. CHAN, M. P. SCHWEIZER and P. O. P. TS'O

Department of Chemistry, University, Riverside, California; California Institute of Technology, Pasadena

S U M M A R Y

I."Monomer-monomer" interactions. The molar osmotic coefficients of aqueous purine, uridine and cytidine have been determined by measurement of vapor pressure lowering, and the activity coefficients of the compounds were calculated. The data indicate that association processes proceed beyond the dimer stage, and that the experimental results are consistent with a successive series of steps having equal equilibrium constants. The tendency for association and interaction between free bases and nucleosides in solution was as follows: purine-purine > purine-pyrimidine > pyrimidine-pyrimidine.

In an attempt to develop a physical model for the association of bases, NMR chemical shifts of purine protons were studied as a function of concentration. The spectral relationships were assigned by deuterium substitution. Again the deviations from ideality involve association beyond dimer stages with equal equilibrium constants. The chemical shift data suggest vertical stacking of purine molecules without direct superposition rather than horizontal hydrogen bonding association.

II. Interactions of polymers with monomers have been measured by the change of melting temperature as a function of concentration of monomer. The increasing order of effectiveness in lowering of T_m was as follows: Adonitol, methyl riboside (both negligible) < cyclohexanol < phenol, pyrimidine, uridine < cytidine, guanidine < purine, adenosine, inosine, deoxyguanosine < caffeine, coumarin, 2.6-dichloro-7-methylpurine. No specificities related to hydrogen bonding in the WATSON-CRICK formulation was noted. It appears likely that the data are best interpreted in terms of hydrophobic interactions. The physical binding of several compounds to poly-A and DNA has been studied by equilibrium dialysis. The equilibrium constant (per mole of base in nucleic acid), over a 20-fold concentration range was as follows for denatured DNA: Phenanthrene, 2,000; testosterone, 220; estradiol, 65; diethyl stilbesterol, 35; adenine and caffeine, 9; and thymine < 1.

III. Polymer-polymer interactions were studied as a function of the nature of the solvent. It was shown that polymers in formamide solution had only slightly different optical properties (rotary dispersion, absorbancy) than did the corresponding monomers. Dimethyl sulfoxide solutions of DNA, however, showed an abnormally low specific rotation. Formamide solutions of DNA prepared below 25° in the presence of at least 0.4 M salt had properties characteristic of undenatured polymer. In general, the phenomena considered in the discussion can be best correlated in terms of hydrophobic interactions.

In the study of organic molecules whose interactions might ultimately be related to information transfer at the biological level, it has been considered of primary importance to include nucleic acids, nucleotides, nucleosides, and simple derivatives of purines and pyrimidines. The molecular relationships are classified in the general categories of "monomer-monomer", "monomer-polymer" and "polymer-polymer", even though some of the species are simple materials such as purine instead of native nucleotides. Recent advances in the relationship of nucleic acid replication to template DNA [1)-8)] have focused attention to specificities that reside in variations in base sequences. As a consequence, the evaluation of an appropriate chemical mechanism for replication could reasonably be initiated through the study of any of the types of interactions indicated above.

It has also been of interest to extend the area of knowledge related to the chemical modification of bases within intact polymeric nucleotide chains. Although such studies were initiated, they immediately diverted attention up to the present to the potential use of organic solvents for nucleic acids and their moeities.

Monomer-Monomer Interactions. Bases and nucleosides that are uncharged in neutral aqueous solutions were chosen for the study of interactions. Thus, complications due to charge effects were avoided.

Osmotic coefficients for aqueous purine, uridine and cytidine at concentrations of up to 1.0 molal were determined by means of thermoelectric measurements of vapor pressure lowering [9)] (Fig. 1, 2). Because of the large concentration effects, it was assumed that association took place and that a general treatment of multiple interactions would be used to describe the system (equation 1).

$$\begin{array}{ll} B_1 + B_1 \rightleftarrows B_2 & K_1 = B_2/(B_1)^2 \\ B_2 + B_1 \rightleftarrows B_3 & K_2 = B_3/(B_2)(B_1) \\ \cdots\cdots & \cdots\cdots \\ B_n + B_1 \rightleftarrows B_{n+1} & K_n = B_{n+1}/(B_n)(B_1) \end{array} \qquad (1)$$

For simplification, it was assumed that all of the K's were equal (equation 2).

$$K_1 = K_2 = K_3 = \ldots = K_n \qquad (2)$$

On the basis of similar assumptions, SCHELLMAN [10)] developed a relationship between K and the osmotic coefficient for the analysis of urea association (equation 3).

$$K = (1 - \emptyset)/m\emptyset^2 \qquad (3)$$

where, m = overall molal concentration

Alternatively, and at the other extreme, it was assumed that only dimer formed. The relationship between K and the osmotic coefficient was again derived according to the SCHELLMAN relationship [10)] (equation 4).

$$K = K_1 = \frac{(1 - \emptyset)}{m(2\emptyset - 1)^2} \qquad (4)$$

The calculation of equilibrium constants based on each of the sets of assumptions was made. Selected values for multiple association and others for dimerization only are given in Table 1. It is evident that the large change in K with change in concentration, especially in the case of purine, indicates that equation 4 is not applicable. On the other hand, the values of K calculated from equation 3 are very nearly constant, and they indicate that the equation can be used with a fair degree of approximation of the association process.

Table 1

Equilibrium Constants, $K = (1-\phi)/m\phi^2$

Molal Conc.	Purine	Uridine	Cytidine
0.1	2.1	0.64	0.74
0.4	2.2	0.60	0.87
0.7	1.9	0.62	0.90
1.1	1.8	----	----
		$K_1 = (1-\phi)/m(2\phi-1)^2$	
0.1	3.1	0.72	0.86
0.4	10.1	0.94	1.66
0.7	34.8	1.35	2.86
1.1	1.2×10^5	----	----

In the problem of the interaction of species related to nucleic acids, it was of interest to determine the nature of the interaction and its association with the inter- and intra-molecular forces within chains of polynucleotides. Thermoelectric measurements gave a fairly precise picture of the extent of aggregation but no clue whatsoever concerning mechanism of aggregation.

The use of proton magnetic resonance to study the structure of purines, pyrimidines and their nucleosides and nucleotides has met with success in the past [11)-14)]. It was thought probable that the chemical shifts of the purine protons would change as a function of concentration as association proceeded. This has indeed been observed [15)].

In aqueous solution, the purine n.m.r. spectrum consists of three lines (Fig. 3). The N-H proton is not observed in water, probably because of exchange; it is readily seen in dimethyl sulfoxide. Although assignments of spectral peaks to the protons have been made [12),16)], these were based solely on theoretical considerations and showed disagreement. Because of the importance of correct assignments in the evaluation of a physical model for the product of interactions, an experimental verification through specific deuteration was carried out.

When purine was treated with deuterium oxide for four hours at 105°, the high field peak disappeared completely [17)], and the other peaks were unaffected. The exchange reaction was found to be 50 % complete in 45 minutes. If the exchange proceeded via protonation of purine to give a charged intermediate, it appeared that the 8-position would be most rapidly affected, for the most stable intermediate would result. In order to verify the interpretation, 8-mercaptopurine was desulfurized with deuterated Raney nickel [18)] in a non-protic solvent containing a small amount of deuterium oxide. The product of the reaction was partially deuterated, and the high field n.m.r. peak was diminshed by 48 %. In the desulfurization of 6-mercaptopurine under identical conditions, the relative area of the low field peak was reduced by 47 %. Better deuterium substitution was accomplished, however, by the reduction of 6-iodopurine with deuterium and platinum in methanol-d solvent. In this case, the low field peak disappeared completely without reducing the intensities of the remaining peaks. The possibility that some exchange might have taken place during the reduction was excluded by noting that purine itself incorporated only trace amounts of deuterium under the reaction conditions. It was concluded from these results that the high field peak originated from the 8-proton, the low field peak from the 6-proton, and the remaining peak from the 2-proton.

The change in chemical shift with purine concentration in water is given in Fig. 4. The values are given in cycles per second measured from external chloroform, and corrections for bulk susceptibility were incorporated as described by POPLE, SCHNEIDER and BERNSTEIN [19a)]. It was assumed that in pi-electron systems such as that occurring in purine the applied magnetic field, H_0, induces ring currents and the resulting diamagnetic anisotropy causes a field reinforcement at the ring protons [19b)] (Fig. 5). The dilution shift may then be accommodated by a ring current model in which partial overlap of the purine rings occurs (Fig. 6).

When partial overlap occurs, the field reinforcement at a proton in the environment of another purine ring is cancelled due to the induced component in the opposite direction (Fig. 7). According to this model, a shift to higher fields as concentration increased would be predicted, for greater numbers of protons experience ring currents from neighboring molecules. On the other hand, the association process can not involve end-on interaction of the purine rings (Fig. 7); for this model predicts a downfield shift as a function of concentration, and the induced components are additive. This is contrary to observation.

If, in the association process, one lets B_n be the moles of species in the nth stage, then nB_n is the total moles of purine in the n-mer sites. The total number of moles of purine in solution is given by equation 5.

$$m = \sum_{n=1}^{\infty} nB_n \qquad (5)$$

The contribution of each species in the associative process to the total screening constant, σ, is given by equation 6, where F_n is the fraction of purine in the n-mer environment.

$$\sigma = \sum_{n=1}^{\infty} \sigma_n F_n = \sum_{n=1}^{\infty} \sigma_n nB_n/m \qquad (6)$$

From equation 1, $B_n = K^{n-1}(B_1)^n$, which on substitution into equation 6 followed by expansion, gives the expression indicated by equation 7.

$$\sigma = \frac{\sigma_1 B_1}{m} + \frac{\sigma_2 2K(B_1)^2}{m} + \frac{\sigma_3 3K^2(B_1)^3}{m} + \ldots\ldots \qquad (7)$$

In equation 7, dividing by B_1/m and extrapolating $\sigma m/B_1$ to zero (equation 8) gives σ_1, the screening constant due to monomer.

$$\frac{\sigma m}{B_1} = \sigma_1 + \sigma_2 2KB_1 + \sigma_3 3K^2 (B_1)^2 + \cdots + \sigma_n nK^{n-1} (B_1)^{n-1} \qquad (8)$$

In order to obtain a limiting value for the screening constant as the associated units approach infinity, assume that at some stage a limit is reached as far as the effect of $_n$ on is concerned. In other words, $_n =$ after n'.

$$\sigma = \sum_{n=1}^{\infty} \sigma_n F_n = \sum_{n=1}^{n'} \sigma_n F_n + \sum_{n=n'+1}^{\infty} \sigma_\infty F_n$$

$$= \sum_{n=1}^{n'} \sigma_n F_n + \sigma_\infty \sum_{n=n'+1}^{\infty} F_n = \sum_{n=1}^{n'} \sigma_n F_n + \sigma_\infty (1 - \sum_{n=1}^{n'} F_n)$$

$$= \sigma_\infty + \sum_{n=1}^{n'} \sigma_n F_n - \sigma_\infty \sum_{n=1}^{n'} F_n$$

$$= \sigma_\infty + \sum_{n=1}^{n'} (\sigma_n - \sigma_\infty)\, nB_n/m$$

$$= \sigma_\infty + \sum_{n=1}^{n'} (\sigma_n - \sigma_\infty)\, nK^{n-1} (B_1)^n/m$$

$$= \sigma_\infty + (\sigma_1 - \sigma_\infty)\, B_1/m + (\sigma_2 - \sigma_\infty)\, 2K(B_1)^2/m + \ldots$$

$$\ldots\ldots + (\sigma_{n'} - \sigma_\infty) n'K^{n'-1}(B_1)^{n'}/m \qquad (9)$$

The resulting expression (equation 9) can lead to an evaluation of σ_∞ by extrapolation to $B_1/m = 0$.

With appropriate algebraic manipulation, the values of σ_2 and σ_3 were also determined. The contribution of higher species to the screening constant was small [15].

The population distribution of purine at each stage of association was calculated by computer from osmotic equilibrium constants [9]. From the relationship $B_n = K^{n-1} \cdot (B_1)^n$, the stoichiometric molality, m, is given by equation 10, and the concentration of monomer, B_1, by equation 11.

$$m = \sum_{n=1}^{\infty} nB_n = B_1 + 2K(B_1)^2 + 3K^2(B_1)^3 + \ldots + nK^{n-1}(B_1)^n$$

$$= \frac{B_1}{(1 - KB_1)^2} \qquad (10)$$

$$B_1 = \frac{2mK + 1 \pm \sqrt{4mK + 1}}{2mK^2} \qquad (11)$$

From the relationships between B and K, and considering the

$$B_2 = K(B_1)^2; \quad B_3 = K^2(B_1)^3; \quad \text{etc.}$$

mole fractions, F_n, the population distribution was calculated.

$$F_1 = B_1/m; \quad F_2 = 2B_2/m; \quad \text{etc.}$$

A weighting of the ring protons at each stage was made on the assumption that there was a statistical distribution of various orientations. For a dimer site, one arrangement was possible and for a trimer site two were possible (Fig. 8). The protons in various environments through the pentamer stage are distributed in the following way:

	M	D	T	Total
Dimer	1	1	0	2
Trimer	5	6	1	12
Tetramer	12	16	4	32
Pentamer	28	40	12	80

Thus, by a treatment of the statistical factors and population distribution in conjunction with equations 8 and 9, a calculation of chemical shift vs. concentration was completed (Fig. 4)[15].

The nature of the intermolecular forces cannot be established precisely at this juncture, but they probably are a composite of VAN DER WAAL's forces, pi-pi electron interactions and dipole-dipole interactions. Purine in polar organic solvents such as dimethyl sulfoxide and dimethyl formamide shows very little chemical shift dependency on concentration. These two solvents have previously been shown to possess marked denaturing powers toward nucleic acids [20]. If dipole-dipole interaction was primarily responsible, it might be expected that association would be greater in the organic solvents than in water because of their lower dielectric constants. However, since the opposite effect was observed, it may be that purine is better solvated in them than in water; in other words, the ratio of purine-solvent interaction to purine-purine interaction is greater in the organic solvents.

A study with 6-methylpurine has shown that its association is much more extensive than that of purine. It would appear that a combination of inductive and steric factors associated with the methyl group may account for the behaviour. A larger dipole moment would favor association and lead to a relatively large value of K, particular at the "dimer" stage where methyl-methyl interactions can be small. At higher degrees of association, the values of K would decrease due to a larger proportion of methyl-methyl steric interactions.

The interaction of different bases was determined by measuring the increase in solubility of a C^{14}-labeled base in the presence of another unlabeled base or other compound [9]. The effect of various materials on adenine-C^{14} is given in Fig. 9. A similar effect was noted for thymine-C^{14}. The mechanism of the interaction does not necessarily involve hydrogen-bonding, for urea should have been highly effective as either a hydrogen bond donor or acceptor. The effectiveness shows a positive correlation with the ability of the same compounds to reduce the stability of the helical form of nucleic acids [21]. This phenomenon, which is attributed to hydrophobic and stacking interactions, will be discussed more fully under "Monomer-Polymer Interactions".

Monomer-Polymer Interactions. In the measurement of the effect of purine and pyrimidine derivatives on T_m of helical structures of polynucleotides, it was necessary to resort to measurements of optical rotation rather than absorbancy because of problems of light transmission in concentrated solutions [21]. The effect of purine on the helix-coil transition for poly A, determined by optical rotatory technique, is given in Fig. 10. In the case of uridine, it was necessary to obtain contributions of poly A by difference because of the optical activity of the interactant (Fig. 11). A compilation of results is given in Table 2 for poly A and in Table 3 for thymus DNA. The effectiveness of added compounds were related in the following manner:

Adonitol, methyl riboside, urea (all negligible) $<$ cyclohexanol $<$ phenol, pyrimidine, uridine $<$ cytidine, thymidine $<$ purine, adenosine, inosine, deoxyguanosine $<$ caffeine, coumarin, 2.6-dichloro-7-methylpurine

Table 2

Compounds	ΔT_m, calcd. for interactant at 1.0 M conc.
Pyrimidine	17-20
Nucleosides	15-40
Purine	40-80
Nucleosides	55-130
2,6-Dichloro-7-methylpurine	300
Coumarin	180
Phenol	16
Cyclohexanol	8
Urea	1
Adonitol	0
Methyl riboside	0

Table 3

Compounds	ΔT_m, calcd. for interactant at 1.0 M
Pyrimidine	17
Thymidine	18
Uridine	17
Purine	30
Inosine	30
Caffeine	60
Urea	3

A large percent error occurs in the last two compounds because of their very limited solubility. The effect of 2.6-diaminopurine was also quite large, but the interaction with poly A was assisted by charge interaction at the experimental pH of 4.85 (pK = 5.1 for the purine derivative [22]).

A study of the physical binding of various compounds to poly A and DNA has been initiated by means of equilibrium dialysis. It was found that the equilibrium constant, nK (per mole of base in nucleic acid), over a 20-fold concentration range was as follows for denatured DNA:

Phenanthrene, 2,000; testosterone, 220; estradiol, 65; diethyl stilbesterol, 35; adenine and caffeine, 9; thymine, 1.

The values were much lower with native DNA, poly A coil, and poly A helix.

It appears that hydrogen bonding, though not excluded, is of secondary importance in the relationship related to lowering of T_m. No correlation is noted between hydrogen bond donor or acceptor ability and the helix-coil transition temperature. However, the higher affinity of DNA for steroids and carcinogens than for bases indicates that the data are best interpreted in terms of hydrophobic interactions [23]. The increase in the number of water-water interactions in transferring a compound from an aqueous disperse system to the interaction species could be the most important factor contributing to the thermodynamic stability.

Polymer-Polymer Interactions. The term polymer-polymer interaction as used here is meant to include those phenomena of inter- and intra-chain effects that arise from interpretation of various types of physical data. Of primary concern here, is the use of optical rotatory properties, particularly of formamide solutions.

Previous publications have given data on optical and physical properties of nucleic acids, synthetic polynucleotides and nucleotides in formamide and aqueous solutions at different temperatures and pH values [20),24),25)]. It was concluded that nucleic acids lost all formal secondary structure in formamide, and the interactions between the constituent bases were minimal. However, in the presence of added electrolyte, the $[\alpha]$ vs. temperature profile indicated the retention of some secondary structure [25)].

When DNA was dissolved in formamide-KCl at temperatures below 25° then reisolated, the melting profiles were a function of the concentration of KCl (Fig. 12) [26)]. If the concentration of KCl was 0.4 molar, the DNA was indistinguishable from that isolated from standard saline citrate buffer. A concentration distribution of reisolated material measured in a cesium chloride density gradient (Fig. 13) was consistent with a picture of increasing denaturation with decreasing KCl concentration. However, if optical rotation measurements were made directly on the formamide solutions, unusually high specific rotations were obtained. A high temperature effect still remained that was attributed to interconformational transition, and the magnitude was related to KCl concentration in the expected manner (Fig. 14).

Previously, it was noted that specific rotations of nucleotides were not considerably different in aqueous and formamide solvents [24)]. Furthermore, optical rotation and dispersion values for polynucleotides in formamide were quite similar to those of corresponding mixtures of nucleotides. It appeared that any specific solvent effects had no significant influence on chain or intimate bond conformations that would be reflected in terms of environments of optically active chromophores. In the case of formamide solutions of DNA in the presence of about 0.3-0.4 M KCl, the rotations were about 1.5 times the value for aqueous SSC solutions (shown by the magnitude of K values from the DRUDE equation), but there was no significant change in λ_c from the position suggested by absorption maxima (Table 4). These data suggested that the conformation of DNA in formamide-KCl was different from that in appropriate aqueous solvents, perhaps in terms of helix pitch or diameter

Table 4

Solution		$K \times 10^6$	λ_c, mμ
Aqueous, SSC		35	231
Formamide,	0.1M KCl	17.5	259
	0.2M KCl	27.5	256
	0.3M KCl	46	250
	0.4M KCl	52	249
	0.3M KCl (5°)	52	249

due to molecules of solvation. The factors related to broadening in the transition zone of DNA in formamide-KCl have yet to be examined.

Acknowledgement:- This work was supported in part by grants GM 8185, RG 5143, GM 10316 and RG 3977 from the National Institutes of Health, U.S. Public Health Service.

REFERENCES

1) WEISS,S.B., T.NAKAMOTO, Proc. natl. Acad. Sci. USA 47, 1400 (1961)
2) GEIDUSCHEK,E.P., T.NAKAMOTO, S.B.WEISS, Proc. natl. Acad. Sci. USA 47, 1405 (1961)
3) BURMA,D.P., H.KRÖGER, S.OCHOA, R.C.WARNER, J.WEILL, Proc. natl. Acad. Sci. USA 47, 749 (1961)
4) FURTH,J.J., J.HURWITZ, M.GOLDMAN, Biochem. biophysic. Res. Commun. 4, 362 (1961)
5) FURTH,J.J., J.HURWITZ, M.GOLDMAN, Biochem. biophysic. Res. Commun. 4, 431 (1961)
6) STEVENS,A., J. biol. Chemistry 236, PC 43 (1961)
7) KORNBERG,A., Science 131, 1503 (1960)
8) JOSSE,J., A.D.KAISER, A.KORNBERG, J. biol. Chemistry 236, 864 (1961)
9) TS'O,P.O., I.S.MELVIN, A.C.OLSON, J. Amer. chem. Soc. 85, 1289 (1963)
10) SCHELLMAN,J.A., C. R. Trav. Lab. Carlsberg, Sér. chim., 29, 223 (1956)
11) GRONOWITZ,S., R.A.HOFFMAN, Ark. Kemi 16, 459 (1960)
12) JARDETZKY,C.D., O.JARDETZKY, J. Amer. chem. Soc. 82, 222 (1960)
13) KOKKO,J.P., L.MANDELL, J.H.GOLDSTEIN, J. Amer. chem. Soc. 84, 1042 (1962)
14) GATLIN,L., J.C.DAVIS,Jr., J. Amer. chem. Soc. 84, 4464 (1962)
15) SCHWEIZER,M.P., S.I.CHAN, G.K.HELMKAMP, P.O.TS'O, J. Amer. chem. Soc., in press
16) REDDY,G.S., L.MANDELL, J.H.GOLDSTEIN, J. chem. Soc. 1414 (1963)
17) The authors are indebted to Professor JARDETZKY for suggesting that H_8 exchanges readily under these conditions.
18) BONNER,W.A., J. Amer. chem. Soc. 76, 6350 (1954)
19) (a) POPLE,J.A., W.G.SCHNEIDER, H.J.BERNSTEIN, "High Resolution Nuclear Magnetic Resonance", McGraw-Hill Boo, Co., Inc., New York, Chapter 2.
(b) ibid, p. 181
20) HELMKAMP,G.K., P.O.P.TS'O, J. Amer. chem. Soc. 83, 138 (1961)
21) TS'O,P.O.P., G.K.HELMKAMP, C.SANDER, Proc. natl. Acad. Sci. USA 48, 686 (1962)
22) ALBERT,A., D.J.BROWN, J. chem. Soc. 2060 (1954)
23) ZIMM,B.H., J. chem. Physics 33, 1349 (1960)
24) TS'O,P.O.P., G.K.HELMKAMP, C.SANDER, Biochim. biophysica Acta 55, 584 (1962)
25) HELMKAMP,G.K., P.O.P.TS'O, Biochim. biophysica Acta 55, 601 (1962)
26) TS'O,P.O.P., G.K.HELMKAMP, C.SANDER, F.W.STUDIER, Biochim. biophysica Acta, in press.

D I S C U S S I O N - Q U E S T I O N S

JORDAN.- Your studies for the denaturation of DNA in formamide are most interesting. Do you demonstrate the observed behaviour is in any way due to the bound water in the DNA, which will still be present in the formamide solutions.

HELMKAMP.- Since the specific rotation of helical DNA is intimately related to its specific secondary structure, it seems likely that the structure of the molecule of solvation would be of importance. I feel that the bound water, which is difficult to remove from solid material, should be in equilibrium with formamide in formamide solution. In this sense, the original water could have an influence on the rotation.

ZIMMER.- Does the interaction of bases depend on the polarity of the organic solvent used in your experiments and are there differences between the four DNA-bases?

HELMKAMP.- Yes. In certain low dielectric media, purine shows an insignificant amount of association. This is the case in dimethylsulfoxide. Only purine and 6-methylpurine were investigated. The limitations because of solubility exclude many or most DNA bases from n.m.r. studies.

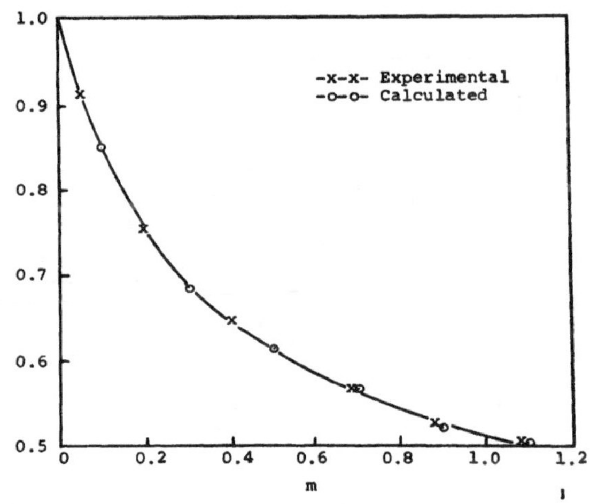

Fig.1 Osmotic coefficients vs. concentration for purine in water [9].

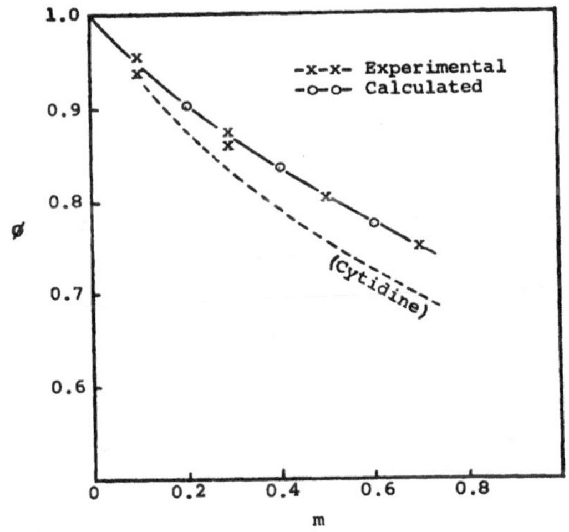

Fig.2 Osmotic coefficients vs. concentration for uridine in water [9].

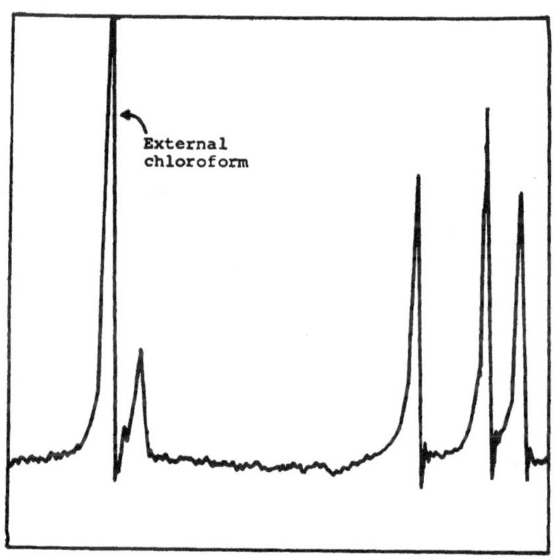

Fig.3 Nuclear magnetic resonance spectrum of purine in water; shifts downfield from external chloroform [15].

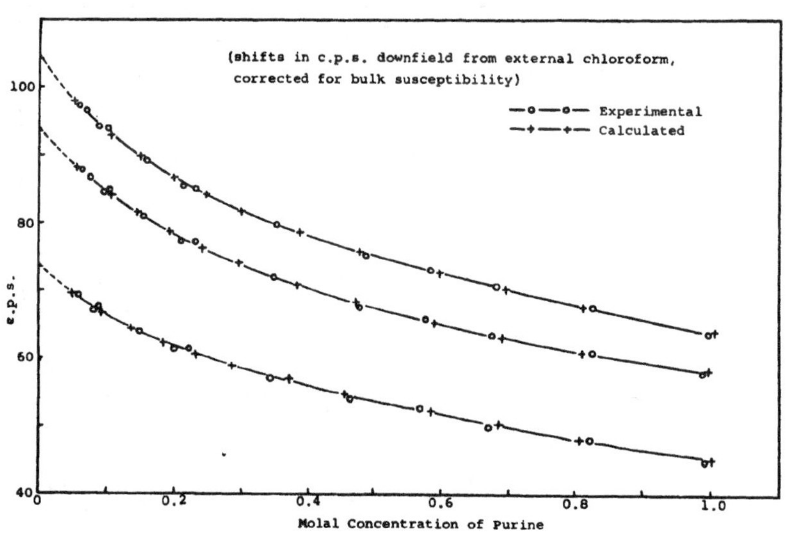

Fig.4 Experimental and calculated chemical shifts for purine in water [15].

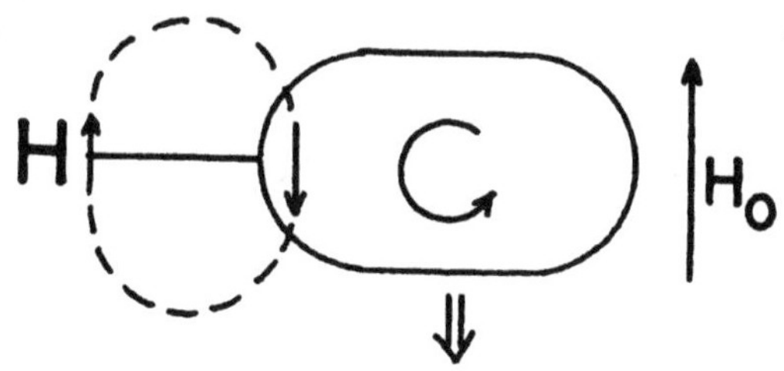

Fig.5 Field reinforcement at ring proton from induced ring current [15].

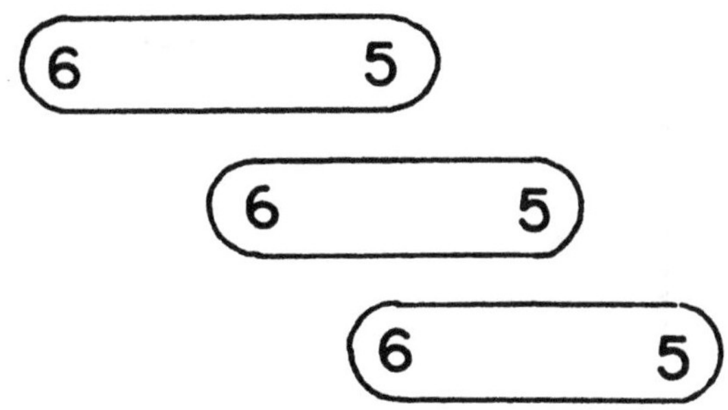

Fig.6 Ring current model with partial overlap [15].

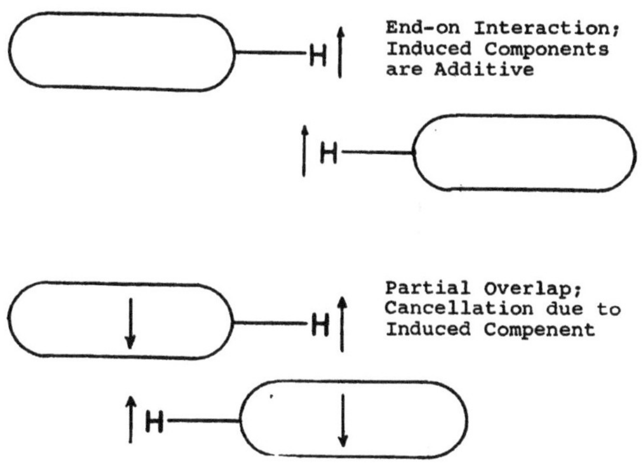

Fig.7 Models for interaction of purine molecules [15].

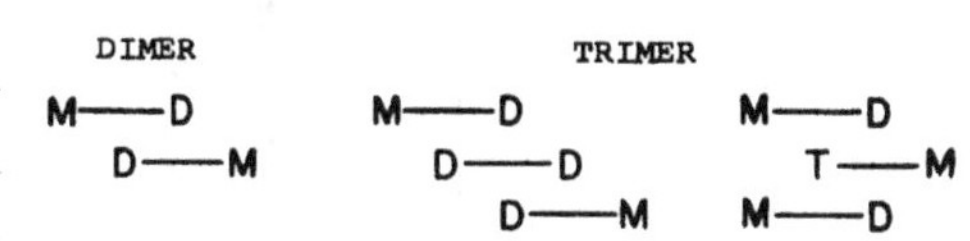

Fig.8 Statistical arrangements in association to form dimer and trimer [15].

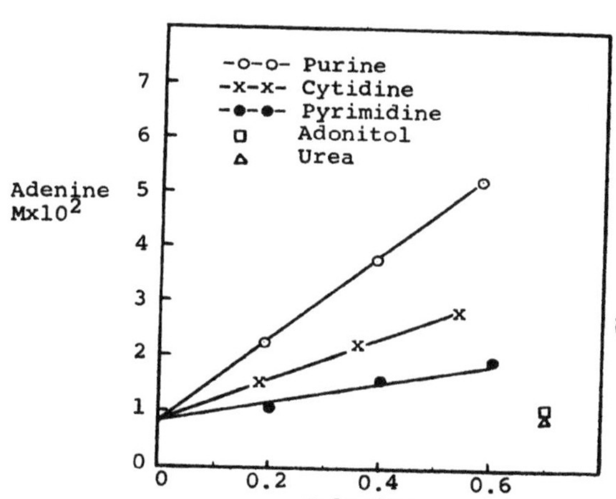

Fig.9 Effects of bases on the solubility of adenine-C^{14} [9].

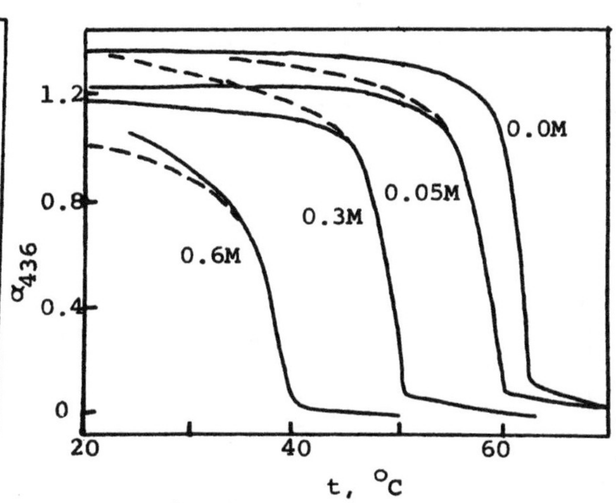

Fig.10 Effect of purine on Tm of poly A [21].

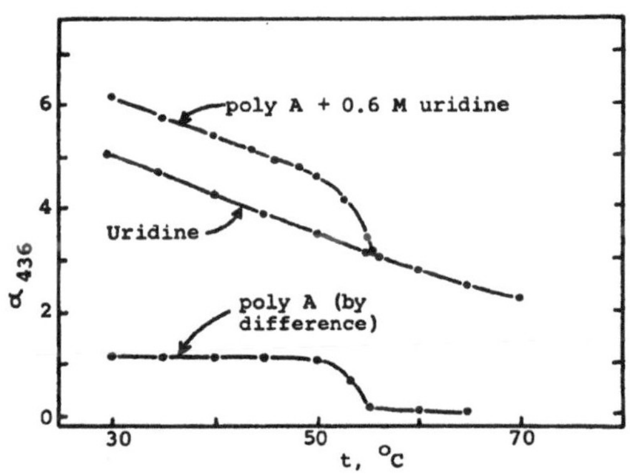

Fig.11 Measurement of Tm in the presence of optically active substances [21].

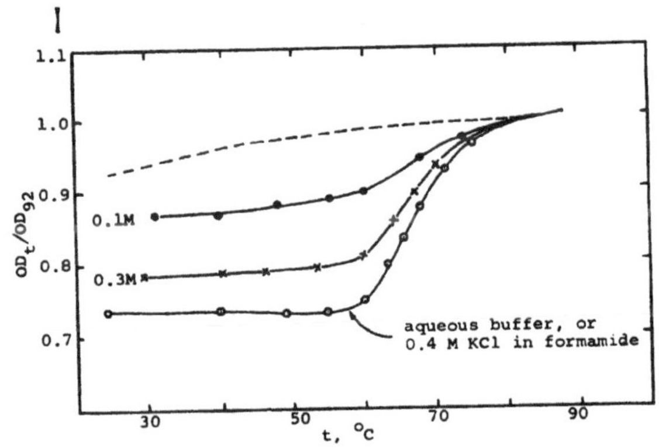

Fig.12 Melting profiles of thymus DNA in HMP after solution in formamide-KCl [26].

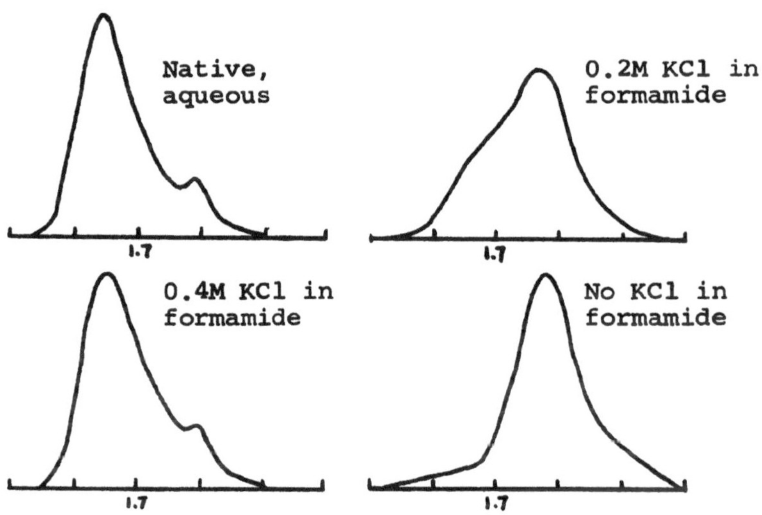

Fig.13 Concentration distribution of thymus DNA in a cesium chloride density gradient at 44,770 rpm [26].

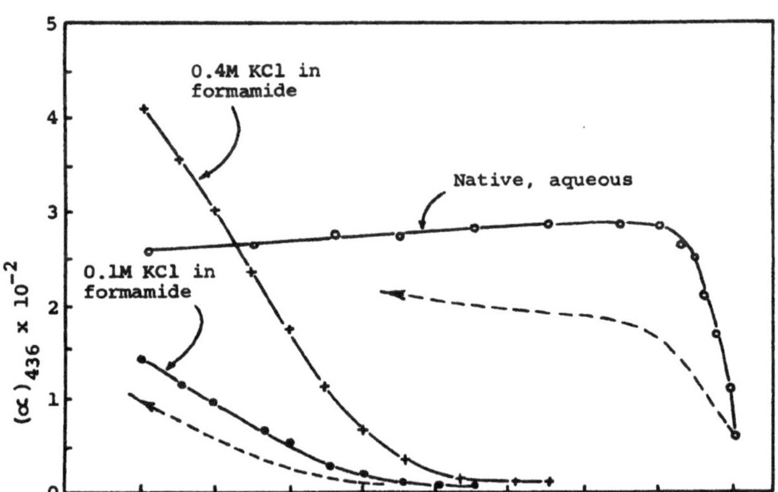

Fig.14 Temperature profiles of optical rotation of thymus DNA in aqueous and formamide solvents [26].

D Nucleinsäuren

Über den Einfluß von Protonen auf die DNA-Konformation

Ch. ZIMMER und H. VENNER

Institut für Mikrobiologie und experimentelle Therapie der Akademie, Jena

S U M M A R Y

Some effects produced by protonation of the deoxyribonucleic acid (DNA) molecule in solution were studied by spectrophotometric and T_m measurements. Below pH 5 the ionization of the bases takes place as indicated by the characteristic absorption ratios and the helix-coil transitions both estimated at different wavelength. Those ratios are determined for various DNAs of different guanine-cytosine contents in the acid range. Protonation of the bases leads to a sharpening of the breadth of the thermal transitions measured at distinct wavelengths. It is concluded, that protonation of the native DNAs takes place due to the basic properties of the corresponding bases. The ionization of the base pairs possibly causes changes in the DNA-conformation and the denaturation process becomes more cooperative.

Die geordnete Sekundärstruktur der DNA stellt in neutraler, wäßriger Salzlösung eine Doppelhelix dar, deren Konformation durch Übereinanderschichtung planar angeordneterBasen und deren Paarung zu wasserstoffbrückengebundenen Basenpaaren zustande kommt [1]. Im wäßrigen System hängt die thermische Stabilität des DNA-Doppelmoleküls wesentlich von der Stärke [2)-3)] und Art[1)3)-4)] der zugesetzten Gegenionen ab.

Unsere Untersuchungen über die Protonisierung der DNA-Molekel beziehen sich vorwiegend auf Veränderungen der DNA-Sekundärstruktur infolge Wechselwirkung der Wasserstoffionen mit den Nucleotidbasen während des thermischen Denaturierungsvorganges [5]. Früher wurde von uns mitgeteilt [4], daß die von MARMUR und DOTY [2),6)] bei pH 7 gefundenen thermischen Stabilitätsunterschiede von DNAs verschiedenen GC-Gehaltes durch Protonenzugabe bei pH 3 nicht vorhanden sind. Die weiteren Messungen stellen einen Versuch dar, die Protonisierungseffekte der Basen im DNA-Molekül zu charakterisieren.

Isolierung, Reinigung [7)-8)] und Messungen [9] erfolgten nach bereits früher gemachten Angaben. Der Einfluß von Wasserstoffionen auf den Molekülzustand läßt sich auch durch Messungen der spezifischen Rotation zeigen, da die optischen Rotationswerte der DNA sehr stark von der helikalen Konformation des Moleküls abhängen [10]. Wie Abb. 1 zeigt, fällt bei Zusatz von Protonen der $T_{m[\alpha]_D}$-Wert für Heringssperma-DNA in SSC [+)] beim Übergang von pH 7 nach pH 4,7 um 9 °C ab. Für diese Veränderung ist offenbar die Protonenwechselwirkung mit den Basenpaaren, die bereits unter pH 5 zur Ionisation der Basen und Spaltung einiger Wasserstoffbrücken führt, verantwortlich. Um die Protonisierungseffekte an den Basenpaaren AT und GC, die sich im UV-

---

+) SSC = Standard-Salz-Citrat = 0,15 m NaCl + 0,015 m Na-Citrat.

Spektrum der DNA charakteristisch bemerkbar machen sollten, zu studieren, wurden die Messungen der thermischen Denaturierung im sauren Milieu auf den gesamten Wellenlängenbereich der UV-Absorption d r DNA ausgedehnt. Das UV-Spektrum der DNA ergab jedoch in SSC unterhalb 240 nm kein normales Minimum, wie aus Abb. 2 hervorgeht. In Phosphatpuffer (Abb. 2) kann dagegen das Minimum bei 228 nm und ein zweites Maximum bei 207 - 208 nm gemessen werden. Der Phosphatpuffer wurde der Ionenstärke von SSC (μ = 0,24) entsprechend gewählt: 0,1 m NaCl + 0,04 m Na_2HPO_4 + 0,02 m NaH_2PO_4 = SPP (Standard-Phosphat-Puffer) (μ = 0,24). Nach Abb. 3 sind in SPP im sauren Gebiet deutliche Veränderungen der bei pH 7 sehr ähnlichen Kurvenprofile der Helix-Knäuel-Übergänge zu beobachten. Es erfolgt eine Änderung des Hyperchromie-Effektes und der thermischen Übergangsbreite Δt bei verschiedenen Wellenlängen. Ein Anstieg der Hyperchromie und eine deutliche Schärfung der thermischen Übergangsbreite sind besonders bei 280 nm festzustellen.

Tabelle 1

DNA	pH	Temp. °C	220	230	240	250	260	270	280	290 nm
Proteus mirabilis GC 42 Mol-%	7,0	T_m	84,7	83,9	84,4	84,3	84,0	83,8	83,9	83,4
		Δt	5,7	5,9	5,0	5,5	5,3	4,8	4,9	17,7
	3,4	T_m	40,2	40,6	40,6	40,8	40,8	40,3	40,9	41,0
		Δt	5,9	5,0	4,5	3,5	2,8	6,5	0,4	1,2

T_m- und Δt-Werte der DNA aus Proteus mirabilis im neutralen und sauren Milieu bei Gegenwart von 0,1 m NaCl + 0,04 m Na_2HPO_4 + 0,02 m NaH_2PO_4.

Tab. 1 zeigt für die Ionenstärke 0,24 (0,1 m NaCl + 0,04 m Na_2HPO_4 + 0,02 m NaH_2PO_4) die T_m- und Δt-Werte von Proteus mirabilis-DNA im neutralen und sauren Bereich. Alle Werte wurden entsprechend der Wasserausdehnung für erhöhte Temperaturen korrigiert. Unter den Bedingungen der von uns benutzten Meßmöglichkeiten ist keine starke Abweichung der T_m-Werte zu beobachten, während bei Messungen bis zu 0,05° C Genauigkeit nach FELSENFELD und SANDEEN [11] offenbar eine charakteristische T_m-Abhängigkeit von λ besteht. Die Werte der thermischen Übergangsbreite Δt, die für die Heterogenität der Zusammensetzung der DNA charakteristisch sind [12], und die thermischen Denaturierungsprofile der Abb. 3 zeigen, daß prinzipiell eine Schärfung der Übergangsbreite durch Zugabe der Protonen auftritt, d. h. Δt wird kleiner. Dies gilt im wesentlichen bei allen Wellenlängen. Für die DNA aus Str. chrysomallus (GC = 71 Mol-%) war Δt bei pH 7 und pH 3 allgemein größer und eine Schärfung der Übergangsbreite konnte nur bei einigen Wellenlängen beobachtet werden: λ = 280 nm Δt pH 7 = 17,1°, Δt pH 3 = 11,2. Die Auswertung der Hyperchromie-Effekte in Abhängigkeit von λ ergab qualitativ vergleichbare Absorptionsänderungen, die entsprechend dem unterschiedlichen GC-Gehalt den betreffenden charakteristischen Absorptionen der Basen zugeschrieben werden können. Die Protonisierung der Basen Cytosin, Guanin und Adenin führt zur Ionisation der N - CO^- bzw. N = C - NH_2-Gruppierungen, was Änderungen im Absorptionsverlauf der DNA bedingt. Desoxycytidin zeigt einen starken Anstieg bei 280 - 285 und einen Abfall bei 230 - 240 nm; Guanosin einen Abfall bei 275, 230 und 250 nm und Adenin einen Anstieg bei 260 - 265 nm bzw. einen Abfall bei 225 - 230 nm [13]. Im Zustand der geordneten DNA-Struktur sind die Ionisationen der Nucleotidbasen infolge der Wasserstoffbrückenbindungen und Schichtung der Basenpaare mehr oder weniger unterdrückt [14]. Die Einwirkung der Protonen auf die reaktiven Zentren der Basen, die entsprechend der Reihenfolge der Basizitätsunterschiede erfolgt, dürfte eine geringe Ionisation bewirken, was eine Verringerung der Stabilität oder Lösung von Wasserstoffbrückenbindungen zur Folge haben muß. Beim thermischen Übergang werden dann auch denaturierte Bereiche im DNA-Helix-Molekül mit vollständig ionisierten Basensequenzen frei. Adenin, Guanin und Cytosin rufen bei ca. 230 - 240 nm einen Absorptionsabfall hervor, was tatsächlich in der Dispersion des Hyperchromie-Effektes der DNA im sauren Milieu stark zum Ausdruck kommt (Abb. 4). Bei 280 nm ist infolge Ionisation der Cytosinreste ein Anstieg im Falle von Proteus mirabilis-DNA zu beobachten. Der AT-Typ sollte infolgedessen eine geringere Steilheit dieses Kurvenprofiles zeigen, da Adenin bei 260 - 265 nm einen Absorptionsanstieg besitzt [13]. Wie aus Abb. 4 zu ersehen, ist dies tatsächlich der Fall (Abb. 4, Proteus mirabilis-DNA bei pH 3,1 im Bereich 260 - 265 nm). Diese Effekte sind in Übereinstimmung mit den von uns früher gemachten Beobachtungen über die Veränderungen des Helix-Knäuel-Überganges der Protonisierung des DNA-Moleküls. Danach kommt es bei pH 3 praktisch zu einer Aufhebung der GC-Abhängigkeit der Denaturierungstemperatur [4].

Zur Erklärung für die konformativen Veränderungen des DNA-Moleküls infolge Protonisierung ist ein Vergleich der Ionisationszustände von DNA verschiedenen GC-Gehaltes nützlich.

Tabelle 2

DNA (Mol-% GC)	p_H 3			p_H 4		
	% $(GC)^+H$	% $(AT)^+H$	$\frac{(GC)^+H}{(AT)^+H}$	% $(GC)^+H$	% $(AT)^+H$	$\frac{(GC)^+H}{(AT)^+H}$
71	35	11	3,18	22	3	7,33
50	25	19	1,36	15	6	2,50
42	21	22	0,95	13	7	1,86

Anteil der Kationformen der Basenpaare der DNA verschiedenen GC-Gehaltes in % der Gesamtmolprozente der Basen und ihre Relationen. Die Abschätzung der Ionisation der DNA wurde summarisch unter Berücksichtigung der Ionisation der Nucleoside im sauren Milieu vorgenommen.

In Tab. 2 haben wir die Ionisationen eines der Basenzusammensetzung verschiedener DNAs äquivalenten Nucleosidgemisches mit Hilfe der pK_a-Werte [+)] berechnet. Das Ergebnis, welches lediglich eine Abschätzung der Ionisationsunterschiede darstellen soll, zeigt, daß bei DNA höheren GC-Gehaltes mehr GC-Basenpaare ionisiert werden sollten, was vor allem auf das Cytidin als stärkste Base zurückzuführen ist. Dies geht auch aus dem Vergleich der in Tab. 2 angeführten Verhältnisse $(GC)^+/(AT)^+$ hervor. Bei pH 4 ist dieses Verhältnis größer, d. h. die vorhandenen Protonen werden vorwiegend vom Cytosin als stärkste Base gebunden. Erst bei pH 3 nimmt auch die Ionisation von Adenin beträchtlich zu. Um die hier wiedergegebene Abschätzung der Ionisation der Basen qualitativ an der DNA zu prüfen, wurde das Auftreten der betreffenden Basenkationform in der DNA bei Protonisierung an Hand der veränderten UV-Spektren im sauren Milieu im Falle einer DNA mit hohem GC-Gehalt (71 Mol-%) untersucht. Wie aus Abb. 5 zu ersehen ist, führt der Zusatz von Säure zu einem Anstieg der Absorption im Gebiet von 265 - 300 nm, wobei sich das Maximum von 258 nm (pH 7) nach 265 nm (pH 2,98) verschiebt, und im Bereich von 230 - 260 kommt es zu einem Absinken der Absorption entsprechend der Ionisation der Cytosinreste. Interessant ist dabei das Auftreten des isosbestischen Punktes bei 253 - 256 nm. Der prozentuale Anstieg bzw. Abfall der charakteristischen Absorptionsverhältnisse bei Protonenzugabe gegenüber pH 7 ist in Tab. 3 für zwei verschiedene DNA-Typen wiedergegeben.

+) Die Berechnung erfolgte summarisch auf der Grundlage bekannter pK_a-Werte [14] der Basen in saurer Lösung.

Tabelle 3

DNA	λ/260	Änderung der Absorptionsverhältnisse mit dem pH-Wert in %			
		pH 4,31	pH 3,62	pH 3,10	pH 2,98
Streptomyces	230/260	+ 5,84	+ 3,20	+ 3,89	- 14,72
chrysomallus	240/260	- 2,13	- 5,62	- 8,53	- 22,49
GC 71 Mol-%	250/260	- 1,47	- 3,15	- 6,50	- 12,48
	270/260	+ 2,77	+ 8,94	+ 18,0	+ 25,56
	280/260	+ 8,22	+ 25,62	+ 51,8	+ 65,39
		pH 4,23	pH 3,68	pH 3,10	pH 2,98
Proteus mirabilis	230/260	+ 8,48	+ 3,34	- 9,25	- 8,23
GC 42 Mol-%	240/260	- 2,79	- 0,74	- 12,29	- 12,66
	250/260	- 0,23	- 0,00	- 7,22	- 7,22
	270/260	+ 1,21	+ 3,63	+ 12,45	+ 12,80
	280/260	+ 2,73	+ 9,11	+ 25,68	+ 26,22

Änderung der Absorptionsverhältnisse der DNA in Abhängigkeit vom pH-Wert der Lösung, in Gegenwart von 0,2 M $NaClO_4$. Die Vorzeichen beziehen sich auf die Absorptionsänderungen bei der betreffenden Wellenlänge im sauren Gebiet, auf eine Herabsetzung (-) oder einen Anstieg (+) der Absorption.

Tabelle 4

DNA	GC Mol-%	Änderung der Absorptionsverhältnisse pH 7 - 3,1 in %				
		230/260	240/260	250/260	270/260	280/260
Str. chrysomallus	71	+ 3,89	- 8,53	- 6,50	+ 18,00	+ 51,80
E. coli	50	- 3,03	- 24,28	- 10,98	+ 16,39	+ 37,67
B. subtilis	43	- 8,99	- 12,78	- 5,84	+ 12,50	+ 34,20
Proteus mirabilis	42	- 9,25	- 12,29	- 7,22	+ 12,45	+ 25,68
Heringssperma	46	- 9,90	- 12,77	- 7,69	+ 13,80	+ 31,84
Weizenkeim	46	- 6,87	- 13,39	- 7,68	+ 11,16	+ 28,60
Kalbsthymus	42	- 18,52	- 18,18	- 8,49	+ 14,30	+ 29,42

Änderung der Absorptionsverhältnisse verschiedener DNAs bei Zugabe von Protonen (~ pH 3,1) in 0,2 m $NaClO_4$-Lösung. Die Vorzeichen beziehen sich auf die Absorptionsänderungen bei der betreffenden Wellenlänge im sauren Gebiet, auf eine Herabsetzung (-) oder einen Anstieg (+) der Absorption.

Die Absorptionsverhältnisse 270/260 und 280/260 steigen mit zunehmender Wasserstoffionenkonzentration der Lösung für DNA hohen GC-Gehaltes (71 Mol-%) stärker an gegenüber DNA mit niederem GC-Gehalt (42 Mol-%). Entsprechend werden die Werte für 230/260 und 240/260 nm kleiner bzw. negativ, da die Extinktion im Gebiet 230 - 240 nm zunehmend absinkt [13]. Ein Vergleich verschiedener DNA-Typen (Tab. 4) ergibt bei Änderung des pH-Wertes von 7 nach 3,1 einen Anstieg der Verhältnisse 270/260 und 280/260 mit wachsendem GC-Gehalt, während die Quotienten 250/260, abgesehen von E. coli-DNA, nahezu in der gleichen Größenordnung liegen, was auf das Vorhandensein eines nahezu ähnlichen isosbestischen Punktes verschiedener DNA-Typen in diesem Wellenlängenbereich hindeutet. Bei heterogenen DNA-Typen sind die Werte weniger übersichtlich. Auf Grund dieser Resultate können wir feststellen, daß bei Protonisierung des DNA-Moleküls eine Beeinflussung und Änderung der Konformation infolge Ionisation der Basenreste bereits im Bereich unter pH 5 erfolgt und bei pH 4 deutlich vorhanden ist. Dies betrifft zunächst das Cytosin, da Cytosin die stärkste der vier DNA-Basen darstellt. Demnach kann infolge einer bevorzugten Protonisierung der Cytosinreste die GC-Bindung im sauren Gebiet stärker labilisiert werden (vgl. Diskussion in Literaturzitat [5]), was bei Raumtemperatur nicht unbedingt die Bildung von denaturierten Strukturen nach sich ziehen muß, wohl aber bereits eine gewisse Veränderung des Konformationszustandes bedeuten kann. Zusätzlich ist auch eine Stabilitätsverminderung des GC-Paares im sauren Milieu (pH 3) durch Protonisierung am N_7 des Guanins [15] nicht ausgeschlossen, was eine Ionisation des Guanins, ähnlich der Methylierung [16] bewirken könnte. Andererseits muß im Temperaturbereich oberhalb des T_m-Wertes bei pH 3 mit einer beginnenden Purinabspaltung [17] gerechnet werden (unveröffentlichte Ergebnisse dieses Laboratoriums).

Einige theoretische Überlegungen zu den Veränderungen des thermischen Schmelzvorganges der DNA bei Protoneneinwirkung sollen kurz angedeutet werden. In einer früheren Arbeit [5], die auf die qualitativen Veränderungen der DNA-Moleküle infolge Wechselwirkung der Protonen mit den Basenpaaren gerichtet war, ist unter den dort angewandten Bedingungen die Änderung der Übergangsbreite des Schmelzvorganges zunächst ungeklärt geblieben. Eine Reihe von Messungen in SPP (Tab. 1, Abb. 3) und in 0,1 SSC (unveröffentlichte Ergebnisse der Autoren) zeigten wiederholt, daß eine deutliche Schärfung der Übergangsbreite, d. h. eine Erniedrigung von Δt erfolgt. Dieses Verhalten trifft für verschiedene Wellenlängen zu, wie aus Tab. 1 hervorgeht, was jedoch bei einer DNA mit hohem GC-Gehalt (Streptomyces chrysomallus 71 Mol-% GC) nicht in derselben Weise eindeutig gefunden werden konnte. Die vorliegenden Ergebnisse stimmen jedoch prinzipiell mit den Beobachtungen anderer Autoren [20] überein. Nach statistisch-mechanischen Betrachtungen von ZIMM [18] hängt der Schmelzvorgang des DNA-Moleküls unter anderem von zwei wesentlichen thermodynamischen Parametern ab, die die Bildung von denaturierten Bezirken in der DNA-Konformation charakterisieren: die Größe s, welche die Gleichgewichtskonstante zwischen der Zahl von wasserstoffbrückengebundenen und ungebundenen komplementären Basensequenzen im Doppelhelixbereich darstellt und der Schichtungsparameter σ_0. Nach DE VOE und TINOCO [19] ist die freie Energie der Helix von der Basenzusammensetzung und der Sequenz der DNA abhängig. Offenbar wird primär die DNA-Helix-Stabilität gegenüber den Einzelsträngen im wäßrigen System hauptsächlich durch elektrostatische Anziehungskräfte stabilisiert [19]. Der negative Enthalpiebetrag der Helixbildung ist bei pH 7 für die stabilere GC-reiche DNA entsprechend größer gegenüber einer AT-reichen. Man kann annehmen, daß die von uns diskutierte Protonisierung der DNA-Basenpaare bzw. deren Ionisation diese bestehenden elektrostatischen Kräfte im Molekül verändert und damit sollte der größere negative Enthalpiebetrag der Helix vom GC-Typ gegenüber einer Helix vom AT-Typ im sauren Milieu niedriger werden. Da Δt kleiner wird (Tab. 1, Abb. 3), bedingt die Protonisierung des DNA-Moleküls an den Basenpaaren offenbar eine Verringerung der innermolekularen Heterogenität. Nehmen wir an, daß die Protonisierung der DNA-Helix-Konformation den freien Energiebetrag erhöht, und zwar sowohl die freie Energie der Wasserstoffbrückenbindung wie die der Schichtungsenergie der Basenpaare. Dies müßte einerseits zu einem Anstieg von s und damit zur Erniedrigung des T_m-Wertes führen, andererseits wird der Schichtungsparameter σ_0 herabgesetzt [18][20], was das Vorhandensein einer kleineren Anzahl längerer helikaler Bezirke und einer geringeren Zahl denaturierter Bezirke in einem teilweise denaturierten DNA-Molekül während des thermischen Überganges bedeutet. Der thermische Übergang wird dadurch schärfer [18], der Ablauf des Vorganges ist stärker kooperativ gegenüber neutralen Bedingungen. Eine klare Verringerung der Übergangsbreite wurde im wesentlichen bei 240, 250, 260, 280 und 290 nm beobachtet. Dies hängt vielleicht davon ab, welche an der Absorptionserhöhung beteiligten protonisierten Basenpaare auf Grund der Basenzusammensetzung der betreffenden DNA bei der jeweiligen Wellenlänge ausschlaggebend sind.

L I T E R A T U R

1) WATSON,J.D., F.H.C.CRICK, Nature 171, 737 (1953)
WILKINS,M.H.F., Angew. Chem. 75, 429 (1963)
2) MARMUR,J., P.DOTY, J. molecular Biol. 5, 109 (1962)
3) DOVE,W.F., N.DAVIDSON, J. molecular Biol. 5, 467 (1962)
4) ZIMMER,Ch., H.VENNER, Naturwissenschaften 49, 86 (1962)
5) ZIMMER,Ch., H.VENNER, J. molecular Biol. 7, 603 (1963)
6) MARMUR,J., P.DOTY, Nature 183, 1427 (1963)
7) ZIMMER,Ch., H.VENNER, Hoppe-Seyler's Z. physiol. Chem. 333, 20 (1963)
8) VENNER,H., Hoppe-Seyler's Z. physiol. Chem. 333, 5 (1963)
9) ZIMMER,Ch., H.VENNER, Hoppe-Seyler's Z. physiol. Chem. (1964), im Druck
10) FRESCO,J.R., Tetrahedron 13, 185 (1961)
TINOCO,J., R.W.WOODY, D.F.BRADLEY, J. chem. Physics 38, 1317 (1963)
11) FELSENFELD,G., G.SANDEEN, J. molecular Biol. 5, 587 (1962)
12) DOTY,P., J.MARMUR, N.SUEOKA, Brookhaven Symp. Biol. 12, 1 (1959)
13) BEAVEN,G.H., E.R.HOLIDAY, E.A.JOHNSON, in "The Nucleic Acids", ed. by E.CHARGAFF und J.N.DAVIDSON, S. 493 (1955), New York: Academic Press, Inc.
14) JORDAN,D.O., in "The Nucleic Acids", ed. by E. CHARGAFF und J.N.DAVIDSON, S. 447 (1955), New York: Academic Press, Inc.
15) JARDETZKY,C.O., O.JARDETZKY, Fed. Proc. 17, 380 (1958)
JARDETZKY,C.O., O.JARDETZKY, J. Amer. chem. Soc. 82, 222 (1960)
PULLMAN,B., A.PULLMAN, Nature 189, 725 (1961)
16) LAWLEY,P.D., P.BROOKES, Nature 183, 1427 (1961)
17) GREER,S., S.ZAMENHOF, J. molecular Biol. 4, 123 (1962)
18) ZIMM,B.H., J. chem. Physics 33, 1349 (1960)
19) DE VOE,H., J.TINOCO, J. molecular Biol. 4, 500 (1962)
20) SUCHORUKOW,B.J., Ju.Sch.MOSCHKOWSKI, T.M.BIRSCHTEIN, W.N.LYSZOW, Biophysik 8, 294 (1963).

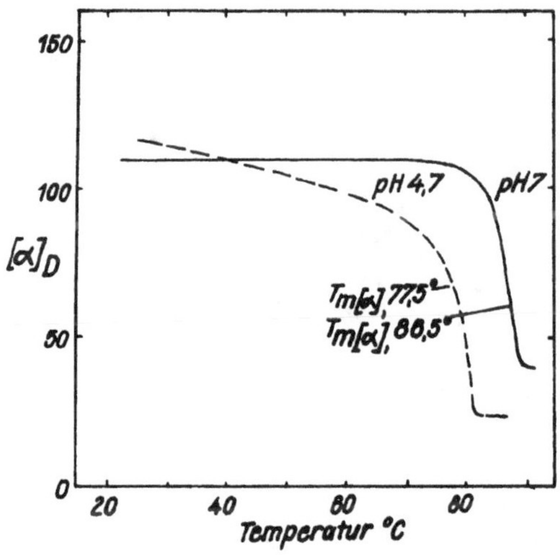

Abb.1 Thermischer Übergang (Änderung der spezifischen Rotation mit der Temp.)der Desoxyribonucleins. aus Heringssperma in SSC.

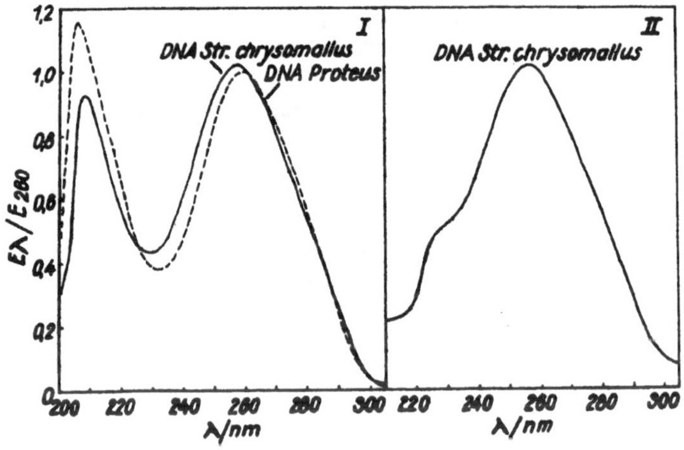

Abb.2 UV-Spektrum mikrobieller Desoxyribonucleinsäuren in Kochsalz-Citrat (SSC) (I) und Phosphatpuffer (SPP) (II).

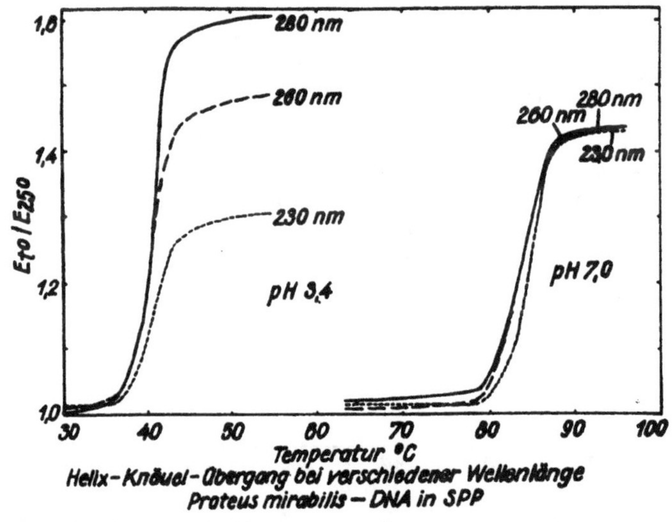

Abb.3 Änderung des Helix-Knäuel-Überganges (Absorptions-Temperatur-Profile) von Proteus mirabilis-Desoxyribonucleinsäure bei 230, 260 und 280 nm; gemessen in SPP für pH 7 und 3,4.

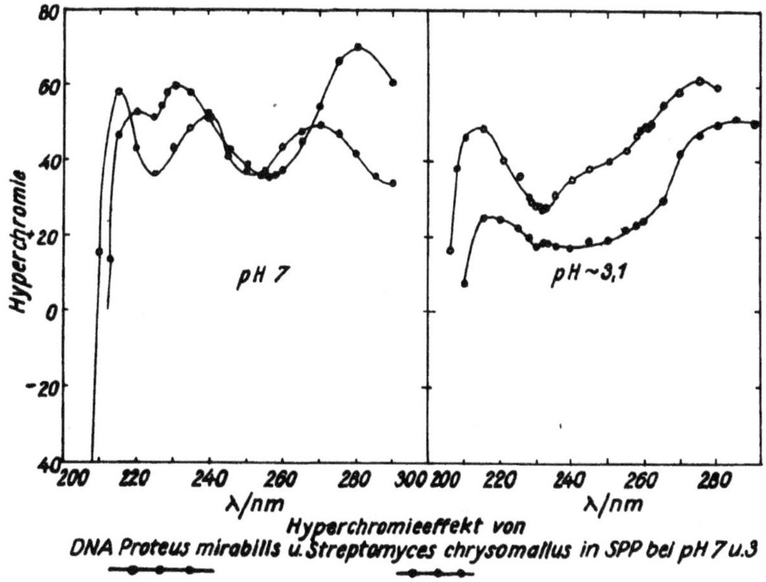

Abb.4 Dispersion des Hyperchromie-Effektes von Desoxyribonucleinsäure aus Proteus mirabilis (Guanin + Cytosin 42 Mol-%) und Streptomyces chrysomallus (Guanin + Cytosin 71 Mol-%), gemessen in SPP im neutralen und sauren Bereich. Die Einstellung des pH-Wertes erfolgte unter den gleichen Bedingungen wie in Abb.3 angegeben; der Hyperchromie-Effekt (h) wurde nach der Gleichung $h = \frac{A}{A_{25^0}} \cdot 100$ berechnet, a = Absorption.

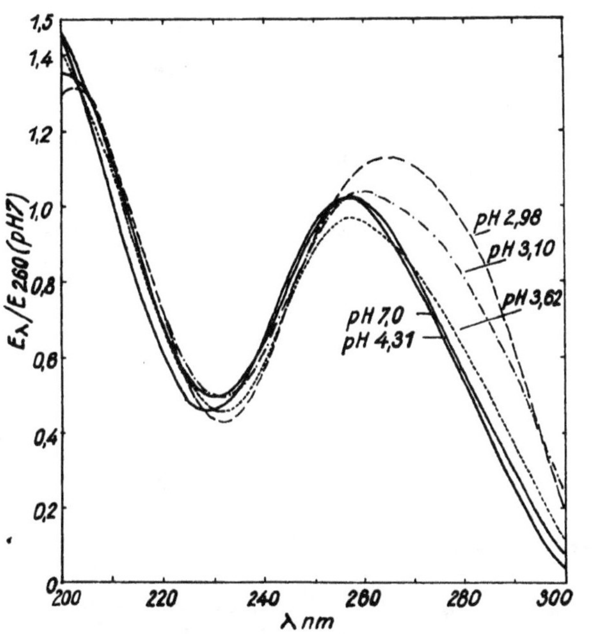

Abb.5 UV-Spektrum von Streptomyces chrysomallus-Desoxyribonucleinsäure in 0,2m $NaClO_4$ bei verschiedenen pH-Werten. Die pH-Einstellung wurde mit 0,2 n HCl vorgenommen.

D Nucleinsäuren

Denaturation of DNA in Heavy Water

V. GRECHKO, R. MASLOVA, L. SHKARENKOVA and J. VARSHAVSKY

Institute of Radiation Biology and Physico-Chemical Biology of the Academy, Moscow

S U M M A R Y

1. A different stability of DNA at the thermal and acid-base denaturation in ordinary and heavy water has been shown.

2. The differences between the temperatures of thermal denaturation and values of pH of acid-base denaturation of DNA in ordinary and heavy water has been determined.

3. Comparison of differences in dissociation constants of -NH-CO- and NH_2 - groups in native DNA (Δ pK DNA $\approx$ 0,4) and those in constants of the same groups of free nucleotides (Δ pK Nucl $\approx$ 0,4) in ordinary and heavy water suggested that the observed differences in the stability of DNA to acid and base denaturation were mainly due to the difference in the dissociation constants of these groups in ordinary and heavy water.

I N T R O D U C T I O N

Investigations of protein solutions in heavy water by LINDERSTROM-LANGE and his coworkers from Karlsberg Laboratory, Dr. SHERAGA with coworkers and others, have shown that incorporation of deuterium into a macromolecule results in stabilization of helical structures. The stability of such structures depends upon the strength of hydrogen bonds. This effect leads to an increase in the temperature corresponding to the helix-coil transition.

From modern structural conceptions concerning nucleic acids it would be expected that substitution of deuterium for hydrogen in the molecules of such biopolymers should result in a similar effect. This problem is of considerable interest for two reasons. Firstly, great attention is paid at present to the role of structural transition in nucleic acids during their vital activity. Secondly, it is closely connected with the elucidation of the mechanism of inhibition of propagation and growth by deuterium.

We studied thermal and acid-base denaturation of DNA in ordinary and heavy water solutions.

The samples of DNA under study were prepared from Bacteriophage T_2 Ech. Coli. In a number of experiments we used also DNA from calf's thymus. The results obtained were in principle similar for both DNA.

To investigate the thermal and acid-base denaturation of DNA we have used two different methods, i. e. firstly the measurement of the hyperchromism at 260 nm, secondly - measurement of the viscosity.

THERMAL DENATURATION

In some preliminary experiments we have compared the temperature dependence of hyperchromic effect of neutral salt solutions of DNA in ordinary and heavy water. Contrary to expectations no essential differences in the shape of curves has been found. This result is in agreement with recent results of several American authors who also measured the temperature curves of hyperchromic effect of neutral salt solutions of DNA, RNA and synthetic polynucleotides in ordinary and heavy water.

However our experimental data on the temperature curves of viscosity of neutral salt solutions of DNA in ordinary and heavy water showed that the melting temperature of native (double stranded) DNA in heavy water is somewhat higher than that of the same DNA in the solution in ordinary water. In fig. 1 you can see experimental curves on viscosity of DNA depending on the temperature between 20 and 95^{0} at pH 7 in 0.15M solutions of sodium chloride in ordinary and heavy water. You can see that the melting temperature in heavy water is somewhat higher then that in ordinary water.

However, viscosity measurements of DNA at pH 3 have shown, that with acid denaturation an opposite effect is observed. That is "melting" of DNA in heavy water occurs at a much lower temperature than in ordinary water. This cannot be explained as has been done with proteins, by the change in strength of hydrogen bridges between DNA strands due to deuterium substitution, because the relative stability of deuterium and ordinary DNA in neutral and acidic solutions should not be markedly affected.

In fig. 2 you can see the temperature curves of viscosity of solutions of DNA in ordinary and heavy water in acidic medium at pH 3.

You can see, that melting temperature of DNA in ordinary water is actualy higher than in heavy water.

We have interpreted the observed effect on the assumption that substitution of ordinary water by heavy water leads to change in some properties of DNA that are essentially dependent on the pH of the medium and markedly affect the DNA denaturation process in the solutions.

To make clear the reasons determining the observed differences in stability of DNA in neutral solutions in ordinary and heavy water, as well as the changes in the character of these differences in acidic solution we had to study the dependence of DNA denaturation upon the pH of the medium at constant temperature for ordinary and heavy water solutions.

That is to say, it was necessary to compare the curves of acid-base denaturation of the same DNA in ordinary and heavy water.

ACID-BASE DENATURATION

If one starts from modern conceptions that denaturation of DNA is closely connected with the cleavage of hydrogen bonds between functional groups - NH_2 and - NH-CO, then the dissociation of these groups taking place at particular values of the pH of the medium must result in the cleavage of hydrogen bonds and thus in denaturation of DNA. The acidic dissociation constants of chemical groups under investigation, whose values are known to be somewhat different in ordinary and heavy water provide a quantitative characteristic of the ability of these groups to undergo dissociation.

Table 1 summarises data given in the literature for the constants of dissociation of NH_2 and NH-CO-groups both for free nucleotides and nucleic acids.

Table 1:

Values of pK dissociation of $-NH_2$ and -NH-CO-groups in nucleotides and DNA

Nucleotide	Functional group	pK of free nucleotides	pK of nucleotides in DNA: denaturated DNA	pK of nucleotides in DNA: nativ DNA
Guanilic acid	-NH-CO-	9.3	10.4; 10.9	11.4
	NH_2	2.4	2.4	2.5
Cytidilic acid	$-NH_2$	4.2	4.8	4.7; 4.4
Adenilic acid	$-NH_2$	3.7	3.6	3.7; 4.4
Thymidilic acid	-NH-CO	9.4	10.4	11.5

It was experimentally shown that the difference in magnitude of acidic dissociation constants in heavy and ordinary water $\Delta pK = pK_{D_2O} - pK_{H_2O}$ for individual groups of chemical compounds is a linear function of pK_{H_2O}.

Acids dissociate in heavy water to a smaller extent than they do in ordinary water. For amino and imino groups the corresponding dependence is as follows:

$$\Delta pK = 0{,}243 + 0{,}417\ pK_{H_2O}\ .$$

Using this equation it is possible to estimate the changes in dissociation constants of $-NH_2$ groups in molecules of native DNA when substituting deuterium oxide for water.

With cytidine and adenine in native DNA molecules ΔpK is about 0,4, whereas with similar group of guanine ΔpK is 0,35.

This means that if a 50 % dissociation of $-NH_2$ of cytidine and adenine groups from DNA is achieved in ordinary water, say at pH = 4.4 in heavy water it should take place at pD = 4,8. Similarly for NH_2 groups of guanine of DNA such a dissociation is achieved in ordinary water at pH = 2,5, whereas in heavy water it should take place at pD = 2,85.

Thus the addition of protons to $-NH_2$ groups and directly related to it the process of acidic denaturation of DNA should take place in heavy water at lower acidity than in H_2O. In other words, DNA in heavy water must be more prone to acidic denaturation than in ordinary water.

There are no data available on the quantitative dependence of ΔpK on pK_{H_2O} for groups - NH-CO-. There are reasons to believe that this group is not an exception and that its acidic dissociation resulting from its ability to donate a proton (just as with acidic dissociation of other chemical groups), should be facilitated in ordinary water as compared to heavy water. If it is indeed so, solutions will be more stable to alkaline denaturation, resulting from the dissociation of -NH-CO-groups of guanine and thymine, in heavy water than in ordinary water. The picture in this case is opposite to that expected with acidic denaturation of DNA. The above considerations about the difference in degree of dissociation of $-NH_2$ and -NH-CO-groups in ordinary and heavy water hold for dissociating groups not linked by hydrogens bonds, for instance for nucleotides. In native DNA molecules such groups are hydrogen bonded. Their dissociation is also to some extent affected by the ordered (parallel stacked) arrangement of purine and pyrimedine rings in the spiral helix structure. Therefore, in addition to the study of dependence of the DNA denaturation on pH it was of interest to find out whether the quantitative dependences of ΔpK on pK_{H_2O} are the same in functional groups of separate free nucleotides and in ordered structures of native nucleic acids in which these groups are hydrogen bonded. Comparing the difference in critical values of pH and pD at acidic and alkaline denaturation of DNA, which is equal to pK of $-NH_2$ and -NH-CO- groups of DNA nucleotides, with ΔpK of the same groups of free nucleotides, it is possible to evaluate the effect of a change in dissociation constants of these groups in heavy water on the change in stability of DNA.

In fig. 3 you can see the experimental data on viscosity and hyperchromism of DNA solutions depending on the acidity of medium at 20°C. These data were obtained by measuring the viscosity and absorption in the ultra-violet spectral region of solutions one hour after the neutral DNA solution had been mixed with the corresponding buffer solution. Along ordinates are plotted the values of

$$\frac{\eta^{20^0}_{max} - \eta^{20^0}_{pH}}{\eta^{20^0}_{max} - \eta^{20^0}_{min}} \quad \text{and for A also} \quad \frac{A\ pH - A\ min}{A\ max - A\ min}$$

These values are respectively the ratio of viscosity and hyperchromism of certain pH to their maximum values observed on transition from the native form of DNA in neutral solutions to the completely denatured state in acidic or alkaline solutions. One should note a rather good agreement between critical values of pH determined both by the change in absorption at 260 nm and in viscosity. In table 2 it will be seen that in the case of acid-base denaturation both these methods give a very good agreement in contrast to the thermal denaturation.

Table 2:

Values of pH and pD for 50 % denaturation of DNA

	Acidic medium		Alkaline medium	
	From Hyperchromic effect	From the change in viscosity	From Hyperchromic effect	From the change in viscosity
pH	2,7	2,8	11,3	11,3
pD	3,1	3,2	11,7	11,8
pD-pH	0,4	0,4	0,4	0,5

It is now generally agreed that there exists a linear dependence between the percentage of broken H-bonds and that of the hyperchromic and viscosity effects upon their maximum values at the completely denatured state of DNA. Hence, the value of pH for 50 % hyperchromic effect (i. e. 50 % ionisation of NH-CO groups of guanine and thymine) should be numerically equal to the exponent of the acidic dissociation constant of these groups (pK).

A l k a l i n e   d e n a t u r a t i o n . From these curves it will be seen that the value of pK in ordinary water is 11.3 and in heavy water 11.7. Thus ΔpK is 0,4.

The comparison of ΔpK for the -NH-CO group in ordered structures of native DNA with that of the same groups in free nucleotides could have enabled to come to a conclusion concerning possible changes in the dissociation character of -NH-CO groups due to their participation in the formation of hydrogen bonds. Such a comparison was however, earlier impossible owing to the absence of data for dependence of ΔpK on the pK_{H_2O} for -NH-CO groups. Therefore to determine the values ΔpK of such groups in guanosine we have titrated guanosine spectrophotometrically in solutions of ordinary and heavy water. In table 3 you can see the results of this experiment.

Table 3:

Spectrophotometric Titration of Guanosine Dissolved in ordinary and heavy water

pH	A	$Lg\frac{C_{A^-}}{C_{HA}}$	pK_{H_2O}	pD	A	$Lg\frac{C_{A^-}}{C_{DA}}$	pK_{D_2O}
1-st series				3-th series			
4.8	0.728	-	-	5.4	1.764	-	-
6.5	0.721	-	-	7.4	1.751	-	-
7.25	0.726	-	-	9.21	1.703	-0.52	9.73
8.55	0.704	-0.63	9.18	9.65	1.656	-0.15	9.80
9.08	0.674	-0.12	9.20	10.10	1.586	+0.32	9.80
9.69	0.632	+0.50	9.19	10.80	1.502	-	-
11.10	0.602	-	-	11.59	1.502	-	-
2-nd series				4-th series			
4.75	1.820	-	-	5.35	1.794	-	-
6.32	1.815	-	-	9.10	1.735	-0.56	9.65
7.11	1.813	-	-	9.50	1.681	-0.15	9.65
8.50	1.765	-0.67	9.17	10.07	1.602	+0.37	9.70
9.62	1.621	+0.34	9.28	11.80	1.520	-	-
10.83	1.534	-	-				
11.86	1.533						

The mean experimental value of ΔpK is 0,5. This value is in good agreement with ΔpK for -NH-CO groups determined from alkaline denaturation of DNA. Such an agreement shows that in going from water to heavy water the change in the exponent of the dissociation constant of -NH-CO groups of hydrogen bonded guanine and thymine in macromolecules of native DNA does not essentially differ from the value of ΔpK for the mentioned groups in free nucleotides. Hence, the isotopic effect displayed in the changing stability of DNA to alkaline denaturation in ordinary water in comparison with that in heavy water is mainly accounted for by the difference in dissociation constants of functional groups in ordinary and heavy water. The change in strength of hydrogen bonds due to deutero substitution either does not seem to affect the process of alkaline denaturation of DNA at all, or is of secondary impotance and is not pronounced when DNA is titrated in the alkaline region.

A c i d i c   d e n a t u r a t i o n . If one suggests, that acidic denaturation of DNA is also closely connected with the rupture of hydrogen bonds caused by ionisation of amino groups (at least of $-NH_2$ groups of guanine), then in acidic medium the value of pH corresponding to 50 % denaturation should be numerically equal to the pK of the dissociation of these groups, as it is the case in alkaline denaturation.

In fig. 3 you can see, that experimentally observed difference between the values pH in ordinary and heavy water for 50 % denaturation in acidic region is 0.4. This value is numerically equal to ΔpK of dissociation of the aminogroups of DNA. It was shown above that for the dissociation of the $-NH_2$ group of guanilic acid in heavy and ordinary water ΔpK is 0,35.

It is, thus, possible to suggest that the Δ pK dissociation values for free and bonded NH_2 groups of guanilic acid coincide within the experimental error. Thus the change in stability of DNA towards acidic denaturation in heavy water in comparison with ordinary water, as is the case with alkaline denaturation is mainly due to a difference in dissociation constants of functional groups in these solvents.

It has already demonstrated that dissociation constants in ordinary and heavy water are different. In the recent monograph by ROGINSKY and SHNOLL and in the review by RITTENBERG it was pointed out that the exponent of dissociation constants of most compounds is lower in heavy water then in ordinary water by 0.3 - 0.5 units. Our value of 0.4 for the dissociation of NH_2 and NH-CO groups of DNA is between these figures.

Considering a vivid interest, displayed at the Conference on possibilities and limits of various physico-chemical methods, we would like to stress, in particular, the above mentioned fact, according to which the difference in the melting points of DNA in neutral solutions of ordinary and heavy water is observed only when studying viscosity and not on measuring hyperchromism. A number of American workers such as have recently published several papers dealing with thermal denaturation of neutral salt DNA solutions in heavy water by method of hyperchromism. They failed thereby to observe any marked difference in the denaturation curves for ordinary and heavy water.

There is one more characteristic difference between the hyperchromic and on acid-base denaturation viscosimetric methods. Namely the first method is much less sensitive to changes in the secondary structure of DNA. Fig. 4 shows the acid denaturation curves of DNA measured hyperchromically (continuous line) and viscosimetrically (dotted line). It will be readily seen, that the intensity of light absorption changes sharply in a narrow range of pH, corresponding to the structural change of DNA, whereas the viscosity of the solutions changes gradually, and this change reflects not only the complete denaturation of DNA but also the preceding processes.

This differences between the two methods has been observed not only with acidic but also with alkaline denaturation.

As I said above, this work was concerned with the elucidation of possible causes of an inhibiting action of heavy water on living organisms. The evidence presented in this work suggests that the acid-base denaturation of DNA is due to the dissociation of $-NH_2$ and -NH-CO groups, in strongly acidic and strongly alkaline solutions. Usually the corresponding conditions do not hold in living organisms and therefore it is doubtful whether the differences in these DNA groups in ordinary and heavy water that we have succeeded in observing in vitro might exert any substantial effect on the process of reduplication of DNA in living organisms. It is, however, not to be excluded, that for other groups of biologically active substances that are markedly dissociated in neutral solutions, the changes in dissociation constants on substitution of D_2O for H_2O might be of importance. Even slight differences in the extent of dissociation of such compounds might affect the carefully regulated system of chemical equilibria in the body and thus depress the normal development of vital activity.

R E F E R E N C E S

1. LINDERSTROM-LANG, K., "Deuterium exchange and protein structure", Symposium on Protein Structure, ed. by A. NEUBERGER, 23 - 34 (1958)
2. CALVIN, M., J. HERMANS, H. SCHERAGA, J. Amer. chem. Soc. 81, 5048 (1959)
3. HERMAN, J., H. SCHERAGA, Biochim. biophysica Acta 36, 534 (1959)
4. SCHERAGA, H., Ann. N. Y. Acad. Sci. 84, 608 (1960)
5. MAHLER, H., B. MEHROTRA, Biochim. biophysica Acta 55, 789 (1962)
6. CRESPI, H., J. KATZ, J. molecular Biol. 4, 65 (1962)
7. MAHLER, H., G. DUTTON, B. MEHROTRA, Biochim. biophysica Acta 68, 210 (1963)
8. MILES, H., Biochim. biophysica Acta 43, 353 (1960)
9. JORDAN, D. O., "The Nucleic Acids", N. Y. 1955, v. I, p. 447
10. SHACK, J., J. M. THOMSETT, J. biol. Chemistry 17, 197 (1952)
11. JORDAN, D. O., A. R. MATHIESON, S. MATTY, J. chem. Soc. 154 (1956)
12. COX, R. A., Biophys. biochimica Acta, Previews, 2, N 8 (1962)
13. LI, N. C., PH. TRANG, R. MATUR, 139 Meeting Amer. Chem. Soc. 6 R (1961)
14. WEN, W. J., I. M. KLOTZ, J. physic. Chem. 65, 1085 (1961)
15. MICHELSON, A. M., Nature 182, 1502 (1958)
16. VINOGRADOVA, E. N., Methody opredeleniya conzentrazii vodorodnich ionov., Moskva (1956)
17. GLASOE, P. K., F. A. LONG, J. physic. Soc. 64, 188 (1960)
18. COX, R. A., A. R. PEACOCKE, J. Polym. Sci. 23, 765 (1957)
19. EHRLICH, P., P. DOTY, J. Amer. chem. Soc. 80, 4251 (1958)
20. ROGINSKY, S. Z., S. E. SHNOLL, "Isotopy v biochimii", Moskva (1963)
21. RITTENBERG, D., J. Chim. physique 60, 318 (1963)

D I S C U S S I O N - Q U E S T I O N S

ZIMMER : For investigations on acid-base properties of DNA in solution it is important to know the ionic strength. Therefore I want to know how was the counter ion concentration in your experiments ?

WARSCHAWSKIJ : 0,015 m NaCl.

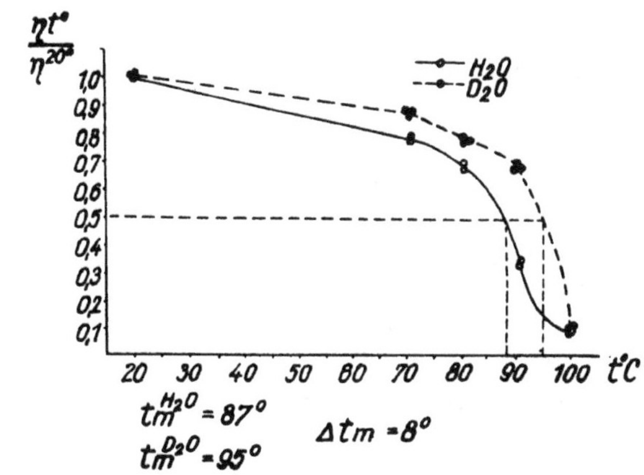

Fig.1 Thermal denaturation of DNA in H_2O and D_2O at pH = 7

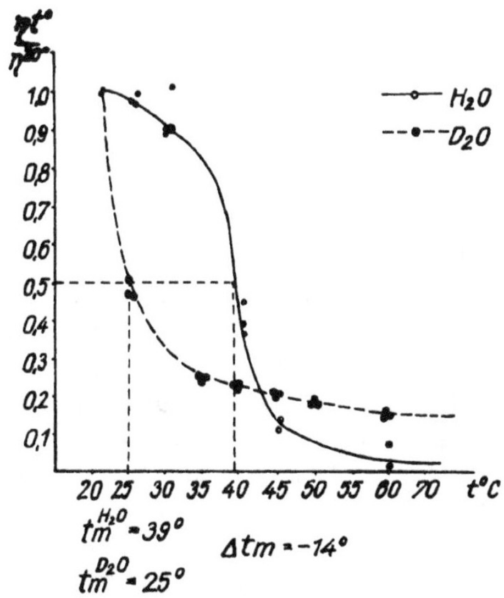

Fig.2 Tho mal denaturation of DNA in H_2O and D_2O at pH = 3

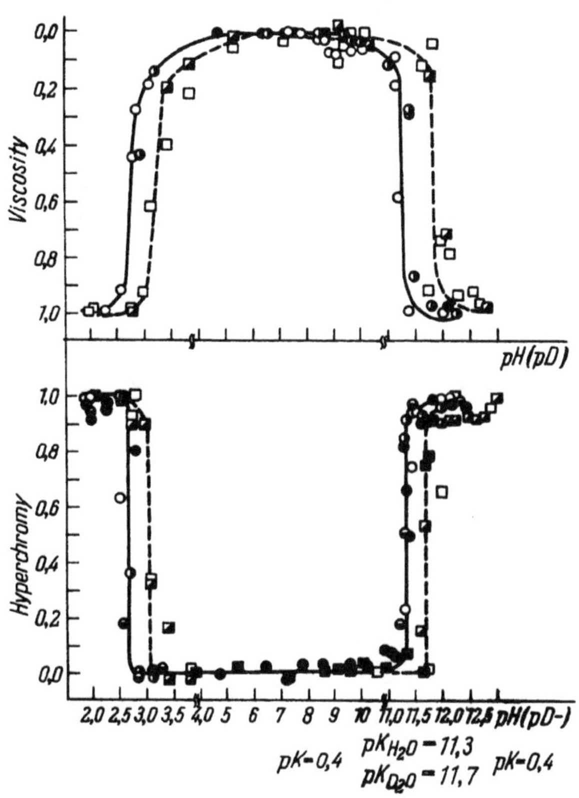

Fig.3 Acid-base denaturation of DNA in H_2O and D_2O

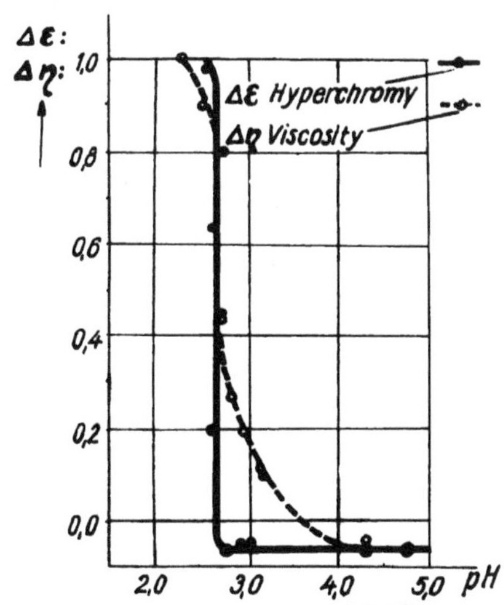

Fig.4 Acid-base denaturation of DNA

D Nucleinsäuren

Die Dispersion der optischen Aktivität von Nucleinsäure-Acridin-Orange-Komplexen

W. J. PERMOGOROW und Ju. Ss. LASURKIN

Kurčatov's Institut für Atomenergie, Moskau

S U M M A R Y

Complexes of acridine orange (AO) with nucleic acids of various types and origin were investigated. The dispersion of optical rotation of the complexes in their absorbtion range was recorded by means of a spectral polarimeter with a sensitivity of 0.0003. The complexes of AO with native T2-phage, E. coli, calf thymus, and salmon sperm DNA's were found to give qualitatively the same curves of the COTTON effect.

The AO-denatured DNA complexes also show a COTTON effect, but of a quite different form. The curves of dispersion of the optical rotation for the AO-RNA complexes - rat liver and yeast RNA - are similar to the corresponding curves for the denatured DNA.

Suggestions are made concerning the connection between the character of the dispersion curves and different possible types of binding of the dye molecules with DNA molecules.

Den Nucleinsäurekomplexen mit Farbstoffen der Acridinreihe ist in der letzten Zeit eine Reihe von Untersuchungen gewidmet worden [1)-5)]. Das läßt sich damit erklären, daß die Acridinfarbstoffe starke Mutagene mit sehr spezifischer Wirkung sind [6)] und daß die Untersuchung ihrer Bindung mit DNA für die Aufklärung des Mechanismus ihrer Mutagenwirkung von Interesse ist. Trotz der vielen Untersuchungen ist der Bindungsmechanismus bis jetzt noch nicht völlig geklärt worden. Als Forschungsmethode wurde die Dispersion der optischen Aktivität gewählt wegen ihrer außerordentlichen Empfindlichkeit für die Konformationsänderungen der Biopolymeren [7)12)]. Da der Kurvencharakter der Dispersion der optischen Aktivität von DNA-Acridinfarbstoff-Komplexen, wie sich aus der folgenden Darlegung ergeben wird, für die Molekülkonformation des Polymers sehr empfindlich ist, kann diese Methode für die Aufklärung der Struktur von Nucleinsäuren selbst angewandt werden. Dabei wird die Untersuchung im sichtbaren Spektralbereich durchgeführt, was sehr bequem ist und die hohe Meßempfindlichkeit der Apparatur garantiert.

Die Menge der für solche Untersuchung notwendigen Nucleinsäure ist zehn- bis hundertmal so klein wie bei der Messung der optischen Aktivität von DNA und RNA selber. Alle Messungen wurden mit einem in unserem Labor entwickelten photoelektrischen Spektralpolarimeter hoher Empfindlichkeit durchgeführt, das noch ausführlich beschrieben werden wird. Im Gerät wurde das Prinzip des sogenannten Halbschattenmodulators angewandt [8]. Als Modulator wird die Faraday-Zelle im Gerät benutzt. Alle Messungen wurden in einer 50 mm langen Quarzküvette durchgeführt. Als Lichtquelle diente die Kryptonröhre ДКсШ-1000. Die Meßempfindlichkeit des Geräts betrug etwa 0,0003°. Bei optischen Dichten der Probe von 1,5 bis 2,0 sank die Empfindlichkeit auf etwa 0,002°. Die Größe des Drehwinkels wurde mit Hilfe der Faraday-Zelle ermittelt.

Gleichzeitig mit den spektralpolarimetrischen Messungen wurden auch Absorptionsspektren von Nucleinsäure-Acridinorange-Komplexen mit einem Spektrophotometer СФ-4 ermittelt. In der Arbeit wurden folgende Präparate verwendet:

1. DNA aus Kalbsthymus, die nach der GULLAND-Methode abgesondert wurde. mit nachfolgender Trennung von Eiweiß durch ein Gemisch von Chloroform und Oktylalkohol.
2. RNA aus Hefe, hergestellt nach MALLTE-LAMANNA-Methode [9], Eiweißgehalt betrug 2,4 %.
3. RNA aus Rattenleber, hergestellt nach der Phenolmethode, Eiweißgehalt betrug 2 %.
4. DNA aus Lachssperma-Handelspräparat (Ungarn).
5. DNA aus T2-Phagen, hergestellt nach der Phenolmethode, Eiweißgehalt war nicht höher als 1 %.
6. DNA aus E. coli, hergestellt nach der Phenolmethode, Eiweißgehalt war nicht höher als 1 %.
7. Lösliche-Transport-RNA (T-RNA) aus E. coli, hergestellt nach der Phenolmethode, Eiweißgehalt wurde nicht bestimmt.

Denaturierte DNA wurde durch 15 minütiges Tempern bei 100° C mit schnellfolgender Kühlung bei 0° C hergestellt.

In der Arbeit wurde das Acridin-Orange (AO) "Extra" ohne nachfolgende Reinigung und das von $ZnCl_2$ gereinigte AO verwendet. Beide Präparate ergaben naheliegende Resultate. Die reale Konzentration von AO "Extra" wurde von uns nach Absorptionsspektren durch einen Vergleich mit denen des von $ZnCl_2$ gereinigten AO ermittelt. Alle Lösungen wurden in Zitratpuffer ($p_H = 6,0$) vorbereitet. Konzentration des Puffers war jeweils 0,001 m. Bei der Untersuchung von AO-Komplexen der nativen DNA verschiedener Herkunft stellte sich heraus, daß sie alle einen ähnlichen COTTON-Effekt aufweisen. Der Abriß der Dispersionskurve, der auftritt, wenn das Verhältnis der Gesamtzahl der Nucleotide P zu der der Farbstoffmoleküle D etwa 1 beträgt ($\frac{P}{D} \sim 1$), ist auf dem Bild 1,1 wiedergegeben. Dieser COTTON-Effekt wird weiter der Kürze halber als ein rechter COTTON-Effekt bezeichnet. Es sei bemerkt, daß der Farbstoff selbst keine optische Aktivität besitzt und die eigene Drehung von DNA bei den Versuchskonzentrationen nicht über 2" steigt.

RNA- und denaturierte DNA-Komplexe mit AO weisen COTTON-Effekte auf, die einander ähnlich sind. Aber sie unterscheiden sich von COTTON-Effekten der nativen DNA-Komplexe bei $\frac{P}{D} \sim 1$ (s. Bild 2,1). Dieser COTTON-Effekt wird als der linke bezeichnet. Das letzte Ergebnis widerspricht dem von BRADLEY [10], der keinen Effekt mit denaturierter DNA gefunden hat. Um den Bindungsmechanismus zu klären, wurden von uns Versuche mit nativer und denaturierter DNA bei verschiedenen $\frac{P}{D}$ -Werten angestellt (Bild 1 und 2). Aus den Bildern läßt sich ersehen, daß der rechte COTTON-Effekt an nativer DNA bei der Vergrößerung von $\frac{P}{D}$ verschwindet und der linke COTTON-Effekt erscheint (Bild 1,4).

Die anfängliche Erhöhung des Effektes bei der Vergrößerung des $\frac{P}{D}$ -Wertes bis 2,5 läßt sich durch das Vorhandensein einer großen Menge von nicht gebundenem Farbstoff in der Lösung, der bei Zusatz des Polymers gebunden wird, erklären. Bei der denaturierten DNA (Bild 2) wird der rechte COTTON-Effekt bei der Vergrößerung des $\frac{P}{D}$ -Wertes allmählich durch den linken ersetzt. Die Versuchsergebnisse kann man folgendermaßen zu deuten versuchen. Wie PEACOCKE und SKERRETT [4] angenommen und LERMAN und LUZZATTI [2][3][11] gezeigt haben, besteht der Grundmechanismus der Bindung von AO mit nativer DNA bei großen $\frac{P}{D}$ -Werten darin, daß sich AO-Moleküle in die Doppelhelix zwischen die

Stickstoffbasen einer DNA hineindrängen. In unseren Versuchen entspricht diesem der linke COTTON-Effekt (Bild 1,4). Nach PEACOCKE und SKERRETT entspricht dieser Bindungsart eine größere Bindungsenergie, aber auf diese Weise können 4-5 Nucleotide nur ein Farbstoffmolekül binden. Ist der Farbstoff im Überschuß vorhanden (kleine $\frac{P}{D}$ -Werte), kann sich einefünfmal größere Menge von Farbstoff mit DNA binden, aber bei kleinerer Bindungsenergie. Offensichtlich findet diese Bindung außerhalb der Helix mit PO_4^--Gruppen statt. Das weist in unseren Versuchen einen rechten COTTON-Effekt auf. Bei kleinen $\frac{P}{D}$ -Werten, wenn alle "inneren" und "äußeren" Bindungsstellen besetzt sind, soll der rechte COTTON-Effekt überwiegen. Bei der Verdünnung von AO mit einem Polymer setzt sich ein immer größeres Teil des AO ins "Innere", was energetisch vorteilhafter ist. So beginnt der linke COTTON-Effekt zu überwiegen. Was die Ergebnisse mit denaturierter DNA anbetrifft, so ist es vielleicht möglich, sie folgenderweise zu deuten. Der Farbstoff dimerisiert sich glatt auf der denaturierten DNA und auf den anderen leichtbeweglichen Strukturen bei kleinen $\frac{P}{D}$ -Werten [5]. Unserer Meinung nach kann dabei der linke COTTON-Effekt, der qualitativ dem linken COTTON-Effekt von nativer DNA bei größeren $\frac{P}{D}$ -Werten etwas ähnlich ist, durch die Bindung von Dimeren mit der denaturierten Struktur der DNA erklärt werden. Dieser linke COTTON-Effekt ist anscheinend kompliziert, und man kann ihn als eine Summe zweier einfacher COTTON-Effekte mit den Übergangspunkten nahe 464 nm und 504 nm darstellen, was der Lage der Absorptionsbanden des Dimers entspricht.

Bei der Vergrößerung von $\frac{P}{D}$ -Werten wird der Farbstoff durch das Polymer verdünnt. Seine Dimerisation wird dabei weniger wahrscheinlich und sein Anteil an der Dispersion der optischen Aktivität wird hauptsächlich durch den Farbstoff bestimmt, der mit den übriggebliebenen Helixteilen der Moleküle gebunden ist.

Um die Natur der Entstehung der optischen Aktivität in der Absorptionsbande des Farbstoffes zu klären, haben wir T-RNA-AO-Komplexe bei größeren $\frac{P}{D}$ -Werten untersucht, wenn nur etwa zwei oder drei Moleküle einem Molekül RNA entsprechen. Dabei beobachteten wir den rechten COTTON-Effekt. Anscheinend kann in diesem Fall von keiner Über-Helix des Farbstoffes auf T-RNA die Rede sein, die der Über-Helix der mit dem Eiweiß [7] gebundenen Farbstoffe ähnlich wäre. Die optische Aktivität des Farbstoffes entsteht durch die Asymmetrisierung von Farbstoff-Molekülen bei ihrer Bindung mit der DNA-Helix oder mit T-RNA. Im Falle der größeren $\frac{P}{D}$ -Werte geht auf der nativen DNA eine Asymmetrisierung der AO-Moleküle bei ihrer Bindung zwischen den flachen Basenpaaren vor sich, die entsprechend der Helix-Struktur der DNA zueinander gewandt sind.

Um die Gültigkeit unserer Annahme über die äußere Bindung des AO mit PO_4^--Gruppen bei dem rechten COTTON-Effekt zu prüfen, haben wir die Wirkung von Mg^{++}-Ionen auf den rechten COTTON-Effekt bei der nativen DNA (Bild 4) und auf den linken bei der denaturierten DNA, wie auch auf den linken bei der nativen DNA bei größeren $\frac{P}{D}$ -Werten ausgenutzt.

Es stellte sich heraus, daß die volle Amplitude des COTTON-Effekts (Abstand zwischen Maximum und Minimum in sek.) für den rechten COTTON-Effekt der nativen DNA mit dem Zusatz von $MgCl_2$ steil absinkt. Bei dem 10 bis 15-maligen Herabsinken des rechten COTTON-Effekts änderte sich die Größe der Stufe im Bereich des Minimums, die offensichtlich der Beimengung des linken COTTON-Effekts entspricht, verhältnismäßig wenig, d. h. der linke COTTON-Effekt war Mg-unempfindlich. Die Versuche mit der nativen DNA bei größeren $\frac{P}{D}$ -Werten, wenn nur der linke COTTON-Effekt beobachtet wird, bestätigen das. Bei der Mg^{++}-Konzentration $2{,}10^{-3}$ m änderte er sich kaum. Diese Versuche sprechen für die Annahme, daß im Falle der nativen DNA der rechte COTTON-Effekt von der äußeren und der linke von der inneren Bindung des Farbstoffes bestimmt wird. Der volle linke COTTON-Effekt der RNA steigt zuerst bei dem Zusatz von Mg^{++} (bis zur Konzentration 10^{-3} m) und beginnt erst dann schnell herabzusinken (Bilder 3,2). Dieser linke COTTON-Effekt scheint folglich für den Mg^{++}-Zusatz empfindlich zu sein, obwohl nicht in so hohem Grade wie bei dem rechten COTTON-Effekt der nativen DNA. Das widerspricht nicht der Annahme, daß der linke COTTON-Effekt bei der denaturierten DNA und der RNA von der Bindung der Farbstoffdimeren mit dem Polymer abhängt. Die Wirkung von Mg^{++}-Ionen auf den AO-RNA-Komplex läßt keine Schlüsse zu, ob sich das dimere AO mit PO_4^--Gruppen oder mit Stickstoffbasen bindet. Um diese Frage zu lösen, werden von uns zur Zeit Versuche mit Formaldehyd angestellt.

Der große Unterschied im Aussehen der Dispersionskurven von AO-nativen- und AO-denaturierten-DNA-Komplexen bei kleinen $\frac{P}{D}$ -Werten kann bei der Ausarbeitung des Testes für den Spiralgehalt der DNA benutzt werden. Dabei hat die Wahl der $\frac{P}{D}$ -Werte, bei denen die Messungen durchgeführt werden, beachtliche Bedeutung, weil das Aussehen der Dispersionskurven (Bild 1,2) für den $\frac{P}{D}$ -Wert sehr empfindlich ist, was Schwierigkeiten bei der Messung bereitet. Es ist möglich, daß sich andere Farbstoffe, welche keine so große Zahl von Bindungsmechanismen mit Nucleinsäuren besitzen (z. B. die, die nicht ins Innere der Helix eindringen), als passender erweisen.

Also ergeben sich aus den durchgeführten Versuchen folgende Annahmen über die Herkunft der anomalen Dispersion der optischen Aktivität von AO-Nucleinsäure-Komplexen:

1. Im Falle der größeren $\frac{P}{D}$ -Werte überwiegt bei den nativen DNA die innere Bindung des Farbstoffes, die den linken COTTON-Effekt aufweist. Bei niedrigeren $\frac{P}{D}$ -Werten bindet sich außerdem eine große Menge von Farbstoffmolekülen bei den PO_4^--Gruppen und der rechte COTTON-Effekt überwiegt. In beiden Fällen findet die Asymmetrisierung der Farbstoffmoleküle statt.
2. Bei der denaturierten DNA und bei der RNA im Falle der niedrigeren $\frac{P}{D}$ -Werte überwiegt die Bindung von AO-Dimeren mit Nucleinsäuren und ihre Asymmetrisierung, was den linken COTTON-Effekt aufweist, der sich von dem linken COTTON-Effekt auf nativer DNA unterscheidet.

An der Durchführung der Messungen, die das vorliegende Referat stützen, hat der Diplomand Ss. S. SCHMURAK teilgenommen.

L I T E R A T U R

1) STEINER,R.F., R.F.BEERS, Arch. Biochem. Biophysics 81, 75 (1959)
2) LERMAN,L.S., J. molecular Biol. 3, 18 (1961)
3) LERMAN,L.S., Proc. nat. Acad. Sci. USA 49, 94 (1963)
4) PEACOCKE,A.R., J.N.H.SKERRETT, Trans. Faraday Soc. 52, 261 (1956)
5) BRADLEY,D.F., G.FELSENFELD, Nature /London/ 184, 1920 (1959)
6) BRENNER,S.L., F.H.BARNET, C.CRICK, A.ORGEL, J. molecular Biol. 3, 121 (1961)
7) BLOUT,E.R., J. Amer. chem. Soc. 83, 1411 (1961)
8) KUSEL,W.A., W.J.PERMOGOROW, Optika i Spektroskopija X, 541 (1961)
9) MALLTE, M.T., C.LAMANNA, Arch. Biochem. Biophysics 47, 174 (1953)
10) NEVILLE,D.M., D.F.BRADLEY, Biochim. Biophysica Acta 50, 397 (1961)
11) LUZZATI,V., F.MASSON, L.S.LERMAN, J. molecular Biol. 3, No. 5 (1961)
12) FRESCO,J.R., Tetrahedron /London/ 13, 185 (1961)

D I S K U S S I O N

LÖBER.- Haben Sie einen Einfluß der Dimerisation des AO auf die Rotationsdispersion beobachtet?

LASURKIN.- Dimere des AO selbst besitzen keine optische Aktivität. Wir haben Rotationsdispersion der AO-Dimere verbunden mit DNA, besonders der denaturierten DNA, beobachtet. Im vollen Text des Vortrages sind diese Resultate angegeben.

ALTMANN.- Können die AO-Komplexe mit DNA verglichen werden mit Benzacridinkomplexen, um etwas über den carcinogenen Mechanismus der letzteren zu erfahren?

LASURKIN.- Wahrscheinlich sind die Untersuchungen über den Mechanismus der Entstehung der Acridinfarbstoff-Nucleinsäurekomplexe nicht nur für die Mutationswirkung, sondern auch für die carcinogene Wirkung solcher Stoffe von Interesse. Vielleicht sind auch beide Mechanismen miteinander verbunden.

DELLWEG.- Können Sie Angaben machen über die Stabilität dieser Komplexe, insbesondere gegenüber sichtbarem Licht? Es ist ja bekannt, daß DNA in Gegenwart von Farbstoffen durch sichtbares Licht unter Zerstörung besonders von Guanin verändert wird.

LASURKIN.- Wir haben die Stabilität der AO-DNA-Komplexe gegenüber Licht nicht untersucht (nur gegen die Wirkung des Lichtes geschützt). Was die Wärmestabilität dieser Komplexe anbetrifft, so sind die Komplexe, die durch "innere" Bindung des AO gebildet sind, mehr temperaturbeständig als die anderen, die durch äußere Bindung gezeichnet sind. Das entspricht einer größeren freien Energie der Bindung im ersten Falle.

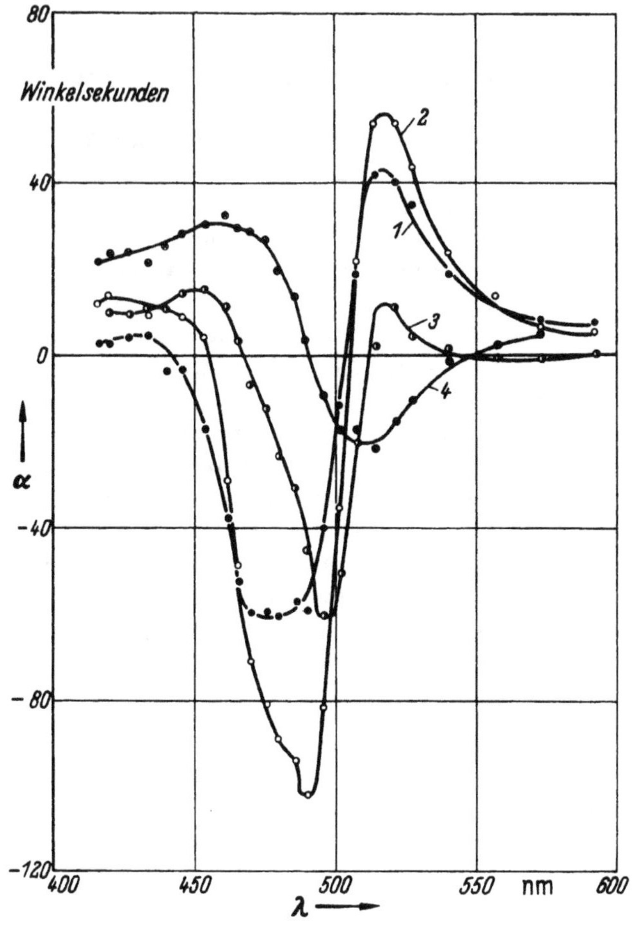

Abb.1 Dispersion der optischen Aktivität von AO-Komplexen mit DNA von T2 Phagen.(Eigendrehung der DNA ist abgezogen). AO-Konz. 0,7.10^{-5}m d-Nucleotiden-Konz.:Kurve 1:1.10^{-5}m; 2: 2.10^{-5}m; 3: 6.10^{-5}m; 4: 3.10^{-4}m

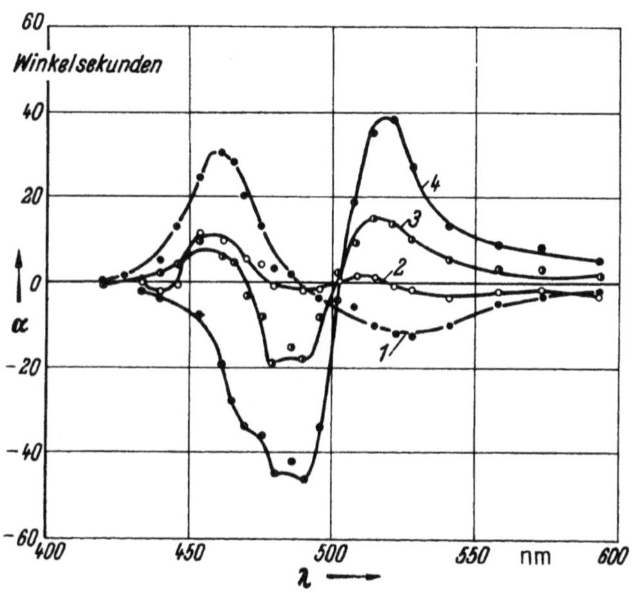

Abb.2 Dispersion der optischen Aktivität von AO-Komplexen mit denaturierter DNA aus Kalbsthymus. AO-Konz. 0,7.10^{-5}m. d-Nucleotiden-Konz.:Kurve 1:0,5.10^{-5}m; 2:1.10^{-5}m; 3:2.10^{-5}m; 4: 6.10^{-5}m.

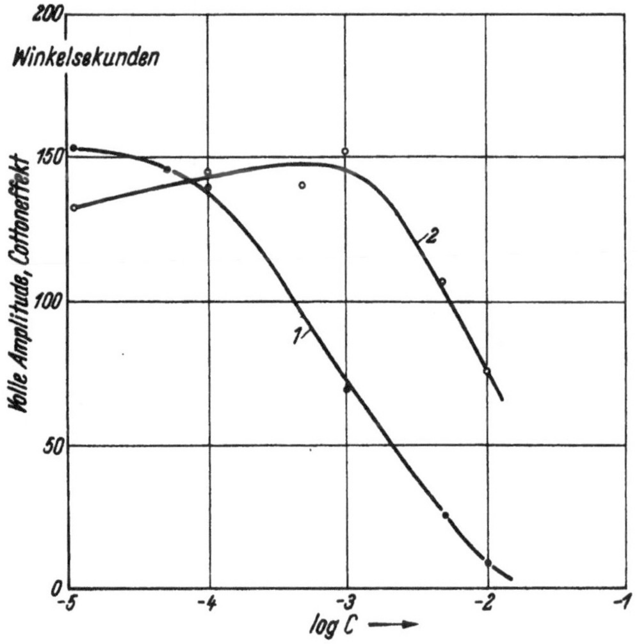

Abb.3 Abhängigkeit der vollen Amplitude des COTTON-Effektes vom Logarithmus der Mg^{++}-Konz.
Kurve 1: AO-Komplexe mit nativer DNA von T2 Phagen. AO-Konz. 0,7.10^{-5}m; d-Nucleotiden-Konz. 2.10^{-5}m
Kurve 2: AO-Komplexe mit RNA aus Hefe. AO-Konz. 1,2.10^{-5}m; r-Nucleotiden-Konz.: 2.10^{-5}m. Linke Punkte entsprechen Nullkonz. von Mg^{++}. Zitratpufferkonz. 10^{-3}m.

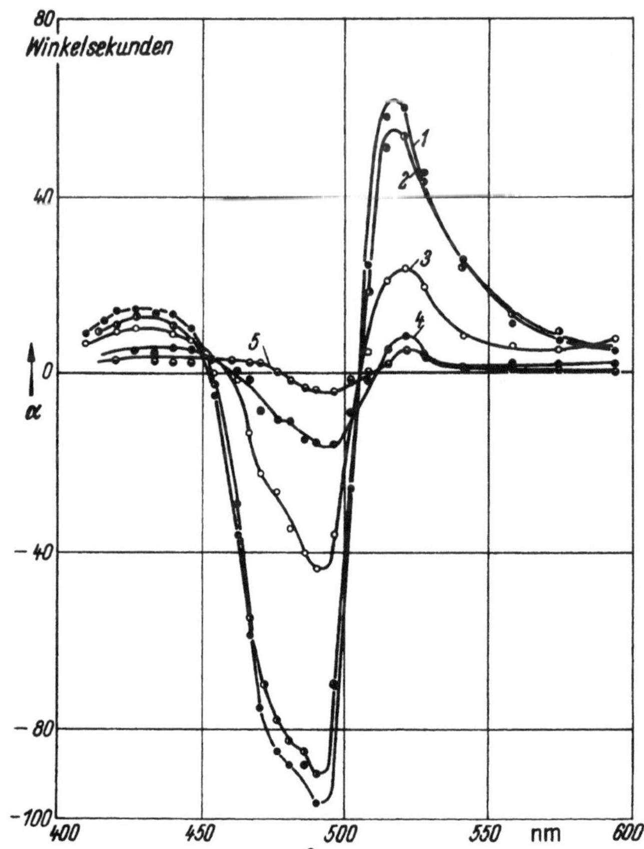

Abb.4 Dispersion der optischen Aktivität von AO-Komplexen mit DNA von T2 Phagen. d-Nucleotiden-Konz. 2.10^{-5}m. AO-Konz. 0,7.10^{-5}m. Mg^{++}-Konz.: 1: keine; 2: 10^{-4}m; 3: 10^{-3}m; 4: 5.10^{-3}m; 5: 10^{-2}m.

D Nucleinsäuren

Hyperchromer Effekt von Nucleinsäure-Farbstoff-Komplexen

Ju. MOSCHKOWSKI

Institut für chemische Physik der Akademie, Moskau

S U M M A R Y

Measurements have been made on the mixtures of cyanine dye in solution with native and thermal denaturated DNA. It is shown that absorption coefficient of dye increases by denaturation of DNA.

Es ist bekannt [1)-5)], daß die optische Dichte bei 260 nm der Lösung von nativer Nucleinsäure im Maximum der Absorptionsbande um 30 - 40 % niedriger als bei einer denaturierten Probe liegt. So eine Senkung des Absorptionskoeffizienten, welche man als hypochromen oder in einigen Fällen als hyperchromen Effekt bezeichnet, ist mit einer regelmäßigen Anordnung der Purin- und Pyrimidinbasen in der Doppelspirale der nativen Nucleinsäure verbunden [6)]. Bei beliebigen Einwirkungen, die zur Störung der regelmäßigen Anordnung der heterozyklischen Kerne führen (Erwärmung, fermentative Zersetzung, Hydrolyse, Veränderung der Ionenstärke der Lösung und dgl.), wächst der Absorptionskoeffizient der Bande bei 260 nm, und auch der kurzwelligen Bande bei 200 nm [7)]. In den meisten Arbeiten, die dem Studium der Strukturänderungen der Nucleinsäuren gewidmet sind, wird die Messung des hyperchromen Effektes als eine Grundmethode zur Bestimmung der denaturierten Form dieses Biopolymers benutzt.

Es ist zu erwarten, daß ein analoger Effekt der Senkung des Absorptionskoeffizienten auch in beliebigen anderen, regelmäßig angeordnete heterozyklische Kerne enthaltenden, absorbierenden Molekülen zu beobachten ist.

Deshalb schien es uns interessant, einen hyperchromen Effekt bei einem am Makromolekül nativer und denaturierter Desoxyribonucleinsäure (DNA) adsorbierten oder als Komplex gebundenen Farbstoff zu erreichen.

Als Substanzen für die Untersuchung wählten wir die Kalbsthymus-DNA, die durch die Phenolmethode [8)] gewonnen war und den Cyaninfarbstoff (I), dessen Absorptionsmaximum bei 424 nm liegt.

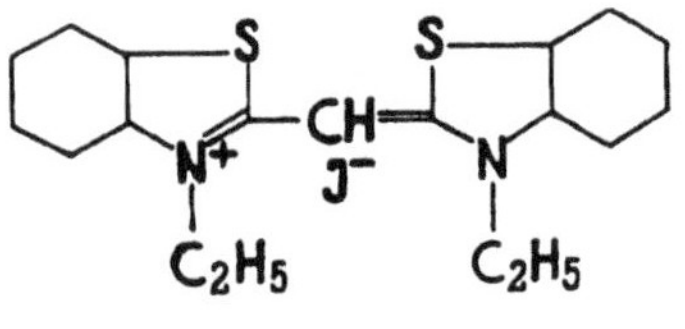

(I)

Abb. 1 zeigt die Absorptionsspektren des Farbstoffs (I) in einer Lösung von DNA bei Zimmertemperatur und bei 95° C. Die Farbstoffkonzentration betrug $8{,}0 \cdot 10^{-6}$ Mol/l, der Gehalt an DNA 0,1 mg/ml. Außerdem befanden sich in der Lösung 0,1 M NaCl. Wie man an den Spektralkurven sieht, wächst die Lichtabsorption des Farbstoffs in einer bis auf 95° C erhitzten Lösung. Die folgenden Versuche zeigen, daß die gefundene Erhöhung des Absorptionskoeffizienten von ca. 35 % beim Erhitzen des Gemisches Farbstofflösung (I) und DNA hervorgerufen ist durch die Änderungen, die im Biopolymer und nicht im Farbstoffmolekül stattfinden. Erstens bewirkt die Erwärmung der Farbstofflösung (I) in einer 0,15 M Lösung von NaCl auf 95° C, daß sich die Spektralkurve praktisch nicht ändert, abgesehen von einer kleinen Senkung der kurzwelligen Schulter der Kurve, bedingt durch die Spaltung der Farbstoffdimeren (Abb. 2). Gibt man den Farbstoff (I) einem anderen Biopolymer, z. B. der Gelatine, zu (1 mg/ml), so führt das auch nicht zur Vergrößerung der optischen Dichte beim Erhitzen. Daraus kann man folgern, daß nur in Lösungen des Farbstoffs (I) und DNA, welche in nativem Zustand ist, die Erwärmung auf 95° C (Wärmedenaturierung) eine Steigerung des Absorptionskoeffizienten des Farbstoffs, d. h. einen hyperchromen Effekt, hervorruft.

In Abb. 3 sind die Schmelzkurven der untersuchten DNA-Probe angegeben. Die eine Kurve (a) wurde aus den Veränderungen der optischen Dichte des Farbstoffs (I) bei der Wellenlänge 424 nm erhalten und die andere (b), wie gewöhnlich, aus der Veränderung der optischen Dichte im Maximum der Absorptionsbande der DNA bei der Wellenlänge 260 nm. Beim Vergleich der Kurven muß man berücksichtigen, daß außer dem Vorhandensein von Farbstoff in einer der DNA-Lösungen sie sich auch durch die DNA-Konzentration unterscheiden. So entspricht der durch die Absorption des Farbstoffes (I) gemessenen Kurve (a) eine DNA-Konzentration von 0,25 mg/ml und der anderen Kurve (b) eine DNA-Konzentration von 10 mg/ml.

Wie man aus Abb. 3 sehen kann, sind in beiden Fällen hinreichend deutliche Schmelzkurven zu erkennen. Jedoch war der Schmelzpunkt der DNA, wie üblich definiert durch die Temperatur, die der Mitte des Übergangs entspricht, in einer Farbstoff (I)-Lösung etwas herabgesetzt (ungefähr um 10° C). Es ist vorläufig noch schwierig zu sagen, wodurch ein solcher Unterschied hervorgerufen wird. Es sei bemerkt, daß im Falle einer anderen aus T2-Phagen gewonnenen DNA die durch die Absorption von Farbstoff und DNA ermittelten Schmelzkurven praktisch zusammenfielen. Der entdeckte hyperchrome Effekt des Farbstoffs in den DNA-Lösungen erlaubt die Strukturumwandlung dieses Biopolymers in einem weiten Konzentrationsbereich zu untersuchen. Dies kann auch in Anwesenheit von Stoffen geschehen, die im Bereich von 260 nm absorbieren, d. h. in Fällen, wo die Messungen der Schmelzkurven auf Grund der Absorption der DNA selbst unmöglich sind.

L I T E R A T U R

1) MAGASANIK,B., E. CHARGAFF, Biochim. biophysica Acta 7, 396 (1951)
2) CAVALIERI,L, J. Am. chem. Soc. 74, 1242 (1952)
3) THOMAS,R, Biochim. biophysica Acta 14, 231 (1954)
4) SPIRIN,A., L.GAWRILOVA, A.BELOZERSKI, Biochimia 24, 600 (1959)
5) DOTY,P., H.BOEDTKER, R.FRESCO, R.HASELKORN, M.LITT, Proc. nat. Acad. Sci. USA 45, 482 (1959)
6) TINOCO,J, J. Am. chem. Soc. 82, 4785 (1960)
7) LYSZOW,W., B.SUCHORUKOW, L.BLUMENFELD, Yu.MOSCHKOWSKI, W.PETUCHOW, Biofisika 7, 662 (1962)
8) GEORGIEW, W.STRUTSCHKOW, Biofisika 5, 745 (1960)

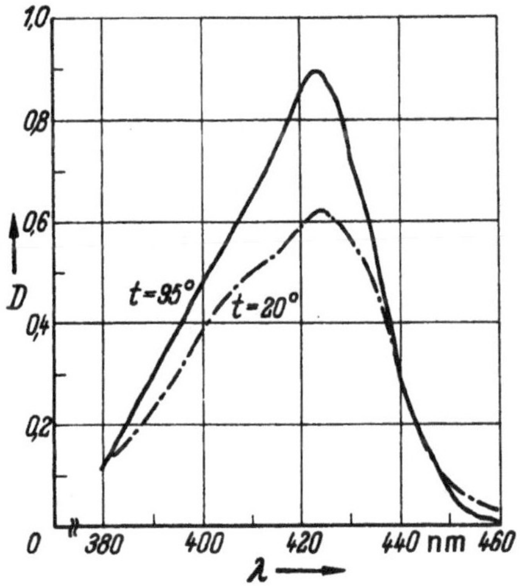

Abb.1 Absorptionskurven des Farbstoffes (I) in Lösung nativer und denaturierter DNA.

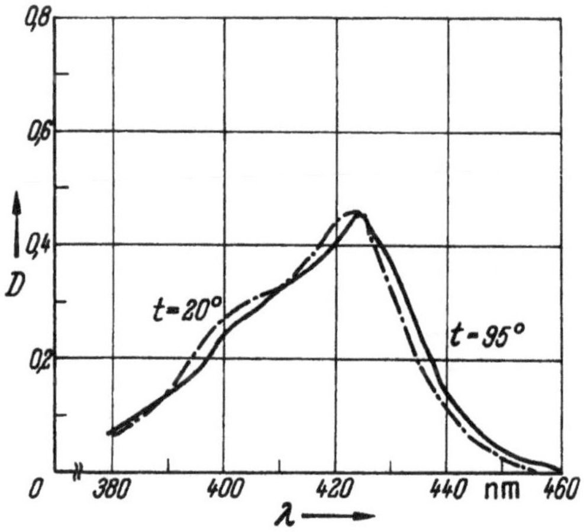

Abb.2 Absorptionskurven des Farbstoffes (I) in Gelatinelösung bei Zimmertemperatur und 95°C.

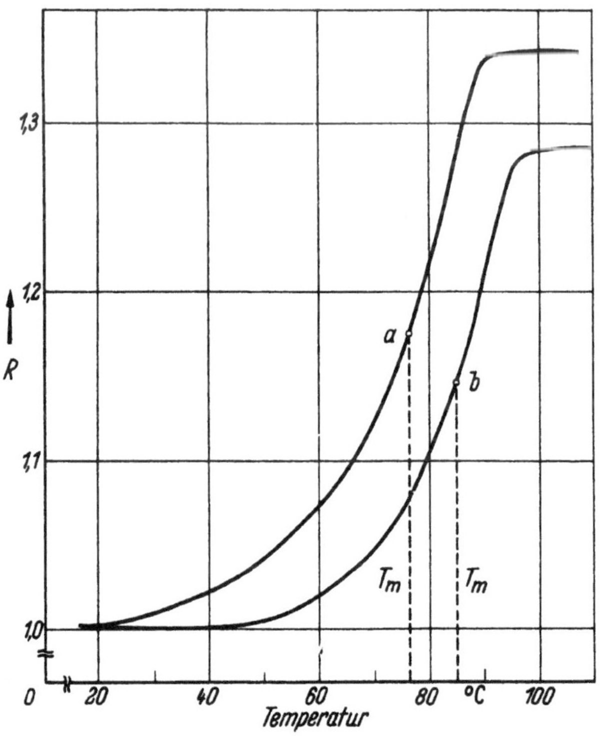

Abb.3 DNA-Schmelzkurven, die durch Messung der optischen Dichte des Farbstoffes (I) (Kurve a) und der DNA (Kurve b) erhalten wurden.

D Nucleinsäuren

A Physico-Chemical Study of the Macromolecular Structure of Deoxyribonucleic Acid (DNA) and Ribonucleic Acid (RNA)

E. V. FRISMAN, V. I. VOROBYEV and L. V. SHCHAGINA

Physical Institute of the State University and Institute of Cytology of the Academy, Leningrad

S U M M A R Y

The dependence of optical anisotropy of DNA macromolecules on molecular weight has been studied by the methods of flow birefringence and viscosimetry. Samples of different molecular weights were obtained by means of ultrasonification of solutions of the native DNA.

It was shown that an increase of molecular weight (M) causes the growth of optical anisotropy of DNA macromolecules.

An analysis of the experimental data permits to conclude that in the range of M from 3×10^5 to 9×10^6 the behaviour of DNA macromolecules can be described by a model of half-rigid chains; at $M = 2 \times 10^5$ one can approximate DNA macromolecules with rigid rods.

Evaluation was made of the geometrical asymmetry of DNA molecules by means of coefficients of rotatory diffusion of macromolecules obtained from measurements of orientation angles and intrinsic viscosity of each fraction.

The influence of thermal denaturation on the structure of DNA macromolecules has been studied.

The investigation of RNA solutions showed that the latter are characterized by a positive double refraction in the flow which is mainly due to form effect.

It was established that a prolonged action of speed field of the flow results in irreversible changes in the structure of RNA molecules which lead to changes of the double refraction sign, orientation angle and viscosity, ultraviolet absorption being unaltered.

The important part the nucleic acids play in the process of the cell division and the biosynthesis of protein has attracted the attention of a wide range of investigators to the study of these substances. Nevertheless the physical and physicochemical properties of the nucleic acid macromolecules and the process of denaturation of DNA remain not yet clear enough.

This fact makes it desirable, while investigating nucleic acids, to use methods sensible to the structure and flexibility of macromolecules.

One can obtain important information on the structure and flexibility of macromolecules during the study of their inner optical anisotropy $(\gamma_1 - \gamma_2)_i$. Indeed, for the Gaussian chain the value $(\gamma_1 - \gamma_2)_i$ is connected with the optical anisotropy of the segment $(\alpha_1 - \alpha_2)$ by the equation

$$(\gamma_1 - \gamma_2)_i = \frac{3}{5} (\alpha_1 - \alpha_2) \qquad (1)$$

The optical anisotropy of the segment can be expressed in the first approximation through the difference of the polarisability of the monomer unit ($a_\parallel - a_\perp$) and the quantity of monomers in the segments S:

$$\alpha_1 - \alpha_2 = (a_\parallel - a_\perp) \cdot S \qquad (2)$$

Thereby $(\gamma_1 - \gamma_2)_i$ does not depend on the molecular weight M of the polymer under investigation what is begin observed experimentally (see for ex. [2]).

At the same time for rigid rods:

$$(\gamma_1 - \gamma_2)_i = (a_\parallel - a_\perp) \cdot Z \qquad (3)$$

where Z is the degree of polymerisation. Here $(\gamma_1 - \gamma_2)_i$ increases proportionally to M.

In such a way the study of the dependence of $(\gamma_1 - \gamma_2)_i$ on M makes it possible to judge about the rigidity of macromolecules.

Experimentally determining $(\alpha_1 - \alpha_2)$ and S one can find the experimental value of $(a_\parallel - a_\perp)_{ex}$. The comparison $(a_\parallel - a_\perp)_{ex}$ with the theoretical value ($a_\parallel - a_\perp$) computed according to the scheme of the tensoric additivity of the polarisability of valent bonds gives the possibility to make out the structure of the monomer.

The optical anisotropy of the macromolecule can be determined from the study of the flow birefringence and the viscosity of the polymer solution. The flow birefringence for the given system is usually characterised by a dynamooptical constant:

$$[n] = \lim_{\substack{g \to 0 \\ c \to 0}} \left(\frac{\Delta n}{gc\eta_0}\right)$$

Here Δn is the value of the flow birefringence of the solution with the concentration c at the velocity gradient g. η_0 is the viscosity of the solvent. Independently of the model of the macromolecule the ratio $[n] / [\eta]$ (where $[\eta]$ is the intrinsic viscosity of the sample) is proportional to the optical anisotropy of the particles solved. So the following equation has been obtained for the Gaussian coils [3],

$$\frac{[n]}{[\eta]} \cdot \frac{27\, n_s\, k\, T}{4\pi (n_s^2 + 2)^2} = \gamma_1 - \gamma_2 \qquad (4)$$

where k is the Bolzman constant, T - the absolute temperature, n_s - the refraction index of the solvent.

Analogical equations differing only in the numeral coefficients are obtained for rigid rods [5] and ellipsoids [6].

One must bear in mind that $(\gamma_1 - \gamma_2)$ in equation (4) includes both the inner anisotropy $(\gamma_1 - \gamma_2)_i = \Theta_i$ and the anisotropy relating from the geometrical asymmetry of the particles $(\gamma_1 - \gamma_2)_f$ - the form anisotropy [4].

The value of $(\gamma_1 - \gamma_2)_i$ is directly determined according to the measured values $[n]$ and $[\eta]$ in the case of very small form anisotropy:

$$\frac{[n]}{[\eta]} \cdot \frac{27\, n_s\, k\, T}{4\pi (n_s^2 + 2)^2} = \Theta_1 \qquad (5)$$

Under these conditions the value Θ_1

$$\Theta_1 = \frac{(\Delta n/g)_{g \longrightarrow 0}}{\eta_0\, (\eta_r - 1)} \; \frac{27\, n_s\, k\, T}{4\pi (n_s^2 + 2)^2} \qquad (6)$$

(where η_r is the relative viscosity of the solution) does not depend on the concentration of the solution and coincides with the value Θ_1 computed from equation (5). I.e. Θ_1 can be found from the data of the flow birefringence and the viscosity in the solution of the finit concentration [7].

In this work we give the results of the study of the optical anisotropy of DNA molecules and its dependence from the molecular weight of the polymer. Both the structure of the DNA molecule wich has undergone a thermal denaturation and the molecular structure of RNA in solutions of different ionic strength have been examined.

The Sample and the Method of Measuring

We have studied the sodium salt of DNA which was obtained from the calf thymus by a preliminary extraction of deoxyribonucleoprotein according to the method of MIRSKY [8] with a subsequent deproteinisation after KIRBY [9]. The contents of nitrogen (N) and phosphorus (P) in the given sample of DNA were: N = 13.6 %; P = 7.8 %, which corresponds to N/P = 1.74, the latter showing that there is a slight admixture of protein in the preparation. The quantity of the atomic extinction coefficient at λ = 260 mμ, E(P) = 6500 speaks about the nativity of the sample under investigation.

The fractions with corresponding molecular weights were obtained by sonification of the DNA solutions with the concentration c = 0.03 g/100 ml in 0.15 M NaCl, using the piecoquartz ultrasonic generator with an oscillation frequency of 420 kc. The sonification time varied from 1 to 60 minutes. In all cases the ultrasonic field did not have any influence upon the value E (P), which testifies to the nativity of all the samples obtained. The concentration of the solutions was determined according to their absorption in the ultraviolet part of the spectrum after the hydrolysis by 0.5 M $HClO_4$ and according to the phosphorus content.

The studied samples of RNA (from the tobacco-mosaic virus (TMV), from E. coli, and from the transplantable strain of the mucouse cancer of the rat liver) were obtained according to the method described by SPIRIN [11]. The nativity of the samples was verified according to the temperature dependence of the viscosity of their solutions [11].

The measurement of the value and orientation of the flow birefringence was carried out on the optical apparatus with a half shadowed elliptical compensator, described in a number of scientific reports [2),12]. A mica plate with a path length difference Δ_0 = 0.0232λ served as compensator. The magnitude of the flow birefringence was computed from the relation

$$\Delta n = \frac{\Delta_0}{l} \cdot \sin 2\Delta\varphi \qquad (7)$$

where l is the thickness of the anisotropic layer, $\Delta\varphi$ - the difference of the half-shadowed compensator azymuth at the rotating and immobile rotor of a dynamooptimeter.

Dynamooptimeters with an inner and outer rotor were used [12]. The magnitudes of Δn and η_r were measured at the temperature of 21° C.

Investigation of DNA Solutions

We shall not describe in detail the process of investigation described in the works [13),14)], but only mention the results obtained. The flow birefringence and the viscosity of solutions of different concentrations of the native DNA have been studied as well as the solutions of DNA heated during 20 minutes at temperatures of 80°, 90° and 100° with a subsequent cooling at 0°.

The results of measurement have shown that the value $\dfrac{\Delta n/g}{\eta_0(\eta_r - 1)}$ practically does not depend on the concentration of the solution.

Thus in spite of the great significance of $(n_k - n_s)$ in this system (n_k is the refractive index of the polymer) the effect of the macroform plays a small role. In these conditions the values Θ_1 can be computed according to equation (5) or (6).

The optical anisotropy of DNA macromolecules proved to be several orders higher than the anisotropy obtained even for extremely anisotropic molecules of synthetic polymers. This may occur for two reasons: either the anisotropy of a monomer is large or the DNA macromolecule is very rigid. By using the CRICK and WATSON model [15),16)] we have computed $(a_\parallel - a_\perp)$. The calculation has shown that the monomers containing pairs of adenin-thymin or guanin-cytosin practically show the same anisotropy:

$$a_\parallel - a_\perp \quad = \quad -15 \times 10^{-24}\ \mathrm{cm}^3,$$

which exceeds only ten times the anisotropy of a monomer unit of the polystyrene molecule. Hence, the great value of Θ_1 obtained for the molecule of DNA depends on its great rigidity. In the state of denaturation the molecule of DNA is less rigid than in the native state, but nevertheless the rigidity is rather considerable.

Further we studied the flow birefringence, the viscosity and absorption in the ultraviolet region of the spectrum ($\lambda = 260\ m\mu$) of DNA in solutions of different ionic strengthes being heated at different temperatures and cooled at 0° C. After cooling the solutions were exposed to the temperature of 21° at which measurements were taken.

In Fig. 1–3 the dependence of the values measured on the heating temperatures for the solution of DNA in 0.15 M NaCl is shown. The diagram shows that the heating at 40° influences markedly the value of anisotropy. At the same time the optical density, the angle of orientation and the specific viscosity of the solutions of DNA don't change with heating to 60°.

For solutions of smaller ionic strength the value $\Delta n/g$ undergoes greater changes at the same heating temperature.

The results of these experiments may serve as indication to the existence of "weak points" in the molecular structure of DNA. It is possible that the heating of solutions of DNA at comparatively low temperatures may stimulate the break of a few bonds which stabilize the structure of a double helix at certain parts of the DNA macromolecule. The break of single bonds is unlikely to occur here. One should rather admit that there occours a destruction of several bonds at certain parts of the macromolecule which can form a winding of the chain in those regions and create conditions for same freedom of rotation of the nitrogen bases.

From the equations (1) and (3) it follows that the study of the dependence of the optical anisotropy of macromolecules on the molecular weight of the polymer enables us to judge about the rigidity of the chain. One may make use of the value $[n]/[\eta]$ proportional to the optical anisotropy independently of the model of a macromolecule.

Let us draw our attention to the fact that in the preceding experiments we used

the value $[\eta]$ measured at the finit g.
While studying the dependence $[n]/[\eta]$ from M the values $[\eta]_o = [\eta]_{g=o}$ were determined for all the fractions studied.

For the measurement of $[\eta]_o$ we improved the magnetic rotational viscosimeter described in work [17]. In our viscosimeter the magnetic field changes in a very simple way, which allows to obtain with the aid of one rotor a wide interval of speed gradients, differing more than a thousand times. The smallest value is $g \approx 0.01\ sec^{-1}$.

It should be mentioned that in the viscosimeter described in the paper [17] each rotor gives a single value of g.

Let us consider the results of the study of the flow birefringence and viscosity in the solutions of different concentrations of DNA fractions at 0.15 M NaCl.

For all the solutions studied Δn increases proportionally to g in the whole gradient region used. In Fig. 4 the dependence $(\frac{\Delta n}{gc\eta_o})_{g \to o} = f(c)$ is shown which allowed us to find the values $[n]$ given in Table 1.

Table 1

Hydrodynamic and optical parameters of DNA molecules of different molecular weights

num. of fractions	$[\eta]_o \cdot 10^{-2}$	$M \cdot 10^{-6}$	$[n] \cdot 10^3$
1	90,0	9,10	- 2,8
2	41,5	4,55	- 1,05
3	32,0	3,60	- 0,83
4	14,4	1,77	- 0,32
5	5,6	0,76	- 0,13
6	1,8	0,28	- 0,03
7	1,0	0,16	- 0,014

The study of η_r at different g enabled us to determine $(\eta_r)_{g=o}$ and by means of Fig. 5 to determine the values $[\eta]_o$ for different fractions of DNA.

The molecular weights of the DNA fractions obtained from the relation [18]:

$$[\eta]_o = 1{,}45 \times 10^{-6} \times M^{1.12}$$

are given in Table 1.

As it has already been mentioned, the macroform anisotropy of the DNA molecules is exceedingly small in comparison with its inner optical anisotropy in spite of a great difference between the refraction indexes of DNA and the solvent. This fact has been confirmed by the results of this work.

The dependence $[n]/[\eta]$ from M shown by the curve 1 in Fig. 6 indicates that the optical anisotropy of the DNA molecule increases with increasing molecular weight, the optical behaviour of the DNA molecules differing from the behaviour both of the Gaussian coils and rigid rods.

TSVETKOV [19] showed that in many cases the optical anisotropy of the microform may play an important role in the anisotropy measurements. The optical anisotropy of the microform is the form anisotropy of the rigid part of the chain, which determines the inner anisotropy as well.

In the case when the anisotropy of the macroform can be neglected the measured value $[n]/[\eta]_o$ should consist of two parts: $[n]_i/[\eta]_o$ characterizing the inner anisotropy, and $[n]_{fs}/[\eta]_o$, connected with the microform anisotropy and equal to:

$$\frac{[n]_{fs}}{[\eta]_o} = \frac{(n_s^2+2)^2\,(n_k^2-n_s^2)\,M_g \cdot S}{180\,\pi\,RT\,n_s^3\,\rho}\;(L_2-L_1) \qquad (8)$$

Here ρ is the density of the dry polymer, n_k is its refractive index, M_g is the molecular weight of the monomer, $L_2 - L_1$ is the shape factor of the rigid part of the chain.

As the macromolecule of DNA is not described by the Gaussian coil we shall use the model of "a wormiform chain" [20)-21)] for determining S and compute the persistence lenght (a) according to the radii of inertia, obtained from the relation $\bar{R}_z \sim M^{0,58}$ given in the work [18)]. This gives the value a = 550 Å which corresponds to S = 325. In this case we can take $L_2 - L_1 = 2\pi$.

The values $[n]_i/[\eta]_o$ given in Table 1 are calculated from the relation:

$$-\frac{[n]}{[\eta]_o} = -\frac{[n]_i}{[\eta]_o} + \frac{[n]_{fs}}{[\eta]_o} \qquad (9)$$

It is seen from Fig. 6 that the calculation of the microform effect does not change the nature of the dependence of the optical anisotropy of DNA molecules from M.

One must notice that the value 2a approximately coincides with the hydrodynamic length L of the DNA molecule of the 6th fraction, and exceeds the value L for the latter fraction. That is why in computing $[n]_{fs}/[\eta]_o$ for the 7th fraction the value S was supposed to be equal to L/3,4, i.e. to the ratio of the hydrodynamic length to the length of the monomer unit. In other words the DNA molecule of the 7th fraction approximates the rigid rod.

The following comparison also helps to draw the same conclusion. If to apply the theory of intrinsic viscosity developed for rigid cylinders [22)], and take the diameter of the cylinder d = 20 Å, and the length $L = \frac{M}{660} \times 3.4$ Å, then the theoretical value $[\eta] = 107\ cm^3/g$ coincides well with the experimental $[\eta] = 100\ cm^3/g$ obtained for the 7th fraction.

The rodlike structure of the DNA molecules of the 7th fraction is also confirmed by the data obtained from the study of the optical anisotropy. Indeed, making use of the theoretical value $(\alpha_\parallel - \alpha_\perp) \cong -150 \times 10^{-25}\ cm^3$ computed according to the scheme of the tensor additiveness of the polarisability of the bonds [13)], one can find the value $(\gamma_1 - \gamma_2)_i$ for the rods: $(\gamma_1 - \gamma_2)_i = (\alpha_\parallel - \alpha_\perp)\frac{M}{M_0} = -3,7 \times 10^{-21} cm^3$, which is in good accordance with the value found from the experimental results by means of the equation connecting $(\gamma_1 - \gamma_2)_i$ with $[n]/[\eta]_o$ for rigid rods [5)].

Thus because of the dependence of $[n]_i/[\eta]_o$ on M (Fig. 6), and subsequent comparisons, one can draw the conclusion that at $M \geq 3 \times 10^5$ the behaviour of DNA molecules (in 0.15 M NaCl) can be described by a model of a halfrigid rod. At the smaller molecular weights the molecule approximates well with a rigid rod.

In order to judge about the geometrical asymmetry of DNA molecules we used the coefficients of rotation diffusion D_r obtained from the measurement of the angles of orientation [6)]. Not fixing our attention on the calculations we shall only point out that the estimation has given the ratio value of the axes ratio p ≅ 20.

Investigation of RNA Solutions

The dependences of $\Delta n = f(g)$ and $\alpha = f(g)$ have been studied as well as the viscosity of solutions with different concentrations of RNA in 10^{-3} M Tris-buffer [23)]. Some of the samples have also been studied at a greater ionic strength (10^{-2} M Tris-buffer).

For all the solutions there has been found a positive double refraction in the flow which increases in the used region of speed gradients proportionally to g as we see in Fig. 7 given as an example.

While investigating RNA from the tumour in the solutions of greater ionic strength the positive value Δn observed at the start of experiment decreases while the solution remained in the speed field and passed through the zero in the negative region. The effect of the hydrodynamical forces of the flow on the RNA molecule during three hours at $g = 2000\ sec^{-1}$ resulted in a stable negative value of Δn. Repeated experiments have shown a complete reproduction of the dependence $\Delta n = f(g)$ shown in Fig. 8. The change of the sign of Δn is followed by substantial changes of η_r and α.

On attaining a stable value of Δn the orientation angle was $\alpha = 45^o$ in the whole interval of g used. All those data show a substantial change of the RNA molecule under the influence of hydrodynamic forces of the flow. The ultraviolet absorption measured for two states of the solution didn't show marked changes and constituted $E\,(P)_{270} = 7200$.

From equation (4) it follows that the value $[n]/[\eta]$ is proportional to the whole anisotropy of the macromolecule. If there exists no anisotropy of fom the inner anisotropy may be defined according to equation (5) or (6). But if the form anisotropy (Θ_f) plays a substantial role, equation (5) will be rewritten:

$$\frac{[n]}{[\eta]} \quad \frac{27\, n_s\, k\, T}{4\pi (n_s^2 + 2)^2} = \Theta_i + \Theta_f \tag{10}$$

In order to judge about the structure and shape of the macromolecule it is necessary to define Θ_i and Θ_f separately. The value Θ_i might be defined by choosing a solvent satisfying the condition $n_k = n_s$ [4]. In this case from the value $\Theta_i + \Theta_f$ obtained for the system possessing the form effect one can find Θ_f. However for RNA the solvent with $n_k = n_s$ is unknown. Moreover, in substituting the solvent the structure of the chain my be changed and, hence, Θ_i may alter as well. The second meaning for dividing the values Θ_i and Θ_f consists in studying a number of fractions [2], which is also difficult while studying RNA.

Let us point out one more possibility of evaluating the role which Θ_f plays with regard to Θ_i in the given system. If the form effect exists, equation (6) will be rewritten as follows:

$$\frac{\Delta n/g}{\eta_0(\eta_r - 1)} \quad \frac{27\, n_2\, k\, T}{4\pi\, (n_s^2 + 2)^2} = \Theta_i + \Theta_f^* \tag{11}$$

where Θ_f^* is the form anisotropy with the given concentration of the solution. As the for effect decreases together with the concentration of the polymer $\Theta^* \longrightarrow o$ [25)-26)], i.e. $\Theta_i + \Theta_f^* \longrightarrow \Theta_i$.

The dependence of $\Theta_i + \Theta_f^* = f(c)$ for the solutions of virus RNA at two ionic strengthes is given in Fig. 9. At a small ionic strength the role of the form effect is very great. The lack of the sample did not allow to investigate solutions with higher concentration, and to approach the value Θ_i connected with the chain structure. But one can make some preliminary conclusions at the basis of these data. If we take the smallest measured value $\Theta_i + \Theta_f^* \approx \Theta_i$ one can evaluate Θ_f. and, hence, also the ratio of the axes which proved to be equal to $A/B \approx 80$. This ratio is close to that obtained by the electron microscopy method [27].

It seems that the great value of A/B does not go with the irregular distribution of the helical parts. Moreover, the small value of Θ_i obtained for RNA in 10^{-3} M Tris-buffer coincides with the model suggested by SPIRIN [28] (Fig. 10).

Indeed it is to be supposed that at the orientation of RNA macromolecules by the flow the nonhelical parts will give a negative anisotropy. It follows from the negative anisotropy for denaturated DNA showing that the nitrogen bases are perpendicular to the chain of the main valence. The helical parts must give positive anisotropy. In solution of low ionic strength the hydrogen bonds are broken and the role of the helical and nonhelical parts is almost the same which gives a small value of Θ_1. At a greater ionic strength the compression of the coil results in the small role of Θ_f. The large positive value Θ_1 is due to the increase of helical parts.

The same results have been obtained during the study of RNA from other sources. There has been mentioned the influence of the hydrodynamic forces of the flow on the solutions of RNA, the latter undergo considerable changes in the optical and hydrodynamic properties without changing ultraviolet absorption. We suppose that this is connected with the break of internucleotide bonds. We think that further investigations of this phenomenon may help to draw a conclusion concerning the details of the structure of the RNA macromolecule.

L I T E R A T U R E

1) KUHN,W., F.GRÜN, Kolloid-Z. 101, 248 (1942)
2) FRISMAN,E.V., V.N.TSVETKOV, Shurn. Tekn. Fix. 25, 447 (1955); 29, 212 (1959)
3) KUHN,W., H.KUHN, Helv. chim. Acta 26, 1934 (1943)
4) TSVETKOV,V.N., E.V.FRISMAN, Doklady Akad. Nauk SSSR 97, 647 (1954)
5) KUHN,W., H.KUHN, P.BUCHNER, Ergebn. exact. Naturwiss. 25, 1 (1951)
6) PETERLIN,A., H.STUART, Z. Physik 112, 1 (1939)
7) PETERLIN,A., J. polymer. Sci. 12, 45 (1954)
8) MIRSKY,A.E., A.W.POLLISTER, J. gen. Physiol. 30, 117 (1946)
9) KIRBY,H.S., Biochem. J. 66, 495 (1957)
10) SPIRIN,A.S., Biochimia 23, 656 (1958)
11) SPIRIN,A.S., Biochimia 26, 511 (1961)
12) FRISMAN,E.V., V.N.TSVETKOV, Shurn. Exper. Theor. Fis. 23, 690 (1952)
13) FRISMAN,E.V., V.I.VOROBIEV, L.V.SCHAGINA, N.K.YANOVSKAYA, Visokomolek. Soed. 4, 762 (1962)
14) FRISMAN,E.V., V.I.VOROBIEV, L.V.SCHAGINA, N.K.YANOVSKAYA, Visokomolek. Soed. 5, 622 (1963)
15) WATSON,J.D., F.H.C.CRICK, Nature (London) 171, 964 (1953)
16) CRICK,F.H., J.D.WATSON, Proc. Roy. Soc. (London), Ser. A 223, 80 (1954)
17) ZIMM,B.H., D.M.CROTHERS, Proc. nat. Acad. Sci. USA 48, 905 (1962)
18) EIGNER,J., Doctoral thesis, Harvard University (1960)
19) TSVETKOV,V.N., Visokomolek. Soed. 5, 740 (1963)
20) KRATKY,O., G.POROD, Recueil Trav. chim. Pays-Bas 68, 1106 (1949)
21) EIZNER,Yu.E., O.B.PTITSYN, Visokomolek. Soed. 4, 1725 (1962)
22) SADRON,C, In book: Flow properties of disperse systems (ed. Hermans). Amsterdam, 1953
23) FRISMAN,E.V., V.I.VOROBIEV, N.K.YANOVSKAYA, L.V.SCHAGINA, Biochimia 28, 137 (1963)
24) FRISMAN,E.V., E.N.ARCHIPOVA, Doklady Akad. Nauk SSSR 115, 491 (1957)
25) FRISMAN,E.V., M.A.SIBILEVA, A.V.KRASNOPEROVA, Visokomol. Soed. 1, 597 (1959)
26) JANESCHITZ-KRIGL,H., Makromolekulare Chem. 33, 55 (1959)
27) KISELEV,N.A., L.P.GAVRILOVA, A.S.SPIRIN, Doklady Akad. Nauk SSSR 138, 692 (1961)
28) BOGDANOVA,E.S., Z.P.GAVRILOVA, G.A.DVORKIN, N.A.KISELEV, A.S.SPIRIN, Biochimia 27, 387 (1962).

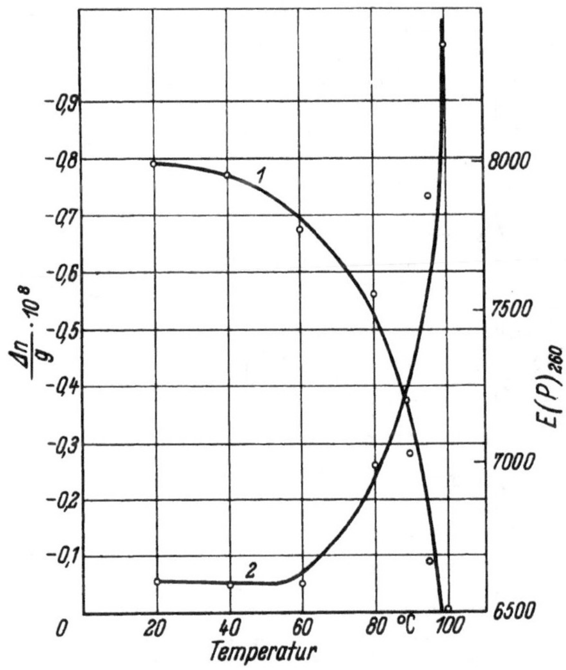

Fig.1 Dependence of $\frac{\Delta n}{g}$ (curve 1) and $E(P)_{260}$ (curve 2) on heating temperature of solution of DNA in 0,15 M NaCl c=0,0155 $\frac{g}{100\ ml}$

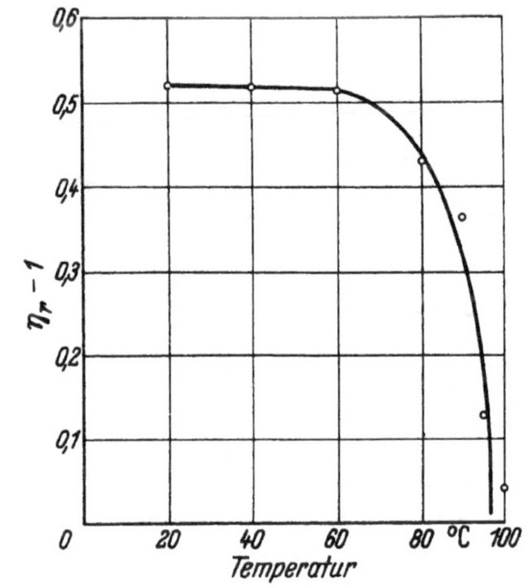

Fig.2 Dependence of η_r -1 on heating temperature of solution of DNA in 0,15 M NaCl c=0,0155 $\frac{g}{100\ ml}$

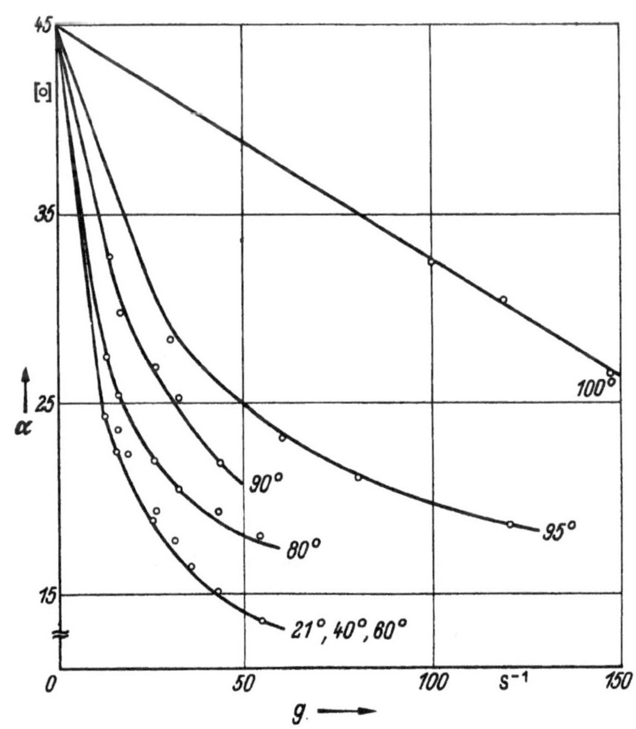

Fig.3 Dependence of $\alpha = f(g)$ for solutions in 0,15 M NaCl at different heating temperatures c=0,0155 $\frac{g}{100\ ml}$

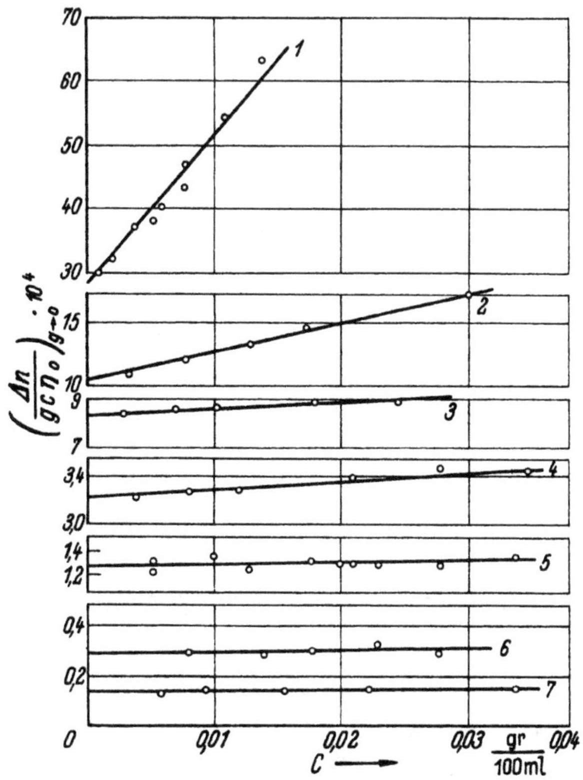

Fig.4 Dependence of $\frac{\Delta n}{(g c \eta_0)_{g\to 0}} = f(c)$ for the solutions of DNA-fractions in 0.15 M NaCl. The figures are fraction numbers.

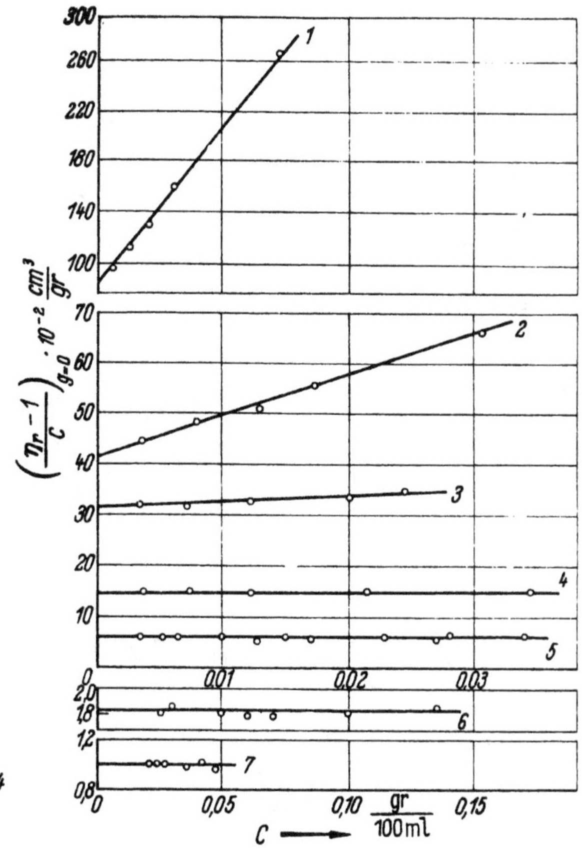

Fig.5 Dependence of $\left(\frac{\eta_r - 1}{c}\right)_{g\to 0} = f(c)$ for the solutions of DNA-fractions in 0.15 M NaCl. The figures are fraction numbers.

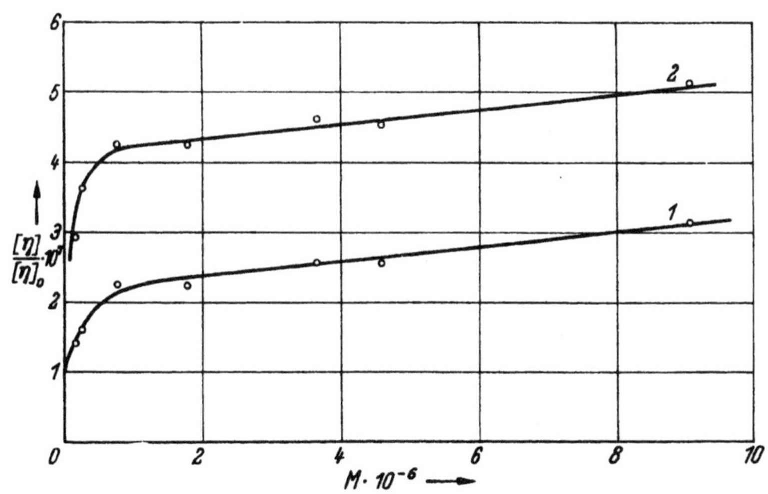

Fig.6 Dependence of $[\eta]/[\eta]_0 = f(M)$ for the DNA molecules in 0.15 M NaCl. Curve 1 without taking into consideration the anisotropy of microform, curve 2 taking into consideration the microform anisotropy.

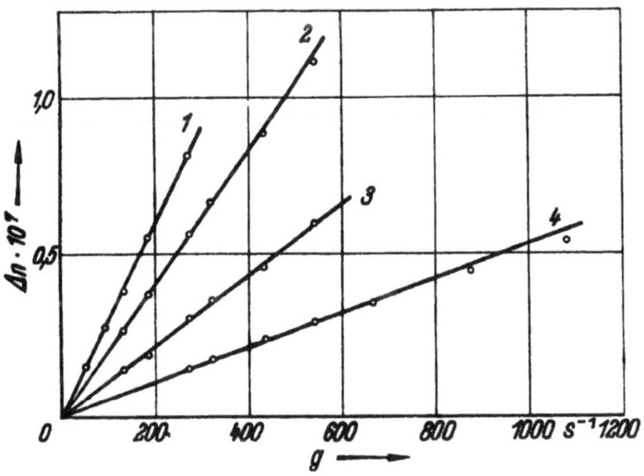

Fig.7 Dependence of $\Delta n = f(g)$ for solutions of RNA from the tumour with a small ionic stength of tris-buffer 10^{-3}M. The quantity of grams of RNA in 100 ml: 1- 0.2; - 0.113; 3- 0.058;4-0.03

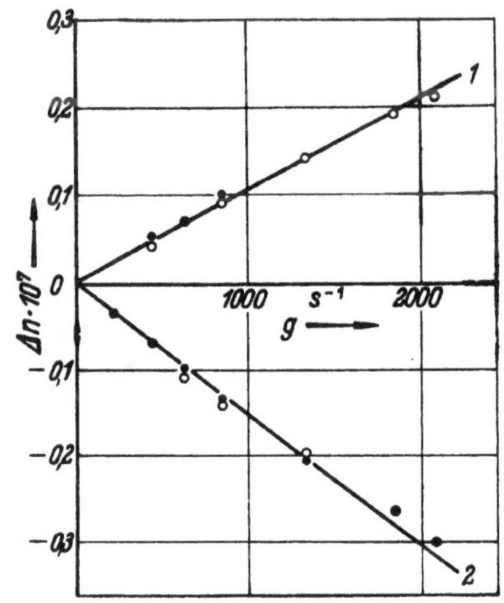

Fig.8 Influence of the hydrodynamical forces of the flow on the value Δn of RNA-solution from the tumour. Tris-buffer 10^{-2}M. 0 and are different experiments. 1 is $\Delta n = f(g)$ at the starting moment. 2 is $\Delta n = f(g)$ after the solution had been in the speedfield.

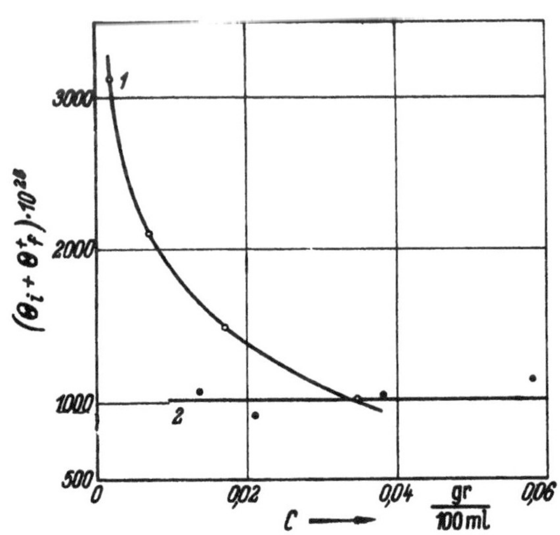

Fig.9 Dependence of the whole optical anisotropy of RNA molecules from TMV on solution concentration. Tris-buffer: $1-10^{-3}$M; $2- 10^{-2}$M

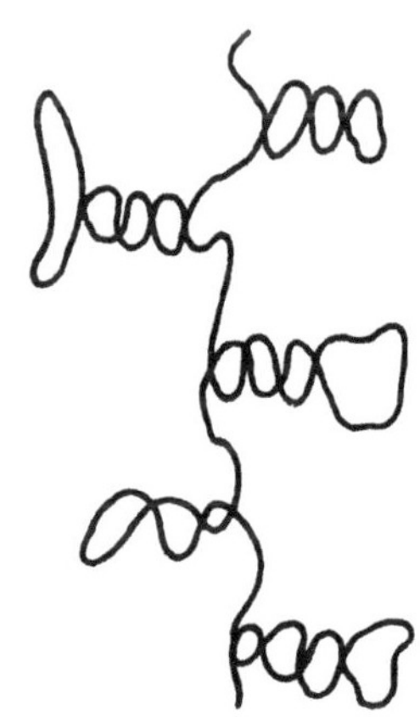

Fig. 10.- Model of a RNA molecule [28].

D Nucleinsäuren

Zur UV-Photochemie der Nucleinsäuren

W. SAUERBIER

Institut für Genetik der Universität, Köln

S U M M A R Y

The present state of experiments on the identification of the lethal UV-photoproduct in DNA is discussed. Whenever a quantitative description of the lethal event - as determined in the biological test - is desired, serious difficulties arise by reactivating mechanisms which erase UV-lesions. Thus far, we are in no case absolutely sure to measure the intrinsic UV-sensitivity of a DNA.

Studies with 5BU-substituted DNA and chemical measurements on the formation and the elimination of UV-photoproducts in parallel with the inactivation and the reactivation of biological functions strongly suggest that the thymine-dimer ($\widehat{TT}$) is the most frequent lethal UV-photoproduct. However, some results with host cell reactivation, and with the competitive inhibition of the photoreactivating enzyme might point into another direction.

The ratio $\widehat{TT}$ / lethal UV-hits equals about 5 with double stranded DNA in which no reactivations are believed to occur. With single stranded DNA the ratio $\widehat{TT}$ / lethal UV-hits is about 0.7. These results suggest that the double strandness <u>per se</u> offers a possibility for by-passing UV-lesions.

In diesem Beitrag will ich zusammenfassen, inwieweit es zur Zeit möglich zu sein scheint, an Nucleinsäuren beobachtete strahlenbiologische Phänomene durch die Ergebnisse der UV-Photochemie zu erklären. Die Betrachtungen werden auf DNA und auf Organismen, deren primäre Informationsträger DNA sind, beschränkt. Wir wollen uns außerdem auf der Seite der Photochemie beschränken, indem wir ausschließlich das Thymin-Dimer ($\widehat{TT}$) betrachten.

Dimerisationsprodukte des Thymins wurden nach Bestrahlung gefrorener, wäßriger Thyminlösungen mit UV von 254 nm von BEUKERS und BERENDS [1] erstmals nachgewiesen. Einen Hinweis, daß das $\widehat{TT}$ auch in der UV-Photobiologie bedeutsam sein könnte, erbrachten dann WACKER et al [20], indem sie aus UV-bestrahlter DNA ein Produkt isolierten, das chromatographisch mit dem $\widehat{TT}$ identisch war.

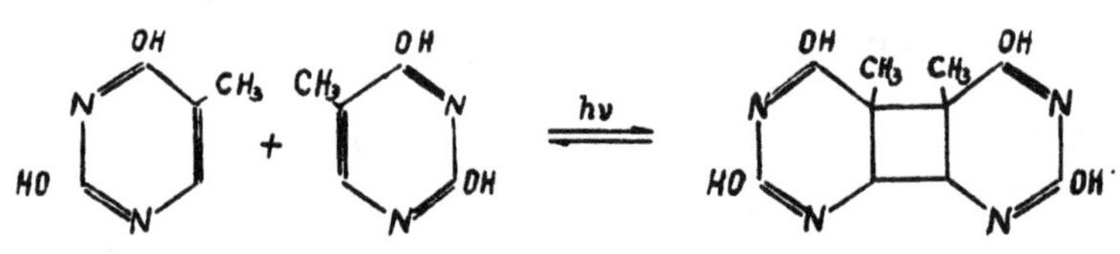

Wir müssen uns zuerst der Strahlenbiologie zuwenden und nachsehen, ob überhaupt eine eindeutige Dosis-Effekt-Zuordnung möglich ist. Als biologisches Objekt wählen wir den Bakteriophagen (oder kurz Phagen), bei dem die Voraussetzung erfüllt ist, daß bei 254 nm Bestrahlung praktisch ausschließlich die DNA geschädigt wird. Als biologischen Effekt wählen wir die Fähigkeit eines Phagen, in einem geeigneten Wirtsbakterium Phagen-Nachkommenschaft zu erzeugen. Wir werden dabei v o r l ä u f i g annehmen, daß für diese Funktion das gesamte DNA-Molekül des Phagen benötigt wird, d. h. daß es im Uberlebensfalle "ungetroffen" ist. Zum Uberleben muß eine DNA zwei Funktionen ausführen können; sie muß eine "richtige" Information an die messenger RNA abgeben, u n d sie muß eine identische Replikation erlauben. Von mutativen Änderungen wollen wir der Ubersichtlichkeit halber absehen, und uns nur mit der biologisch leicht meßbaren Inaktivierung befassen. Dabei müssen wir zur Beschreibung der strahlenbiologischen Situation mindestens die 4 folgenden Phagentypen vorstellen (Tab. 1). (Bakterienwirte sind Stämme von Escherichia coli.)

Tabelle 1

Phagentyp	ungef. Mol.Gew. der DNA	ungef. MG/MG_{T2}	DNA-Typ	ungef. Verhältnis der UV-Empfindlichkeiten k/k_{T2}
T2 bzw. $T4v^-$	$1{,}3 \times 10^8$	1	2-strängig	1
$T4v^+$	$1{,}3 \times 10^8$	1	2-	1/2
T1	4×10^7	1/3	2-	1/2 bzw. 1/10
ØX 174	$1{,}7 \times 10^6$	1/60	1-	1/5

Die Basenzusammensetzung der DNA ist bei den 4 Phagen nicht wesentlich unterschiedlich mit der Ausnahme, daß bei T2 und T4 5-Hydroxymethylcytosin an Stelle von Cytosin vorhanden ist. Da wir uns auf die Betrachtung des Photoprodukts $\widehat{TT}$ beschränken, dürfen wir diese Phagen als vergleichbare Objekte ansehen und das MG der DNA v o r e r s t als allein bestimmend für die UV-Empfindlichkeit annehmen.

Wie aus Abb. 1 hervorgeht wird ØX 174 durch UV (254 nm) nach einer Eintrefferfunktion inaktiviert. Es ist also:

1. $\log S = - kD$,

wobei S die Zahl der überlebenden Phagen, D die UV-Dosis und k eine Inaktivierungskonstante darstellen. Für T2 und T4 gilt 1. nicht streng. Diese Phagen zeigen UV-Uberlebenskurven (Abb. 1), die einen Mehrtreffercharakter andeuten. Die Ordinaten-Extrapolationswerte liegen zwischen 1,4 und 1,8. Wir wollen aber vorläufig auch diese Kurven als Eintrefferkurven behandeln. Im Falle von T1 (s.Abb. 1) läßt sich die Uberlebenskurve durch die Annahme zweier Fraktionen mit unterschiedlicher Empfindlichkeit hinreichend annähern:

2. $S/S_0 = pe^{-k_1 D} + (1-p)e^{-k_2 D}$,

wobei p die Fraktion mit der Empfindlichkeit k_1 und (1-p) die Fraktion mit der Empfindlichkeit k_2 bedeuten ($k_1:k_2 = 5$). Die Aufspaltung der T1-Population in zwei Fraktionen unterschiedlicher UV-Empfindlichkeiten ist aber keine primäre Eigenschaft der Phagen, sondern sie wird durch den physiologischen Zustand der Wirtsbakterien bestimmt (SAUERBIER [12]). Durch geeignete Vorbehandlung der Wirtsbakterien können also auch für T1 UV-Uberlebenskurven vom Typ 1. erhalten werden, wobei alle Phagen die geringe UV-Empfindlichkeit k_2 annehmen und außerdem die Ordinaten-Extrapolationswerte wie bei T2 um 1,6 liegen. (Außerdem kann durch den Wirt coli B_{s-1} (HILL [7]) für alle T1-Phagen die Empfindlichkeit k_1 erreicht werden.)

Gehen wir zurück zur Tab. 1. Ein Vergleich der k-Werte von T2 bzw. $T4v^-$ und $T4v^+$ zeigt, daß $T4v^+$ bei gleicher DNA-Menge nur die halbe UV-Empfindlichkeit besitzt wie $T4v^-$. Untersuchungen von STREISINGER [19] und von HARM [5] zeigten, daß in beiden Phagen bei gleicher UV-Dosis gleichviel Treffer entstehen, daß aber im Falle des $T4v^+$ nach dem Eindringen der Phagen-DNA in die Wirtszelle wieder die Hälfte der letalen UV-Treffer ausgelöscht wird. Verantwortlich für dieses Auslöschen ist vermutlich ein Enzym. Die Information zur Synthese des Enzyms liegt im v-Gen des T4. Die Mutation von v^+ nach v^- bewirkt also eine scheinbare Steigerung der UV-Empfindlichkeit um den Faktor 2, die in Wirklichkeit aber durch die genannte Reversion der UV-Schäden zustandekommt. Den Mechanismus bezeichnen wir als v - G e n r e a k t i v i e r u n g , ohne zu wissen wie er im einzelnen abläuft.

Bei der Betrachtung von T1 (Tab. 1) sehen wir, daß das Verhältnis k/k_{T2} der empfindlichen Fraktion etwa MG/MG_{T2} entspricht, so daß wir uns nur um die Empfindlichkeitsunterschiede z w i s c h e n den b e i d e n Fraktionen kümmern wollen. Aus Versuchen von GAREN und ZINDER [3] war bekannt, daß dieser Unterschied durch einen reaktivierenden Mechanismus der Wirtsbakterien verursacht wird, dem ein Genom-Austausch zwischen Phage und Bakterium zugrunde liegen sollte. Inzwischen ist aber gezeigt (SAUERBIER [10][11][13]), daß eine direkte Reversion der UV-Schäden - vermutlich durch einen enzymatischen Mechanismus ohne Genom-Austausch - diese Reaktivierung bewirkt. Da die Information für diesen Mechanismus im Bakteriengenom enthalten ist, und da außerdem UV-Schäden im Bakteriengenom selbst durch diesen Mechanismus reaktiviert werden, sprechen wir von W i r t s r e a k t i v i e r u n g (WR).

Ich habe diese beiden Typen der Reaktivierung hier vorgestellt, um zu zeigen, daß zwar die Zuordnung biologischer Effekt zur UV-Dosis möglich ist, daß damit aber noch keine Zuordnung primäre UV-Empfindlichkeit zur Dosis erreicht ist. Diese ist nur dann möglich, wenn keine Mechanismen im biologischen Milieu vorhanden sind, die UV-Schäden beheben oder sonstwie umgehen können. Im Augenblick können wir bei keinem unserer biologischen Objekte mit absoluter Sicherheit behaupten, daß im Überlebenstest nicht ein unbekannter Reaktivierungsmechanismus eine zu geringe UV-Empfindlichkeit vortäuscht. Mit anderen Worten: Wir wissen nicht, ob uns ein Objekt zur Verfügung steht, an dem wir die primäre UV-Empfindlichkeit messen können. Es ist weiter wichtig darauf hinzuweisen, daß die Reaktivierungsmechanismen mit unterschiedlicher Wirksamkeit reaktivieren, so daß wir nicht mit einem für alle Objekte gültigen Korrekturfaktor die primären Empfindlichkeiten errechnen können. Wir wollen aber im Folgenden annehmen, daß die für $T4v^-$ und für die empfindliche Fraktion von T1 gemessenen k-Werte die primären Empfindlichkeiten der DNA dieser Phagen angeben. (Allerdings muß ich darauf hinweisen, daß einige Ergebnisse von HARM [6] diese Annahme für $T4v^-$ in Frage stellen.)

Einen dritten Reaktivierungstyp muß ich noch vorstellen, die Photoreaktivierung (PhR) (s. JAGGER [8]). Sie ist sowohl i n v i v o , als auch bei geeigneten Objekten i n v i t r o möglich. Für letzteres sind eine hochgereinigte Proteinfraktion (Enzym), ein Cofaktor (?) und langwelliges UV-Licht (360 nm) notwendig (s. RUPERT [9]). Dabei können abhängig vom biologischen Objekt 20 bis 90 % der primären UV-Schäden reaktiviert werden.

Die drei genannten Reaktivierungstypen haben neben ihrer Bedeutung bei der Bestimmung der primären UV-Empfindlichkeit noch eine wesentliche Aufgabe bei der Identifizierung des letalen, reaktivierbaren UV-Photoprodukts. Für diese Identifizierung stehen prinzipiell zwei Wege zur Verfügung:

1. Substitution natürlicher Basen durch Basenanaloga in der DNA.
2. Der chemische Nachweis von UV-Photoprodukten in der DNA.

S U B S T I T U T I O N

Die Substitutionsmethode lieferte folgende Ergebnisse: Wird in der DNA der Phagen T2 und T4 Thymin durch 5-Bromuracil (5BU) ersetzt, so führt das bei T2 zu einem Anstieg der biologisch gemessenen UV-Empfindlichkeit um den Faktor 2, bei T4v$^+$ hingegen zu einem Empfindlichkeitsanstieg um den Faktor 4 (STAHL et al. [18]). Dieser Unterschied wurde als ein Verlust der v-Genreaktivierung interpretiert. In den gleichen Versuchen wurde festgestellt, daß die UV-Photoprodukte in den 5BU-substituierten T2 und T4v$^+$ nicht mehr photoreaktivierbar waren. Diese Ergebnisse legten die Annahme nahe, daß die v-Gen- und die photoreaktivierbaren UV-Photoprodukte in thyminhaltiger DNA photochemische Veränderungen des Thymins darstellen. Quantitativ mußte es sich dabei um etwa 60 % aller letalen UV-Treffer handeln. Wird das Thymin der DNA von T1 durch 5BU ersetzt, dann ergibt sich im biologischen Test ein Anstieg der UV-Empfindlichkeit der sensiblen T1-Fraktion um einen Faktor von annähernd 2, während die UV-Empfindlichkeit der resistenten Fraktion um etwa den Faktor 10 ansteigt (SAUERBIER [10]). Gleichfalls ist auch hier die Photoreaktivierbarkeit der UV-Schäden verloren. Wir interpretierten die Ergebnisse wie STAHL et al., nämlich, daß in diesem Falle die wirtsreaktivierbaren und die photoreaktivierbaren UV-Schäden etwas mit dem Thymin zu tun haben müssen. Quantitativ bedeutet das aber bei T1, daß etwa 93 % aller letalen UV-Treffer im Thymin gelegen sein müssen, da durch WR und durch PhR z u s a m m e n 93 % der o h n e Reaktivierung biologisch gemessenen UV-Treffer ausgelöscht werden.

Einen weiteren Hinweis, daß es sich beim letalen UV-Schaden um ein Photoprodukt des Thymins handelt, und zwar vermutlich um das Dimerisierungsprodukt zweier benachbarter Thymine, lieferten Versuche von GÜNTHER und PRUSOFF [4]) und von WACKER und JACHERTS [21]). In 6-Azathymin ist das 6-Kohlenstoffatom durch Stickstoff ersetzt.

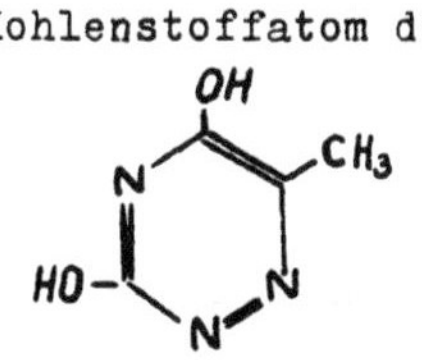

Das führt dazu, daß zwischen C_5 und dem Atom in 6-Stellung keine Doppelbindung mehr besteht. Deswegen kann auch der für die Dimerisierung zweier Thymine erforderliche Zyklobutanring nicht mehr geschlossen werden. Falls das Thymin-Dimer das letale Photoprodukt ist, müßte die Einführung von 6-Azathymin zu einer Abnahme der UV-Empfindlichkeit führen. Eine solche Abnahme wurde von GÜNTHER und PRUSOFF bei teilsubstituierten Streptococcus faecalis und von WACKER und JACHERTS ebenfalls bei teilsubstituierten Enterococcus stei beschrieben. Da aber die UV-Empfindlichkeiten von Bakterien stark vom physiologischen Zustand abhängen, lassen diese Ergebnisse nur unsichere Schlüsse zu.

C H E M I S C H E R   N A C H W E I S

Mit dem chemischen Nachweis von UV-Photoprodukten in der DNA allein ist noch keine Aussage möglich, ob ein Photoprodukt biologisch letal, mutativ oder überhaupt bedeutungslos ist, falls es sowohl bei der Informationsablesung als auch bei der Replikation wie das Ausgangsprodukt behandelt wird. Zuverlässigere Hinweise werden nur erhalten, wenn außer der Parallele der Bildung des UV-Photoprodukts zur Inaktivierung zusätzlich die Reversion des UV-Photoproduktes parallel mit der biologisch gemessenen

Reaktivierung beobachtet wird. Der erste verläßliche Versuch wurde von WULFF und RUPERT [22] unternommen, die zeigten, daß durch PhR i n v i t r o $\widehat{TT}$ monomerisiert wird. Wenn auch nicht an der gleichen DNA-Präparation gemessen, so wurde doch durch die gleiche PhR-Enzym-Präparation transformierende DNA photoreaktiviert, wie sich im biologischen Test zeigte.

Aus den Arbeiten von R. SETLOW und J. SETLOW ergab sich dann eine zweite Möglichkeit der Monomerisierung von TT durch eine Nachbestrahlung von 254 nm bestrahlter DNA durch 240 nm. Dabei zeigte sich, daß parallel mit der Spaltung des $\widehat{TT}$ eine Reaktivierung der biologischen Aktivität transformierender DNA einherging. Aus der Inaktivierungskinetik bei nachfolgender 254 nm Bestrahlung der 240 nm reaktivierten DNA ließ sich außerdem schließen, daß durch die 240 nm-Reaktivierung das biologisch letale 254 nm Photoprodukt zum Ausgangsprodukt zurückverwandelt worden war. Mit der Beobachtung stand natürlich die Monomerisierung des $\widehat{TT}$ in Einklang, so daß mit diesen Versuchen, neben den Befunden der PhR, die zweite strenge Parallele vorlag. Für die v-Gen-Reaktivierung gibt es entsprechende Versuche noch nicht. Für die Wirtsreaktivierung haben wir versucht, die gleiche Parallele nachzuweisen. Das Ergebnis war, daß durch WR $\widehat{TT}$ n i c h t monomerisiert wird (SAUERBIER, unveröffentlicht). Dieser Befund bedeutet entweder, daß trotz der aufgezeigten Parallelen das $\widehat{TT}$ doch nicht das letale Photoprodukt ist, oder - was ich für wahrscheinlicher halte - daß die Letalität des $\widehat{TT}$ in der DNA durch WR anderweitig behoben wird, o h n e das $\widehat{TT}$ z u m o n o m e r i s i e r e n. Die genannten Parallelen zwischen Dimerisierung und Inaktivierung und Monomerisierung und Reaktivierung sind in Tab. 2 nochmals zusammengefaßt.

Tabelle 2

Einwirkung	Ausgangsmaterial	Chem. Befund	Biol. Befund	Autor
254 nm	unbestr. DNA	T zu $\widehat{TT}$	Inaktiv.	WACKER et al. [20]
240 nm	254 nm best. DNA	$\widehat{TT}$ zu T	Reaktiv.	SETLOW [14]
PhR	254 nm best. DNA	$\widehat{TT}$ zu T	Reaktiv.	WULFF und RUPERT [22]
v-Gen	254 nm bestr. DNA	?	Reaktiv.	RUPERT [9] STREISINGER [19] HARM [5]
WR	254 nm bestr. DNA	$\widehat{TT}$ bleibt $\widehat{TT}$	Reaktiv.	SAUERBIER, unveröffentlicht

Die Betrachtungen über Inaktivierung und Reaktivierung einerseits und Thymin-Dimerisierung und Monomerisierung andererseits waren so weit nur qualitativ. Die ersten zuverlässigen quantitativen Untersuchungen wurden von WULFF [23] am Phagen T4v⁻ durchgeführt, indem er in einem niederen UV-Dosisbereich an den gleichen Proben das Verhältnis $\widehat{TT}$ pro biologisch gemessenem Letaltreffer bestimmte. Das Ergebnis war 4,8 $\widehat{TT}$/Letaltreffer. Eigene Versuche ergaben für T1 etwa 6 $\widehat{TT}$/Letaltreffer (gemessen unter Bedingungen, unter denen WR ausgeschaltet war). Das wirft die Frage auf, wieso bei einer angenäherten Eintrefferkinetik der Inaktivierung nur etwa jedes fünfte $\widehat{TT}$ letal ist. Ein einfacher Mehrtreffermechanismus kann dem nicht zugrunde liegen, so daß als Erklärung entweder besondere Nachbarschaftsstellungen oder die Degeneration des genetischen Codes oder, was am plausibelsten erscheint, die Überbrückung von $\widehat{TT}$ durch die Zweisträngigkeit der DNA vorgebracht werden muß. Falls diese letzte Interpretation richtig ist (sie wurde ausführlich von STAHL [17] diskutiert), müßte ØX 174, bezogen auf den DNA-Gehalt im Vergleich zu T2, wesentlich sensibler sein. Wie aus Tab. 1 hervorgeht, ist das auch der Fall. Weiterhin müßte bei ØX 174 das Verhältnis $\widehat{TT}$/Letaltreffer ≤ 1 sein. Auf Anregung von M. DELBRÜCK wurde dieses Verhältnis an unserem Institut durch C. DAVID bestimmt. Das vorläufige Ergebnis ist: 0,7 $\widehat{TT}$/ ØX 174-Letaltreffer.

Kommen wir noch einmal zur Photoreaktivierung zurück. ØX 174 zeigt einen photoreaktivierbaren Sektor von etwa 24 %. Durch Versuche von J. SETLOW[16)] ist andererseits bekannt, daß alle $\widehat{TT}$ durch PhR monomerisiert werden können. Von den 100 letalen UV-Treffern in ØX 174 lassen sich durch PhR aber nur 24 reaktivieren. Nehmen wir an, das $\widehat{TT}$ ist das letale UV-Photoprodukt, dann bedeutet das entweder: N i c h t   a l l e $\widehat{TT}$ lassen sich so monomerisieren, daß sie wieder das Ausgangsprodukt Thymin darstellen, oder aber auch im einsträngigen ØX 174 sind   n i c h t   a l l e   letal.

Vielleicht ist es ein Zufall, daß 30 % der letalen UV-Schäden in ØX 174 nicht dem $\widehat{TT}$ zugeschrieben werden können, und daß auch etwa 30 % (genau 24 %) der Letalschäden photoreaktivierbar sind. Vielleicht ist es aber kein Zufall.

Um am Ende noch einige Ungereimtheiten zu sagen, die nicht ganz in das sonst so schöne Bild vom $\widehat{TT}$ als letalem UV-Schaden passen, möchte ich noch Befunde von RUPERT (pers. Mitt.) zitieren. Das Dimer in dem synthetischen Dinucleotid $\widehat{TpT}$ wirkt nicht als kompetitives Substrat, während UV-bestrahltes Poly-GC kompetitiv hemmt.

L I T E R A T U R

1) BEUKERS,R., W.BERENDS, Biochim. biophysica Acta 41, 550 (1960)
2) DAVID,C.N., (1963) pers. Mitt.
3) GAREN,A., N.D.ZINDER, Virology 1, 347 (1955)
4) GÜNTHER,H.L., W.H.PRUSOFF, Biochim. biophysica Acta 55, 778 (1962)
5) HARM,W., Naturwissenschaften 45, 391 (1958)
6) HARM,W., Virology 19, 66 (1963)
7) HILL,R.F., Biochim. biophysica Acta 30, 636 (1960)
8) JAGGER,J., Bacteriol. Rev. 99 (1958)
9) RUPERT,C.S., J. gen. Physiol. 45, 703; 725 (1963); pers. Mitt.
10) SAUERBIER,W., Virology 15, 465 (1961)
11) SAUERBIER,W., Virology 16, 358 (1962)
12) SAUERBIER,W., Virology 17, 164 (1962)
13) SAUERBIER,W., Z. Vererb.-Lehre 93, 220 (1962)
14) SETLOW,R., Biochim. biophysica Acta 49, 237 (1961)
15) SETLOW,R.B., J.K.SETLOW, Proc. Natl. Acad. Sci. USA 48, 1250 (1962)
16) SETLOW,J., (1963) pers. Mitt.
17) STAHL,F.W., (1958) pers. Mitt.
18) STAHL,F.W., J.M.CRASEMANN, L.OKUN, E.FOX, C.LAIRD, Virology 13, 98 (1961)
19) STREISINGER,G., Virology 2, 1 (1956)
20) WACKER,A., H.DELLWEG, D.WEINBLUM, Naturwissenschaften 47, 477 (1960)
21) WACKER,A., D.JACHERTS, J. molecular Biol. 4, 413 (1962)
22) WULFF,D.L., C.S.RUPERT, Biochem. biophysic. Res. Commun. 7, 237 (1962)
23) WULFF,D.L., (1963) pers. Mitt.

DISKUSSIONSFRAGEN

ZIMMER.- Mir scheint der Vergleich TpT + Photoreakt. Enzym mit Poly GC + PhE, was in einem Falle zu keiner kompetitiven Hemmung, im anderen zu einer solchen führt, nicht ganz beweiskräftig. Es spielen für die Bildung des Enzym-Substratkomplexes auch konfigurative bzw. konformative Zentren eine Rolle. Vielleicht wäre Poly AT + PhE als Vergleich sicherer.

SAUERBIER.- Sicher ist das Ergebnis mit TpT nicht beweiskräftig. Ich möchte es nur als einen Hinweis verstanden wissen, der uns vorsichtig machen sollte, das $\widehat{TT}$ schon als das "bewiesene" letale UV-Photoprodukt anzusehen. Von RUPERT sind außerdem das Thymidylsäure-Okta-Nucleotid und Apurinsäure UV-bestrahlt worden. In beiden Fällen ließ sich mit dem UV-bestrahlten Produkt keine kompetitive Hemmung erreichen.

DOSE.- Ich möchte mir gestatten, 3 Fragen zu stellen.

1. Durch welche Methoden haben Sie die Thymindimeren quantitativ bestimmt?
2. Welche Lichtquelle haben Sie für die UV-Bestrahlung bei 240 nm benutzt?
3. Wie groß waren die eingestrahlten Energien und wie haben Sie diese bestimmt?

SAUERBIER.- Zur Frage 1: Die Thymin-Dimere wurden nach HCOOH-Hydrolyse (WYATT und COHEN, Biochem. J. 55, 774 (1953)) UV-bestrahlter T1-DNA, die Methyl-H^3-Thymin markiert war, durch Papierchromatographie in Butanol-Wasser (86:14) und Ammoniumsulfat-Natriumacetat-Isopropanol (40:9:1) vom Thymin getrennt. Zur Kontrolle der Chromatographie-Verluste wurde C^{14}-Thymin-Dimer zugegeben. Nach 4-maligem Rechromatographieren erlaubt die Methode 1 $\widehat{TT}/10^5$ T zu messen (WULFF (1963) unveröffentlicht). Die Bestimmung der Radioaktivität erfolgte mit dem PACKARD "tricarb" Szintillationszähler.
Zur Frage 2: Zur Bestrahlung steht ein leistungsstarker UV-Monochromator zur Verfügung (Johns und Delbrück, unveröffentlicht).
Zur Frage 3: In den Experimenten, in denen die Monomerisierung des TT durch Wirtsreaktivierung des T1 gemessen werden sollte, betrug die UV-Dosis 400 erg/mm^2. Das sind etwa 20 T1-Letaltreffer. Die Bestimmung der UV-Dosen erfolgt photochemisch durch Umwandlung vom Leuko-Malachitgrün (s. JOHN, RAPOPORT und DELBRÜCK, J. molecular Biol. 4, 104 (1962)).

DELLWEG.- Ich möchte darauf hinweisen, daß nach jüngsten Versuchen von A. WACKER auch eine Photoreaktivierung von UV-bestrahlten Bakterien durch Wasserstoffperoxyd und sichtbares Licht möglich ist. Eine Kultur von E. coli B wurde am Ende der logarithmischen Wachstumsphase in starker Verdünnung mit 10 erg/mm^2 bei 253,7 nm bestrahlt. Danach wurde die Zahl der keimfähigen Zellen zu 60 % bestimmt. Setzte man der Kultur nach der UV-Bestrahlung 3 μg/ml Wasserstoffperoxyd zu und bestrahlte mit einer Tageslichtlampe (5×10^5 erg/mm^2), so stieg die Zahl der keimfähigen Zellen auf etwa 99 % an. Es wäre daher möglich, daß der Mechanismus der Photoreaktivierung mit der Bildung von H_2O_2 zusammenhängt bzw. mit der Entstehung von sauerstoffhaltigen Radikalen, die durch die Enzyme der biologischen Oxydation in Gegenwart von Licht gebildet werden.

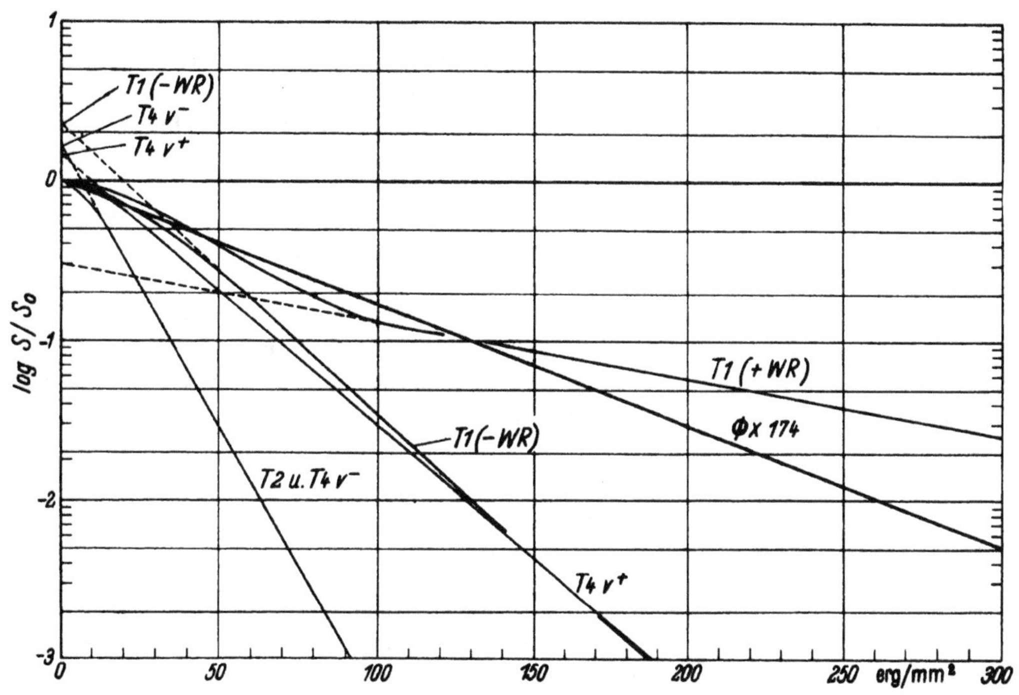

Abb.1 Inaktivierungskurven

D Nucleinsäuren

Strahlenchemische Veränderungen in Nucleinsäuren. Über eine gegen Nucleasen resistente Fraktion aus UV-bestrahlter Desoxyribonucleinsäure

H. DELLWEG

Institut für therapeutische Biochemie der Universität, Frankfurt/M.

S U M M A R Y

By enzymatical degradation of UV irradiated DNA with DN'ase I and snake venom a nuclease-resistent oligonucleotide fraction was formed. This fraction is not splitted by these enzymes. It contains the whole of thymine dimers and in addition the other four bases and was separated by chromatography on Sephadex G 25. The degradation of this fraction by 66 % formic acid containing 3 % diphenylamine according to BURTON led to a number of unusual pyrimidine nucleotides which cannot be found in non irradiated DNA. The value is rendered of this method for analysis of UV lesions in DNA.

Bei unseren früheren Versuchen über die strahlenchemischen Veränderungen in Nucleinsäuren wurde die mit Thymin-[2-^{14}C] markierte DNA nach UV-Bestrahlung (253,7 nm) mit heißer Perchlorsäure hydrolysiert. Nach chromatographischer Trennung der freien Basen bestimmten wir die Menge an dimerem Thymin aufgrund seiner Radioaktivität. Bei diesem Abbau besteht die Möglichkeit der Bildung von sekundären Hydrolysenprodukten, die dann irrtümlich als Bestrahlungsprodukte angesprochen werden. Um diese Fehlerquelle auszuschließen, haben wir versucht, die DNA nach der UV-Bestrahlung mit Enzymen abzubauen.

Wir gingen dabei von Bakterien (Enterococcus Stei) aus, die in Gegenwart von Thymin-[2-^{14}C] kultiviert worden waren und isolierten aus ihnen hochmolekulare DNA nach der Methode von MARMUR [1]. Durch nacheinanderfolgende Behandlung mit DN'ase I und dem Gemisch von Phosphomono- und Phosphodiesterase aus ungereinigtem Schlangengift wurde diese DNA zu über 99 % in die Desoxyribonucleoside gespalten. Nach einer UV-Bestrahlung (253,7 nm) mit 12,9 x 10^4 erg/mm^2 konnte ein Teil der DNA nicht mehr durch das Enzymgemisch abgebaut werden. Von 100 Molen Basen lagen nach der Bestrahlung und Spaltung 91,4 Mole als Nucleoside und 8,6 Mole als Nucleotide vor, aus denen auch durch erneute Behandlung mit dem Enzymgemisch kein Phosphor mehr abgespalten wurde.

Etwa um die gleiche Zeit erschien eine kurze Notiz von BOLLUM und SETLOW [2], in der ebenfalls über den Abbau von UV-bestrahlter DNA mit DN'ase und gereinigter Phosphodiesterase berichtet wurde. Danach konnten etwa 10 % der bestrahlten DNA nicht mehr gespalten werden. Durch weitere Bestrahlung des resistenten Anteils bei 239 nm wurde dimeres Thymin wieder in monomeres Thymin gespalten. Die nachfolgende Enzymbehandlung führte nun zu Mononucleotiden und zwar im Verhältnis dXMP zu TMP wie 1:2.

$$\mathrm{DNA} \xrightarrow{270\ nm} \xrightarrow[\text{Phosphodiesterase}]{\text{DN'ase}} \text{MONONUCLEOTIDE (etwa 90 \%)} + \text{OLIGONUCLEOTIDE (etwa 10 \%)}$$

$$\text{OLIGONUCLEOTIDE} \xrightarrow{239\ nm} \xrightarrow[\text{Phosphodiesterase}]{\text{DN'ase}} n\ dXMP + 2n\ TMP$$

Wir fanden, daß unter den abgespaltenen Nucleosiden kein dimeres Thymidin nachzuweisen war. Da sich das gesamte dimere Thymin in der nuclease-resistenten Fraktion befand, führten wir weitere Versuche zur Charakterisierung dieser Fraktion durch.

Die Trennung der Produkte aus dem enzymatischen Abbau gelang an Sephadex G 25 [3]. Wie Abb. 1 zeigt, erhielten wir vier Fraktionen. Fraktion IV, die als erste aus der Säule eluiert wurde, enthielt den phosphorhaltigen und gegen weitere Enzymbehandlung resistenten Anteil. Fraktion III bestand aus unbedeutenden Mengen an Mononucleotiden. Bemerkenswert ist die überraschend gute Trennung zwischen Pyrimidinnucleosiden (Fraktion II) und Purinnucleosiden (Fraktion I).

Verschiedene Versuche, die Nucleotide der Fraktion IV weiter aufzutrennen, schlugen fehl. Eine Sephadex-Säule (G 25) von 2 Metern Länge ergab wiederum nur eine sehr scharfe Bande. Aus Sephadex G 200 wurde die Substanz ebenfalls in einer einheitlichen, wenn auch etwas verbreiterten Bande eluiert. Von Dowex 1 x 10 war keine Elution mit einem Ammoniumacetat-Gradienten möglich. Auch die Hochspannungselektrophorese führte zu keiner Fraktionierung. Wir glauben daraus den Schluß ziehen zu dürfen, daß es sich bei Fraktion IV um ein Gemisch von Oligonucleotiden handelt mit verhältnismäßig einheitlichen Molekulargewichten.

Die Basenzusammensetzung der einzelnen Fraktionen geht aus Tabelle 1 hervor. Die Werte für Guanin und Adenin lagen in der Nucleosidfraktion etwas höher, der Wert für Thymin wesentlich niedriger, als in der DNA. In Fraktion IV lagen die Werte der Purine und besonders der des Cytosins sehr viel tiefer als in der DNA vor der Bestrahlung. Der Anteil an Thymin in der monomeren und dimeren Form in Fraktion IV betrug über 55 % aller Basen. Bei der Addition der Werte in den Fraktionen I bis IV finden wir zu wenig Cytosin. Dies ist erneut ein Hinweis für die strahlenchemische Umwandlung von Cytosin in der DNA, über die wir schon früher berichtet haben [4].

Zur weiteren Charakterisierung des nuclease-resistenten Anteils der Fraktion IV wählten wir die von BURTON und PETERSEN angegebene Spaltung mit Ameisensäure [5]. Dabei werden bekanntlich die Purine als freie Basen abgespalten und die Pyrimidine erscheinen entsprechend ihrer Sequenz auf der DNA-Kette als Nucleotide der allgemeinen Formel Py_np_{n+1}. Es war uns jedoch unmöglich, die gesamten Spaltprodukte dieses Abbaues zu identifizieren. Erstens enthielt die in unseren Versuchen eingesetzte DNA auch nach Behandlung mit RN'ase noch etwa 2,5 % RNA, die von Schlangengift offenbar nicht abgebaut wurde und daher in Fraktion IV enthalten war. Zweitens vergrößerten die durch Thymin-Dimerisierung entstandenen zusätzlichen Produkte die Zahl der Spaltstücke doch sehr erheblich und erschwerten dadurch die Trennung und Identifizierung.

Für den Abbau standen etwa 3 mg Substanz mit einer Radioaktivität von 0,45 µC zur Verfügung. Die Trennung dieses Hydrolysates wurde an einer Säule mit 2 ml Dowex 1 x 2 ausgeführt [6]. Abb. 2 zeigt den Verlauf der Elutionskurve. Die Aktivitätskurve hat acht mehr oder weniger gut getrennte Banden, nach denen das Eluat in einzelne Fraktionen geteilt wurde. Jede Fraktion wurde nochmals durch Papierchromatographie getrennt. Zur Identifizierung wurden die einzelnen Flecken eluiert und im Spektrophotometer ausgemessen. Danach wurden die Eluate zur Spaltung von dimerem Thymin mit UV bestrahlt, nochmals chromatographiert und mit Perchlorsäure oder Ameisensäure hydrolysiert. Die so erhaltenen freien Basen konnten nach einer weiteren papierchromatographischen Trennung aufgrund ihres Spektrums bestimmt werden. Die Phosphorbestimmung erfolgte nach P.S. CHEN et al. [7], wobei wir mit einem Endvolumen von 0,4 ml arbeiteten. Die ersten vier Fraktionen konnten vollständig aufgeklärt werden, bei den letzten vier Fraktionen mußten wir uns mit der Bestimmung der Basenzusammensetzung begnügen.

Von den aufgefundenen Hydrolysenprodukten seien hier nur kurz die wichtigsten erwähnt: Wir fanden Thymidin und dimeres Thymidin in Fraktion I, wofür wir keine Erklärung haben, da diese Verbindungen durch den Abbau theoretisch nicht entstehen können. Fraktion II enthielt ein Monophosphat des dimeren Thymidin, das am Ende einer Oligonucleotidkette gestanden haben muß. In Fraktion III fanden wir neben Uridyl- und Cytidylsäure, die zum Teil von der RNA herrühren, auch noch Thymidylsäuren, die ebenfalls vom Ende einer Oligonucleotidkette stammen muß. In Fraktion IV, die insgesamt neun verschiedene Verbindungen enthielt, konnten wir folgende radioaktive Verbindungen nachweisen: Thymidylsäure, Thymidin-diphosphat, ein Triphosphat des dimeren Thymidins, ein Tetraphosphat des dimeren Thymidins und ein gemischtes Triphosphat, welches Thymidin und dimeres Thymidin enthielt [+].

Über die Wirkung der Nucleasen auf UV-bestrahlte DNA geben uns die vorliegenden Versuche keine erschöpfende Auskunft. Wenn die Fermente vom Ende der Kette an abbauen würden und ihre Wirkung bei der ersten Thymin-Dimerisierung zum Stillstand käme, müßte der nuclease-resistente Anteil bei einer Dimerisierungsrate von 8,5 % wesentlich größer sein. Umgekehrt ist es aber auch ausgeschlossen, daß die Fermente alle Nucleotide außer den strahlengeschädigten abspalten, da wir ja alle vier Basen im nuclease-resistenten Anteil gefunden haben. Wir haben uns also vorzustellen, daß durch die UV-Bestrahlung bestimmte Abschnitte auf der Kette so geschädigt werden, daß sie von den Enzymen einfach übersprungen werden.

Mit der Methode des enzymatischen Abbaues hat man ein Mittel in der Hand, zwischen den strahlengeschädigten und den nichtgeschädigten Bereichen der DNA zu unterscheiden und die beiden voneinander zu trennen. Die Konzentrierung der für die Untersuchung der strahlenchemischen Veränderungen interessierenden Bereiche der Nucleinsäure auf einen Bruchteil der gesamten DNA bedeutet eine wesentliche Erleichterung der Untersuchungen und eine Vergrößerung der Nachweisempfindlichkeit. Damit kommen wir dem Ziel des Chemikers näher, die Versuche in einem für die Genetik interessanten Dosisbereich durchzuführen, wobei man bekanntlich bisher noch auf erhebliche experimentelle Schwierigkeiten gestoßen ist.

Tabelle 1
Basenzusammensetzung der einzelnen Fraktionen nach Trennung an Sephadex G 25

	Mol-% Base	Guanin	Adenin	Cytosin	Thymin Dimer	Thymin	% Spaltung
1	DNA vor Bestrahlung	20,6	29,4	20,8	-	29,2	
2	Nucleoside Frakt. I, II + III	21,3	31,4	20,4	-	26,9	91,4
3	Oligonucleotide Frakt. IV	16,7	18,5	9,1	29,3	26,4	8,6
4	Summe aus Zeile 2 u. 3	20,9	30,2	19,5	2,5	26,9	100,0

+) Fräulein Ute Güngerich möchte ich für die Durchführung zahlreicher Versuche und die technische Assistenz herzlich danken.

L I T E R A T U R

1) MARMUR,J., J. molecular Biol. 3, 208 (1961)
2) BOLLUM,F.J., R.B.SETLOW, Federat. Proc. 21, 374c (1962) 46th Ann. Meetg.
3) ZADRAZIL,S., Z.SORMOVÁ, F.SORM, Coll. czechoslov. chem. Commun. 26, 2643 (1961)
4) DELLWEG,H., A.WACKER, Z. Naturforschg. 17b, 827 (1962)
5) BURTON,K., G.B.PETERSEN, Biochem. J. 75, 17 (1960)
6) DOSKOCIL,J., F.SORM, Biochim. biophysica Acta 55, 953 (1962)
7) CHEN,P., T.Y.TORIBARA, H.WARNER, Analytic. Chem. 28, 1756 (1956)

D I S K U S S I O N S B E M E R K U N G E N

SAUERBIER.- Ich glaube, Sie können den unvollständigen Abbau Ihrer UV-bestrahlten DNA ohne weiteres erklären, wenn Sie annehmen, daß in Ihrer DN'ase noch eine geringe Endo-Nuclease-Aktivität vorhanden war und daß die Exo-Nuclease nur bis zum $\widehat{TT}$ abspaltet.

DELLWEG.- Auch die Annahme einer Endo-Nuclease erklärt nicht die Tatsache, daß Spaltstücke entstehen, die außer dimerem Thymin auch noch die anderen Basen enthalten. Offenbar kann auch die Endo-Nuclease nur dort spalten, wo zwei Dimerisierungen auf dem Strang einen bestimmten Mindestabstand überschreiten.

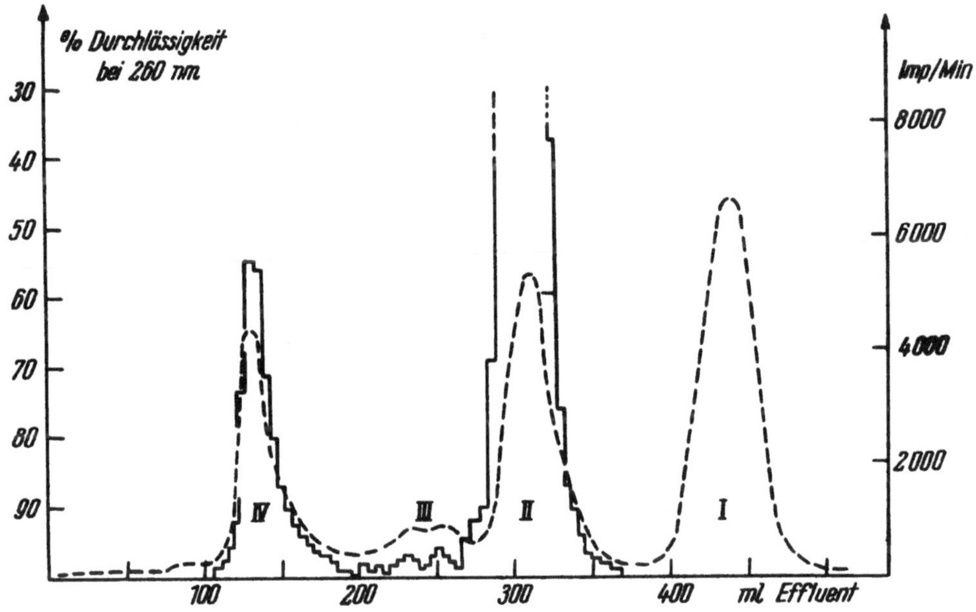

Abb.1 Trennung der Spaltprodukte des enzymatischen Abbaues an Sephadex G 25 (2,8.66 cm). Elution mit 0,01%-iger Ammoniumcarbonatlösung, pH 7,5 [3].
——— Radioaktivität in Imp./Min.
– – – – Prozentuale Durchlässigkeit bei 260 nm.

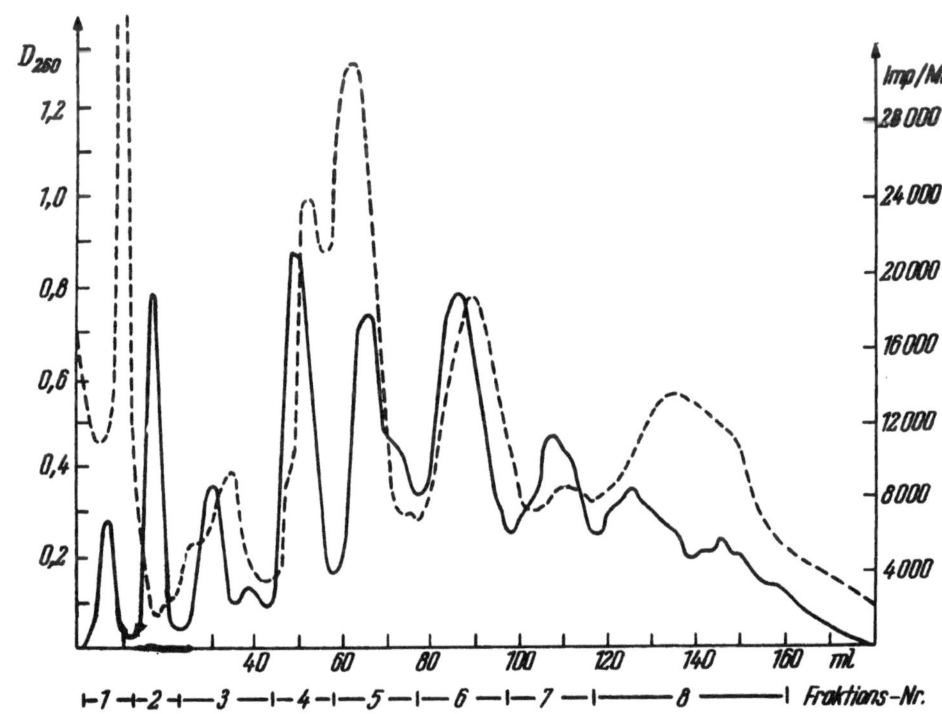

Abb.2 Hydrolyse der Fraktion IV mit 66%-iger Ameisensäure in Gegenwart von 3% Diphenylamin, 19 Std. bei 30°C [5]. Trennung der Nucleotide an Dowex 1.2 (0,36.19 cm). Elution mit einem linearen Gradienten aus Ammoniumformiat, 0,1 bis 1,2 m, pH 4,65 [6]. ——— Radioaktivität in Imp./Min.
– – – – Extinktion bei 260 nm.

D Nucleinsäuren

Zum Mechanismus der durch UV-Bestrahlung hervorgerufenen Veränderungen an DNA

H. VENNER und Ch. ZIMMER

Institut für Mikrobiologie und experimentelle Therapie der Akademie, Jena

S U M M A R Y

DNAs of various sources of different GC-content were irradiated and changes of the thermal transition were found by T_m-technique. T_m decreases with progressing irradiation time as a result of thymine dimerization. The plot of T_m versus irradiation time shows some maxima, which will be explained by cross linking-dimerization or perhaps by appearance of non-helical regions in the DNA helix. The T_m differences of equal irradiation doses exhibit a non-linear relation between the T_m difference and GC-content of the DNAs. The observed increase of the breadth of the transitions is consistent with the suggested reaction mechanism.

Bei der Bestrahlung von Desoxyribonucleinsäure (DNA) mit UV-Licht erleiden die Pyrimidine charakteristische Veränderungen, während die Purinbasen praktisch beständig sind. Wie von BEUKERS und BERENDS [1] festgestellt wurde, treten zwei Moleküle Thymin unter Ausbildung eines Zyklobutanringes zwischen Stellung 5 und 6 zu einem Dimeren zusammen, während Cytosin in das Hydrat überführt wird [2)-3]. Ähnlich wie DNA verhalten sich auch die synthetischen Polynucleotide [4]. Durch UV-Bestrahlung von Thymin im eingefrorenen Zustand gewonnenes Thymindimeres erwies sich als säurestabil, dagegen vermag Alkali in der Hitze oder kurze Einwirkung von UV-Licht das Thymindimere in wäßriger Lösung wieder in die monomere Form zurückzuverwandeln [1]. Die Dimerisierung und Entdimerisierung erwies sich nach SETLOW [4] als wellenlängenabhängig in dem Sinne, daß bei 280 nm lediglich Dimerisierung erfolgt, während bei 240 nm nur Trennung in die Monomeren zu beobachten ist. Bei der üblicherweise angewendeten Wellenlänge von 254 nm besteht ein Gleichgewicht zu Gunsten der Dimerisierung. In der DNA ist die Möglichkeit zur Dimerisierung für das Thymin dann gegeben, wenn entweder auf einem Faden der Doppelhelix zwei Thyminbasen in der Nucleotidsequenz benachbart angeordnet sind oder aber wenn je eine Thyminbase von Faden und Gegenfaden einem benachbarten Basenpaar angehören (Abb. 1). Die Thymindimerisierung im DNA-Molekül ist möglich bei vorangehender oder gleichzeitiger Lösung der Wasserstoffbrückenbindung zwischen dem entsprechenden Adenin-Thymin(AT)-Basenpaar, wenn die Dimerisierung innerhalb eines Fadens erfolgt (Fall 1). Eine Dimerisierung von Thyminmolekülen, die in Faden und Gegenfaden lokalisiert sind (cross-linking), ist nur denkbar, wenn zuvor eine Auflockerung der Helix-Struktur der benachbarten Bereiche erfolgt ist (Fall 2). In beiden Fällen kann nebenher eine Aufspaltung der GC-Wasserstoffbrücke erfolgen, die jedoch im ersten Fall ohne Zusammenhang mit der

Thymin-Dimerisierung bleibt. Bei der Thymin-Dimerisierung innerhalb eines Fadens (Fall 1) ist eine Abnahme der Helix-Stabilität zu erwarten. Im zweiten Fall kann es nach einer vorangehenden Labilisierung durch die mögliche Aufspaltung von GC-Wasserstoffbrücken zu einer Stabilitätserhöhung durch die 'cross-linking'-Dimerisierung kommen. Dabei erfolgt räumliche Deformation der ursprünglichen Helix. Im Verlaufe längerer Bestrahlung können die Thymindimeren wieder aufgespalten werden, was im Fall 2 dann zu einer weiteren Abnahme der "Helix"-Stabilität führen sollte. Bei Thymin-Dimerisierung innerhalb eines Fadens ist kein Effekt einer nachträglichen Entdimerisierung auf den bereits vorhandenen molekularen Zustand hinsichtlich der Helix-Stabilität zu erwarten. Bereits von MARMUR [5] wurde die Veränderung der Helix-Stabilität der DNA bei UV-Bestrahlung mit Hilfe der Messung der thermischen Stabilität (T_m-Technik) untersucht. Dabei wurde rasche Abnahme der Stabilität gefunden. Ziel der vorliegenden Arbeit war es, nachzuprüfen, ob evtl. Hinweise für eine im Verlaufe der Bestrahlung auftretende Thymin-Dimerisierung von Faden zu Gegenfaden (cross-linking) zu beobachten war. Demgemäß wurden Nucleinsäuren verschiedenen GC-Gehaltes aus Mikroorganismen und Heringssperma sowie Kalbsthymus-DNA mit UV-Licht der Wellenlänge 254 nm bestrahlt [6] (Quecksilberniederdruckbrenner Sterisol-Lampe, Heraeus Hanau; Dosis ähnlich den Untersuchungen von MARMUR [5] bei 1,89 x 10^4 erg/mm^2/min). Nach steigenden Bestrahlungszeiten wurde dann der T_{mUV}-Wert bestimmt. Die verwendeten Nucleinsäuren hatten folgende Gehalte an Guanin + Cytosin:

Herkunft	Mol-% G + C
Proteus mirabilis	42
Streptomyces chrysomallus	71
Heringssperma	46
Kalbsthymus	41

In Übereinstimmung mit MARMUR [5] zeigten alle Nucleinsäureproben einen Abfall des T_m-Wertes, der vor allem bei Nucleinsäuren mit höherem Gehalt an Adenin + Thymin stärker ist. Jedoch beobachteten wir im Bereich von etwa 4 - 12 Minuten Anomalien (Maxima) im Kurvenverlauf, die gut reproduzierbar waren (Abb. 2). Das Auftreten dieser Maxima ist nur erklärbar, wenn man annimmt, daß entweder eine Thymindimerisierung zwischen Faden und Gegenfaden (cross-linking) auftritt und/oder in der DNA bereits zuvor helikale und nichthelikale Bezirke vorhanden waren, die zu einer unterschiedlich raschen Thymin-Dimerisierung Anlaß geben können. Weiterhin sollte man erwarten, daß mit zunehmendem Thymingehalt der DNA-Proben die Dimerisierungsreaktion bei gleichen Bestrahlungsbedingungen zunimmt, wobei die Beziehung auf Grund des Zusammenhanges - mögliche Thyminbenachbarung (Sequenz-benachbartes Thymin) in Abhängigkeit vom Thymingehalt - nicht linear sein kann. In Abb. 3 ist der T_m-Wert ebenfalls für DNA-Proben verschiedenen GC-Gehaltes bei jeweils gleichen Bestrahlungsdosen in Abhängigkeit vom GC-Gehalt aufgetragen, wobei aus dem von MARMUR in [5] angegebenen Diagramm die entsprechenden T_m-Differenzen errechnet und mit in Abb. 3 einbezogen wurden. Wie ersichtlich, existiert eine nicht lineare Beziehung, die durch einen raschen Anstieg der T_m-Differenz im Bereich von niederen GC-Gehalten (hoher AT-Gehalt) gekennzeichnet ist. Die von MARMUR [5] und die in der vorliegenden Arbeit beobachteten Werte zeigen dabei eine gute Korrelation. Weiterhin wurde die Heterogenität (2 σ) aus den thermischen Übergängen errechnet. Bei einer Auftragung dieser Kenngröße gegen die Bestrahlungszeit resultieren für die verschiedenen Nucleinsäureproben Kurvenzüge unterschiedlicher Steilheit, die wiederum eine deutliche Abhängigkeit vom GC-Gehalt aufweisen. Charakteristisch ist der geringe Endanstieg bei einer DNA extrem hohen GC-Gehaltes (Streptomyces chrysomallus). Auch die Heterogenitätszunahme, insbesondere bei Nucleinsäure mit hohem Thymingehalt, spricht für die Thymin-Dimerisierung als Ursache der Abnahme der DNA-Helix-Stabilität. Die Versuche sollen fortgesetzt werden, um weitere Einblicke in den Reaktionsmechanismus zu erhalten.

L I T E R A T U R

1) BEUKERS,W., W.BERENDS, Biochim. biophysica Acta 49, 181 (1961)
BEUKERS,W., W.BERENDS, Rec. trav. Chim. 77, 729 (1958)
2) SHUGAR,D., in "The Nucleic Acids", edited by E. CHARGAFF and J.N.DAVIDSON, Academic Press New York, 1960, Vol. 3, S. 40
3) WACKER,A., in "Progress in Nucleic Acid Research", edited by J.N.DAVIDSON and W.E.COHN, Academic Press New York, 1963, Vol. 1, S. 369
4) SETLOW,R., J.SETLOW, Proc. Nat. Acad. Sci. USA 48, 1250 (1962)
5) MARMUR,J., W.F.ANDERSON, L.MATTHEWS, K.BERNS, E.GAJEWSKA, P.DOTY, J. cellular comp. Physiol. 58, Suppl. 1, 33 (1961)
6) VENNER,H., Ch.ZIMMER, R.HORN, Naturwissenschaften 49, 542 (1962)

D I S K U S S I O N S B E M E R K U N G

ROWND bestätigt auf Grund eigener, unveröffentlichter Befunde das Auftreten geringer Anteile nicht helikaler Bezirke in der DNA als mögliche Ursache für das Auftreten von Maxima der T_m-Werte in Abhängigkeit von der Dosis. Weitere Hinweise dafür erhielt ROWND auf Grund der Untersuchung der DNA mit Hilfe der CsCl- und Cs_2SO_4-Dichtegradienten-Sedimentation.

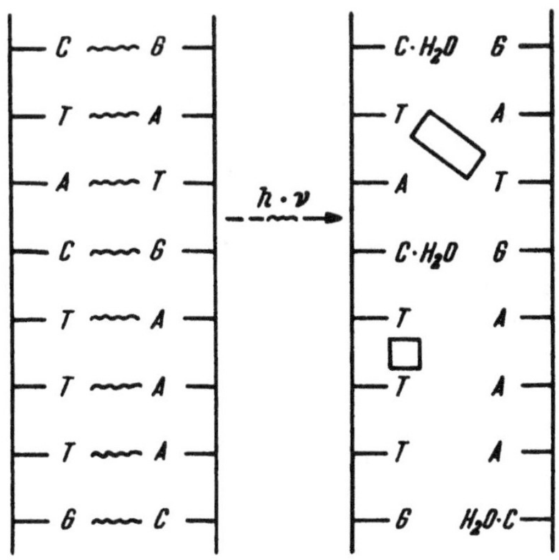

Abb.1 Schematische Darstellung der Wirkung von UV-Strahlen auf einen Abschnitt des DNA-Doppelfadens.

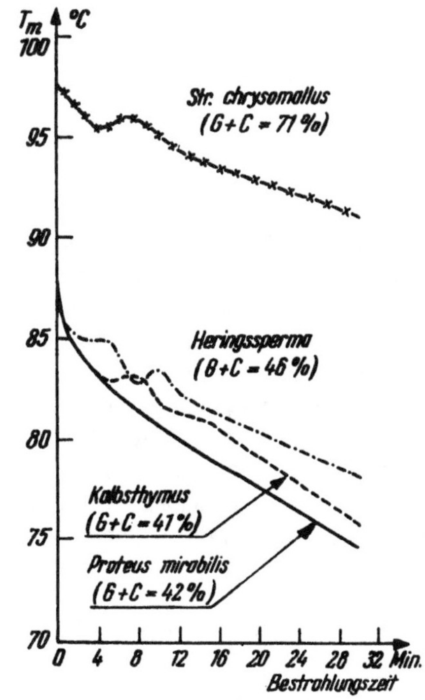

Abb.2 Wirkung von UV-Licht (254 nm) auf den T_m- Wert von DNA-Proben verschiedenen GC-Gehaltes.

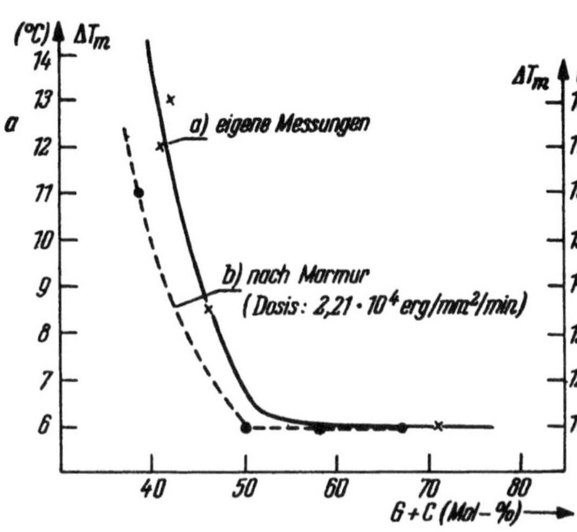

Abb.3 Abnahme der T_m-Werte (ΔT_m) nach UV-Bestrahlung mit gleichen Dosen. Intensität und Bestrahlungszeit waren in allen Fällen gleich.

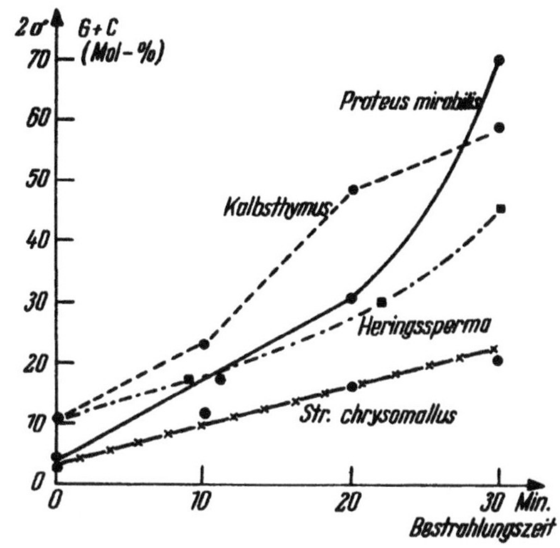

Abb.4 Wirkung von UV-Strahlen auf die Heterogenität (2σ) der Zusammensetzung der DNA verschiedenen GC-Gehaltes.

D Nucleinsäuren

Free Radicals in γ-Irradiated Deoxyribonucleic Acid

G. LÖFROTH, L. EHRENBERG and A. EHRENBERG

Institute of Radiobiology of the University, Stockholm and Karolinska Institute, Stockholm

S U M M A R Y

γ-irradiation of pure, dry and O_2-free DNA gives an ESR spectrum with a high degree of similarity to those of thymine, thymidine, and thymidylic acid, but different from those of other simple nucleic acid constituents and their analogues. In the presence of O_2 a simple spectrum is obtained.

I N T R O D U C T I O N

From some points of view, especially the slowness of certain initial reactions, plant seeds represent a suitable subject of investigation in radiobiological studies. In this material water content, oxygen concentration and temperature have been found to affect strongly the degree of radiation damage (growth inhibition, chromosomal aberration, mutation, etc). At about 12 percent water, corresponding to about one layer of hydration water in proteins and perhaps other hydrophilic macromolecules, a radiosensitivity minimum is found. The higher sensitivity at lower water contents seems to be due chiefly to an oxygen conditioned after effect. At these lower water contents free radicals produced by irradiation are fairly stable. The fact that a nitric oxide treatment, decreasing the radical concentration to zero, also extinguishes the influence of water content on the radiosensitivity indicates the importance of free radicals in the production of lesions [1].

Among other methods electron spin resonance (ESR) has therefore been introduced as a tool in the evaluation of the mechanisms of action of radiations. Apart from living tissue, macromolecules of biologic importance (proteins, nucleic acids and polysaccharides) as well as their simple building stones are studied in this respect [2)-4)].

In the present communication the results from some preliminary experiments with DNA and its constituents will be reported.

M A T E R I A L S   A N D   M E T H O D S

A calf thymus DNA of high purity ($<$0.1 percent protein) was prepared according to HAMMARSTEN [5]. Before irradiation samples were evacuated and washed repeatedly with oxygen-free nitrogen. ^{60}Co γ-irradiation at a dose rate of 0.4 Mrad/h was used. All irradiations and measurements were performed at + 20° C. ESR measurements were made with a Varian spectrometer having a 100 kc/s modulation unit [3]. The ESR spectrum of DNA was compared with those of all commercially available bases, nucleosides, and nucleotides.

R E S U L T S

The fine structure of the ESR spectrum of DNA shows great similarities with the spectra of thymine, thymidine, and thymidylic acid, whereas all other simple components, adenine, adenosine, deoxyadenosine, adenylic acid, guanine, guanosine, deoxyguanosine, cytosine, cytidine, cytidylic acid, deoxycytidylic acid, uracil, uridine, deoxyuridine, uridylic acid, 5-hydroxymethylcytosine, 5-methylcytosine, 5-methyldeoxycytidine, 5-hydroxymethyluracil, and 5.6-dimethyl-2.4-dioxypyrimidine, gives fundamentally deviating spectra.

A similar spectrum was obtained with the sodium salt of a highly polymeric, pure DNA (from Calbiochem. Calif., USA), whereas a fairly degraded, impure (2-5 percent protein) DNA from herring sperm gave a simple absorption line spectrum.

The radicals in DNA are sensitive to water (not yet studied in detail) and oxygen.

Earlier investigators have found different and often more or less simple ESR spectra in irradiated DNA [6)-10)]. It is possible that the reasons for this have to be ascribed to such factors as: impurities, oxygen, water, and high dose. On the other hand the spectra of crystallizable components agree with those of other authors.

C O N C L U S I O N S

The investigation indicates that thymine is the main centre of stabilization of radicals. This is in agreement with calculations of PULLMAN and PULLMAN [11], and is also of interest considering the fact that thymine dimerization has been found as the main change in DNA of UV-irradiated bacteria [12][13]. In a recent publication TOKARSKAYA [14] "gives confirmation to the assumption that thymine is the point of injury after having studied effects of ionizing radiation on dry seeds and isolated DNA".

A C K N O W L E D G E M E N T

The present work was supported by grants from Riksföreningen mot Cancer, Statens Råd för Atomforskning, Knut och Alice Wallenbergs Stiftelse, and a Public Health Service grant AM-05895 from the National Institute of Arthritis and Metabolic Diseases.

L I T E R A T U R E

1) SPARRMAN,B., L.EHRENBERG, A.EHRENBERG, Acta chem. scand. 13, 199 (1959)
2) EHRENBERG,A., L.EHRENBERG, Ark. Fysik 14, 133 (1958)
3) LÖFROTH,G., L.EHRENBERG, A.EHRENBERG, The Fifth International Symposium on Free Radicals, July 6-7, 1961. Almqvist & Wiksell, Stockholm. Paper No. 38

4) LÖFROTH,G., Acta chem. scand. (in press)
5) HAMMARSTEN,E., Biochem. Z. 144, 383 (1924)
6) SHIELDS,H., W.GORDY, Proc. nat. Acad. Sci. USA 45, 269 (1959)
7) GORDY,W., Radiat. Res., Suppl. 1, 491 (1959)
8) ALEXANDER,P., J.J.LETT, M.G.ORMEROD, Biochem. biophysica Acta 51, 207 (1961)
9) SHEN-PEI-GUEN, L.A.BLUMENFELD, A.E.KALMANSON, A.G.PASYNSKI, Biofisika 4, 265 (1959)
10) SHEN-PEI-GUEN, L.A.BLUMENFELD, A.E.KALMANSON, A.G.PASYNSKI, Biofisika 5, 534 (1961)
11) PULLMAN,B., A.PULLMAN, in M.BURTON et al. (eds.) Comparative Effects of Radiation, John Wiley Sons, Inc., New York, p. 105 (1960)
12) WACKER,A., H.DELLWEG, D.WEINBLUM, Naturwissenschaften 47, 477 (1960)
13) WACKER,A., H.DELLWEG, E.LODEMAN, Angew. Chem. 73, 64 (1961)
14) TOKARSKAYA,V.I., Second Intern. Congress of Radiat. Res., Abstracts of papers, p. 15 (1962)

D I S C U S S I O N - Q U E S T I O N S

BLUMENFELD. - What was the separation between hfs components in your spectra?

LÖFROTH. - About 135 Gauss between the outermost satellites.

BLUMENFELD. - Have you any structural interpretation for the hfs pattern?

LÖFROTH. - No.

D Nucleinsäuren

Physikochemische Untersuchung der Strahlenwirkung auf DNA, Proteine und Fermente und der Schutzmechanismus mit Hilfe von Inhibitoren von Radikalprozessen

N. M. EMANUEL, K. KRUGLJAKOWA, E. BURLAKOWA, I. SAPESHINSKI, N. NIKOLAJEWA, W. LUSIKOW und J. SILAJEW

Institut für chemische Physik der Akademie, Moskau

S U M M A R Y

1. The protective degrees of few inhibitors of the free-radical reactions with respect to DNA radiation damage have been obtained by the viscosity measurements. The oxypyridines and the gallates are shown to possess the therapeutic and protective abilities.

2. The direct dependence between the protective degree of DNA and the inhibitor antiradical activities has been demonstrated.

3. The changes in the configuration and the molecular weight with and without the propylgallate addition in DNA irradiated with x-rays (300 kr) have been investigated by the electron microscopy and sedimentation methods. The propylgallate is shown to prevent the DNA decay with 32 per cent efficiency.

4. The chemiluminescence of irradiated DNA in the temperature range above 45° C has been discovered.

The velocity constant and activation energy of the first stage of chemiluminescence reaction agrees well with those of the DNA hydroperoxides decay.

5. The inhibition of the heat inactivation of x-ray irradiated trypsin with propylgallate and the absence of that in the case of UV-light irradiation have been demonstrated.

6. The chemiluminescence of the recombination of radicals of irradiated protein has been studied. The protein radicals R transformation into the peroxy radicals RO_2 at the presence of oxygen have been proved by the chemiluminescence and ESR methods. The interaction of some inhibitors of the free radical reactions with the biological peroxy-radicals has been demonstrated by ESR method.

Die Suche rationeller Prinzipien zur Schaffung wirksamer Strahlenschutzpräparate auf der Grundlage von Ergebnissen über den detaillierten Mechanismus der Strahlenschädigung ist eine der Hauptrichtungen der modernen Radiobiologie. Die Tatsache der Bildung freier Radikale bei der Strahlenwirkung auf die wesentlichsten biologisch bedeutsamen Verbindungen wurde in vielen experimentellen Arbeiten unterstrichen[1)-5)]. Es ist sehr wahrscheinlich, daß gerade diese Zustände freier Radikale im Organismus zur Entwicklung pathologischer Prozesse führen, wie sie der Strahlenschädigung eigen sind.

Aus diesem Grunde wurden seit Beginn 1957 von uns Untersuchungen durchgeführt, die auf die Klärung der möglichen Anwendung wenig toxischer Hemmstoffe von Radikalprozessen (in erster Linie Phenole) für die Unterdrückung oder Abschwächung der akuten

Strahlenkrankheiten bei Tieren gerichtet waren, darüber hinaus wurde die Wirkung von Inhibitoren auf Modellsysteme, die biochemische Verschiebungen im bestrahlten Organismus darstellen sollten, untersucht [6)-8)]. Wir verfolgten mit verschiedenen physikochemischen Methoden den Strahlenabbau der DNA, die Wirkung der Strahlung auf Proteine und Fermente wie auch den Einfluß von Inhibitoren der Radikalreaktionen auf diese Prozesse.

Bei der Untersuchung des Strahlenabbaues der DNA wurde gefunden, daß die Inhibitoren wesentlich das Ausmaß der strahlenbedingten Schädigung der DNA verringern können, sowohl bei Zusatz vor als auch nach der Bestrahlung [8)]. Es wurden chemische Verbindungen aus den Klassen der Phenole, Amine und Oxypyridine untersucht. In den Versuchen wurde native, hochpolymere DNA verwendet, die aus Rattenmilz nach der Methode von ZUBAY und DOTY [9)] bei Deproteinisierung nach KAY [10)] isoliert wurde. Alle verwendeten Inhibitoren wurden in äquimolaren Konzentrationen zur Konzentration des Propylgallats genommen, das einen 100 %igen Schutz der DNA bei Bestrahlung gewährt. Für alle Verbindungen wurden kinetische Kurven der veränderten relativen Viskosität der DNA-Lösungen nach der Bestrahlung in Gegenwart der Präparate und bei ihrem Zusatz nach der Bestrahlung erhalten. Derartige kinetische Kurven für die Derivate des 3-Oxypyridins sind in Abb. 1 und 2 dargestellt. Sie stimmen gut mit der Kontroll-Kurve bei einfacher Veränderung des Ordinatenmaßstabs überein. Dies weist darauf hin, daß alle Kurven infolge eines einheitlichen Wirkungsmechanismus der untersuchten Verbindungen ähnlich sind. Für die Einschätzung des Schutzgrades der DNA vor radioaktivem Zerfall wurde die Veränderung der Größe der relativen Viskosität δ bei Zusatz des Inhibitors verwendet. Es ist zu sehen, daß alle untersuchten Hemmstoffe bei Zusatz vor der Bestrahlung die DNA im Mittel um 50 - 100 % schützen, während die Derivate der Gallensäure und des 3-Oxypyridins eine Schutzwirkung auch bei Zusatz nach der Bestrahlung äußern. Diese Ergebnisse wurden in Versuchen mit Versuchstieren bestätigt [11)-13)].

Da die Aktivität der verwendeten Inhibitoren verschieden war, mußte geklärt werden, ob eine Beziehung zwischen dem Schutzgrad und der Antiradikal-Aktivität des Inhibitors besteht. Um so mehr als in einer Arbeit unseres Laboratoriums [13)] eine lineare Abhängigkeit zwischen der Überlebensrate der Versuchstiere und der Antiradikalwirksamkeit des Inhibitors A gefunden wurde ($A = \epsilon \cdot c$, wo ϵ die relative Wirksamkeit des Hemmstoffs und c seine Konzentration sind). Die gleiche Abhängigkeit wurde von uns in Versuchen mit DNA bei Zusatz der Inhibitoren vor und nach der Bestrahlung erhalten (Abb. 3). Die Ergebnisse bezüglich des Wirkungsmechanismus der Hemmstoffe auf molekularem Niveau wurden bei der Untersuchung der Wirkung von Pyrogallat auf den Strahlenzerfall der DNA mit Hilfe der Elektronenmikroskopie und der Sedimentation in der Ultrazentrifuge erhalten. Bei der Herstellung der für die Elektronenmikroskopie verwandten Proben wurde eine von uns in einer unserer Arbeiten beschriebene Methode verwendet [15),16)]. Auf der Mikrofotografie der nativen DNA (Abb. 4a) sind lange DNA-Moleküle mit einer Stärke von etwa 20 Å zu erkennen. Nach Bestrahlung mit $3 \cdot 10^5$ r zerfallen die DNA-Moleküle in eine große Anzahl kurzer Stücke, gleichzeitig bilden sich Klümpchen im Ergebnis der Denaturierung (Abb. 4b). Bei Zusatz von Propylgallat verringert sich der Grad der Schädigung der DNA-Moleküle. Neben kleinen Fragmenten von DNA-Molekülen kann man auch eine mehr oder weniger große Anzahl größerer Moleküle erkennen (Abb. 4c).

Äußerlich unterscheiden sich diese Moleküle von dem Molekül der nativen DNA durch einige Krümmungen. Man kann im Moment schwer sagen, ob diese mit starken Schädigungen des Ausgangsmoleküls verbunden sind oder ob sie das Ergebnis der Vereinigung kleinerer DNA-Fragmente darstellen.

Für die Untersuchung der Veränderung des DNA-Molekulargewichts bei der Bestrahlung mit und ohne Propylgallat wurde die Sedimentationsmethode in der Ultrazentrifuge angewendet. Die Arbeit wurde mit der analytischen Ultrazentrifuge "Spinco", Modell E, bei 59 780 U/Min (259 700 g) durchgeführt. Der Verlauf der Sedimentation wurde mit Hilfe der UV-Absorption (258 nm) registriert. Daraus wurden die Sedimentationskoeffizienten errechnet und ihre Verteilung bei der Konzentration 0 graphisch dargestellt (Abb. 5). Aus den Sedimentationskonstanten errechneten wir die mittleren Molekulargewichte nach der Formel von EIGNER und stellten die Verteilung der Molekulargewichte graphisch dar. In Abb. 6 ist die Verteilung der Molekulargewichte für die in Gegenwart von Propylgallat bestrahlte und unbestrahlte sowie die native DNA aufgezeichnet. Man kann sehen, daß der Zusatz von Propylgallat vor der Bestrahlung der polymeren DNA mit einem mittleren Molekulargewicht von $6{,}9 \cdot 10^6$ das Molekulargewicht bis zu 38 % erhält, während die ohne Hemmstoff bestrahlte DNA ein Molekulargewicht aufweist, das nur 12 % dessen der Ausgangs-DNA beträgt.

In der Beziehung ist die Tatsache enthalten, daß die Strahlungsnachwirkung vom Zerfall von Peroxydverbindungen abhängt. Daher begannen wir die Untersuchung der Kinetik bei der Bildung und dem Zerfall von Peroxydverbindungen mit Hilfe von chemischen Methoden und Chemolumineszenzmethoden. Hierbei setzten wir voraus, daß, wenn der Zerfall der Peroxydverbindung der DNA unter Teilnahme freier Radikale abläuft, die Entstehung einer Lumineszenz in den bestrahlten DNA-Lösungen möglich sein müßte. Eine derartige Lumineszenz beobachteten wir tatsächlich in unseren Versuchen (Abb. 7). Dieses Leuchten, das in bestrahlten DNA-Lösungen bei Erwärmung entsteht, wurde mit dem Photomultiplier FEU-29 registriert (400 - 600 nm). Es ist zu sehen, daß der schnelle Abfall der Intensität durch einen wesentlich langsameren abgelöst wird, was auf den mehrphasigen Charakter dieses Prozesses hinweist. Man kann die Konstante K_1 des schnelleren Prozesses errechnen, sie hat bei 75° den Wert $1{,}2 \pm 0{,}3 \cdot 10^{-2}\ s^{-1}$ und stimmt gut mit dem mit Hilfe der chemischen Methode gefundenen überein. Die zweite Phase des Intensitätsabfalls der Lumineszenz ist wahrscheinlich damit verbunden, daß in den Reaktionen Wasserstoffperoxyd teilnimmt, dessen Konzentration sich im Verlauf längerer Zeit langsam verändert (80 - 100 Min).

Die Aufnahme von Kurven der Lumineszenzveränderung bei verschiedenen Temperaturen erlaubte die Werte der Aktivierungsenergie der ersten Phase zu beurteilen. Nach den Ergebnissen der Chemolumineszenz wurde der Wert $E_1 = 11 \pm 2$ kcal/Mol gefunden, der gut mit der spektrophotometrisch gefundenen Größe für die Aktivierungsenergie übereinstimmt. Ausgehend von den Vorstellungen über die Natur der latenten Schädigungen durch freie Radikale untersuchten wir die Wirkung des Inhibitors bei Reaktionen freier Radikale (Propyläther der Gallensäure) auf den Prozeß der Wärmeinaktivierung des Trypsins, das mit Röntgen und UV bestrahlt wurde.

In Abb. 8 sind die kinetischen Inaktivierungskurven bei $73{,}5^\circ$ C vom nativen (I) und vom röntgenbestrahlten (5) Trypsin angeführt. Die Kurve (5) hat im Unterschied zur Kurve (I) eine komplizierte Form. Auf dem gleichen Bild ist der Einfluß des Propylgallats (PG) auf die Wärmenachwirkung gezeigt, das zur bestrahlten Lösung vor dem Erwärmen zugesetzt wurde (4). Die Konzentration des PG war in der Lösung 10^{-3} m (in dieser Konzentration beeinflußt der PG praktisch die Wärmedenaturierung des nativen Trypsins nicht). Wie auf der Abbildung zu sehen ist, kommt der Schutzeffekt des PG auf die Wärmenachwirkung in röntgenbestrahlten Trypsinlösungen klar zum Ausdruck. Analoge Ergebnisse wurden für eine Temperatur von $96{,}8^\circ$ C erhalten. Im Falle einer Bestrahlung mit UV wurde dieser Schutz nicht beobachtet.

Die Wirkung des Propylgallats kann man auf zweierlei Art und Weise erklären: entweder reduziert das PG die latente Schädigung oder es verringert in den bestrahlten

Proteinlösungen den Einfluß langlebiger aktiver Produkte auf die Wärmeinaktivierung des nativen Fermentes. Die Untersuchung von nach der Bestrahlung in flüssigem Stickstoff eingefrorenen Trypsinlösungen mit Hilfe der EPR-Methode zeigte, daß, auch wenn unter den erwähnten Bedingungen Radikale vorkommen, ihre Anzahl um ein Vielfaches geringer ist als die auf Grund der erhaltenen kinetischen Werte über die Nachwirkung zu erwarten wären.

Für die Beweisführung zur zweiten Überlegung wurden folgende Versuche durchgeführt. Die Trypsinlösungen wurden mit einer Dosis von $5 \cdot 10^5$ r unter den oben erwähnten Bedingungen bestrahlt. Hierbei betrug die Restaktivität weniger als 2 (der pH-Wert der Lösung veränderte sich unwesentlich bei der Bestrahlung). Die Lösungen des bestrahlten und des nativen Trypsins wurden in die Kälte gebracht und das Gemisch bei einer bestimmten Temperatur inkubiert. Die Versuchsergebnisse sind in Abb. 9 dargestellt. Es erwies sich, daß die mit Röntgenstrahlen behandelten Trypsinlösungen stark die Wärmeinaktivierung des nativen Proteins beeinflussen können. Dabei zeigt es sich, daß eine zweistündige Erwärmung der bestrahlten Trypsinlösung bei 96,8° C vor Zusatz derselben zum nativen Trypsin wesentlich den erwähnten Effekt verringert. Noch deutlicher wirksam wird die Erwärmung in Gegenwart von PG (Konzentration 10^{-3} m). Das vorherige Zusammenbringen der mit einer Dosis von $5 \cdot 10^5$ r bestrahlten Trypsinlösung mit Katalase bis zur völligen Entfernung des Wasserstoffperoxyds (bei 0° C) zeigte, daß ein Teil des beobachteten Effektes mit der Wirkung des bei der Bestrahlung gebildeten Peroxyds auf das native Trypsin bei Erwärmen erklärt werden kann, ein anderer Teil mit der Wirkung des bestrahlten Trypsins (Kurve 5, Abb. 9). Wir beobachteten auch, daß bestrahlte RNase-Lösungen in 10^{-3} n HCl wirksamer das native Trypsin inaktivieren als - 10^{-3} n HCl bei der gleichen Strahlendosis. Abb. 8 (Kurven 2 und 3) zeigt den Schutz von bestrahlten Trypsinlösungen durch Propylgallat in Abwesenheit von Wasserstoffperoxyd. Die Schutzwirkung ist hier gering. Wahrscheinlich ist sie mit der Beseitigung der aktiven Gruppen im bestrahlten Trypsin verbunden. Die Schutzwirkung wird wesentlich deutlicher, wenn in der Lösung der Gehalt dieser aktiven Gruppen (bei Erhöhung der Strahlungsdosis) erhöht wird. Zur Klärung der Rolle der Hydroperoxyde und stabiler freier Radikale, die sich im Protein bei Bestrahlung bilden können, führten wir in der Reaktion der Wärmenachwirkung Versuche mit dem Hydroperoxyd des tertiären Butyls und stabilen Radikalen folgender Strukturen durch. Die Ergebnisse dieser Versuche sind in Abb. 10 dargestellt. Es ist ersichtlich, daß die Radikale und Hydroperoxyde an der Thermoinaktivierung des Trypsins beteiligt sein können, wobei man sie auf Grund ihrer Wirksamkeit in folgende Reihe einordnen kann: Wasserstoffperoxyd, Hydroperoxyd, stabiles Radikal. Die EPR-Ergebnisse zeigten, daß unter den Versuchsbedingungen zur Trypsin-Inaktivierung die Radikale P-1 und P-2 nicht auf Kosten einer Rekombination oder anderer Radikal-Reaktionen verschwinden. In Bezug auf die Auswahl zwischen Radikal- und Molekularmechanismen der Inaktivierung kann man im Moment noch keinerlei Schlußfolgerungen treffen.

Insgesamt kann der Effekt der Wärmenachwirkung nicht nur der Inaktivierung des nativen Trypsins bei Erwärmen mit Wasserstoffperoxyd und bestrahltem Eiweiß zugeschrieben werden, da die Reaktionskinetik der Nachwirkung sich von der Kinetik dieser beiden Reaktionen unterscheidet. Es ergeben sich also im Prozeß der Wärmenachwirkung in verdünnten röntgenbestrahlten Trypsinlösungen mindestens drei Reaktionen: die eigentliche Nachwirkung (oder "wahre" Nachwirkung), die wahrscheinlich nicht vom typischen Inhibitor der Reaktionen freier Radikale - dem Propylgallat - beeinflußt wird, die Inaktivierung des nativen Trypsins durch Wasserstoffperoxyd [26] und durch bestrahltes Eiweiß ("Pseudo-Nachwirkung"). Die beiden letzten Reaktionen werden durch Propylgallat unterdrückt. Wahrscheinlich stellen die Hydroperoxyde in der dritten Reaktion das aktive Prinzip dar, da sie wie bekannt [25] bei Bestrahlung des Trypsins stark angereichert werden.

Die angeführten Ergebnisse zeugen davon, daß die Inhibitoren von Radikal-Reaktionen biologische Makromoleküle vor der Strahlenwirkung schützen. Vom Gesichtspunkt des Mechanismus der Schutzwirkung muß man feststellen, ob die Inhibitoren von Radikalprozessen in Austauschreaktionen mit den Radikalen der bestrahlten Biopolymere treten, ähnlich wie dies bei chemischen Radikalreaktionen der Fall ist. Es wurden Versuche mit bestrahltem Serumalbumin unter Anwendung zweier paralleler Methoden - EPR und Chemolumineszenz - durchgeführt. In Vorversuchen war das Entstehen einer schwachen Lumineszenz beim Lösen der bestrahlten Biopolymere festgestellt worden. Die Versuchsdurchführung bestand in folgendem. Zum bestrahlten Serumalbumin wurden die Lösungsmittel zugesetzt und die Verringerung der Radikale mit Hilfe des EPR-Gerätes sowie die Dämpfungskinetik der Lumineszenz mit Hilfe eines Gerätes zur Registrierung schwacher Lumineszenz untersucht. Das Protein wurde in vorher evakuierten Ampullen in der γ-Strahlenanlage GUT-Co-400 mit Dosen von 5 - 20 mrad bestrahlt. Bei der Versuchsdurchführung der EPR wurde das bestrahlte Protein in eine Ampulle überführt und das Spektrum aufgenommen (EPR). Danach wurde in der Ampulle das Lösungsmittel zugesetzt (Wasser oder Eisessig) und die Verringerung der Anzahl der Radikale zeitlich registriert. Zur Untersuchung der Kinetik der Chemolumineszenz wurde das bestrahlte Protein in ein vor dem Photomultiplayer befindliches Reaktionsgefäß gebracht. Der Photostrom wurde verstärkt und mit Hilfe eines elektronischen Potentiometers die kinetischen Kurven der nach Zusatz des Lösungsmittels zum bestrahlten Protein beginnenden Lumineszenz registriert. Die Möglichkeit der Versuchsdurchführung sowohl im Vakuum als auch an der Luft war vorgesehen gewesen. Die kinetischen Kurven, die mit dem EPR-Gerät erhalten wurden, und die Chemolumineszenzkurven, die in Vakuumversuchen erhalten wurden, zeigten eine Abnahme der Proteinradikale entsprechend den kinetischen Gleichungen der Reaktion 1. Ordnung:

$$R = R_0 \cdot e^{-K_1 t} \quad \text{und} \quad I = I\,max \cdot e^{-K_2 t}$$

Dabei war in verschiedenen Versuchen K_1 gleich:
$1 - 2{,}3 \cdot 10^{-2}$ Sek^{-1}, der Wert von K_2 war $1{,}3 - 2{,}5 \cdot 10^{-2}$ Sek^{-1}.
Die Übereinstimmung der mit Hilfe der EPR-Methode und der Chemolumineszenz erhaltenen Werte zeugt davon, daß für die Lumineszenz die Radikale des bestrahlten Proteins, die bei der Lösung frei werden, verantwortlich sind (die Lumineszenz entsteht bei ihrer Rekombination). In der Arbeit [27] konnte gezeigt werden, daß bei Austausch der Radikale R' durch RO_2' im Prozeß der Kohlenwasserstoffoxydation die Intensität der Chemolumineszenz stark ansteigt.

Man könnte erwarten, daß auch in unserem Falle bei Zusatz von Sauerstoff sich die Lumineszenzintensität verändert. Aus Abb. 11 ist die Abhängigkeit der Chemolumineszenzintensität von der Zeit nach Zusatz von Wasser zum bestrahlten Protein im Vakuum dargestellt. Im Verlaufe des Prozesses wurde das System belüftet. Wie man auf der Abbildung sieht, erhöhte sich die Lumineszenzintensität um das mehrfache, was zugunsten der Annahme über den Austausch der Radikale R' durch RO_2' spricht. Bei Vorhandensein der Radikale RO_2' im System könnte man erwarten, daß die Inhibitoren der Prozesse freier Radikale vom Typ des Ionols (2.6-Di-tert.-butyl-4-methylphenol) mit solchen Radikalen reagieren werden. Es erwies sich, daß das zum System bestrahltes Eiweiß - Eisessig zugesetzte Ionol stark die Intensität der Chemolumineszenz verringert.

Ein direkter Beweis für die Existenz der Reaktion

$$RO_2' + InH \longrightarrow In' + RO_2H$$

wäre im EPR-Spektrum das Auftreten des Ionolradikals nach Verschwinden des Radikals des bestrahlten Serumalbumins. In der Tat konnte in einer Reihe von Fällen von uns das EPR-Spektrum des Ionols aufgenommen werden [28].

Die Gegenüberstellung des EPR-Spektrums des Ionols nach der Reaktion mit den Proteinradikalen mit dem Spektrum des im Bleitetraacetat gewonnenen Ionolradikals [24] ist auf Abb. 12 dargestellt.

Die Verwendung von Inhibitoren von Radikalreaktionen als Strahlenschutzmittel gibt also die Möglichkeit, den primären Mechanismus der Strahleneinwirkung auf wichtigste biochemische Zellkomponenten zu untersuchen.

R E F E R E N C E S

1) GORDY,W., W.ARD, H.SHIELDS, Proc. nat. Acad. Sci. USA 41, N 11, 283 (1955)
2) BLUMENFELD,L.A., Dokl. Akad. Nauk SSSR 117, N 1, 72 (1957)
3) MCCORMIC,J., W.GORDY, J. physic. Chem. 62, N 7, 784 (1958)
4) GORDY,W., Proc. nat. Acad. Sci. USA 46, N 8, 1124 (1960)
5) PEN-GEN, Shen, L.A.BLUMENFELD, A.E.CALMANSON, A.G.PASINSKY, Biophysika 4, B. 3, 263 (1959)
6) BURLAKOVA,E.B., D.G.DZANTIEV, G.B.SERGEEV, N.M.EMANUEL, Reports of Conference on problem: "Biochemical and physico-chemical basis of biological effects of the ionizing radiation" 25-28 Febr. 1957, M. 1957
7) EMANUEL,N.M., Proceedings of Moscow Society of Naturalists, 7, 73 (1963)
8) EMANUEL,N.M., K.E.KRUGLYAKOVA, N.A.ZAKHAROVA, I.I.SAPEZHINSKII, Dokl. Akad. Nauk SSSR 13, 6, 1451 (1960)
9) DOTY, P., ZUBAY,G., J. Amer. chem. Soc. 78, 6207 (1956)
10) KAY,E.R.M., N.S.SIMMONS, D.Z.DOUNCE, J. Amer. chem. Soc. 74, 1724 (1952)
11) ERSHOFF,D.H., C.W.STEERS, Proc. Soc. exper. biol. Med. 104, 2, 274 (1960)
12) POZZA,F., G.VERZA, Boll. Soc. ital. biol. sperim. 36, 20, 1120 (1960)
13) BURLAKOVA,E.B., V.D.GAINZEVA, L.V.SLEPUKHINA, N.G.KHRAPOVA, N.M.EMANUEL, in press
14) YUSUPOVA,L.S., V.S.BALABUKHA, Radiobiologija 3, B. 2, 256 (1963)
15) KISELEV,N.A., N.V.NIKOLAEVA, Biochimija 26, N 5, 956 (1961)
16) NIKOLAEVA,N.V., K.E.KRUGLYAKOVA, N.A.KISELEV, B.K.VAINSHTEIN, N.M.EMANUEL, Dokl. Akad. Nauk SSSR 142, N 3, 713 (1962)
17) BUTLER,J.A.V., Les peroxydes organiques en radiobiologie, Masson et Cie, Editeurs, Paris, 1958, 36.
18) DANIELS,M., G.SCHOLES, J.WEISS, C.M.WHEELER, J. chem. Soc. 226 (1957)
19) EIDUS,L.Kh., E.E.GANASSY, Biofisika 4, 215 (1959)
20) EIDUS,L.Kh., L.P.KAYUSHIN, Dokl. Akad. Nauk SSSR 142, 3, 615 (1960)
21) SCHOLES,G., J.WEISS, Rad. Res., Suppl. 1, 177 (1959)
22) MCDONALD,M.R., E.C.MOORE, Rad. Res. 2, 426 (1955)
23) VASILIEV,R.F., A.A.VICHUTINSKY, Dokl. Akad. Nauk SSSR 142, 3, 615 (1962)
24) BOGDANOV,G.P., M.S.POSTNIKOVA, N.M.EMANUEL, Izv. Akad. Nauk SSSR OKhN, 1 (1963)

D I S K U S S I O N S B E M E R K U N G E N

DOSE.- Sie haben die Trypsinlösungen mit sehr hohen Dosen bestrahlt. Sofern Sie hierbei einen Sauerstoffeffekt untersucht haben, möchte ich Sie fragen, wie Sie hierbei den durch die strahlenchemischen Prozesse verbrauchten Sauerstoff wieder ersetzt haben. Einfaches Durchleiten von Gasen durch Proteinlösungen führt bekanntlich infolge der Schaumbildung zu weiterer Denaturierung der Proteine.

NIKOLAJEWA.- Der Sauerstoffeffekt war nicht erforscht worden. Die Trypsininaktivierung war in Luftatmosphäre durchgeführt worden.

DAMERAU.- Von verschiedenen Autoren wurden Beziehungen zwischen freien Radikalen und der Carcinogenese (und auch Altern) vermutet. Wurden Versuche mit Inhibitoren von Radikalprozessen auch in vivo gemacht?

NIKOLAJEWA.- Unsere Ergebnisse in vitro stimmen gut mit solchen in vivo überein (siehe Literatur zu Vortrag: [11],[12],[14]).

BERG.- Wird die DNA unverändert zurückgewonnen? Propylgallat inhibiert nur die Peroxydation und müßte nach Ihrer Gleichung selbst nach $2\,\mathrm{In} \longrightarrow \mathrm{In}^{-}\,\mathrm{In}$ verbraucht werden.

NIKOLAJEWA.- Die Inhibitorenwirkung verläuft nach Schema: $\mathrm{RO_2{}' + InH \longrightarrow RO_2H + In}$. Die Rekombination der wenig aktiven Inhibitorradikale ist nicht ausgeschlossen, geht aber zu langsam.

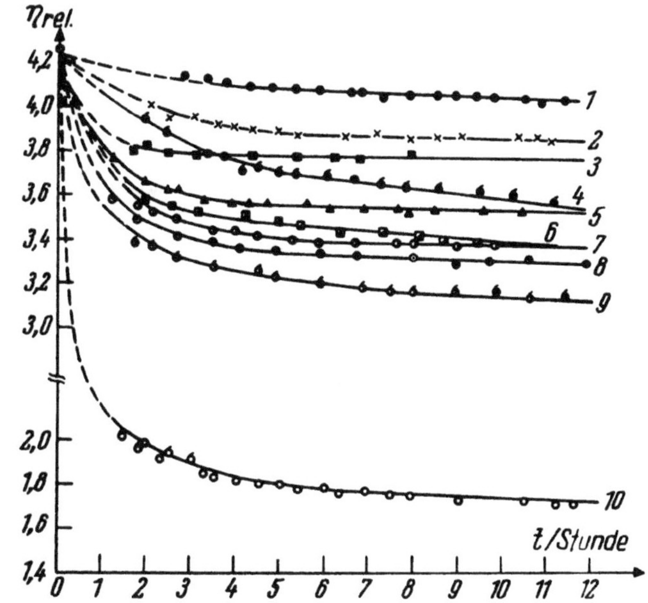

Abb.1 Kinetische Kurven veränderter relativer Viskosität der DNA-Lösgn. nach Bestrahlung mit $6 \cdot 10^4$r bei verschiedenem Inhibitorzusatz.
1. Propylgallat; 2. 2-Propyl-6-methyl-3-oxypyridinchlorhydrat; 3. 2-Äthyl-6-methyl-3-oxypyridinchlorhydrat; 4. 2-Äthyl-3-oxy-4-morpholinmethyl-6-methylpyridindichlorhydrat; 5. Pyridoxin (Vitamin B_6)-chlorhydrat; 6. 2-Äthyl-6-methyl-4-dimethylaminomethyl-3-oxypyridindichlorhydrat; 7. 2-Isobutyl-6-methyl-3-oxypyridinchlorhydrat; 8. 2,6-Dimethyl-3-oxypyridinchlorhydrat; 9. 2-Äthyl-3-oxy-4-piperidinmethyl-6-methylpyridindichlorhydrat; 10. Ohne Inhibitor.

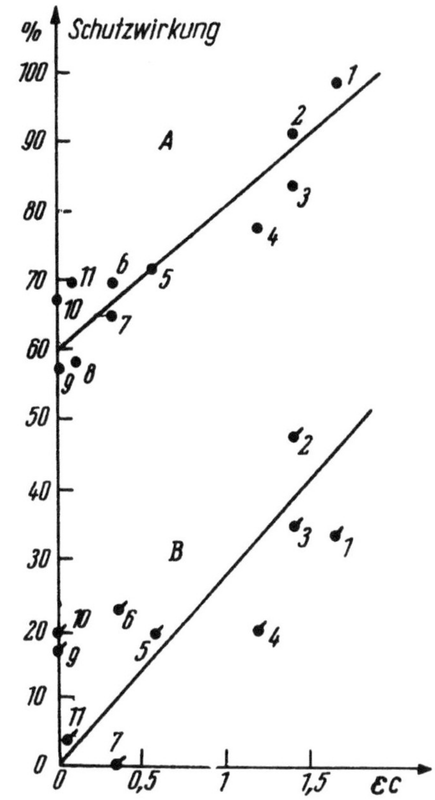

Abb.3 Verhältnis Inhibitorschutzwirkung - Antiradikalwirkung.

A. Zusatz vor Bestrahlung; B. Zusatz nach Bestrahlung.

1. Propylgallat; 2. Isopropylgallat; 3. Gallensäure; 4. Äthylgallat; 5. 2-Äthyl-6-methyl-3-oxypyridinchlorhydrat; 6. 2-Propyl-6-methyl-3-oxypyridinchlorhydrat; 7. 2-Isobutyl-6-methyl-3-oxypyridinchlorhydrat; 8. 2,6-Dimethyl-3-oxypyridinchlorhydrat; 9. 2-Äthyl-3-oxy-4-piperidinmethyl-6-methylpyridinchlorhydrat; 10. 2-Äthyl-3-oxy-4-morpholinomethyl-6-methylpyridindichlorhydrat; 11. 3-Methoxy-4-oxyphenyläthylaminchlorhydrat.

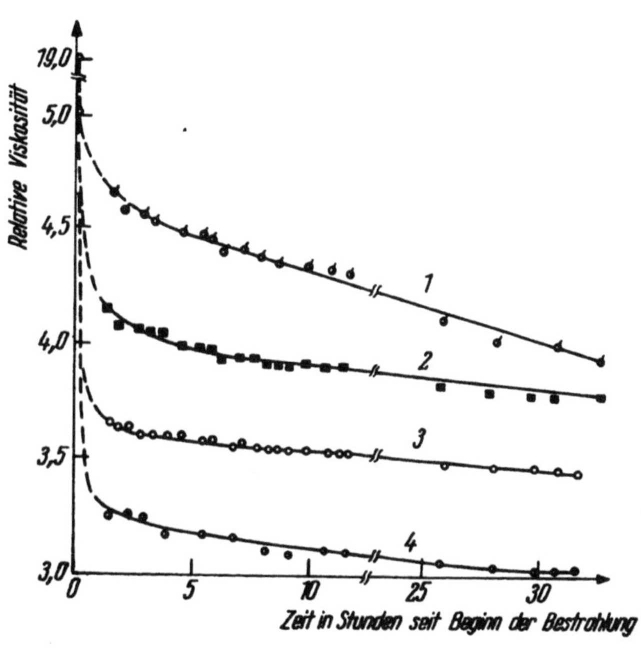

Abb.2 Kinetische Kurven veränderter Viskosität der DNA-Lösungen, Inhibitorzusatz nach der Bestrahlung.

1. 2-Äthyl-3-oxypyridinchlorhydrat; 2. 2-Äthyl-6-methyl-3-oxypyridinchlorhydrat; 3. Ohne Inhibitor; 4. 2-Isobutyl-6-methyl-3-oxypyridinchlorhydrat.

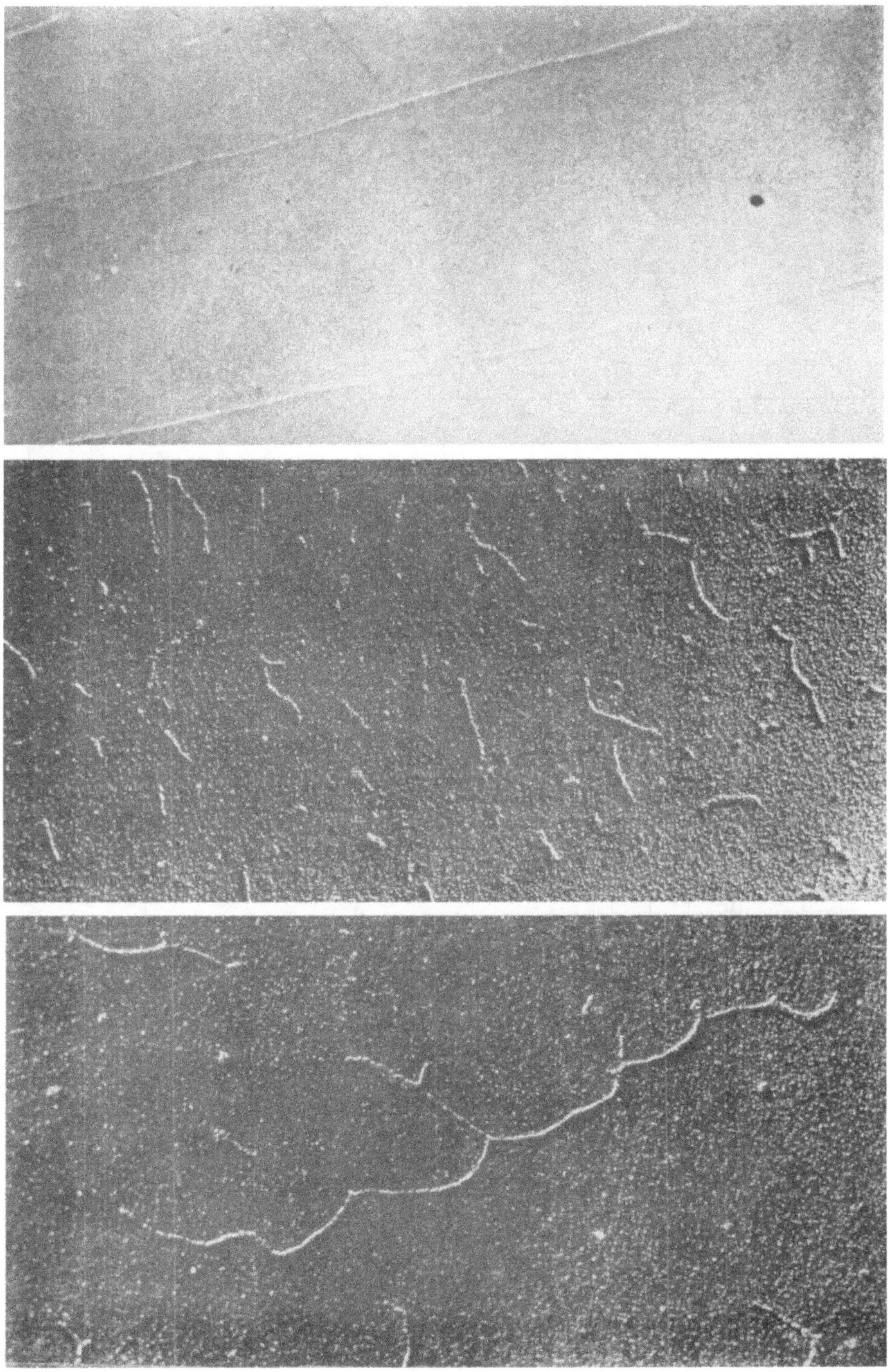

A

B

C

0,1 μ

Abb. 4 DNA-Mikrofotografie (elektronenmikroskopisch). A. nativ; B. Bestrahlungsdosis 3.10^{5} r; C. Bestrahlung 3.10^{5} r bei Propylgallatzusatz.

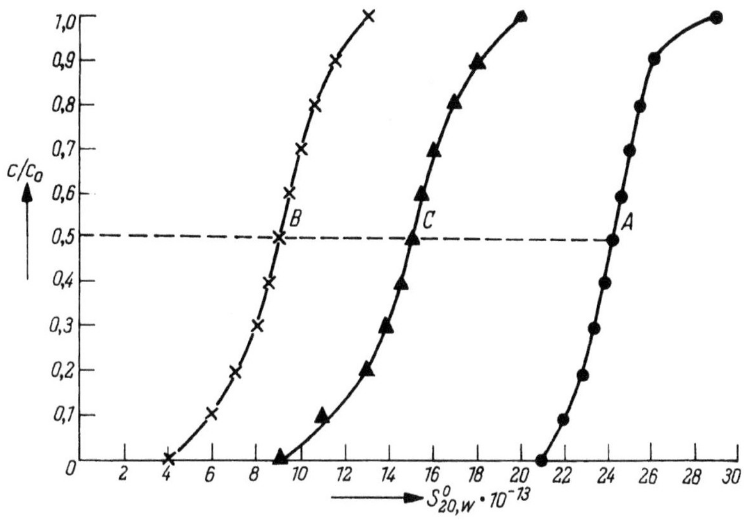

Abb. 5 Sedimentationskoeffizientenverteilung der DNA bei der Konzentration 0. Bezeichnungen wie auf Abb. 4.

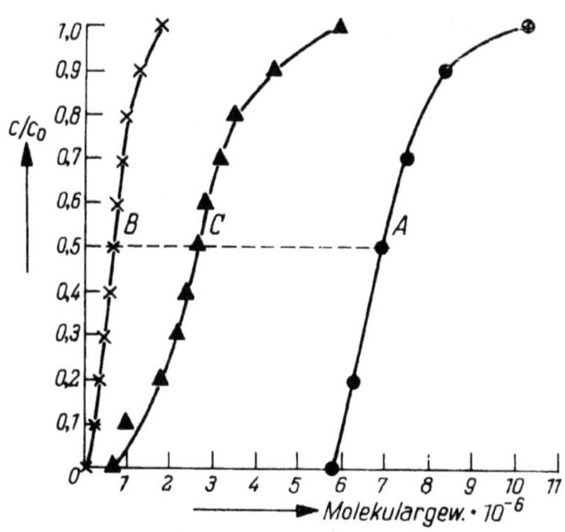

Abb. 6 Molekulargewichtsverteilung der DNA. Bezeichnungen wie auf Abb. 4.

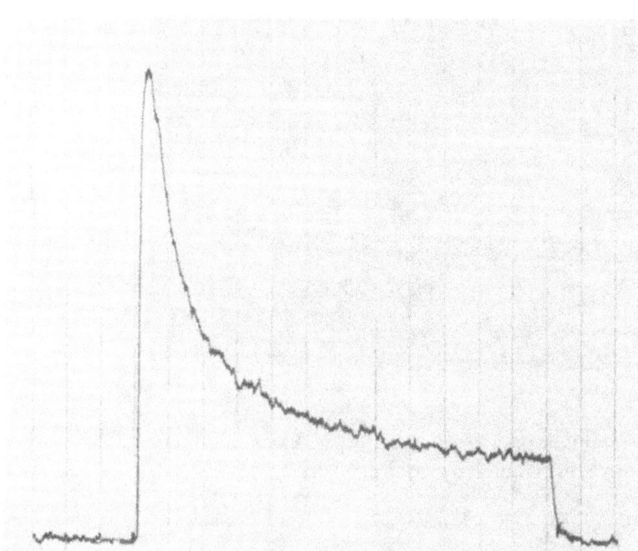

Abb. 7 Zeitverlauf der Lumineszenzintensität der bestrahlten DNA-Lösungen bei der Erwärmung (I: Fotostrom in relativen Einheiten).

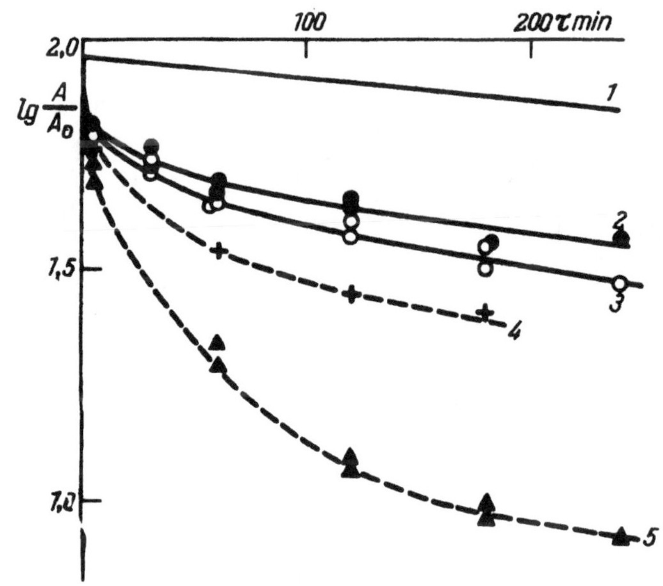

Abb.8 Inaktivierung der Trypsinlösungen bei Bestrahlung mit 10^5r (Temp. 73,5°C). 1.Trypsin (Konzentration 1 mg/ml); 2. Trypsinlösung mit 10^5r bestrahlt; H_2O_2 mit Katalase entfernt, danach mit HCl-Lösung (10^{-3}n) 1:1 verdünnt und Propylgallatlösung (bis zur Konzentration 10^{-3}m) zugegeben; 3. wie in 2., aber ohne Propylgallat; 4. Trypsinlösung (2,0 mg/ml) mit 10^5r bestrahlt, mit HCl-Lösung (10^{-3}n) verdünnt und danach Propylgallat (bis 10^{-3}m) zugegeben; 5. wie in 4., aber ohne Propylgallat. A_0. Spezifische Trypsinwirksamkeit unmittelbar nach der Bestrahlung; A.Spezifische Wirkung bei Erwärmung.

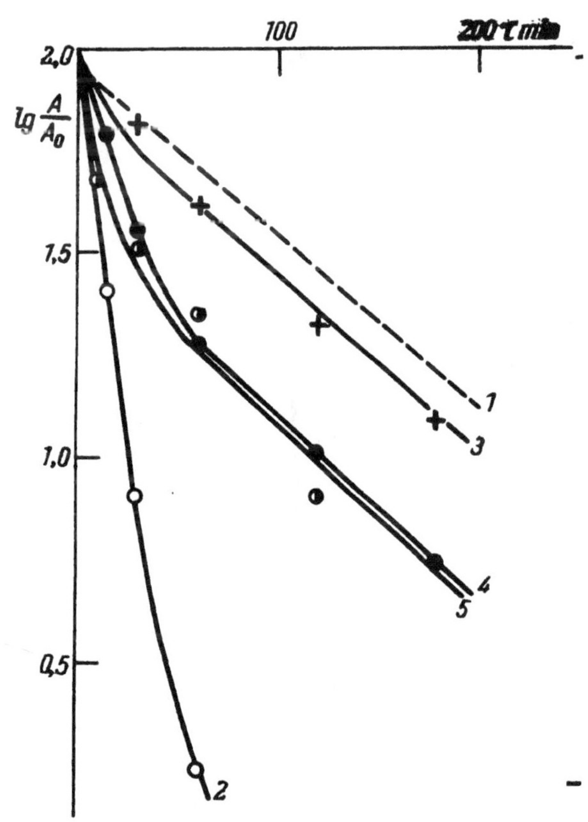

Abb.9 Inaktivierende Wirkung bestrahlter Trypsinlösungen bei 96,8°C.

1. Natives Trypsin; 2. Trypsinlösung (1 mg/ml) mit 5.10^5r bestrahlt und mit nativer Trypsinlösung (1 mg/ml) 1:1 verdünnt; 3. Wie in 2.; die bestrahlte Lösung aber vor der Verdünnung 120 Minuten bei 96,8°C mit Propylgallat erwärmt; 4. Wie in 3., die Erwärmung aber ohne Propylgallat; 5. Wie in 2., vor Verdünnung H_2O_2 mit Katalase entfernt.

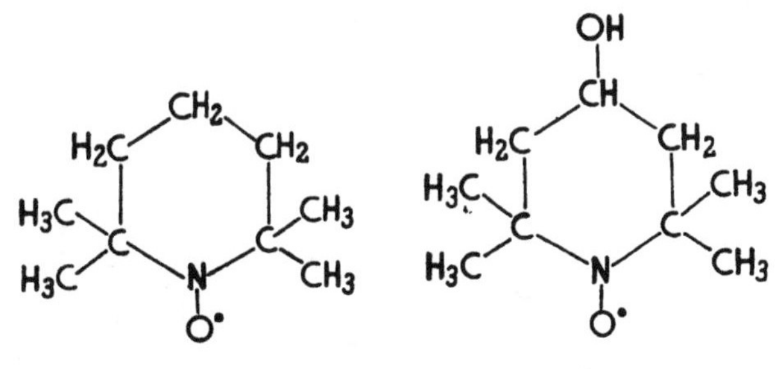

Abb.10 Stabilradikalformeln

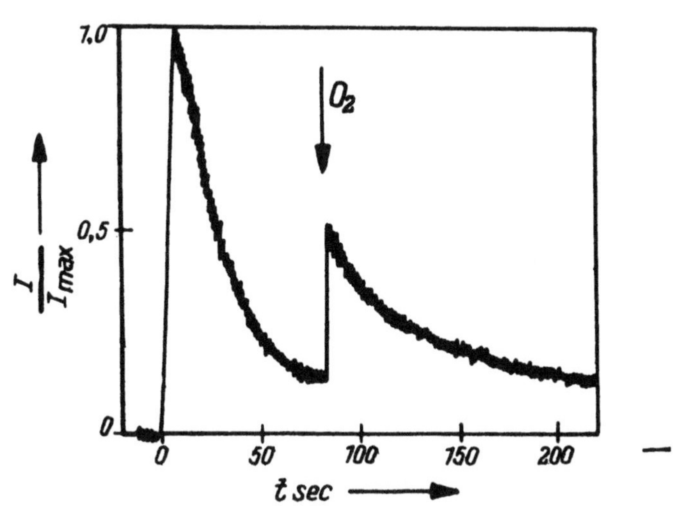

Abb.11 Zunahme der Chemolumineszenzintensität mit O_2 während des Prozesses.

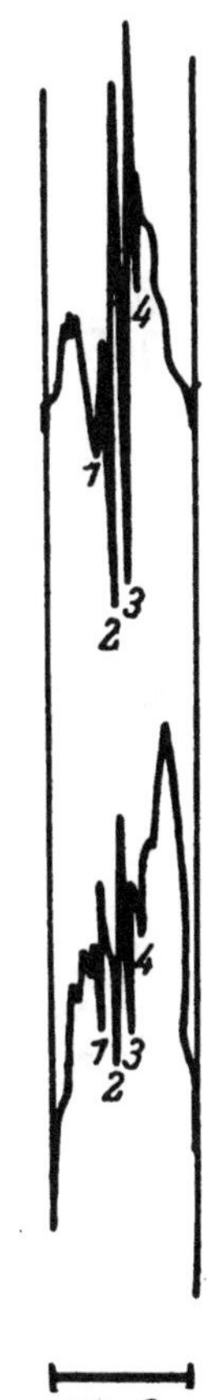

Abb.12 ESR-Spektren der Phenoxylradikale von Ionol; a) nach Reaktion mit Proteinradikalen; b) nach Reaktion mit Bleitetraacetat.

D Nucleinsäuren

Elektronenspinresonanz-Untersuchungen an Nucleinsäuren

W. DAMERAU

Institut für Biophysik der Akademie, Berlin-Buch

S U M M A R Y

A review is given on electron spin resonance investigations on irradiated and non-irradiated nucleic acids and their constituents. The results are compared, and possibilities and limitations in further studies are discussed.

E I N L E I T U N G

Während der letzten fünf Jahre wurde im Rahmen physikalischer Untersuchungen an Nucleinsäuren auch die Elektronenspinresonanz (ESR) eingesetzt. In der ESR liegt eine Methode vor, die mit hoher Empfindlichkeit den Nachweis paramagnetischer Spezies gestattet. Unter den zahlreichen Anwendungen der ESR sind es vor allem die Erfassung von Übergangsmetallionen und der Nachweis freier Radikale, die bei der ESR-Untersuchung von Nucleinsäuren in Frage kommen.

Im folgenden soll über bisherige Ergebnisse bei der ESR-Untersuchung von a) unbestrahlten und b) γ- oder röntgenbestrahlten Nucleinsäuren berichtet werden.

M E S S U N G E N

U n b e s t r a h l t e   N u c l e i n s ä u r e n

1959 wurden von BLUMENFELD und Mitarb. [1)2)] in Nucleinsäuren breite ESR-Signale beobachtet. Die Linienbreite zwischen den Wendepunkten der Absorptionskurve betrug 600 - 1500 Gauß, der g-Faktor 2,2 - 3,0. Bei Messungen bei tiefen Temperaturen wurde ein Ansteigen der g-Werte und eine Verbreiterung der Kurve festgestellt. Die Spinkonzentration lag in der Größenordnung von 10^{20} u.s./g, war also relativ hoch, jedoch wurde darauf hingewiesen, daß bei den Nucleinsäuren die ferromagnetische Resonanz eine große Rolle spielt und die gemessene Spinzahl nur als Vergleich dienen kann.

Die Herkunft der Nucleinsäuren und die Art ihrer Herstellung ist für die Form des ESR-Signals von großer Wichtigkeit. So wurde bei Hitzedenaturierung von DNA und Denaturierung mittels HCl bei 37° C und nachfolgender Dialyse und Lyophilisierung eine Änderung des ESR-Signals beobachtet [3)]. Andererseits wurde eine Schutzwirkung von schwachen Salzlösungen bei optimal 0,42 m NaCl auch am ESR-Signal angezeigt [3)].

Auffällig war die starke Erhöhung der Spinkonzentration in den Nucleinsäuren, wenn diese in Eiweißkomplexe überführt wurden [3]. Die Intensitätssteigerung war im Falle des Eialbumin-Komplexes bei RNA größer als bei DNA.

Nach diesen Arbeiten wurden von anderen Arbeitsgruppen ebenfalls ESR-Messungen an Nucleinsäuren durchgeführt [4]-[8]. BLOIS und MALING [4] erhielten von Herings- und Lachssperm-DNA und Hefe-RNA ESR-Signale, die sich in der Form bei Denaturierung veränderten, in ihrer Intensität jedoch etwa gleich blieben. Kühlung auf 70° K brachte die Signale nicht zum Verschwinden, Erhitzen auf 125° - 175° C zerstörte die Signale. Interessant war, daß die gemessene ESR-Absorption durch Druckanwendung bis optimal 3000 - 6000 at erhöht wurde. Da Cytosin und Nucleoside bei Druckanwendung ebenfalls ein DNA-ähnliches Signal gaben, das jedoch bis zu 15 000 at von der Höhe des Druckes abhängig war, wird von BLOIS [4] vermutet, daß die ESR-Signale "von elektronischen Wechselwirkungen zwischen Basenmolekülen stammen, wenn die letzteren von ihren normalen kristallographischen Positionen im Gleichgewicht entfernt werden", und daß das ebenfalls im intakten DNA-Molekül der Fall ist.

WALSH und Mitarb. [5] vertreten dagegen die Ansicht, daß die beobachteten Signale von ferromagnetischen anorganischen Einschlüssen herrühren, die wegen der Stärke des Signals nicht dispers, sondern in konzentrierter Form vorliegen. Solche Aggregate ließen sich elektronenmikroskopisch nachweisen. Die Natur dieser Einschlüsse wurde nicht geklärt. Da sie in ihrer Intensität bis 900° K wenig geändert wurden, können es keine Fe_3O_4-Einschlüsse (Curiepunkt 855° K) sein. α-Fe_2O_3 kann es ebenfalls nicht sein, da dessen Magnetismus zu gering ist, um den gefundenen Paramagnetismus zu erklären. Da in den Nucleinsäuren paramagnetische Ionen von Natur aus enthalten sind [9] und nach längerem Aufbewahren von Lachsspermatozoenköpfen und E. coli-Präparationen die Intensität des Signals steigt, nehmen WALSH und Mitarb. an, daß die Herausbildung der ferromagnetischen Einschlüsse eines Fe^{II}-/Fe^{III}-Oxyd-Hydroxydtyps erst bei Alterung erfolgt.

Das wird durch Arbeiten von PTAK, ROPARS und DOUZOU [10] unterstrichen, in denen beim Altern frisch hergestellter DNA-Präparationen, die 0,02 % Eisen enthielten, eine ähnliche Veränderung hinsichtlich Intensität, g-Faktor, Linienform und -breite festgestellt wurde wie bei Eisenkolloiden, die durch Hydrolyse von Fe^{II}-Salzen hergestellt wurden.

Andere Autoren [11][12] kommen ebenfalls zu der Ansicht, daß die ESR-Signale durch ferromagnetische Einschlüsse zu erklären sind. In metallarmen RNA-Präparationen wurden keine ESR-Signale festgestellt [6], ebenfalls nicht in Nucleoprotein aus Heringssperm [13].

In einer neuen, kritischen Übersicht wird von BLUMENFELD [14] jedoch darauf hingewiesen, daß die beobachteten Effekte nicht auf geringfügige Verunreinigungen zurückgeführt werden können, sondern angenommen werden muß, "daß die ferromagnetischen Bereiche und ihre Eigenschaften mit der organischen Struktur eng verbunden sind".

Bestrahlte Nucleinsäuren

1959 berichtete GORDY [15] über ESR-Untersuchungen an bestrahlten Nucleinsäuren, Nucleotiden, Nucleosiden und Purin- bzw. Pyrimidinbasen, deren Spektren miteinander verglichen wurden.

In Adenin und Guanin wurden nach Röntgenbestrahlung keine Signale gefunden, während sie in den entsprechenden Nucleosiden auftraten, und die Pyrimidinbasen ebenfalls strahlenempfindlicher waren. DNA und RNA waren relativ strahlenresistent, doch war das von der Herkunft der Proben abhängig. Bei den RNA und DNA traten einfache Linien mit schwacher Andeutung von nichtaufgelöster Hyperfeinstruktur (Hfs) auf, die in der

Gesamtbreite dem Uridin, Cytidin und Adenosin glichen. Eine Andeutung der Thymidin- bzw. Desoxyguanosinstrukturen wurde dagegen nicht beobachtet.

Von GORDY wurden diese Ergebnisse so interpretiert, daß das durch die Bestrahlung induzierte ungepaarte Elektron in dem Nucleinsäuremolekül migrieren und innerhalb der intakten Kette mit einem anderen ungepaarten Elektron rekombinieren kann. Beim Auseinanderbrechen der Kette ist das jedoch nicht mehr der Fall, und das ungepaarte Elektron wird in dem Bruchstück "getrapt" und bildet ein freies Radikal. Dieser Elektroneneinfang sollte dann eher an Cytosin bzw. Adenin stattfinden als an Guanin und Thymin.

Tatsächlich wurden später durch andere Methoden in den Nucleinsäuren Halbleitereigenschaften entdeckt [16]. Die Ergebnisse von GORDY wurden später auch durch Berechnungen von PULLMAN [17] gestützt, wonach die Elektronenacceptoreigenschaft bei Guanin, Thymin, Adenin und Cytosin in dieser Reihenfolge zunimmt. Andererseits sollte aus theoretischen Gründen die Strahlenresistenz von Adenin, Guanin, Cytosin und Thymin in dieser Reihenfolge abnehmen [17], was experimentell belegt ist [18] und in gewissem Grade auch von GORDY [15] nachgewiesen wurde.

Andererseits fanden VAN DE VORST und Mitarb. [27], daß abweichend von der Theorie [17] bei γ-Bestrahlung von Adenin, Guanin, Cytosin und Thymin die ESR-Absorption im Adenin am größten war.

Spektren bestrahlter Nucleinsäuren oder ihrer Basen wurden auch in anderen Arbeitsgruppen aufgenommen [6][8][13][28]. Ein sehr interessantes Problem ist die Frage nach Schutzmechanismen bei der Bestrahlung von DNA. ORMEROD und ALEXANDER [13] untersuchten die ESR-Spektren von Nucleoprotein, das mit oder ohne Zusatz von Cysteamin in Abwesenheit von O_2 γ-bestrahlt wurde. Bei -196^0 C wurde eine symmetrische einzelne Linie beobachtet, die beim Erwärmen auf Zimmertemperatur eine schlecht aufgelöste Hfs zeigte. Zusatz von Cysteamin veränderte jedoch das Spektrum, wobei bei Zimmertemperatur ein Radikal vorherrschend war, das nach der für S-Radikale typischen g-Faktoranisotropie aus dem Cysteamin entstanden war. Während die Cysteaminradikale zugenommen hatten, war die Anzahl der DNA-Radikale gesunken. Als Schutzmechanismus wird ein Wasserstofftransfer vom Cysteamin zum DNA-Radikal angenommen, während ein Schutzmechanismus durch intermolekularen Energietransfer für gering gehalten wird.

MÖGLICHKEITEN UND GRENZEN IN DER ESR-UNTERSUCHUNG VON NUCLEINSÄUREN

A. Zu Messungen an unbestrahlten Nucleinsäuren

Über die Herkunft des ESR-Signals in unbestrahlten Nucleinsäuren konnte bisher noch keine endgültige Klarheit erreicht werden. Die Unterschiede in den Ergebnissen verschiedener Arbeitsgruppen liegen großenteils darin begründet, daß anfangs handelsübliche Präparate oder Nucleinsäurepräparationen nicht genau bekannten Metall- und Wassergehalts benutzt wurden. Wie nachgewiesen wurde, spielt das Alter der Proben ebenfalls eine große Rolle. Auch meßtechnisch müssen wegen der im Verhältnis zur Breite sehr geringen Amplitude größere Anforderungen gestellt werden.

Der Metallgehalt der Proben spielt eine maßgebende Rolle. Nach RUSSEL und WYARD [19] ist für das ESR-Signal in Adenosintriphosphat (ATP) nicht ein vermuteter innerer charge-transfer [20] verantwortlich, sondern ein Kupfer-ATP-Komplex, der durch Verunreinigung gebildet wird. Diese Befunde ergänzen die Ergebnisse an Nucleinsäuren.

Ein innerer charge-transfer in den Nucleinsäuren oder Nucleotiden ohne Beteiligung von Metallionen, der so stark ist, daß er zur Herausbildung von paramagnetischen Zentren führt, scheint überhaupt fraglich. Bei der Untersuchung von Donatoreigenschaften von biologisch wirksamen Verbindungen haben wir auch die Elektronendonatorbereitschaft von Adenin, Guanin, Cytosin sowie RNA gegenüber dem starken Acceptor Jod untersucht und keine ESR-Signale erhalten [21].

Im geringen Maße sind freie Radikale von der Beteiligung am ESR-Signal unbestrahlter Nucleinsäuren nicht ganz ausgeschlossen. Dieser Anteil sollte aber abhängig von der Art der Lagerung der Nucleinsäuren und nicht a priori in diesen enthalten sein. So fanden wir in handelsüblichem Guanin ein verhältnismäßig schmales Signal von ca. 15 G und eine Spinkonzentration von $6 \cdot 10^{15}$ u.s./g [21]. Nach 8stündigem Kochen mit Alkohol/Wasser verschwand dieses Signal. Da Signale ähnlichen Typus von uns mitunter auch in aromatischen Aminen gefunden wurden [22], nehmen wir an, daß es sich hier um radikalische Oxydationsprodukte dieser Verbindungen handelt. Das von THOM und NICOLAU [8] berichtete schmale Signal in Hefe-RNA könnte hierin seine Erklärung finden.

Bei weiteren ESR-Untersuchungen an unbestrahlten Nucleinsäuren muß daher auf die sorgfältige Darstellung unter möglichst gleichbleibenden Bedingungen und eine möglichst metallarme Präparation Wert gelegt werden.

B. Zu Messungen an bestrahlten Nucleinsäuren

Diesen Untersuchungen kommt große Bedeutung zu, da sie Modelle sind für das Auftreten irreversibler Strahlenschäden in vivo und deren Verhinderung.

Von Interesse sind Vergleiche zwischen den Radikalausbeuten (G-Werten) in Basen, Nucleosiden, Nucleotiden und den Nucleinsäuren unter Beachtung der Dosisabhängigkeit und des Einflusses von Feuchtigkeit, O_2, Resteiweiß und Metallgehalt und des Alters der Nucleinsäurepräparationen.

Von Wichtigkeit ist bei der Bestrahlung wäßriger Nucleinsäurelösungen die Frage, in welchem Maße die Nucleinsäuren direkt durch die Strahlung oder indirekt durch Radikale, die aus der Umgebung stammen (z. B. OH-Radikale), geschädigt werden. Im wäßrigen Milieu sind die entstehenden Radikale wegen ihrer schnellen Rekombination ESR-spektroskopisch nicht nachweisbar, jedoch liefert die Untersuchung eingefrorener Lösungen, die schrittweise aufgetaut werde können, wertvolle Informationen. Entsprechende Untersuchungen sind an Aminosäuren bereits durchgeführt worden [23][24].

In der gleichen Richtung liegen ESR-Untersuchungen über die Schutzwirkung von Strahlenschutzmitteln [13] und den Einfluß von Eiweißen mit oder ohne -SH-Gruppen auf die Radikalausbeute. Diese Messungen dürfen allerdings nicht überschätzt werden und benötigen ein besonderes Maß an Vorsicht bei der Interpretation, da ein Verschwinden der Radikale nicht immer bedeutet, daß die Strahlenschädigung aufgehoben ist. So können z. B. die Radikale von Nucleinsäuren miteinander unter Vernetzung rekombinieren [13]. Während also das ESR-Signal verschwindet, bleibt im biologischen Sinne eine irreversible Schädigung zurück.

Bei der Untersuchung der Strahlenwirkung in Nucleinsäuren kann außer der γ- oder Röntgenbestrahlung auch die innere Bestrahlung in Tritium-markierten Nucleinsäuren untersucht werden, wie das bereits in unserer Arbeitsgruppe bei einer Reihe von Aminosäuren der Fall ist [25]. Die Tritium-Substituierung bietet, wie wir zeigen konnten, in verschiedenen Fällen Vorteile [26] und kann Informationen liefern über die Wirkung der Inkorporation ^{3}H-markierter Nucleotide in Nucleinsäuren.

Auch bei Untersuchungen bestrahlter Nucleinsäuren ist die Einhal ung reproduzierbarer Präparationsmethoden und definierter Metallgehalte von großen Wert. Vor allem beeinflussen Metallkonzentration und Art der Metallionen die Intensität des ESR-Signals [6]. Durch genaue Beobachtung des Milieueinflusses können Fehlinterpretationen weitgehend vermieden werden.

L I T E R A T U R

1) BLUMENFELD,L.A., Biofisika 4, 515 (1959)
2) BLUMENFELD,L.A., A.E.KALMANSON, SEN-PEJ-GEN, Doklady Akad. Nauk SSSR 124, 1144 (1959)
3) SEN-PEJ-GEN, L.A.BLUMENFELD, A.E.KALMANSON, Biofisika 5, 645 (1960)
4) BLOIS,M.S., J.E.MALING, Biochem. biophysic. Res. Commun. 4, 252 (1961)
5) WALSH,W.M., R.G.SHULMAN, R.D.HEIDENREICH, Nature (London) 192, 1041 (1961)
6) LINDIGKEIT,R., B.E.WAHLER, G.SCHOFFA, O.RISTAU, Hochfrequenzspektroskopie, S. 176, Akademieverlag Berlin 1961
7) MÜLLER,A., G.HOTZ, K.G.ZIMMER, Biochem. Biophysic. Res. Commun. 4, 214 (1961)
8) THOM,H.G., Cl.NICOLAU, "Probleme und Ergebnisse aus Biophysik und Strahlenbiologie", S. 67, Akademieverlag Berlin 1962
9) WACKER,W.E., B.L.VALLEE, J. Biochemistry 234, 3257 (1959)
10) PTAK,M., Cl.ROPARS, P.DOUZOU, J. Chim. Phys. 59, 659 (1962)
11) ALEXANDROV,A.A., V.Yu.GAVRILOV et al., Dokl. Akad. Nauk SSSR 141, 1483 (1961)
12) DORFMAN,J.G., Dokl. Akad. Nauk SSSR 142, 815 (1962)
13) ORMEROD,M.G., P.ALEXANDER, Radiat. Res. 18, 495 (1963)
14) BLUMENFELD,L.A., Dokl. Akad. Nauk SSSR 148, 361 (1963)
15) GORDY,W., Radiat. Res. Suppl. 1, 491 (1959)
16) DUCHESNE,J., J.DEPIREUX, A.BERTINCHAMPS, N.CORNET, Nature (London) 188, 405 (1960)
17) PULLMAN,B., Mem. Acad. roy. Belg. 33, 174 (1961)
18) HEMS,G., Nature (London) 181, 1721 (1958)
19) RUSSEL,D.B., S.J.WYARD, Nature (London) 191, 65 (1961)
20) ISENBERG,I., A.SZENT-GYÖRGYI, Proc. nat. Acad. Sci. USA 45, 1232 (1959)
21) DAMERAU,W., G.LASSMANN, unveröffentlicht
22) DAMERAU,W., G.LASSMANN, Arch. Geschwulstforsch. 22, im Druck
23) HENRIKSEN,Th., Radiat. Res. 17, 185 (1962)
24) LASSMANN,G., H.G.THOM, Vortrag auf der Arbeitstagung Biophysik, Berlin, Nov. 1962
25) LASSMANN,G., H.HERZMANN., W.DAMERAU, unveröffentlicht
26) DAMERAU,W., G.LASSMANN, H.G.THOM, Z. phys. Chem. 223, 99 (1963)
27) VAN DE VORST,A., Cl.WILLIAMS-DORLET, Bull. Ampère, 17e année, fasc. special 1962, 398
28) SEN-PEJ-GEN, L.A.BLUMENFELD, A.E.KALMANSON, A.G.PASSINSKI, Biofisika 4, 263 (1959)

D Nucleinsäuren

Über die Bedeutung von Metallionen in Nucleinsäuren

H. ALTMANN

Institut für Biologie und Landwirtschaft, Reaktorzentrum, Seibersdorf

S U M M A R Y

The metal ion content of various nucleic acids was estimated by means of neutron activation analysis. The significance of these cations for the structural and physiological stability is described. Trace elements in nucleic acids have been cleaved by irradiation, and the resulting radiation-damages are discussed.

Bereits 1525 hat PARACELSIUS im "Buch de Mineralibus" Metallionen mit der Medizin in Verbindung gebracht. Aber erst um 1930 wurde erstmals von einem Mineralstoffwechsel gesprochen [1]. Bei vielen Enzymen wurde die Rolle der Metalle, die diese auf Struktur und Funktion ausüben, bisher genauestens untersucht. Aber obwohl die ersten Literaturzitate über Nucleinsäure (NS)-Isolierungen aus dem späten 19. Jahrhundert stammen [2][3], weiß man über die Rolle von Spurenelementen in NS sehr wenig. Bei Proteinen wird die tertiäre Struktur, die mit der biologischen Aktivität in Zusammenhang steht, hauptsächlich durch S-S-Brücken, H-Brücken, VAN DER WAALS Kräfte, aber auch durch Metallbrücken, wie im Trypsin [4], oder der Amylase [5] aufrechterhalten. Durch UV-Dichroismus [6][7] und elektronenmikroskopische Studien [8] wurde die tertiäre Struktur von Ribonucleinsäure (RNS) wahrscheinlich gemacht. FUWA et al. [9] schlossen aus T_M-Kurven der RNS, die entweder durch den Anstieg der Extinktion bei 258 mμ oder durch die Abnahme der spezifischen optischen Drehung gemessen wurden, daß paramagnetische Metallionen in der Erhaltung der tertiären Struktur sowie der Stabilisierung der sekundären Struktur eine Rolle spielen. DAVE und DAVIDSON [10] fanden eine Stabilisierung der Desoxyribonucleinsäure (DNA) gegen Hitzedenaturierung durch Metallionen und YAMANE und DAVIDSON [11][12] bemerkten eine Änderung der "intrinsic viscosity" und UV-Absorption von DNA nach Metallionenzugaben. Über den Sitz der Metallionen besteht Unklarheit, doch ist die Affinität von Cu^{++} und Zn^{++} zu N-Liganden sehr groß. Co^{++}, Fe^{++} und Ni^{++} bilden Komplexe, sowohl mit O als auch mit N, während Mn^{++}, Mg^{++} und Ca^{++} hauptsächlich O-Donoren bevorzugen. Zur Aufrechterhaltung der tertiären Struktur dürfte das N_7 des Guanin mit seiner großen Elektronendichte eine bevorzugte Bindungsstelle von Metallionen sein. Die qualitative und quantitative Bestimmung von Spurenelementen in Nucleinsäuren bildet vielfach eine Hürde zur intensiven Forschung auf dem beschriebenen Gebiet, da die zu analysierenden Nucleinsäuremengen meist relativ klein sind. Obwohl die Analysenmethoden immer empfindlicher wurden, so ist doch die Verunreinigungsgefahr besonders bei der chemischen Trennung der kleinsten Mengen der Elemente sehr stark. Wir verwendeten für unsere Proben die Neutronen(n)-Aktivierungsanalyse, die sich besonders gut für die von uns untersuchten Übergangselemente eignet.

Die Anwendung der n-Aktivierung erfolgte erstmals 1936 von HEVESY, der mit Hilfe einer Ra-Be-Quelle Aktivierungen von seltenen Erden, die einen sehr hohen Einfangquerschnitt haben, vorgenommen hat. Aber erst nach dem zweiten Weltkrieg, als Kernreaktoren mit hohen n-Flüssen zur Verfügung standen, wurden Spurenelemente in biologischem Material bestimmt. Mit Hilfe der Aktivierung kann die Erfaßbarkeit der meisten Elemente gegenüber den bis jetzt empfindlichsten Methoden mindestens um den Faktor 10 unterboten werden. Neben der hohen Genauigkeit der Methode besteht aber ihr größter Vorteil darin, daß alle Elemente, die uns interessieren, in einem Arbeitsgang bestimmt werden können. Mit Hilfe von Koinzidenzmethoden und Rechenmaschinen hat man versucht, radiochemische Manipulationen zu vermeiden, was aber nur bei Bestimmung von Elementen mit höherer Konzentration und Vernachlässigung der Empfindlichkeit möglich ist. Biologische Proben enthalten aber meistens einen großen Komplex verschiedener Elemente, deren Gammaspektren nach der Aktivierung sich stark überschneiden, so daß man eine radiochemische Trennung vornehmen muß. Nur bei einigen sehr kurzlebigen Elementen wie Se und V gelang uns eine einwandfreie Messung ohne vorherige Abtrennung störender Elemente. Auch die Bestrahlung im epithermischen Bereich (CdB-Abschirmung) brachte nur bei Mn^{55} und Co^{59} eine stärkere Aktivität im Gegensatz zu K, Na und Cl.

Die Bestrahlung der Proben erfolgte in Quarzröhrchen bei einem integrierten thermischen n-Fluß von 10^{16} bzw. 10^{18} n/cm^2. Es wurde ein Verfahren ausgearbeitet, diese Spurenelemente nach Aktivierung und nasser Veraschung des biologischen Materials chemisch aufzutrennen. Dabei wurden nach Trägerzusatz die einzelnen Elemente durch spezifische Komplexbildner bei verschiedenen pH-Werten mit Hilfe von Flüssigkeits-Flüssigkeitsextraktionen voneinander getrennt.

Durch Vergleich der erhaltenen Gammaspektren mit Spektren von Standards, die unter den gleichen Bedingungen gewonnen wurden, konnten diese Spurenelemente quantitativ bestimmt werden [13].

Weiter wurde eine neue Isolierungsmethode von RNS und DNA entwickelt und mit den herkömmlichen Methoden verglichen, im Hinblick auf eine möglichst quantitative Extraktion aus dem Gewebe ohne gleichzeitige Abspaltung der Spurenelemente. Fast keine Spurenelement-Verluste wurden bei der Extraktion der Nucleinsäure mit Ammonformiat [14] gefunden.

Da einerseits Nucleinsäuren Strahlenresistenzfaktoren sind [15], andererseits Metallionen in Nucleinsäuren die Strahlenresistenz beeinflussen können [16], wurden Untersuchungen über die Entfernung der Metallionen sowohl durch ionisierende Strahlung als auch durch Komplexbildner durchgeführt. In der folgenden Tabelle sind die erhaltenen Werte angeführt :

Tabelle 1 Spurenelementgehalt verschieden behandelter Nucleinsäuren :

Probenmaterial	ppm Cu	ppm Mn	ppm Zn
RNS (Hefe)	155 ± 7	57 ± 5	1430 ± 40
RNS - einmal umgefällt	153 ± 7	56 ± 5	1320 ± 40
RNS - dreimal umgefällt	148 ± 7	35 ± 3	781 ± 28
RNS + EDTA - einmal umgef.	2,0 ± 0,2	0,3 ± 0,1	< 5
RNS + DTPA - einmal umgef.	3,9 ± 0,3	0,6 ± 0,1	< 5
RNS - acetyliert	96 ± 10	20 ± 3	850 ± 32
RNS - in vitro mit 10.000 rad bestrahlt	65 ± 3	25 ± 1,5	493 ± 10
RNS - aus in vivo bestrahlter Hefe :			
3.000 rad	135 ± 8	35 ± 4	435 ± 34
8.000 rad	55 ± 3	18 ± 3	330 ± 28
24.000 rad	75 ± 4	8 ± 1	135 ± 8

Mit Hilfe von Komplexbildnern kann man die untersuchten Spurenelemente fast vollständig aus der RNS entfernen. Während Mn^{++} und Zn^{++} mit steigender Strahlendosis stetig vermindert werden, erfolgte bei Cu^{++} ein Wiederanstieg der Werte bei 24.000 rad, was mit Rekombinationsvorgängen bei höheren Strahlungsdosen erklärt werden kann. Bei Untersuchungen mit anderen Bestrahlungseinrichtungen und Komplexbildnern soll untersucht werden, ob eine selektivere Abspaltung möglich ist. Eine Stabilitätsreihe von Komplexen zweiwertiger Ionen wurde von IRVING und WILLIAMS [17)] aufgestellt. Die Reihe $Mn^{++} < Fe^{++} < Co^{++} < Ni^{++} < Cu^{++} > Zn^{++}$ gilt für nahezu alle Komplexe, ungeachtet der Art der Liganden. Bei vorläufigen Untersuchungen mit sehr niedrigen Strahlendosen konnten wir auch Spurenelementabspaltungen feststellen. Dies deutet darauf hin, daß Energiewanderungen innerhalb der Nucleinsäureketten stattfinden. Der physikalische Vorgang des Energietransportes ist noch weitgehend ungeklärt - es werden sowohl Elektronen-Wanderungsmechanismen als auch Übertragungen über angeregte Zustände diskutiert [18)19)]. Energie aus direkten Treffern oder aber auch Anregungsenergie wird so schnell verteilt, daß zu wenig Zeit für Reaktionen wie z.B. die Radikalbildung vorhanden ist. Die Untersuchungen wurden sowohl bei Zimmertemperatur als auch bei sehr niedrigen Temperaturen durchgeführt [20)]. Der Energieauffänger kann vom Treffer relativ weit entfernt sein. Paramagnetische Ionen bilden in polymeren Nucleinsäuren einen Schutzfaktor gegen ionisierende Bestrahlung [16)]. Sie können sehr leicht Elektronen aufnehmen oder abgeben. Die dadurch bedingte räumliche Veränderung im Komplex kann zu einer leichten Abspaltung führen. In bezug auf Übertragung angeregter Zustände können paramagnetische Ionen verbotene Übergänge erleichtern und so für eine frühere Überführung eines Triplettzustandes in einen Grundzustand verantwortlich sein. Paramagnetische Ionen steigern aber auch die Zahl der Übergänge zwischen den einzelnen Zuständen des Kernspins der Protonen des H_2O. Der Effekt ist bei der Messung der longitudinalen kernmagnetischen Relaxationszeit erkennbar. So wird die Relaxationszeit für H_2O um den Faktor 10.000 verkürzt, wenn einmolare Lösungen von Mn^{++} untersucht werden [21)]. Die Resonanzenergie in Molekülen ist ein Strahlenresistenzfaktor [22)]. Durch die Chelonatbildung entsteht ein neuer metallhaltiger Heterocyklus von pseudoaromatischem Charakter und dadurch ein Resonanzenergiegewinn. Die Elektronen des Chelons werden durch die Komplexbildung disloziert und die Ladungen verteilen sich je nach der Natur des Metallions auf alle Atome des Ringes [23)]. Über eine katalytische Wirkung der Nucleinsäure-Metallkomplexe ist bisher nichts bekannt, doch zeigen zahlreiche Metallkomplexe im Gegensatz zu den einfachen hydratisierten Kationen eine katalytische Wirkung.

Ein Hinweis, daß durch die Spurenelementabspaltung die tertiäre Struktur von Nucleinsäuren verändert wird, ist auch die Änderung der Viskosität bei 22.000 rad. In diesem Dosisbereich finden aber noch keine Änderungen im Absorptionsspektrum oder Stickstoff- und Phosphorgehalt statt [24)]. Ob die Spurenelementabspaltung aus Nucleinsäuren bei niedrigen Dosen und damit ein teilweiser Verlust der tertiären Struktur bei Strahlenkrebs eine Rolle spielt, müßte eingehend untersucht werden. Nach DRUCKREY hat die carcinogene Wirkung (Strahlung oder chemische Substanzen) ein Maximum an "biologischer Entropie" d.h. sie führt den Organismus aus einem Zustand höherer Ordnung schrittweise in den einer größeren Unordnung.

An Hefe durchgeführte Untersuchungen zeigten, daß ein Teil der abgespaltenen Spurenelemente im "niedermolekularen Teil" der Zelle auftritt [25)].

Diese Metallionen sind für viele Enzyme Hemm- bzw. Stimulationssubstanzen und können Enzymaktivitätsverschiebungen hervorrufen. Auch Nucleasen werden von Spurenelementen beeinflußt. Bestrahlte RNS wird von Ribonucleasen im Vergleich zur unbestrahlten RNS leichter abgebaut [26)]. Durch vergleichsweise Inkubation von Kupfer und Silberionen, einmal mit RNase und einmal mit RNS, vor der Abbaureaktion konnten wir zeigen, daß die Metallionen die RNS gegen den RNase-Angriff stabilisieren. Aus kinetischen Messungen wurde geschlossen, daß die Hemmung der Aktivität der Ribonuclease durch Zn^{++} auf der Bildung eines tertiären Komplexes mit dem Reaktionsprodukt als Liganden beruht [27].

Die Stabilsierung von Nucleinsäuren durch Metallionen dürfte bei Viruserkrankungen eine Rolle spielen, aber es ist bis jetzt die Frage noch nicht geklärt, warum die Virus-RNS stabiler als die Wirts-RNS ist. Bei Untersuchungen des Spurenelementgehaltes von TMV-RNS und Tabakpflanzen-RNS zeigte sich, daß der Cu^{++}-Gehalt der Virus-RNS wesentlich höher ist [28] und deshalb von Tabakpflanzen-RNase schlechter angegriffen werden kann als die Wirts-RNS.

SINSHEIMER et al. [29] beobachteten, daß Metallionen die Infektiosität der TMV-RNS stark erhöhen und daß bereits die Konzentration von 1 Ni^{++} auf 700 Nucleotide eine signifikante Stabilisierung der Infektiosität bewirkt.

Strahlendosen von 100 - 1.000 rad bewirken nur sehr wenig signifikante Veränderungen an DNS, doch wird die Transformationsaktivität zum Großteil zerstört [30]. Ob diese Erscheinungen ebenfalls mit Spurenelementen und Veränderungen der tertiären Struktur in Zusammenhang stehen, müßte noch überprüft werden.

L I T E R A T U R

1) KLINKE,K., In "Der Mineralstoffwechsel, Physiologie und Pathologie"; Leipzig-Wien,1931
2) MIESCHER,F., Medicin-chemische Untersuchungen; 1871, Seite 441
3) ALTMANN,R., Archiv für Anatomie und Physiologie; 1889, Seite 524
4) GORINI,L., Biochim.biophysica Acta 7, 318 (1952)
5) STEIN,E.A., Fed.Proc. 16, 254 (1957)
6) DVORKIN,G.A, A.S.SPIRIN, Doklady Akad.Nauk SSSR 135, 987 (1960)
7) BOGDANOVA,E.S., L.P.GAVRILOVA, G.A.DVORKIN, N.A.KISSELEV, A.S.SPIRIN, Biochimica 27, 387 (1962)
8) KISSELEV,N.A., L.P.GAVRILOVA, A.S.SPIRIN, J.Mol.Biol. 3, 778 (1961)
9) FUWA,K., W.E.C.WACKER, R.DRUYAN, A.F.BARTHOLOMAY, B.L.VALLEE, Proc.nat.Acad.Sci. USA 46, 1298 (1960)
10) DOVE,W.F., N.DAVIDSON, J.molecular Biol. 5, 467 (1963)
11) YAMANE,T., N.DAVIDSON, Biochim.biophysica Acta 55, 609 (1962)
12) YAMANE,T., N.DAVIDSON, J.Amer.chem.Soc. 83, 2599 (1961)
13) STEHLIK,G., H.ALTMANN, Mh.Chem. - in Druck
14) ALTMANN,H., G.STEHLIK, Atompraxis 8, 471 (1962)
15) BACQ,Z.M., P.ALEXANDER, Grundlagen der Strahlenbiologie. Thieme Verlag, 1958, S.232
16) ALTMANN,H., G.STEHLIK, K.KAINDL, Atompraxis 9, 255 (1963)
17) IRVING,H., R.J.P.WILLIAMS, J.chem.Soc. London 3192 (1953)
18) GOODMAN,J., J.STEIGMAN, In Rad.Effects in Physics, Chemistry and Biology ed. M.EBERT and A.HOWARD, S. 264; North-Holland Publ.Comp.Amsterdam 1963
19) YUSTER,P., S.J.WEISSMAN, J.chem.Physics 17, 1182 (1949)
20) LETT,J.T., M.G.ORMEROD, Biochim.biophysica Acta 51, 207 (1961)
21) COHN,M., J.S.LEIGH, Nature London 193, 1037 (1962)
22) PULLMAN,B., A.PULLMAN, In Comparative Effects of Radiation, John Wiley and Sons Inc., (1960) Seite 112
23) BERSIN,T., In "Biochemie der Mineral- und Spurenelemente"; Akad.Verlagsges. Frankfurt/Main, (1963), Seite 18
24) TAYLOR,B., J.P.GREENSTEIN, A.HOLLANDER, Arch.Biochemistry 16, 19 (1943)
25) ALTMANN,H., G.STEHLIK, K.KAINDL, Nature London 199, 823 (1963)
26) ALTMANN,H., II.Wissenschaftliche Tagung der Deutschen Gesellschaft für Biophysik, 25./26.4.63 Baden-Baden

27) FINDLAY,D., Nature London 190, 781 (1961)
28) ALTMANN,H., G.STEHLIK, K.KAINDL, IIIE Colloque International de Biologie de Saclay, 26/28.9.63
29) CHEO,P.C., B.S.FRIESEN, R.L. SINSHEIMER, Proc. nat. Acad.Sci. USA 45, 305 (1959)
30) GUILD,W.R., Radiat.Res. Suppl. 3, 257 (1963)

D I S K U S S I O N S B E M E R K U N G E N

ECKART : Vom Bakteriophagen T5 ist bekannt, daß seine Aktivität die Anwesenheit von Ca^{++}-Ionen im Nährmedium erfordert. Diese werden nicht zur Adsorption des Phagen an die Bakterienmembran benötigt, sondern für anschließende Stadien. Dies ergibt sich aus der Tatsache, daß man auch nach erfolgter Adsorption durch Ca-Gaben die Aktivität wiederherstellen kann, d.h. den Ablauf des kompletten Vermehrungszyklus des Phagen gewährleisten kann. Halten Sie für wahrscheinlich, Sie haben ja das Calcium in diesem Zusammenhang erwähnt, daß es noch nachträglich in der Bakterienzelle einen stabilisierenden Effekt auf die Phagen-DNA ausübt ?

ALTMANN : Es ist möglich; TMV-RNA kann auch durch einige Metallionen nach der Trennung vom TMV-Protein stabilisiert werden.

ZIMMER : Als Erklärung der von Ihnen beobachteten Schutzwirkung von divalenten Kationen in RNA gegen UV-Strahlung könnte vielleicht die Beobachtung von KRATZ über DNA und deren Reaktion mit Hg^{2+} dienen. In diesem Falle wird angenommen, daß Hg^{2+} mit Thymin reagiert.

ALTMANN : Vielleicht wirkt auf diese oder irgend eine andere Weise das Metallkation der Veränderung von Uracil bzw. anderer Basen in der RNA entgegen.

EINS : Kann man von den genannten Affinitäten erwarten, daß es gelingt, bestimmte Orte im DNA-Molekül elektronenmikroskopisch darzustellen ?

ALTMANN : Es ist möglich, aber es kommt doch leicht zu Spurenelementabspaltungen und damit möglicherweise zur Veränderung einer tertiären Struktur.

D Nucleinsäuren

Stabilitätskonstanten einiger Purin-Metallkomplexe

H. REINERT

Institut für Mikrobiologie und experimentelle Therapie der Akademie, Jena

S U M M A R Y

Formation constants of some purines and pyrimidines have been determined from potentiometric measurements applying the BJERRUM pH-technique. The stability sequence for divalent metal ions is: Cu > Ni ≥ Zn > Co according to IRVING's and WILLIAMS's rule. For the copper complexes of purines and pyrimidines the order of stability is: uracile < cytosine < 9-methyladenine < 9-methylhypoxanthine ≮ 7-methyladenine < 7-methylhypoxanthine < purine ≤ hypoxanthine < adenine ≤ 2.6-diaminopurine. From potentiometric measurements and in some cases from spectrophotometric measurements a ratio metal:ligand = 1:2 was found. The binding sites are discussed.

Purine und Pyrimidine sind als wesentliche Bestandteile der Nucleinsäuren und vieler Coenzyme schon seit langem Gegenstand intensiver chemischer, biochemischer und biologischer Untersuchungen. Es lag daher nahe, das Verhalten dieser Verbindungen auch gegenüber Metallionen zu untersuchen.

Erste Ergebnisse über Untersuchungen des Komplexbildungsvermögens von Purinen und Nucleinsäuren wurden von PARNAS und KLIMEK sowie von SCHÜTZ und UMSCHWEIF in den dreißiger Jahren veröffentlicht. Aus ihren Untersuchungen folgerten sie, daß Komplexbildung mit Purinen nur auftreten kann, wenn diese am Imidazolring unsubstituiert sind. WEISS stellte fest, daß man durch Variieren der Konzentration des Metallsalzes und der Purin- bzw. Pyrimidinbase einerseits und des pH-Wertes andererseits eine große Zahl von Komplexen darstellen kann. Auch im Imidazolring methylierte Purinbasen reagieren auf diese Weise mit Metallsalzen unter Komplexbildung. Von einigen Autoren wurde auch die Stabilität von Komplexen des Adenins und verschiedener Purine gemessen. Jedoch sind die sich aus diesen Messungen ergebenden Daten und Schlußfolgerungen teilweise widersprüchlich und es ist auch nicht ohne weiteres möglich, diese Ergebnisse auf biologische Systeme anzuwenden, da die Messungen fast durchweg in Wasser/Äthanol- bzw. Wasser/Dioxan-Mischungen durchgeführt wurden.

Unseren Versuchen lag der Gedanke zu Grunde, durch Messungen der Stabilität von Purin- und Pyrimidinkomplexen Schlüsse auf die Bindung bzw. auf die mögliche Struktur der Komplexe zu ziehen. Ein Blick auf den Molekülbau der Purine läßt nämlich erkennen, daß in diesen Verbindungen mehrere Orte für eine Haftung der Zentralatome in Frage kommen können. Aus diesem Grunde wurden folgende Verbindungen auf ihr Verhalten gegenüber Metallionen untersucht: die Nucleinsäurebausteine Uracil, Cytosin und Adenin. Zur Ermittlung der Haftstellen wurden noch Hypoxanthin, Purin-2.6-diaminopurin sowie die am Stickstoffatom N_7 und N_9 methylierten Derivate des Adenins und des Hypoxanthins in die Untersuchungen einbezogen. Als Zentralatome wurden vorwiegend Metalle der 1. Übergangsreihe des periodischen Systems wie Kupfer, Kobalt, Nickel und Zink gewählt. Daneben wurden auch Komplexe des Quecksilbers untersucht. Da alle untersuchten Purine und Pyrimidine Protonenakzeptoren sind, war es möglich, die Stabilität der Komplexe durch potentiometrische pH-Messungen nach der Methode von BJERRUM zu untersuchen. Grundlage der pH-Methode ist, daß man das zu untersuchende Metallion mit dem Proton und dem Komplexbildner in Konkurrenz treten läßt, gemäß der Gleichung:

$$M + n\,HpL \rightleftharpoons ML_n + npH \qquad p = \text{Anzahl der Protonen} \quad n = 1, 2, 3 \ldots n \tag{1}$$

Die Meßgröße $[H^+]$ erlaubt dann sowohl die Berechnung des Ausmaßes, mit dem die Reaktion abläuft, als auch die Ermittlung der Konzentration des an das Metallion gebundenen Liganden. Die einfache und sichere Messung der Wasserstoffionenkonzentration mit Hilfe der Glaselektrode gestattet dann eine im Vergleich zu anderen Methoden ziemlich genaue Berechnung der Komplexstabilitätskonstanten.

Zur Berechnung dieser Konstanten nach der oben skizzierten pH-Methode ist die Kenntnis der Dissoziationskonstanten der Liganden erforderlich. Ihre Bestimmung muß unter gleichen Bedingungen durchgeführt werden, wie sie bei der Untersuchung der Komplexe angewendet werden. Aus den pH-Werten, die man durch Titrieren von annähernd $2 \cdot 10^{-3}$ molaren Lösungen der Ligandenperchlorate mit Carbonat-freier Lauge erhält sowie aus der Konzentration des Liganden und der Konzentration an zugesetzter Lauge werden die Dissoziationskonstanten nach einem von MARTELL und Mitarbeitern angegebenen Verfahren berechnet. Eine Zusammenstellung der Dissoziationskonstanten zeigt Tabelle I.

Tabelle I

	pK_1	pK_2	pK_3
Uracil	-	9,53	-
Cytosin	4,62	-	-
Purin	2,50	-	8,89
Adenin	4,22	-	9,80
7-Methyladenin	4,05	-	-
9-Methyladenin	3,98	-	-
Hypoxanthin	-	8,83	-
7-Methylhypoxanthin	-	9,32	-
9-Methylhypoxanthin	-	9,22	-
2.6-Diaminopurin	4,96	-	10,45
4-Aminobenzimidazol	5,35		10,31
6-Diäthyladenin			

$t = 25{,}0^\circ$ C $\qquad \mu = 0{,}05$ $NaClO_4$

pK_1 entspricht der Dissoziation des protonisierten Teilchens
pK_2 entspricht der Dissoziation des Substituenten in Stellung 6
pK_3 entspricht der Dissoziation der Imidazolgruppe im Purinkern.

Wie im Falle der Dissoziationskonstanten wurden zur Bestimmung der Stabilitätskonstanten $2 \cdot 10^{-3} - 5 \cdot 10^{-3}$ molare Lösungen der Liganden in Gegenwart von Metallionen titriert. Die Auswertung der auf diese Weise gewonnenen Titrationskurven erfolgte nach einer von HARKINS und FREISER ausgearbeiteten Methode. Aus den Werten der gemessenen Wasserstoffionenkonzentrationen, der Gesamtkonzentrationen des Liganden und des Metallions sowie aus der Menge der zugesetzten Lauge berechnet man zunächst den BJERRUMschen Komplexbildungsgrad, das ist die durchschnittliche Zahl von Liganden, die in einer Lösung bestimmter Zusammensetzung an ein Metallion gebunden sind, und die Konzentration des freien nicht im Komplex gebundenen Liganden. Diese Werte wurden benutzt, um die Bildungskurven der Liganden-Metallionen-Systeme zu zeichnen. An Hand dieser Kurven läßt sich eine vorläufige Abschätzung der Komplexkonstanten durchführen, deren Beträge dann durch Anwendung von Näherungsverfahren verbessert werden können.

Aus der Komplexbildungskurve ergibt sich auch die maximale Zahl der an das Metallion gebundenen Ligandenmoleküle, da sich die Kurven, die in den meisten der hier untersuchten Fälle S-förmige Gestalt besitzen, dem Wert der maximalen Koordinationszahl asymptotisch nähern.

Abbildungen 1 und 2 zeigen die Bildungskurven des 7-Methyladeninkupferkomplexes und des Adeninkupferkomplexes. Die vollständige Bildungskurve läßt sich nicht immer erhalten. Beginnende Hydrolyse der Metallionen bzw. die Ausfällung schwerlöslicher Komplexe setzen der vollständigen Auswertung der Titrationskurven Grenzen, da durch diese Prozesse die Gleichgewichtseinstellung gestört ist. Hydrolyseerscheinungen treten vermutlich bei Komplexen des Kobalts und des Nickels auf, weil die Stabilität der Komplexe dieser Metalle im Vergleich zu denen des Kupfers gering ist. Beim Kupfer fallen häufig schwerlösliche Komplexe aus sehr verdünnten Lösungen aus.

In diesen Fällen wurde nur der Wert der 1. Stabilitätskonstante und die Gesamtkomplexkonstante berechnet. Die Kenntnis der 1. Stabilitätskonstanten genügt, um das komplexchemische Verhalten der Liganden miteinander vergleichen zu können. In ihrem Betrag spiegeln sich sowohl die Stabilitätsverhältnisse als auch strukturelle Verhältnisse wider.

In einigen Fällen versagte jedoch die pH-Methode. So bei starker Komplexbildung, d. h. wenn das in (1) gezeigte Gleichgewicht extrem nach rechts verschoben ist oder bei zu schwacher Komplexbildung, wenn das Gleichgewicht sehr stark nach links verschoben ist. Daher war es nicht möglich, Stabilitätskonstanten der Quecksilberkomplexe von Purinen durch potentiometrische pH-Messungen zu bestimmen. Infolge der hohen Affinität der Purine gegenüber Quecksilber-II-Ionen bildeten sich schon in stark saurer Lösung schwerlösliche Komplexe. Umgekehrt versagte die Methode bei den Pyrimidinen wegen ihrer zu schwachen Komplexbildungstendenz gegenüber den Ionen der Metalle der 1. Übergangsreihe des periodischen Systems. In diesen Fällen ermittelten wir die Stabilitätskonstanten durch optische Messungen. Die nach beiden Methoden berechneten Stabilitätskonstanten sind in Tab. II zusammengestellt.

Optische Methoden gestatteten auch die Ermittlung der stöcheometrischen Zusammensetzung der Komplexe. Hierbei wurden verschiedene bekannte Methoden, wie die Methode der kontinuierlichen Variation nach JOB, die Methode der molaren Verhältnisse nach HARVEY und MANNING und die Geradenmethode nach ASMUS verwendet.

Die Auswertung nach diesen Methoden und die Analyse der Titrations- und Bildungskurven ergab, daß alle untersuchten Purine zweizählige Liganden darstellen und somit zwei Koordinationsstellen der Metallionen besetzen können. Die Zahl der maximal durch Metallionen gebundenen Ligandenmoleküle ist in allen untersuchten Fällen zwei. Im Falle der Pyrimidine waren keine eindeutigen Aussagen möglich.

Vergleicht man die Übergangsmetalle Kupfer, Nickel, Kobalt und Zink miteinander im Hinblick auf ihr komplexchemisches Verhalten gegenüber den Purin- bzw. den Pyrimidinbasen, so stellt man fest, daß Kupfer-II-Ionen stets die stabilsten Komplexe bilden. Erst dann folgen Nickel, Zink und Kobalt. Offensichtlich gehorchen die Metalle in ihrem Verhalten der von WILLIAMS und IRVING aufgestellten Regel. Anschaulich wird diese bevorzugte Stellung des Kupfers in Diagrammen dargestellt, in denen

Tabelle II

	potentiometrisch								optisch			
	Kupfer		Zink		Nickel		Cobalt		Kupfer		Quecksilber	
	pK_1	pK_2	pK_1	pK_2	pK_1	pK_2	pK_1	pK_2	pK_1	pK_2	pK_1	pK_2
Uracil	-	-	-	-	-	-	-	-	-	-	-	-
Cytosin	1,0	-	-	-	-	-	-	-	-	-	-	-
Purin	6,20	-	4,90	-	4,20	-	-	-	5,85	-	9,40	-
Adenin	6,86	5,60	5,28	-	5,70	-	5,50	-	6,90	-	10,60	-
7-Methyladenin	3,20	-	-	-	-	-	-	-	-	-	4,40	-
9-Methyladenin	-	-	-	-	-	-	-	-	-	-	-	-
Hypoxanthin	6,20	-	4,75	-	4,10	-	-	-	-	-	-	-
7-Methylhypoxanthin	5,78	-	-	-	-	-	-	-	-	-	-	-
9-Methylhypoxanthin	-	-	-	-	-	-	-	-	-	-	-	-
2.6-Diaminopurin	7,10	6,58	5,90	5,60	6,57	4,30	6,04	-	7,15	-	11,0	-

$t = 25{,}0^{\circ}$ C $\qquad \mu = 0{,}05$ $NaClO_4$

die Beträge der 1. Stabilitätskonstanten gegen die Ordnungszahlen der Elemente bzw. in denen die pH-Werte der beginnenden Ausfällung der Komplexe gegen die Ordnungszahlen aufgetragen sind. Bild 3 zeigt derartige Diagramme der Komplexe des 2.6-Diaminopurins.

Bezogen auf die Stabilität der Kupferkomplexe ergibt sich für das Komplexbildungsvermögen der Liganden folgende Reihenfolge:

Uracil $<$ Cytosin $<$ 9-Methyladenin $<$ 9-Methylhypoxanthin $\ll$ 7-Methyladenin $<$ 7-Methylhypoxanthin $<$ Purin $\leq$ Hypoxanthin $<$ Adenin $\leq$ 2.6-Diaminopurin.

Auffallend ist die geringe Affinität der Pyrimidine gegenüber Metallionen, ganz im Gegensatz zu den Purinen, die verhältnismäßig stabile Komplexe bilden. Man erhält bei der Titration von Lösungen, die Pyrimidine und Metallionen enthalten, die Hydrolysekurven der entsprechenden Metallionen. Bild 4 zeigt die Titrationskurve des Systems Cytosin - Kupfer-II-Ionen.

Im Imidazolkern methylierte Purine zeigen ebenfalls eine geringere Neigung sich mit Metallen zu kombinieren als ihre unsubstituierten Grundkörper. Interessant ist hierbei das Verhalten der beiden isomeren Paare 7- und 9-Methyladenin und 7- und 9-Methylhypoxanthin. Während die am Stickstoffatom N_9 methylsubstituierten Verbindungen keinerlei Komplexbildungsvermögen zeigen, ist bei den am Stickstoffatom N_7 methylierten Verbindungen die Fähigkeit zur Bildung von Komplexen noch deutlich ausgeprägt. Wie die Komplexe des Purins, dem Grundkörper dieser Verbindungsklasse, zeigen, ist die Anwesenheit von Substituenten am Kohlenstoffatom C_6 nicht unbedingt zur Bildung stabiler Komplexe erforderlich. Obwohl im Purin keinerlei Substituenten vorhanden sind, die mit freien Elektronenpaaren als zweite Haftstellen für das Metallion in Frage kämen, erweist sich das Purin als zweizähliger Ligand und vermag Komplexe von relativ hoher Stabilität zu bilden.

Diesen Sachverhalt zeigt Tab. III an Hand der Kupferkomplexe des Purins, Adenins und des 2.6-Diaminopurins.

Tabelle III

	Purin	Adenin	2.6-Diaminopurin
log K_1	6,28	6,98	7,10

$t = 25{,}0^{\circ}$ C $\qquad \mu = 0{,}05$ $NaClO_4$

Die Einführung der Aminogruppen in das Purinmolekül bringt also keine wesentliche Steigerung der Stabilität der Kupferkomplexe mit sich. Die Purine besitzen zwei

Zentren, die infolge ihrer hohen Elektronendichte für eine Anlagerung von Metallionen in Betracht gezogen werden können. Diese beiden Möglichkeiten sind in Bild 5 dargestellt. Nach der Formulierung I, die der allgemeinen Vorstellung über die Chelatstruktur entspricht, würde das Metallion mit dem Substituenten am Kohlenstoffatom 6 und dem Stickstoffatom des Imidazolrings einen spannungsfreien Fünfring bilden. Eine derartige Struktur erklärt jedoch nicht das unterschiedliche Verhalten der am Imidazolring methylsubstituierten isomeren Purine. Infolge der Raumbeanspruchung durch die Methylgruppe sollte man erwarten, daß diese bei der Komplexbildung durch sterische Hinderung eine Abschwächung der Komplexstabilität hervorruft. Wie die Beträge der 1. Stabilitätskonstanten der Kupferkomplexe des 7-Methyladenins und des 7-Methylhypoxanthins zeigen, ist die Komplexbildungstendenz dieser Liganden geringer als die der nicht methylierten Verbindungen. Aus dem gleichen Grund sollten aber die am Stickstoffatom N_9 methylierten Purine eine größere Komplexbildungstendenz als die 7-Methylpurine zeigen. Das Verhalten dieser beiden Verbindungsgruppen ist jedoch gerade entgegengesetzt.

Man kann diesem Sachverhalt gerecht werden, wenn man eine Bindung der Metallionen über die Stickstoffatome N_9 und N_3 annimmt, gemäß der Formulierung II in Bild 5. Eine derartige Verknüpfung erklärt auch am besten, warum Purin als zweizähliger Ligand fungiert, denn es wäre denkbar, daß sich Purin, wie das Benzimidazol, Metallionen gegenüber wie ein einzähliger Ligand verhält und z. B. mit Kupfer-II-Ionen Komplexe der Zusammensetzung $[Cu(Purin)_4]$ bilden könnte. Vom Benzimidazol sind derartige Komplexe bekannt.

L I T E R A T U R

PARNASS,J.K., R.KLIMEK, Biochem. Z. 252, 392 (1932)
KLIMEK,R., J.K.PARNASS, Hoppe-Seyler's Z. physiol. Chem. 218, 30 (1933)
SCHÜTZ,A.F., B.UMSCHWEIF, Biochem. Z. 268, 326 (1934)
ALBERT,A., Biochem. J. 54, 646 (1953)
HARKINS,Th.R., H.FREISER, J. Amer. chem. Soc. 80, 1132 (1958)
CHENEY,G.E., H.FREISER, Qu. FERNANDO, J. Amer. chem. Soc. 81, 2611 (1959)
BJERRUM,J., Metal Ammine Formation in aqueous Solution, Copenhagen 1941
JOB,P., Ann. Chem. 10 9, 113 (1928)
HARVEY,E., D.L.MANNING, J. Amer. chem. Soc. 72, 4488 (1950)
ASMUS,E., Z. analyt. Chem. 178, 104 (1960)
WILLIAMS,R.J.P., H.IRVING, J. chem. Soc. London 3192 (1953)

D I S K U S S I O N S B E M E R K U N G E N

BLUMENFELD.- Why you have shown us only one pK value for the dissociation of uracil. Uracil has really two dissociation constants. The pK values for tauthomeric equilibrium of two ionic forms of uracil are 9.54 and 9.66.

REINERT.- Die Dissoziationskonstanten der Purine und Pyrimidine interessierten nur in Zusammenhang mit der Berechnung von Komplexstabilitätskonstanten. Da Uracil mit Metallionen keine Komplexe bildet, wurde das Protonengleichgewicht des Uracils nicht näher untersucht.

ALTMANN.- Es ist erfreulich, daß die von uns untersuchten verschiedenen Spurenelementabspaltungen sehr gut in Zusammenhang zu bringen sind mit den von Ihnen gebrachten Stabilitätskonstanten verschiedener Metallionen, die an Nucleinsäurebasen gebunden sind. Auch bei unseren Untersuchungen war Cu gegenüber den anderen paramagnetischen Ionen am stärksten gebunden.

DELLWEG.- Haben Sie eine Vorstellung über den Aufbau der Metallkomplexe von Pyrimidinen?

WEISS.- Das unsubstituierte Pyrimidin bildet einen Kupferkomplex, indem es als einzähliger Ligand auftritt. Daraus ist ersichtlich, daß die Heterostickstoffatome zur Komplexbildung Anlaß geben können. Bei Pyrimidinderivaten, die als zweizählige Liganden auftreten, scheint der Substituent in 2-Stellung zur Chelatbildung beizutragen. Es würde dann ein dreigliedriger Ring anzunehmen sein.

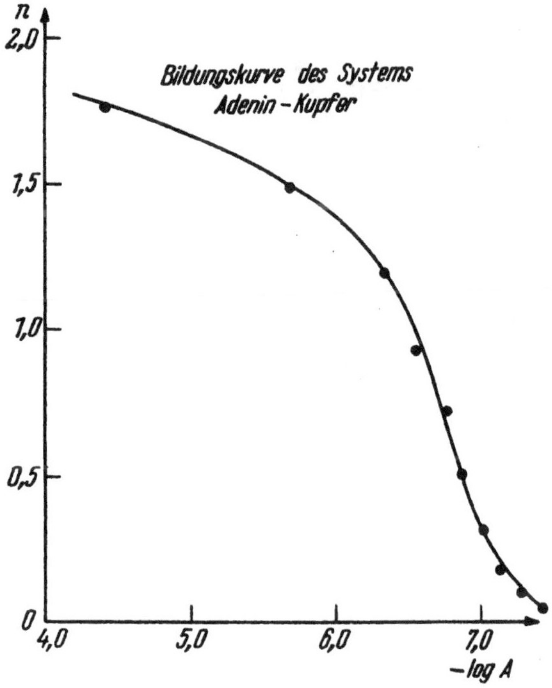

Abb.1 Bildungskurve des Systems Adenin-Kupfer(II)-ionen 2.10^{-3}m Adeninperchlorat. 2.10^{-4}m $Cu(ClO_4)_2$ in $NaClO_4$ μ=0,05 t = 25°C

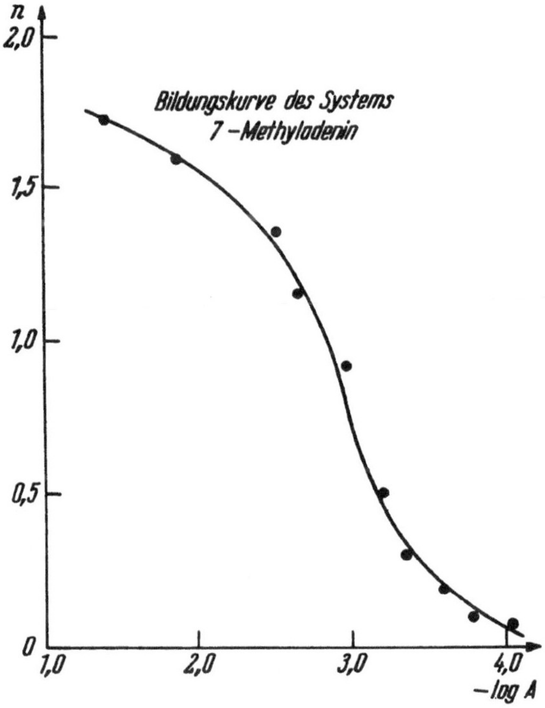

Abb.2 Bildungskurve des Systems 7-Methyl-adenin-Kupfer(II)-ionen 5.10^{-3}m 7-Methyl-adeninperchlorat 5.10^{-4}m $Cu(ClO_4)_2$ in $NaClO_4$ u=0,05 t = 25°C

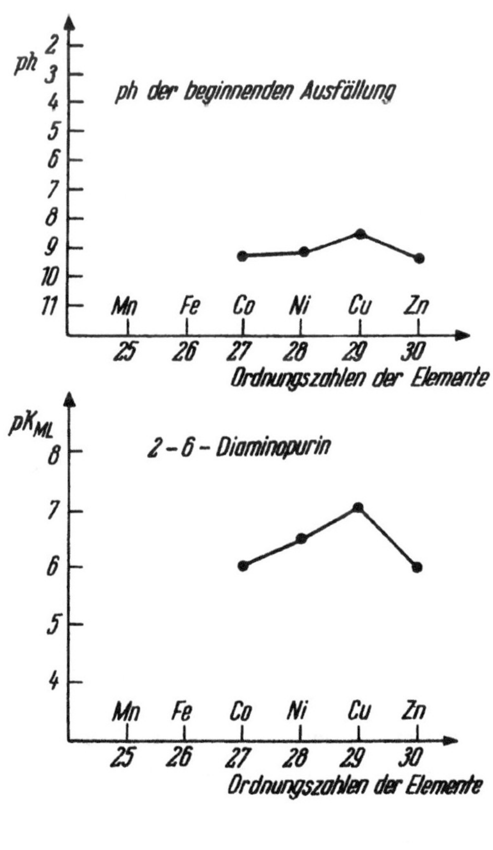

Abb.3 Relative Stabilitäten der Metallkomplexe des 2.6-Diaminopurins. a) pH bei beginnender Ausfällung aufgetragen gegen die Ordnungszahl der Elemente Co; Ni; Cu; Zn. b)Die erste Stabilitätskonstante der Metallkomplexe des 2.6-Diaminopurins aufgetragen gegen die Ordnungszahl der Elemente Co; Ni; Cu; Zn.

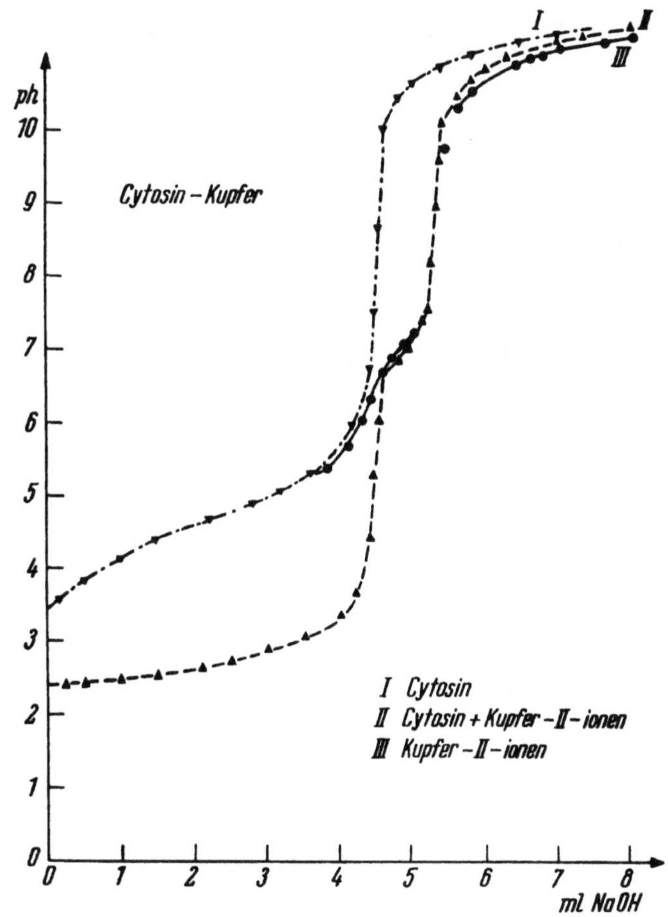

Abb.4 System Cytosin-Kupfer(II)-ionen 5.10^{-3}m Cytosinperchlorat 5.10^{-4}m $Cu(ClO_4)_2$ in $NaClO_4$
$\mu = 0{,}05$ $t = 25^oC$ titriert mit 0,1m NaOH.

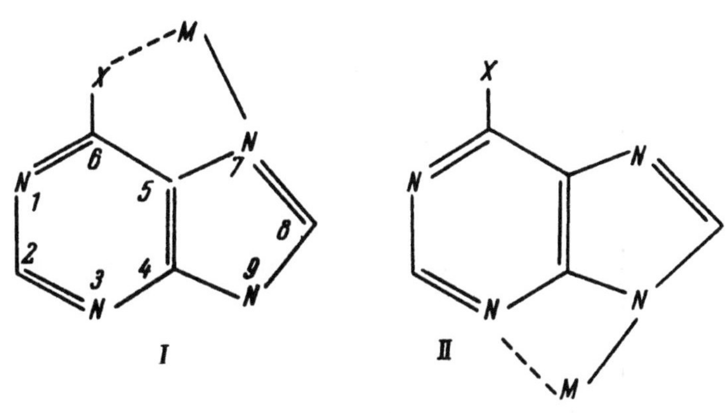

Abb.5 Bindungszentren der Purine, die für eine Anlagerung von Metallionen in Frage kommen können.

D Nucleinsäuren

Transfer of Protons between Bases of DNA

B. SUCHORUKOW, V. POLTEW and L. BLUMENFELD

Institute of Chemical Physics, Moscow

S U M M A R Y

The ionisation and protonation constants of the components of nucleic acids have been measured spectrophotometrically in the ground and the lowest excited state. The thermodynamical parameters (ΔH, ΔF, ΔS) of these processes have been estimated.

It tourns out that electron excitation leads to large changes in the acid base properties of the components of the nucleic acids; pK values vary about some units. With these above data the thermodynamical parameters of the proton transfer between N-bases in a mixture of free nucleotides and in double helical DNA are calculated. It results that in the ground state of isolated nucleotides the proton transfer is thermodynamically unfavourable. In their lowest excited state this does be favorable. In double helical DNA this also holds in the ground state of the complementary N-bases. The charge distribution in the double helical parts is an unambiguous function of the nucleotide sequence. This conclusion is confirmed by results of infrared spectroscopy of nucleic acids and of synthetic polynucleotides.

In this work, the thermodynamic parameters of ionisation of nucleic acids and of their components were measured for the ground and excited states. The pK values of corresponding equilibria for adenine, hypoxanthine, guanine, cytosine, inosine, uracyl, guanosine, adenosine were determined spectrophotometrically over large pH and temperatures ranges. The thermodynamic ionization parameters: ΔH, ΔF, and ΔS were determined as well. The results of measurements for $\mu = 0.1$ solutions at 25° C are summarized in table 1.

The presence of several tautomeric molecular and ionic forms makes the calculations somewhat more complicated. The molecules investigated proved to be capable of tautomerism, but the tautomeric equilibrium for all components shown in the table was shifted essentially towards one form. The concentration of molecules and respective ionic forms may be obtained spectrometrically when the absorption coefficients of these forms are different and their amount is within the method's sensitivity. For all compounds studied except uracyl, we succeeded in observing only one molecular and corresponding ionic form. The uracyl spectrum showed one molecular, but two singly charged ionic forms. This was revealed by analysis of spectra of 1-methyluracyl, 3-methyluracyl, and uracyl [1]. At pH = 7.2 the λ and ε values are spectral characteristics of these molecular forms, since pK is 9.95, 9.75, and 9.5, respectively [2]. In a solution with pH = 12 almost all molecules of these compounds are singly-ionized.

Table 1
The thermodynamic parameters of ionization of bases and ribosides in ground and excited states

Compound	pK	pK^*	ΔH	ΔH^*	ΔF	ΔF^*	ΔS
Cytosine	4,5	6,78	5 000	8 100	6 100	9 200	- 3,7
	11,82	7,78	11 000	5 500	16 100	10 600	- 17,1
Adenine	4,12	4,71	4 200	5 000	5 600	6 400	- 4,7
	9,72	7,3	9 500	6 200	13 200	9 900	- 12,4
Guanine	9,42	-	10 100	-	12 800	-	- 9,1
Hypoxanthine	8,8	6,45	7 200	3 200	12 000	8 800	- 16,1
Cytidine	4,22	6,86	4 400	8 000	5 700	9 300	- 5,0
Adenosine	3,55	2,6	3 800	2 500	4 800	3 500	- 3,4
Guanosine	1,6	3,07	1 000	3 000	2 200	4 200	- 4,0
	9,24	7,77	3 600	6 600	12 600	10 600	- 13,0
Inosine	8,9	7,57	7 200	5 600	12 100	10 300	- 16,4
Uridine	9,51	9,81	8 000	8 400	13 000	13 400	- 16,8

pK^*, ΔH^* and ΔF^* refer to the excited state.
The error in the pK values is $\pm$ 0,02; in $\Delta H \rightarrow \pm$ 500 cal.

Ionization of 1-methyluracyl proceeds by abstraction of protons from the N_3 form, λ_{max} of the absorption band being displaced from 259 nm to 283 nm, and ε_{max} increasing from 7 300 to 10 700. Ionization of 3-methyluracyl is due to abstraction of protons from N_1. This induces no change in λ_{max}, but ε_{max} decreases by a factor of 1.4.

Two overlapping absorption bands with λ_{max} = 260 and 284 nm are characteristics of the singly-charged ions of uracyl. These values are very close to λ_{max} for ions of N-methyl substituted derivatives of uracyl, which seems to be evidence for proton abstraction from uracyl both for the N_3 and N_1 forms. A scheme for ion-molecular conversions of uracyl and respective characteristics is summarized in table 2.

Table 2
The thermodynamic characteristics of protolitic reactions and tautomeric equilibrium of uracyl

Reaction	pK	pK^*	ΔH	ΔH^*	ΔF	ΔF^*	ΔS
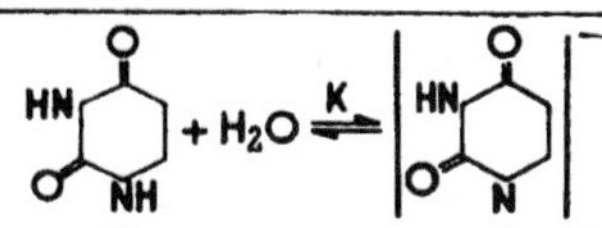 $+H_3O^+$	9,66	2,78	9 100	- 200	13 100	3 300	-13,4
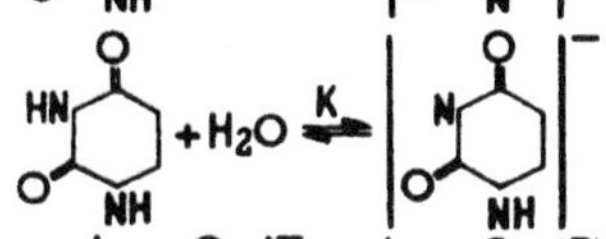 $+H_3O^+$	9,54	10,17	7 600	8 500	13 000	13 900	-13,1
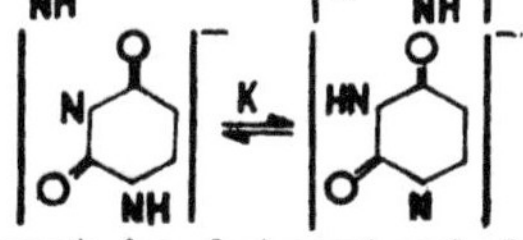	0,12	-7,39	1 500	-8 700	100	-10 100	4,7

ε cannot be determined for every individual tautomeric-ionic form of uracyl, since the latter cannot be isolated. Consequently we have used the ε value for ions of N-methyl substituted derivatives of uracyl allowing for the spectrum shift and the variations in ε_{max} induced by the CH_3 group.

RAWITCHER and STURTEVANT [2)] made calorimetric determinations of the heat of ionization for certain bases, nucleosides and nucleotides in acid media. The ΔH values obtained by us show good agreement with their results. Within our knowledge determination of ΔH for ionization of nucleic acid components in alkali media was not made before. It may be seen from the table that ΔH values for dissociation in alkali media are high compared to those for acid media. This might obtain a qualitative explanation by the fact that abstraction of protons from a neutral molecule would require more energy than in the case of a positive ion.

The ΔH for ionization of purine and pyrimidine bases contributes to the heat of "helix - random coil" transitions in nucleic acids. STURTEVANT and GEIDUSCHEK [3] suggest that the heats of ionization of bases both in native and denaturated DNA are negligeably small. However, this cannot be considered as justifiable. Certainly the heats of ionization of bases in DNA may differ from those for free bases or ribosides, but shifts of the titration curves for denaturated DNA [4] are indications of ΔH values different from zero.

The experimental ΔH value was obtained for the transfer of a proton from an A molecule to a water molecule; the initial species and reaction products were in a hydrated state. This reaction may be considered as proceeding by several steps and the heat of reaction would be then the sum of heats for elementary steps

$$\Delta H = \Delta H^{A}_{dhr} + \Delta H^{H_2O}_{dhr} + \Delta H^{A}_{ion} + \Delta H^{H_2O}_{prot} + \Delta H^{H_3O^+}_{hdr} + \Delta H^{A^-}_{hdr} \qquad (1)$$

the subscripts meaning dehydration, ionization, protonation, hydration, respectively, and the superscript: the molecules of ions participating in the reaction. By grouping the additives it may be written

$$\Delta H = \Delta H_1 + \Delta H_2 \qquad (2)$$

Here

$$\Delta H_1 = \Delta H^{A}_{ion} + \Delta H^{H_2O}_{prot} \qquad (3)$$

is the difference in proton bond energies for an A molecule and a hydroxonium ion, and

$$\Delta H_2 = \Delta H^{A}_{dhr} + \Delta H^{H_2O}_{dhr} + \Delta H^{H_3O^+}_{hdr} + \Delta H^{A^-}_{hdr} \qquad (4)$$

is the difference in hydration energies for initial species and reaction products. ΔH_2 is negative and would vary considerably with temperature due to variations in physical properties of water. The fact that the ΔH values obtained from our results are temperature independent seems to show that the amount of the additives is small. The ΔH_2 value was calculated for the ionization of arginine and glycine using the KIRKWOOD theory [5]. It appeared to be low for the secondary dissociation (pK for arginine was 9.1). We made no such calculations in view of absence of data on the distribution of charges and interatomic separations for ions of nucleic acid components.

The difference in heats of ionization of guanine and guanosine will be noted. It exceeds the possible experimental error and may be explained by different distributions of the electron density in purine bases. The latter seems to be due to the fact that the hydrogen of the imidazole ring of guanine is added to nitrogen in the N_7 position, while the ribose ring of guanosine is added to the N_9 position, i.e. to different tautomeric states of guanine.

The ΔF and ΔS values of table 1 are composite, i.e. are connected both with dissociation as such and with reactions of the solvent with the initial species and reaction products. The negative values of entropy variations, as obtained by us, may be due to ordering of water molecules in the protolytic reaction. This is in agreement with the fact that entropy variations for a reaction of the type $AH^+ \rightleftharpoons A + H^+$ are higher in absolute magnitude than those for a reaction of the type $A \rightleftharpoons A^- + H^+$. This may obtain an explanation thereby that the first reaction is isoionic, while the second reaction yields two ions and each ion may represent a centre of ordering for the water molecules by ion-dipole coupling.

The thermodynamic functions of dissociation in an excited state may be found from the spectrum shift in an ion-molecular conversion. It will be seen from fig. 1 that

$$\Delta E + D^{*} = \Delta E' + D \qquad (5)$$

where ΔE and $\Delta E'$ are energy quanta absorbed by the molecular and ionic form, respectively. D and D^{*} are dissociation energies of (ΔH^{A}_{ion}) for the ground and excited states.

Regarding (2), (3), and the low value of ΔH_2, (5) may be rewritten as

$$\Delta H^{*} = \Delta H + (\Delta E' - \Delta E) \qquad (6)$$

where ΔH^{*} is the heat of dissociation in an excited state. It will be assumed by calculating thermodynamic functions for an excited state that the absorption maxima for a molecule and for an ion correspond to the same vibrational level. Possible shifts of vibrational levels are within experimental error. The values obtained for ΔH^{*} are shown in table 1 and may be seen to differ from ΔH values for the ground state.

The change in the dissociation energy due to excitation seems to be accounted for by rearrangement of the electron density in a molecule (changes in the bond orders and charges on the atoms). In a protonated cytosine structure, for instance, this rearrangement is such that the electronic density increases on nitrogen which is responsible for proton dissociation. In future it might be possible to determine dipole moments for the excited states from excitation-induced changes in ΔH. Assuming that the entropy change in dissociation is similar both for the ground and the excited state, an equation will be obtained for pK in the excited state

$$pK - pK^{*} = \frac{\Delta E - \Delta E'}{2.3RT} \qquad (7)$$

The pK values obtained and the resulting ΔF values are shown in table 1. The basicity of bases and the acidity of acids usually increase with excitation and may amount to several pK units. Effects of this kind seem to play an essential part in the photochemistry of purines, pyrimidines and other conjugated molecules of importance for biology. It will be noted that perturbation with a similar redistribution of the electron density may occur in exothermic dark reactions and this would also change the acid-base properties. With tautomeric conversions excitation may bring about a considerable shift in the tautomeric equilibrium. An equilibrium shift of this kind is shown for uracyl in table 2.

Proton transfer may occur between an acid and a base molecule at certain acidity and basicity values. For instance, the reaction of proton transfer from uracyl to adenine

$$U + A \underset{}{\overset{K}{\rightleftharpoons}} U^{-} + AH^{+} \qquad (8)$$

may be conceived as proceeding by two steps

$$U \overset{K_1}{\rightleftharpoons} U^{-} + H^{+} \qquad (9)$$

$$AH^{+} \overset{K_2}{\rightleftharpoons} A + H^{+} \qquad (10)$$

Then the equilibrium constant for proton transfer will be

$$k = \frac{k_1}{k_2} \qquad (11)$$

The constants for formation of ionic pairs are shown in table 3.

Table 3

The pK and ΔH values of the proton transfer between bases and ribosides in ground and excited states.

Ion pair	pK	pK^*	ΔH	ΔH^*
$CH^+ + A^-$	5,22	0,52	4 500	- 1 900
$CH^+ + U^-$ (1)	5,16	- 3,0	4 100	- 8 300
$AH^+ + U^-$ (1)	5,42	- 1,93	4 900	- 5 200
$RCH^+ + RU^-$ (2)	5,29	2,95	3 600	400
$RSH^+ + RS^-$	5,02	0,91	4 200	- 1 400
$RAH^+ + RU^-$	5,96	7,21	4 200	6 100

(1) These data are for the uracyl in ionic form with hydrogen in position N_1.
(2) R - riboside.

It will be seen that in the mixture of bases (or ribosides) there will be practically no proton transfer. The calculated heats of these reactions show that the latter are unprofitable from the energy standpoint, and this has obtained experimental confirmation [6]. The excited state would make the proton transfer profitable even for mixtures of bases and ribosides.

While proton transfer between two free bases is unprofitable, it may occur for purine and pyrimidine bases ordered along a double helical DNA chain, due to the additional energy from electrostatic interaction between the charges formed. For free charged nucleotides this energy would be low due to the high dielectric constant value ε and the intermolecular separations r. For ordered regions of DNA or RNA molecules, where the separation between charges arising by proton transfer is close to 3 Å, the electrostatic interaction energy $E = -\frac{e_I e_2}{r}$ would be higher in absolute magnitude than the ΔH value for proton transfer at an effective $\varepsilon < 20$. The ε value inside DNA obviously satisfies these conditions. Proton transfer between bases in DNA proceeds through one of the hydrogen bonds, while the second seems to become stronger due to rearrangement of charges in the molecule. The idea of proton transfer between bases was suggested by Japanese scientists [7]. A 1710 cm^{-1} band was shown to occur in the infrared spectrum of DNA; it disappeared by destruction of secondary structure, while a 1690 cm^{-1} band appeared in the spectrum of denaturated DNA. Destruction of the secondary DNA structure effects changes in other bands as well. The temperature dependence of the 1710 cm^{-1} band intensity follows the changes in the 260 nm band in the UV region: the decrease in intensity of the 1710 cm^{-1} band corresponds to increasing absorption for the 260 nm band. The authors suggest that the absorption in the 1710 cm^{-1} region is due to vibration of C = O and C = N bonds of the protonated bases. They do not estimate the probability of proton transfer from the energy standpoint, and the only argument in favour of their suggestion are the results of the investigations on infrared spectra of complex nucleoside crystals. However, it was shown [6] that this was not the right argument in favour of proton transfer in nucleic acids. Proton transfer could obtain confirmation in the infrared spectra of synthetic nucleotide complexes. A double-helical complex was shown to form by mixing polyuridinic (poly-U) and polyadenilic (poly-A) acids in a ratio of 1:1 [8]; its spectrum was different from the additive spectrum shown by components of this complex. A

1672 cm^{-1} band appeared in the high-frequency region of multiple bonds. Before the complex was formed the highest frequency of this region was 1661 cm^{-1}. The poly-U + poly-A complexing results in a 39 per cent decrease in intensity of the 1727 cm^{-1} band and to its shift towards higher frequencies. When complexes are formed by polyinosinic and polycytidilic acids, there appears a 1697 cm^{-1} absorption band, instead of the high frequency band in the multiple bond region. The appearance of absorption bands in the spectra of double-helical synthetic polynucleotide complexes may be considered as evidence for protonation of bases as a result of a proton transfer between the purine and pyrimidine bases. Indeed, the highest frequencies in the multiple bond regions of adenosine and cytidine spectra are 1660 and 1654 cm^{-1}. The addition of proton to these bases is accompanied by the appearance of intensive 1724 and 1685 cm^{-1} bands for cytidine and adenosine respectively [9)-10)]. These frequencies are due to the vibrations of C = O and C = N bonds of the protonated bases. Ionization of the uracyl and inosine bases results in lower frequency in the multiple bond region.

We have studied the infrared spectra of DNA as a function of pH and it was found that the 1712 cm^{-1} band appears at neutral and acidic values of pH and disappears at alkalinic pH values. This seems to be another confirmation of the proton transfer concept. The ideas discussed on the structure of DNA as consisting of two charged chains kept together not only by H-bonds, but by electrostatic interaction as well, may obtain support in the high energy of DNA denaturation per one pair of nucleotides. This value was shown to be 11 kcal/mole [10)]. Without the assumption on the ionic structure of DNA it would be difficult to conceive how are these 11 kcal spent, since rupture of the ordered net of hydrogen bonds in nucleic acids is followed by the formation of disordered H-bonds linking bases with bases, and bases with water molecules, and their number may be no less than that in the original DNA molecule. The energy of denaturation per one hydrogen bond of protein molecules does not exceed 1.5 kcal [11)].

The molecules of nucleic acids involve also disordered regions where proton transfer does not occur as the separation between bases is great and occupied by water. A certain part of disordered DNA regions may be in dynamic equilibrium with the ordered charged regions, i.e. mutual conversion of the two types is possible. Some regions where proton transfer does not occur may be structured by helating effect of metals. Individual regions of the DNA structure suggested may perform different biological functions.

It will be emphasized that the pattern of charges in ordered regions is an unambiguous function of the sequence of nucleotides.

R E F E R E N C E S

1) SHUGAR,D., J.J.FOX, Biochim. biophysica Acta 9, 199 (1952)
2) RAWITCHER,M., J.M.STURTEVANT, J. Amer. chem. Soc. 82, 3739 (1960)
3) STURTEVANT,J.M., E.P.GEIDUSCHEK, J. Amer. chem. Soc. 80, 2911 (1958)
4) COX,R.A., A.R.PEACOCKE, J. chem. Soc. 1957, 4724
5) DATTA,S.P., A.K.SRZYBOWSKI, Biochem. J. 78, 289 (1961)
6) SUKHORUKOV,B.I., G.I.MATKHANOV, Biofisika 8, 134 (1963)
7) KUGOKU,J., M.TSUBOI, T.SHIMANOUCHI, J.WATANABE, Nature 189, 120 (1961)
8) TODD MILES,H., Nature 183, 1814 (1959)
9) SUKHORUKOV,B.I., Biofisika 7, 664 (1962)

10) SUKHORUKOV,B.I., Yu.Sh.MOSHKOVSKII, T.M.BIRSTEIN, V.N.LYSTZOV, Biofisika 8, 294 (1963)
11) LEECH,S.J., Usspechi Chimii 31, 1104 (1962).

DISCUSSION QUESTION

LÖBER.- From what place of the uracyl molecule you suppose, **that** proton dissociates?

BLJUMENFELD.- I think - from the N-atom.

LÖBER.- In consideration the theory of FÖRSTER we should suppose, that proton leaves the molecule from oxygroup, because in excited state this group is more acid than in groundstate. In contrary, the ring nitrogen is more basic.

BLJUMENFELD.- Yes, I think you are right. It can be true for the excited state. Thank you.

FRITZSCHE.- Wie erklären Sie die Diskrepanzen Ihrer infrarotspektroskopischen Messungen mit denen von TSUBOI, der bereits im Gemisch der Nucleoside eine 1710 cm^{-1}-Bande fand, die er dem Protonentransfer zuordnete?

BLJUMENFELD.- I do'nt know. The data of SUKHORUKOV and MATKHANOV are in agreement with the data of TODD-MILES.

ZAHRADNIK.- What do you mean when you are speaking about the excited state? Is it a state formed by the transition of an electron from the highest occupied π-molecular orbital (MO) to the lowest free π-MO?

BLJUMENFELD.- Yes, this is an excited states that corresponds to the absorption band at 260 nm. It is essentially a π-π-transition.

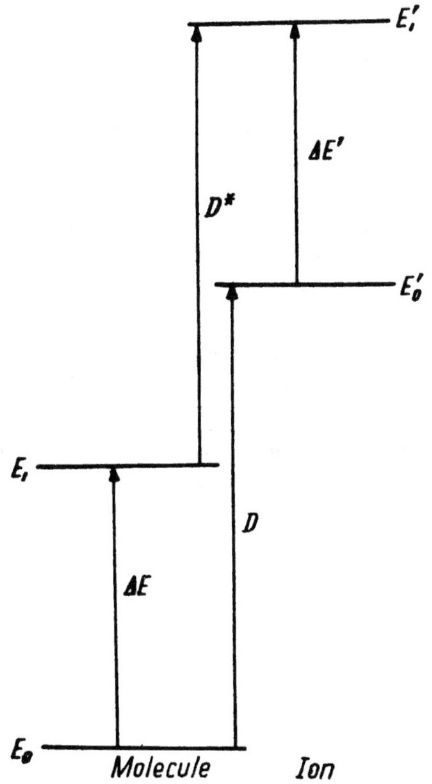

Fig.1 The energies of dissociation in the ground and excited state

D Nucleinsäuren

Eine Bemerkung zu den Elektronenspektren von Pyrimidinen, Purinen und Pteridinen

R. ZAHRADNIK

Institut für physikalische Chemie der Akademie, Prag

S U M M A R Y

A fairly close correlation has been found between the excitation energies ($\tilde{\nu}$) of the first intense bands in electronic spectra of pyrimidines, purines, and pteridines and the $N \longrightarrow V_1$ energies ($E(N \longrightarrow V_1)$) previously computed by A. and B. PULLMAN by means of the simple MO-LCAO method. The data for 32 compounds satisfy the following equation:

$$\tilde{\nu}\,(\text{kcm}^{-1}) = 21.760\ E(N \longrightarrow V_1)\ (\beta) + 6.355, \qquad (1)$$

the correlation coefficient being 0.955. Besides the theoretical $N \longrightarrow V_1$ energies mentioned, the values of these energies for two other sets of empirical parameters were computed for some representatives. It was found that Eq. 1 remains essentially unchanged.

Im Zusammenhang mit der Vorbereitung des Studiums der Elektronenstruktur der Nucleinsäuren haben wir die Anwendbarkeit der einfachen Methode der Molekülbahnen (MO-LCAO, Hückel'sche Methode, HMO) für eine Abschätzung der Anregungsenergien der ersten intensiven Banden der Nucleinsäurebasen, nämlich von Pyrimidinen und Purinen, untersucht. Zu den studierten Verbindungen haben wir Pteridine zugereiht, weil diese Stoffe strukturähnlich den Purinen sind. Unter den Verbindungen, mit welchen wir uns befaßt haben, waren nicht nur die Stoffe, welche aus der Natur bekannt sind, sondern auch einige synthetische Analoge, die vom therapeutischen Standpunkt interessant sind.

An erster Stelle war es notwendig, die folgenden Fragen zu entscheiden: 1) die tautomere Form der untersuchten Stoffe, 2) die Zuordnung der Banden zum $N \longrightarrow V_1$-Übergang, 3) die Werte der empirischen Parameter der HMO-Methode.

In den letzten Jahren [1)-3)] wurde nachgewiesen, daß in neutralen Lösungen das Laktam-Laktim-Gleichgewicht in der Richtung der Laktam-Form verschoben ist, im Falle des Imin-Enamin-Gleichgewichts die Enamin-Form preferiert ist. Bei vielen studierten Verbindungen sind die $n \text{---} \pi^*$ -Banden durch $\pi \text{---} \pi^*$ Banden überlappt; wo das nicht der Fall ist (z. B. beim Thiouracil) wurde die relativ wenig intensive $n \text{---} \pi^*$-Bande durch den Lösungsmitteleffekt entdeckt [3)4)]. Weil die Absorptionskurven der studierten Stoffe keine Feinstruktur aufweisen, wurde im allgemeinen die erste Bande ($\log \varepsilon$ gewöhnlich höher als 3,2) als $N \longrightarrow V_1$-Bande betrachtet. Schon früher haben wir nachgewiesen, daß beim Vergleich der experimentellen Daten mit den theoretischen Angaben die Auswahl von numerischen Werten der empirischen Parameter der HMO-Methode keine kritische Frage ist. Die weiter erwähnten Resultate sind im Einklang mit dieser Erfahrung.

Aus der Tab. I und Abb. 1 wird klar, daß zwischen den experimentellen ($\tilde{\nu}_{exp}$) und theoretischen ($E(N \longrightarrow V_1)$)-Angaben eine Parallelität existiert.

Tabelle I

Anregungsenergien $\tilde{\nu}_{exp}$ der ersten Banden in Pyrimidinen, Purinen, Pteridinen und einigen verwandten Verbindungen und die theoretischen $N \longrightarrow V_1$-Energien

	Verbindung	$\tilde{\nu}_{exp}$ (kcm^{-1})	Lit.	$E(N \longrightarrow V_1)$ (nach 1)
1	Pyrrol	47,6	8	1,812[a]
2	Imidazol	>46,8	8	1,820
3	Pyridin	50	9	1,871[a]
4	Pyrimidin	>48	8	1,883
5	Indol	34,8	8	1,397
6	Benzimidazol	35,8	8	1,482
7	Cytosin	36,5	12	1,390
8	5-Methylcytosin	35,4	2	1,326
9	Orotsäure	35,8	2	1,403
10	Thymin	37,9	2	1,468
11	Uracil	38,6	12	1,557
12	5-Azauracil	42,5	15	1,670
13	6-Azauracil	38,6	3	1,405
14	2,4-Dithiouracil	29,3[b]	8	1,228[b]
15	2,4-Dithio-6-azauracil	31,2	4	1,091[a]
16	Adenin	38,0	11	1,351
17	Purin	38,5	13	1,428
18	2-Aminopurin	31,9	13	1,257
19	2,6-Diaminopurin	35,8	11	1,315
20	6-Amino-8-hydroxypurin	36,8	8	1,394
21	Guanin	36,4	11	1,357
22	Hypoxanthin	40,4[b]	11, 12	1,284[b]
23	Isoguanin	35,4	11	1,318
24	Harnsäure	35,4	10	1,366
25	Xanthin	37,4	11, 12	1,594
26	Pteridin	33,6	8	1,250
27	2-Hydroxypteridin	32,6	8	1,083
28	4-Hydroxypteridin	32,2	8	1,213
29	7-Hydroxypteridin	33,0	8	1,134
30	2-Aminopteridin	27,0	8	1,039
31	4-Aminopteridin	29,8	8	1,082
32	6-Aminopteridin	27,6	8	1,010
33	2-Hydroxy-4-aminopteridin	29,8	8	1,189
34	2-Amino-4-hydroxypteridin	29,4; 32,0	8	1,139
35	2-Amino-4,6-dihydroxypteridin	26,0	8	0,801

[a] Eigene Berechnung für Parameter A (Tab. III).

[b] Eine kritische Überprüfung der Angaben ist notwendig; die Werte waren nicht in Betracht gezogen für die Berechnung der Regressionsgeraden.

Es ist ja so, daß der Korrelationskoeffizient der Abhängigkeit (1) (a und b bezeichnen die Konstanten) kleiner ist als bei anderen Gruppen konjugierter Verbindun-

$$\tilde{\nu}_{exp}\ (\mathrm{kcm}^{-1}) = a \cdot E(N \longrightarrow V_1)(\beta) + b \tag{1}$$

gen, die wir früher untersucht haben, doch ist die Abhängigkeit von $\tilde{\nu}_{exp}$ von $E(N \longrightarrow V_1)$ statistisch hoch signifikant (Tab. II).

Tabelle II

Konstanten a und b der Gl. (1) und Korrelationskoeffizient r

Verbindungen	n^{+}	a	b	r
Pyrimidine, Purine und Pteridine	32	21,760	6,355	0,955
Benzenoide Kohlenwasserstoffe [16]	44	19,207	10,234	0,982
Pyridinartige heterocyclische Verbin. [9]	22	24,031	5,584	0,994
Derivate benzenoider Kohlenwasserstoffe [17]	43	20,446	10,953	0,982

$^{+}$ Anzahl der Verbindungen.

Aus Tab. II folgt, daß die Angaben für Pyrimidine, Purine und Pteridine in einem Gebiet liegen nahe den Daten für benzenoide Kohlenwasserstoffe und verwandte Verbindungen. Hier möchten wir die Arbeit von ISENBERG und SZENT-GYÖRGYI [5] erwähnen, in welcher ein Versuch unternommen wurde, die Anregungsenergien von biologisch interessanten Verbindungen abzuschätzen. Die erwähnte Arbeit ist aber mit gewissen Ungenauigkeiten belastet, die typisch für alle älteren Arbeiten dieser Art waren, d. h. die Anzahl von untersuchten Stoffen war klein, die Verbindungen waren strukturell nicht genügend verwandt und letztens die Verfasser setzten voraus, daß die Abhängigkeit der experimentellen und theoretischen Angaben den Nullpunkt schneidet, was nicht der Fall ist.

Tabelle III

Die Coulomb'schen ($\alpha_X = \alpha + \delta_X\beta$) und Resonanz-($\beta_{XY} = \rho_{XY}\beta$)-Integrale (Definition siehe z. B. Ref. 6)

Bindung	ρ_{XY}			δ_X^a		
	A^{1,b}	B^{6,b}	C^{7,c}	A^{1,b}	B^{6,b}	C^{7,c}
C − N<	0,9	0,8	1	1,0	1,5	1,0
C = N—	1,0	1,0	1	0,4	0,5	0,5
C = O	2,0	1,0	$\sqrt{2}$	1,2	1,0	2,0
C = S	1,2	0,9^{d}	0,9	0	0,5^{d}	0,5

^{a} Betrifft das Heteroatom der angeführten Bindung.

^{b} Induktiver Effekt nicht in Betracht gezogen.

^{c} Induktiver Effekt in Betracht gezogen ($\delta_X = 0{,}1 \sum_i \delta_i$).

^{d} Nach der Arbeit 18).

In Tab. III sind die Werte der Coulomb'schen (α_X) und Resonanz-(β_{XY})-Integrale der HMO-Methode angeführt. Neben den Werten von PULLMAN und PULLMAN [1] (A) haben wir auch die Werte nach STREITWIESER [6] (B) und die Werte, die wir in einer Studie über Azauracile benutzt haben [7] (C), untersucht. Wir haben gefunden, daß der Charakter der Abhängigkeit (1) nicht von den angeführten Werten der Parameter α_X und β_{XY} abhängig ist.

Man kann folgendermaßen zusammenfassen: Die HMO-Methode ist für Abschätzungen der Anregungsenergie der ersten Banden der Nucleinsäurebasen geeignet und deshalb kann man annehmen, daß die Berechnungen der π-Elektronen-Struktur der Nucleinsäuren mit derselben Methode nicht unrealistisch sein werden.

L I T E R A T U R

1) PULLMAN,B., A.PULLMAN, "Quantum Biochemistry", Interscience Publ., New York, 1963
2) SHUGAR,D., J.J.FOX, Biochim. biophysica Acta 9, 199 (1952)
3) JONÁŠ,J., J.GUT, Collect. czechoslov. chem. Commun. 26, 2155 (1961)
4) JONÁŠ,J., J.GUT, Collect. czechoslov. chem. Commun. 27, 1886 (1962)
5) ISENBERG,I., A.SZENT-GYÖRGYI, Proc. Nat. Acad. Sci. USA 45, 519 (1959)
6) STREITWIESER,A., Jr. "Molecular Orbital Theory for Organic Chemists", S. 135, J. WILEY, New York, 1961
7) ZAHRADNÍK,R., J.KOUTECKÝ, J.GUT, J.JONÁS, Collect. czechoslov. chem. Commun. 28, 1499 (1963)
8) "Organic Electronic Spectral Data", Vol I, Interscience Publ., New York, 1960
9) KOUTECKÝ,J., R.ZAHRADNÍK, Collect. czechoslov. chem. Commun. 28, 2089 (1963)
10) CAVALIERI,L.F., A.BENDICH, J. Amer. chem. Soc. 72, 2587 (1950)
11) CAVALIERI,L.F. und Mit., J. Amer. chem. Soc. 70, 3878 (1948)
12) HOTCHKISS,R.O., J. biol. Chemistry 175, 315 (1948)
13) MASON,S.F., J. chem. Soc. London 1954, 2071
14) WYATT,G.R., S.S.COHEN, Biochem. J. 55, 774 (1953)
15) JONÁŠ,J., M.HORÁK, A.PISKALA, J.GUT, Collect. czechoslov. chem. Commun. 27, 2754 (1962)
16) KOUTECKÝ,J., J.PALDUS, R.ZAHRADNIK, J. chem. Physics 36, 3129 (1962)
17) ZAHRADNIK,R., J.KOUTECKÝ, Collect. czechoslov. chem. Commun. 28, 904 (1963)
18) ZAHRADNIK,R., J.KOUTECKÝ, Collect. czechoslov. chem. Commun. 28, 1117 (1963)

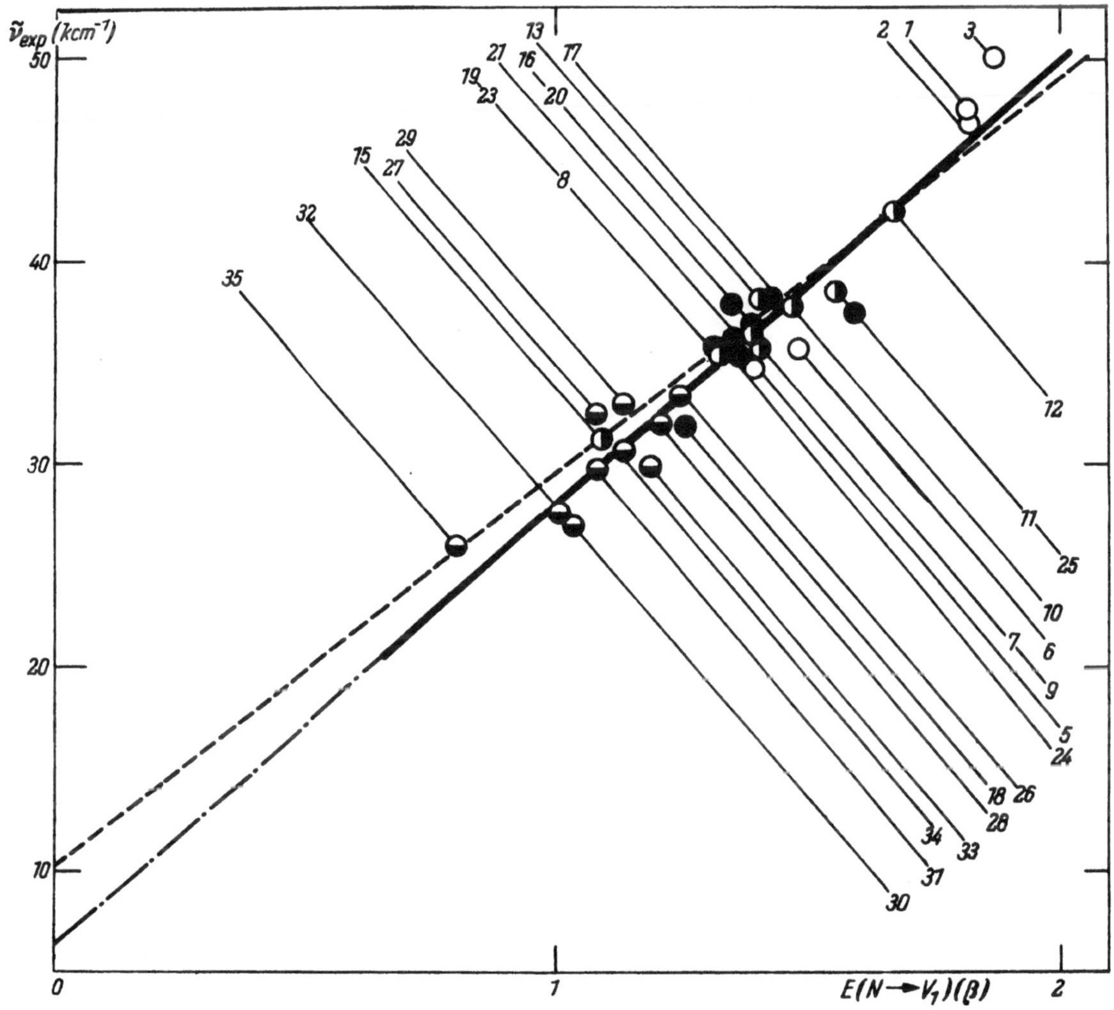

Abb.1 Abhängigkeit der Anregungsenergien der ersten Banden von den $N \rightarrow V_1$ Energien (s.Tab.1!). O Pyrimidine, ● Purine, ◒ Pteridine O Grundsubstanzen. Für Vergleichszwecke wurde die Regressionsgerade (- - - -) für benzoide Kohlenwasserstoffe angeführt.

D Nucleinsäuren

Correlation between the Electronic Structure and the Properties of DNA

J. LADIK and G. BICZÓ

Central Research Institute for Chemistry of the Academy, Budapest

S U M M A R Y

The LCAO MO-s and their energies have been computed in the HÜCKEL approximation neglecting overlap in the single nucleotides, base pairs, dinucleotides and polynucleotides. An outline is given of the way in which from these data the energy band structure of a real DNA molecule should be approximated. Results of these calculations and of a CI-calculation of uracil have opened the way towards the quantum chemical interpretation of the spectroscopic properties of DNA (the positions of the absorption maxima and the hypochromicity of DNA).

The results of the energy band calculation allow for a qualitative interpretation of the semiconductivity of DNA found experimentally.

On the basis of the results obtained it was possible to establish a working hypothesis for the mechanism of DNA duplication and tumor development and to give a possible interpretation of the mutagenic effect of UV-radiation. The connection between the present hypothesis and the new mutation hypothesis of LÖWDIN, based on double proton tunneling in the hydrogen bonds of the base pairs is discussed.

Finally the measured effect of electrostatic field on mitosis is considered.

DNA, as it is well-known, is the material carrier of genetic information. The sequence of its nucleotide bases determines - through RNA - the sequence of the amino acids in proteins; through the proteins thus coded, the DNA particles of a cell determine all biochemical reactions of the cell and so its structure and function.

The WATSON-CRICK stereomodel of DNA having revealed the positions of the atomic nuclei in this macromolecule, it was possible to set about the calculation of the electronic structure of DNA with the aid of quantum chemical, respectively solid state physical methods.

The quantum mechanical calculation of the electronic structure of DNA can be built up of the following steps :

1) Calculation of the energies and wave functions of the delocalized π electron systems of the single bases in their tautomeric form, stable under *in vivo* conditions (neutral pH).
2) Calculation of the delocalized π electrons in a whole base pair (assuming a π electron interaction through the hydrogen bonds).
3) The treatment of all the 16 possible dinucleotides, as common π electron systems. Between two superimposed nucleotide bases there is namely, as shown by LADIK, a

non negligible π electron overlap causing a π electron interaction between them [1]. The detailed investigation of these systems has shown the amount of this interaction to be a function of the sequence of the bases [1) 2)].

4) Calculation of the common π electron systems of two interacting base pairs.
5) Determination of the energy band structure of an infinite chain, containing the same base (polynucleotide).
6) Calculation of the energy bands of a double helix, containing the same base pair as elementary cell (see Fig.1).
7) Approximation of the real band structure of DNA.

Using the simple LCAO MO (HÜCKEL) approximation neglecting overlap, steps 1), 2), 3) and 5) have been already done [2)3)], by solving the appropriate matrix eigenvalue problems with the aid of the Ural I. and M3 electronic computers in Hungary, respectively the Alwax III.E. computer of the Quantum Chemistry Group at Uppsala. In all calculations the real geometry of the molecule was taken into account [4)5)]. The HÜCKEL $\beta_{i,j}$ parameters between atoms belonging to different superimposed bases have been estimated on the basis of the appropriate overlap integrals. It should be mentioned that by using a somewhat different set of parameters, PULLMAN and PULLMAN have also performed steps 1) and 2) [6)7)]. Calculations needed for steps 4), 6) and 7) are in progress.

In the case of U, a comparison of the results of different higher approximations [8)9)] has shown that the semiempirical configuration interaction (CI) method without selfconsistency, gives an excellent agreement with the experimentally found first two absorption maxima of this compound [10)]. (Experimental values : $3{,}88 \cdot 10^4$ cm^{-1}, resp. $4{,}93 \cdot 10^4$ cm^{-1}; theoretical values : $3{,}78 \cdot 10^4$ cm^{-1} and $4{,}94 \cdot 10^4$ cm^{-1}). A wider application of this type of calculations to the other nucleotide bases and to all the 16 possible dinucleotides, is being dealt with at present.

The calculation in the HÜCKEL approximation of the oscillator strenght values of the first four one-electron transitions of the dinucleotides, has shown that as a consequence of the interaction between the superimposed bases, in most of the 16 cases a decrease has occured in the absorption maxima. This fact gives the first step in the quantum mechanical interpretation of the well-known hypochromicity of DNA. It is to be expected that the above mentioned extension of the semiempirical CI calculations will allow a fuller theoretical interpretation of the spectra of the bases and of DNA hypochromicity.

Taking into account that in DNA there exists a π electron interaction between all neighbouring superimposed bases within one chain, we can calculate the band structure of this one-dimensional solid [3)]. This gives immediately a qualitative interpretation of the semiconductivity of DNA found experimentally by DUCHESNE [12)], resp. by ELEY and coworkers [13)]. The experimentally found forbidden band-width between the highest filled and the lowest unfilled band (1,80 eV as found by DUCHESNE and 2,40 EV as found by ELEY) does not agree with the values obtained in the energy band calculation for single chains[3)]. It is, however, possible with the aid of simple HÜCKEL-calculations to be yet performed for a DNA double helix to show that the theoretical value lies between 1,80 and 2,40 eV[2)]. It is, therefore, to be expected that the above mentioned extension of the energy band calculations and the improvement of the approximations will also lead to a quantitative interpretation of the semiconductivity of DNA.

Measurements have proved DNA to possess conductivity also in its ground state [12)13)]. In other words, there are always electrons, resp. positive holes in its conductivity, resp. valence band, due to excitation caused by heat or natural radiation (u.v. sunshine radiation, radioactive and cosmic radiation) or to the effects of electron donor or acceptor molecules present in the cell. It may be assumed further that in a cell there are always strong local electrostatic fields (as for instance the inner field of a nucleoprotein and the local field arising from other molecules with permanent dipoles, charge fluctuations in the solution, etc.). These local electrostatic fields may cause the ends of the DNA double helix to polarise and in consequence the two charged ends of the molecule which have the same charge will repel each other. A classical estimation of the repulsion energy of two netto elementary charges in a base pair, carried out together with HOFFMANN, has shown this energy to be sufficient to bring about the breakage of the hydrogen bonds of two or three base pairs [14)2)]. Since, according to the WATSON-CRICK semiconservative duplication mechanism of DNA, the unwinding of the two helices is a condition necessary for DNA duplication to take place, according to the above mechanism the polarization of the double helix may induce DNA duplication. It would thus seem justified to introduce the hypothesis that DNA duplication and through it the frequency of mitoses in the cells are being regulated by the number of charge carriers present in DNA and by the electrostatic field inside it, or in other words, by the joint effects of the temperature, the natural radiation level, the concentration of electron donor or acceptor molecules in the cell, as well as of the concentration of molecules producing electrostatic fields.

To test this hypothesis normal tissue cultures (chicken-embryo heart cultures prepared by MAXIMOVs method) have been placed into an electrostatic field of a nominal field strength of 15 KV/cm for 8 hours. The field has been found to have a significant mitosis-increasing effect (fixing of the cultures was performed after 18 hours exposure to the field) [15)]. In testing our theory this was, of course, only the first step and though further investigations in this direction are in progress, it can be stated that the first results are at least not in contradiction with our assumption.

In the case of excitation or ionization by radiation or by carcinogens, which according to recent investigation are rather strong electron donors or acceptors [16)], the number of charge carriers in the valence, respectively conductivity band will be increased. This may cause a more intensive polarization of the ends of DNA and with it, according to the above mentioned mechanism, an increase in the number of DNA-duplications. If, owing to the effect of radiation or carcinogens, a large number of DNA duplications takes place in the cells of a tissue, the result may be a new stationary state in the metabolic equilibrium of the cell, in which the rate of DNA synthesis remains faster over a longer period. Perhaps the growth-inducing effect of low doses of radiation may be interpreted in this way.

On the other hand it would seem probable that tumors are formed in consequence of specific changes in the genetic code, changes which may have a modifying influence on the synthesis of those proteins which are the enzimes controlling the growth-cycle of the cell [17)]. These "cancerous DNA sequences" can be produced by the accumulation of errors entering into the genetic code during DNA duplication. This would explain why the occurrence of spontaneous tumors is more probable in old age.

Radiations and carcinogens may have thus a twofold tumor-inducing effect. By increasing the rate of DNA duplication they increase the probability of "spontaneous tumor development". A cell, namely, in which DNA duplication proceeds for a longer time at an increased rate, becomes "old" more quickly than a normal cell, so that the accumulation of errors in the genetic code will proceed also more rapidly.

On the other hand it is most likely that also the probability of the appearance of errors in the course of one DNA duplication-cycle is increased if DNA is ionized or excited. In this connection it is of interest to mention the mutation mechanism recently proposed by LÖWDIN [18)]. According to this author there is a possibility for a nucleotide base pair to go over by double proton tunneling in its unusual tautomeric form ($A{-}T \rightarrow A^{*} - T^{*}$, $G{-}C \rightarrow G^{*} - C^{*}$) (see Fig.2). Should this happen during the duplication of DNA, the two members of a base pair may remain in their unusual tautomeric form (denoted by a star) in both chains. Since for bases in their unusual tautomeric form the relations A^*-C, T^*-G, G^*-T, C^*-A are valid [18] instead of the base pairing relations A-T, G-C, errors will be introduced in the genetic code of the new growing chains.

If a base is ionized by the effect of radiation or by carcinogen attack, there is also the probability that both protons of the two hydrogen bonds get located at the same base[18]. If this happens again during duplication, one base will have no proton, while the other will have them both. Such bases exclude the possibility of the specific binding of an other nucleotide base, so that the genetic information carried by them ($_2\log 4 = 2$ bit) gets lost. The loss of genetic information taking place in this way, represents perhaps the molecular mechanism [18] of the deletion hypothesis of cancer [17].

Calculations of the π electron desnsities of the nucleotide base pairs in their several excited states (performed again in the simple HÜCKEL approximation) show for the case of a very probable transition of the G-C base pair that a charge redistribution occurs which makes the LÖWDIN-type simultaneous tautomeric shift of a base pair more probable than it is in the ground state. This would offer a possible interpretation of the mutagenic and tumor-inducing effects of u.v. radiation [20]. On the other hand, evidence obtained from detailed investigations has shown that both in the A-T and the G-C base pairs there are excited states in which the probability to have both protons of the two hydrogen bonds at the same bases will increase. In the light of the above described mechanism this fact gives a further possible interpretation of the tumor inducing effect of u.v. radiation.

L I T E R A T U R E

1) LADIK,J., Acta physica Acad.Sci.hung. 11, 239 (1960)
2) HOFFMANN,T.A., J.LADIK, Adv. in Chem.Phys. (in the press)
3) LADIK,J., K.APPEL, Preprint N° 78, Quantum Chemistry Group, University of Uppsala (1962)
4) SPENCER,M., Acta crystallogr. [London] 12, 59 and 66 (1959)
5) LANGRIDGE,R., H.R.WILSON, C.W.HOOPER, M.H.F.WILKINS, L.D.HAMILTON, J.molecular Biol. 2, 38 (1960)
6) PULLMAN,A., B.PULLMAN, Bull.Soc.chim. France, 766 (1958); 594 (1959)
7) PULLMAN,A., B.PULLMAN, Biochim.biophysica Acta, 36, 343 (1959)
8) VEILLARD,A., B.PULLMAN, Comptes Rendus 253, 2277 (1961); J.theoretical Biol. 4,37(1963)
9) LADIK,J., K.APPEL, Preprint N° 79, Quantum Chemistry Group, University of Uppsala (1962)
10) LANG,L., J.LADIK, Über die Absorptionsspektren der Nukleotid-basen. Lecture at the VIIth European Congress on Molecular Spectroscopy, Budapest, 1963
11) HOFFMANN,T.A., J.LADIK, J.theoretical Biol. (in the press)
12) DUCHESNE,J., J.DEPIREUX, A.BERTINCHAMPS, N.CORNET, J.M.van der KAA, Nature 188, 405 (1960)
13) ELEY,D.D., D.I.SPIVEY, Trans.Faraday Soc. 58, 411 (1962)
14) HOFFMANN,T.A., J.LADIK, Cancer Research 21, 474 (1961)
15) BOZOKY,L., Gy.KISZELY, T.A.HOFFMANN, J.LADIK, Nature (in the press)
16) ALLISON,A.C., T.NASH, Nature 197, 758 (1963)
17) J.Carcinogenesis, Mechanism of Action. Ciba Found.Symp.,1958.(Churchill, London,1959)
18) LÖWDIN,P.O., Preprint N° 85,Quantum Chemistry Group,University of Uppsala (1962)
19) WATSON,J.P., F.H.C.CRICK, Nature 171, 964 (1953)
20) LADIK,J., J.theoretical Biol. (1963) (in the press).

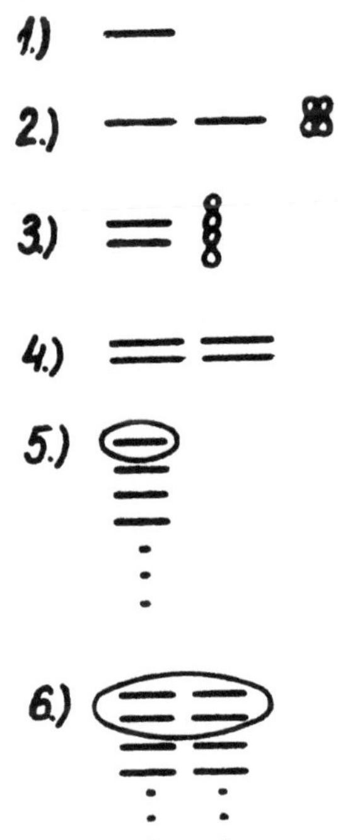

Fig.1 The different steps of the approximation of the electronic structure of DNA

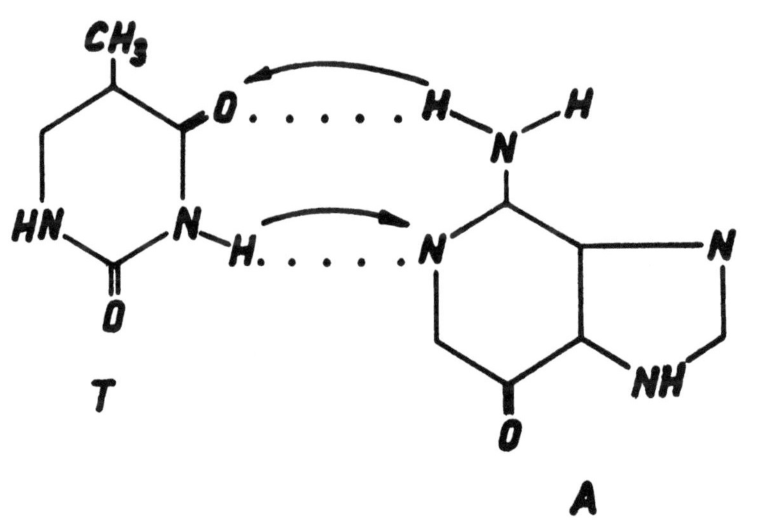

Fig.2 Tautomeric rearrangement by double proton-tunneling of a nucleotide base pair

D Nucleinsäuren

Die Natur der „superpolymeren" DNA und Methoden ihrer Untersuchung

V. A. STRUČKOV

Institut für Biophysik der Akademie, Moskau

S U M M A R Y

The nature of superpolymeric rat thymus and loach sperm DNA (SP-DNA) isolated by means of soft phenolic method was investigated by aid of electronic microscopic and viscosimetric methods. It is proved that SP-DNA structure is a derivate one from the supermolecular structure of DNA in the cell. It presents itself in the form of fibrilles with a diameter of 50-70 Å associated to asymmetrical nets. There is much in common between the SP-DNA fibrilles and DNA fibrilles in the cell nucleus. That is why it was assumed that the soft isolating method makes it possible to preserve the native supermolecular DNA structure. Those DNA structures must play an exceptionally important roll in the cell vital activity as their radiosensitivity is very high and as there is a connection between the physiological state of the whole organism and changes in a cell supermolecular structures. The properties of DNA supermolecular structures in vivo and in vitro were investigated by means of our "elastoviscosimetric" method.

Die Hauptaufgabe der Molekularbiologie ist gegenwärtig die Untersuchung des Zusammenhangs von Struktur und Funktion biogener Makromoleküle. Im Problemkreis der Nucleinsäuren nimmt diese Frage einen zentralen Platz ein, und die Anstrengungen vieler Forscher sind darauf gerichtet, den Regulationsmechanismus aufzuklären, den die Struktur der DNA mit ihrer Rolle im allgemeinen Stoffwechsel der Zelle verbindet.

Für die Lösung dieses Problems hat die Gewinnung nativer DNA-Strukturen erstrangige Bedeutung. Diese Aufgabe ist äußerst kompliziert, da es die Molekularbiologie mit einem bestimmten Zustand der Makromoleküle oder Gruppen von Makromolekülen zu tun hat, die in der Zelle in übermolekulare Strukturen organisiert sind, deren Zustand durch bestimmte räumliche Anordnung einzelner Elemente dieser Strukturen bestimmt wird. Um die Funktion biogener Systeme zu erforschen, müssen die Anstrengungen des Forschers auf die Erhaltung ihrer Nativität im Untersuchungsprozeß gerichtet sein.

In meinen kurzen Darlegungen möchte ich einige Ergebnisse und Vorstellungen anführen, die einen der möglichen Wege zur Lösung dieses großen Problems darstellen könnten. Es wird von der Natur der superpolymeren DNA (SP-DNA) die Rede sein, die wahrscheinlich den nativen Zustand der DNA in der Zelle widerspiegelt und mit der ihre spezielle Funktion verbunden ist. Was ist superpolymere DNA?

Ich will kurz einige Arbeiten anführen, die dieser Untersuchung zugrunde liegen. Im Jahre 1959 berichteten DAVISON [1)] und danach auch andere [2)-4)] über die Möglichkeit der Gewinnung von DNA-Präparaten aus dem T2-Phagen mit einem wesentlich größeren Molekulargewicht als gewöhnlich. Das Molgewicht dieser DNA-Präparate erreichte nach den Angaben dieser Autoren 120 - 140 Millionen. Derartig hochpolymere Präparate von Phagen-DNA wurden mit verschiedenen Methoden erhalten, jedoch war hierbei die Hauptbedingung eine schonende Aufarbeitung des Materials. Gegenwärtig sind diese Präparate der Phagen-DNA ausreichend physikochemisch charakterisiert und die Tatsache der Existenz eines derartig großen und einzelnen DNA-Moleküls in Phagenteilchen ruft keine Zweifel hervor. Im Jahre 1960 isolierten GEORGIEV und STRUCKOV [5] unter Anwendung der Phenolmethode ähnliche hochpolymere DNA-Präparate aus dem Ehrlich'schen Ascites-Carcinom und dem Thymus der Ratte. Diese Arbeit fand ihre weitere Entwicklung in den Untersuchungen von KUSIN, STRASHEVSKAJA und STRUCKOV [6]-[8], die auch der Untersuchung der Natur der "superpolymeren" DNA zugrunde gelegt wurden.

Es konnte gezeigt werden, daß die "superpolymere" DNA viele physikochemische Eigenschaften mit der Phagen-DNA gemeinsam hatte: hohe Werte einer charakteristischen Viskosität und Sedimentationskonstanten. Sie wurde leicht bei mechanischer Behandlung zerstört (Schütteln, Durchfluß durch eine enge Kapillare). All dies unterschied diese DNA-Präparate von den gewöhnlichen Präparaten, die Schmelztemperatur der Wasserstoffbindungen war die gleiche wie bei gewöhnlichen Präparaten. Außerdem veränderten die DNA-Präparate ihre Eigenschaften in einem breiten pH-Bereich (von 4,5 - 13) und bei verschiedener Ionenstärke (0 - 1 M) nicht.

Obwohl diese Präparate zusammen mit RNA isoliert wurden, deren Menge in Abhängigkeit vom Untersuchungsobjekt variierte, wurden ihre Eigenschaften nur von denen der DNA bestimmt. Dies ging aus der Gegenüberstellung von nach RN'ase-Behandlung gewonnenen Ergebnissen im Vergleich mit DNA-Präparaten hervor, die stark von RNA bei der Isolierung gereinigt waren. Der wesentliche Unterschied dieser aus tierischem Gewebe schonend isolierten DNA gegenüber normalen DNA-Präparaten und superpolymeren Präparaten von Phagen-DNA ist das eigentümliche Verhalten im Ostwald-Viskosimeter in Abhängigkeit vom Gradienten, das wesentlich klarer ausgeprägt ist in einem Viskosimeter mit einer besonderen Eingangsöffnung (Abb. 1), das von uns wegen dieser konstruktiven Besonderheit "Elasto-Viskosimeter" genannt wurde.

Die Versuchsergebnisse dieser ungewöhnlichen DNA sind in den Abb. 2 dargestellt. Auf diesen ist zu erkennen, daß die DNA aus Thymus und Leber ein Maximum aufweisen, das bei der Phagen-DNA und den DNA-Präparaten mit einem Molgewicht von 4,5 Millionen, die auf gewöhnliche Weise isoliert wurden, fehlt. Bei bestimmten Werten des Gradienten werden die Struktur der Thymus- und Leber-DNA zerstört, wodurch ein Abfall der Kurve erfolgt. Bei völliger Zerstörung der DNA-Struktur, die durch 3maliges Durchfließen einer Kapillare des Elasto-Viskosimeters bei G = 23.000 Sek^{-1} erreicht wird, verhielt sich die superpolymere DNA ähnlich wie die Phagen-DNA (Abb. 2, e). Eine härtere Bearbeitung - starkes Schütteln - läßt diese ungewöhnliche DNA dem gewöhnlichen niederpolymeren DNA-Präparat ähneln (Abb. 2, f).

Die erhaltenen Ergebnisse verstärkten unsere Meinung darüber, daß eine derartige Besonderheit im Verhalten der DNA im Elasto-Viskosimeter vom Vorhandensein einer eigenartigen Struktur zeugt und die Äußerung ihrer elasto-viskosen Eigenschaften darstellt, die im Eingangseffekt des Kapillar-Elasto-Viskosimeters zum Ausdruck kommt. Das Vorhandensein dieser besonderen Struktur bestimmt das Maximum bei gewissen kritischen Gradienten, wonach ein Abfall einsetzt, der mit einer irreversiblen Deformation der DNA vor dem Übergang in die stationäre Strömung verbunden ist.

Das ist in kurzen Zügen die Besonderheit der DNA-Präparate aus tierischen Geweben, die mit der schonenden Phenolmethode isoliert wurden und die Besonderheit der mit der elasto-viskosimetrischen Untersuchungsmethode erhaltenen Eigenschaften dieser DNA. Diese Methode erwies sich als sehr empfindlich und besaß Vorteile gegenüber der gewöhnlichen Viskosimetrie, da sie die Möglichkeit gab, geringfügige Veränderungen im Zustand der DNA in der Zelle nach Einwirkung verschiedener modifizierender Agentien zu fixieren (ionisierende Strahlung, physiologisch wirksame Stoffe).

Die Meßergebnisse drückten wir in Größen der angeführten Viskosität aus (η_{sp}/c), die die elasto-viskosen Eigenschaften der DNA charakterisieren.

Derartige DNA-Präparate wurden von uns "superpolymere" (SP-DNA) genannt.

Auf Grund der Untersuchung der Viskositäts- und Elastoviskositäts-Eigenschaften entstand die Vorstellung, daß die SP-DNA eine poly-Molekularstruktur besitzt, die ihre elastoviskosen Eigenschaften bestimmt und die ihrer Natur nach den übermolekularen Strukturen des Interphasenkerns der Zelle ähnlich ist [9].

Zur Überprüfung der Richtigkeit dieser Hypothese verwendeten wir die Elektronenmikroskopie, wobei beim Auftragen der DNA-Lösung auf die Objektträger die Adsorptionsmethode mit nachfolgender Kontrastierung der DNA mit 2 %igem Uranylacetat angewendet wurde [10].

Die elektronenmikroskopische Untersuchung der SP-DNA aus Rattenthymus und Fisch-Sperma bestätigten die Richtigkeit unserer Annahme darüber, daß diese DNA eine polymolekulare Struktur besitzt. Die Ergebnisse sind in Abb. 3 dargestellt. Es konnten asymmetrische Netze (Abb. 3 a) gefunden werden, die aus DNA-Fäden mit einem Durchmesser von 50 bis 70 Å (Abb. 3 b) bestehen. Folglich stellen die SP-DNA-Präparate DNA des Kerns dar, die zu netzförmigen Strukturen organisiert ist und deren Elemente Fibrillen darstellen. Eine derartige Fibrille kann nach unseren Errechnungen nicht weniger als 2 DNA-Moleküle enthalten, die miteinander lateral verbunden sind. Die Fibrillen in den Netzen sind möglicherweise durch Resteiweiß, das in unseren Präparaten 3 % ausmacht, vereinigt.

Die an Zellen durchgeführten elektronenmikroskopischen Untersuchungen nach der modifizierten Methode von GALL überzeugten uns davon, daß derartige DNA-Strukturen auch im Kern des Interphasenstadiums der Zelle auftreten (Abb. 4).

Auf Abb. 5 wird die charakteristische Veränderung der Struktur der SP-DNA in Lösung nach γ-Bestrahlung gezeigt.

Nach Gegenüberstellung der elektronenmikroskopischen Bilder isolierter DNA-Präparate mit den zu beobachtenden Bildern im Zellkern kommen wir zum Schluß, daß die schonende Isolierungsmethode erlaubt, native DNA zu gewinnen, die bestimmten übermolekularen Strukturen der Interphasenzelle entspricht. Dies stimmt gut mit den Ergebnissen von RIS über die Ultrastruktur des Interphasenkerns überein [12]. Die Anwendung der gewöhnlichen Isolierungsmethoden für DNA [13]-[16], die mit einer groben Behandlung verbunden ist, führt zur Zerstörung der nativen fibrillären DNA-Struktur, wodurch ein heterogenes DNA-Gemisch entsteht, dessen Zusammensetzung je nach der Isolierungsmethode variiert [17]-[19]. Es ist nicht zufällig, daß die Frage über das Molekulargewicht der DNA für verschiedene Objekte, außer bei Phagen-DNA, offen bleibt. Wenn die Struktur der SP-DNA möglicherweise eine Widerspiegelung des nativen Zustands der DNA in der Zelle darstellt, dann kann man mit Hilfe der elasto-viskosimetrischen Methode eine Verbindung zwischen dem Zustand der DNA-Struktur in der Zelle und dem physiologischen Zustand des Organismus verfolgen. So konnten z. B. bestimmte dynamische Veränderungen im Zustand übermolekularer DNA-Strukturen in Rattenthymus nach Einwirkung von Adrenalin (Abb. 6) und ionisierender Strahlung und eines gonadotrophen Hormons [20] gefunden werden. Dank der Erhaltung nativer DNA-Strukturen im Isolierungsprozeß gelingt es, primäre Strahlenschädigungen über molekulare DNA-Strukturen in der Zelle nachzuweisen [6]-[8]. Unsere Methode gestattet Schädigungen der übermolekularen DNA-Struktur von Rattenthymus sofort nach Bestrahlung des Versuchstieres schon bei Dosen von 200 r genau festzustellen, während bei den gewöhnlichen Isolierungsmethoden, die mit der Zerstörung der nativen Struktur verbunden sind, selbst bei einer Bestrahlung des Tieres mit um das 2fache höheren Dosen, keinerlei Veränderungen der DNA beobachtet werden können. Wir glauben, daß den übermolekularen DNA-Strukturen eine wesentliche Rolle bei der Strahlenschädigung der Zelle zukommt.

Zusammenfassend kann man also sagen, daß die Struktur der SP-DNA eine Widerspiegelung des nativen Zustands der DNA in der Zelle darstellt, mit der ihre bestimmte biogene Funktion verbunden werden kann. Es ist durchaus möglich, daß die Fibrillen des Interphasenkerns die funktionellen Einheiten darstellen, die die spezifische übermolekulare Organisation der DNA-Makromoleküle ist, die für die Verwirklichung der Regulationsfunktion von Stoffwechselprozessen in der Zelle notwendig ist, d. h. den Regulationsmechanismus darstellt, mit dessen Hilfe die Übertragung der in der DNA angelegten Information in der Zelle erfolgt. Selbstverständlich sind die hier dargelegten Vorstellungen, insbesondere im letzten Teil, in bestimmtem Maße hypothetisch, sie können jedoch unserer Meinung nach von Interesse sein als einer der möglichen Wege des komplizierten Problems der Regulationsfunktion übermolekularer DNA-Strukturen des Zellstoffwechsels und können als Diskussionsgrundlage dienen.

L I T E R A T U R

1) DAVISON,P.F., Proc. Nat. Acad. Sci. USA 45, 1510 (1959)
2) THOMAS,C.A., A.T.MATHESON, J. molecular Biol. 2, 5 (1960)
3) HERSHEY,A.D., E.BURGI, J. molecular Biol. 2, 143 (1960)
4) BURGI,E., A.D.HERSHEY, J. molecular Biol. 3, 458 (1961)
5) GEORGIEW,G.P., W.A.STRUČKOV, Biofisika 5, 745 (1960)
6) STRUČKOV,W.A., N.B.STRASCHEWSKAJA, Radiobiologia 1, 172 (1961)
7) KUSIN,A.M., N.BARANOWSKAJA, N.B.STRASCHEWSKAJA, W.A.STRUČKOV, Radiobiologia 1, 476 (1961)
8) STRASCHEWSKAJA,N.B., W.A.STRUČKOV, Radiobiologia 2, 9 (1962)
9) STRUČKOV,W.A., Biofisika 7, 538 (1962)
10) HUXLEY,H.E., G.J.ZUBAY, J. molecular Biol. 2, 10 (1960)
11) GALL,J., Science 139, 120 (1963)
12) RIS,H., Canad. J. Genetics Cytology 3, 95 (1961)
13) SEVAG,M.G., D.B.LACKMAN, J.SMOLENS, J. biol. Chemistry 124, 425 (1938)
14) KAY,E.R.M., N.S.SIMMONS, A.L.DOUNCE, J. Amer. chem. Soc. 74, 1724 (1952)
15) KIRBY,K.S., Biochem. J. 66, 495 (1957)
16) GEORGIEW,G.P., Biochimia 24, 472 (1959)
17) BUTLER,D., K.SHOOTER, Chemische Grundlagen der Vererbung I.L.M. 435 (1960)
18) EPHRUSSI-TAYLOR,G., Chemische Grundlagen der Vererbung I.L.M. 255 (1960)
19) KONDO,N., S.OSAWA, Nature [London] 183, 1602 (1959)
20) BEKETORA,A.A., A.M.KUSIN, Dokl. Akad. Nauk SSSR 146, 1201 (1962)

D I S K U S S I O N S B E M E R K U N G E N

MUNK.- In welchem Geschwindigkeitsgefällegebiet arbeitet Ihr Viskosimeter und bei welchem Gefälle liegt das Maximum?

STRUČKOV.- Der Gradient im "Elastoviskosimeter", der für 0,14 M NaCl-Lösung bei T = 25° C gemessen wurde, beträgt 2100 s^{-1}. Das Maximum für die untersuchten DNA-Lösungen lag bei den genannten Werten P für die Rattenthymus-SP-DNA bei G ~ 1500 s^{-1}; für die Rattenleber-SP-DNA bei ~ 3500 s^{-1}.

ALTMANN.- 1. Welchen Dosisbereich haben Sie bei den Bestrahlungsversuchen untersucht?

2. Könnte man die Änderungen der Viskosität sowie die elektronenmikroskopischen Untersuchungen mit den von uns in diesem Dosisbereich gefundenen Spurenelementabspaltungen in Zusammenhang bringen?

STRUČKOV.- 1. In den Untersuchungen, die sich mit der Frage über die Wirkung der ionisierenden Strahlung auf die übermolekulare Struktur der DNA befassen und am Thymus von weißen Mäusen durchgeführt wurden, bestrahlten wir die Tiere mit verschiedenen Dosen von 200 r bis 50 000 r.

Die genannte Methode zur Extraktion der SP-DNA und die elastoviskosimetrische Methode zu ihrer Untersuchung ermöglichen bereits bei einer Dosis von 200 r die Feststellung von Strahlungsschäden der übermolekularen Strukturen der DNA in der Zelle.

2. Die Resultate der von Ihnen durchgeführten Untersuchungen über die Bedeutung der Ionen zweiwertiger Metalle für die tertiären Strukturen der RNA sind äußerst interessant. Die Möglichkeit, die von uns beobachteten Konfigurationsänderungen in der Struktur der SP-DNA bei Bestrahlung mit der Abspaltung dieser Metalle in Verbindung zu bringen, ist sehr verlockend.

Wir untersuchten die Rolle der Ionen der zweiwertigen Metalle auf die übermolekularen DNA-Strukturen. Aber bei der Behandlung der SP-DNA-Lösungen mit 0,2 %igem Versen und die Ausfällung der SP-DNA bei Gegenwart von Versen ergaben nicht die erwarteten Resultate. Le ROY KUEHL behandelte die analogen Präparate der DNA mit Versen und stellte auch keine Veränderung ihrer Eigenschaften fest.

Trotzdem ist es nicht ausgeschlossen, daß die Ionen der zweiwertigen Metalle an dem Aufbau der übermolekularen DNA-Struktur beteiligt sind. Die Unwirksamkeit des Versens kann in den vorliegenden Versuchen durch die festere Bindung des Metalls in der SP-DNA erklärt werden.

Mir scheint es, daß durch die von Ihnen dargestellte Methode zur Untersuchung der tertiären Struktur neue Daten in bezug auf den Aufbau der übermolekularen Struktur der DNA gewonnen werden können und daß durch diese Methode die Ursachen ihrer hohen Strahlenempfindlichkeit verständlich werden.

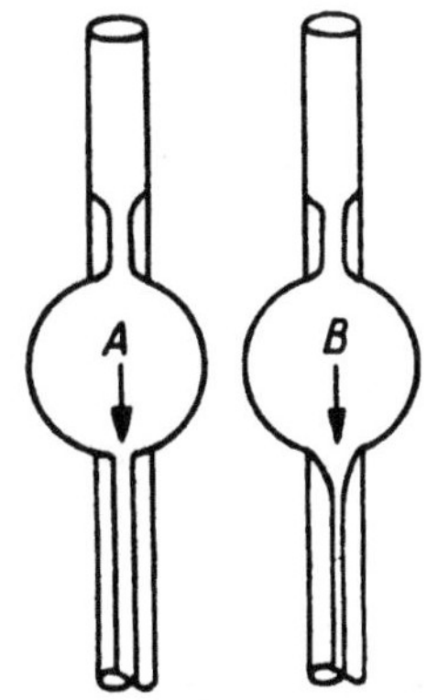

Abb.1 Eingangsformen der Kapillare von der oberen Kugel. A OSTWALD-Viskosimeter, B Elastoviskosimeter

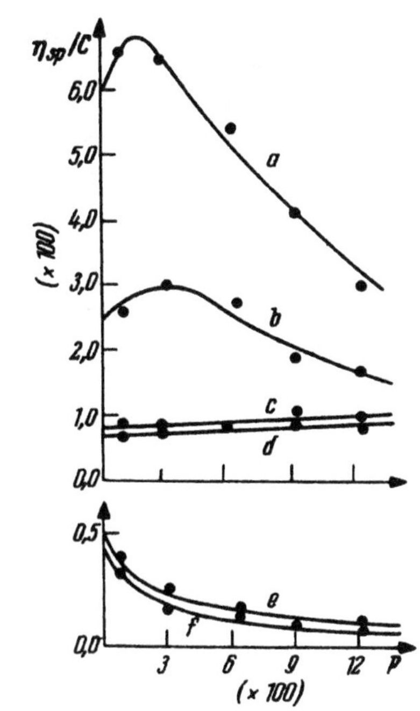

Abb.2 Abhängigkeit η_{spez}/c der Konzentrationen der SP-DNA vom Druck P[cm Wassersäule] im Elastoviskosimeter. a)SP-DNA von Rattenthymus, c=0,002%;b)SP-DNA von Rattenleber,c=0,002%;c)Hochpolymere DNA von T2-Bakteriophagen ([η] =170), c=0,002%;d)a,b,c,nach Abbau der SP-DNA durch 3-maliges rasches Durchlaufen durch die Kapillare des Elastoviskosimeters bei einem Gradienten von 23000 sec^{-1};e)DNA von Kalbsthymus nach GEORGIEW [16] mit $M=4,5.10^6$ bei c=0,035%;f)a,b nach intensivem Schütteln.

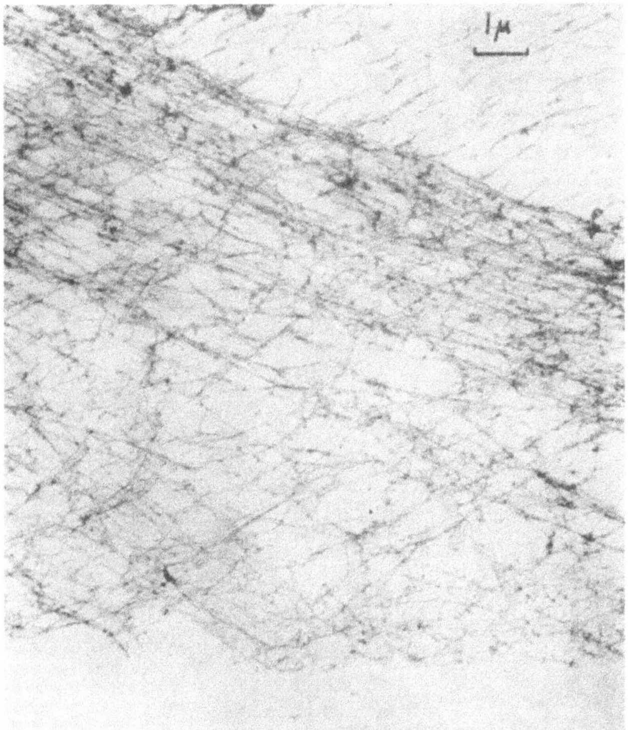

Abb. 3a Vernetzte SP-DNA aus Rattenthymus; Vergrößerung 9000. Eine ähnliche Struktur ist auch bei der Untersuchung der SP-DNA aus Spermien der Schmerle zu finden.

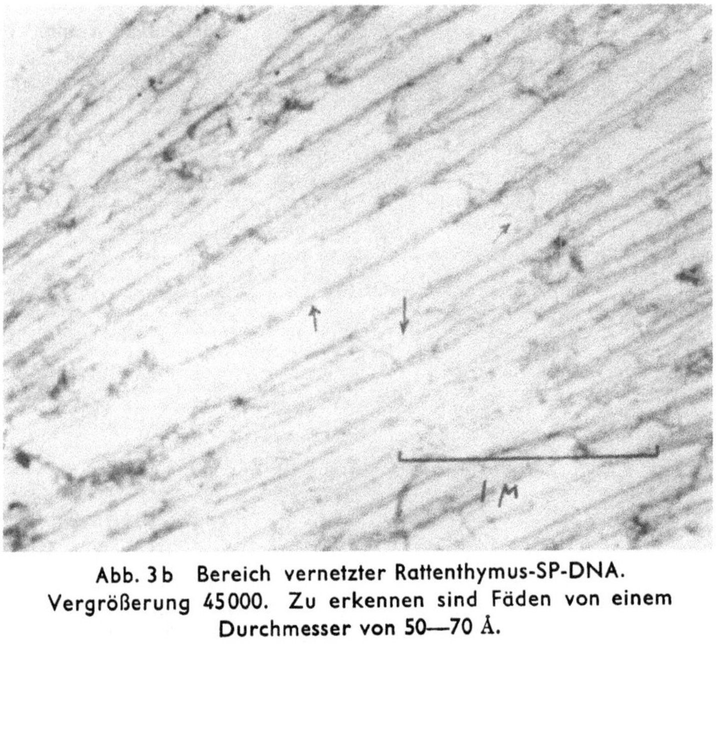

Abb. 3b Bereich vernetzter Rattenthymus-SP-DNA. Vergrößerung 45000. Zu erkennen sind Fäden von einem Durchmesser von 50—70 Å.

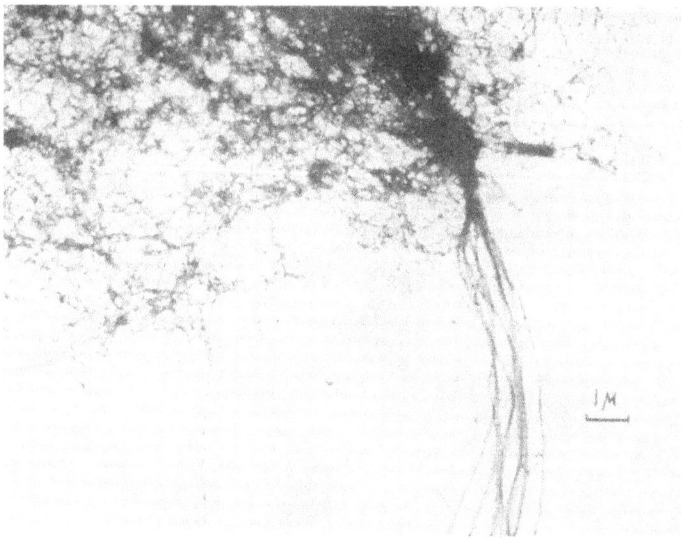

Abb. 4 Ia Kopfbereich eines Schmerlen-Spermiums, dessen wesentlicher Bestandteil Nucleoproteid-Fibrillen sind. Vergrößerung 7500.

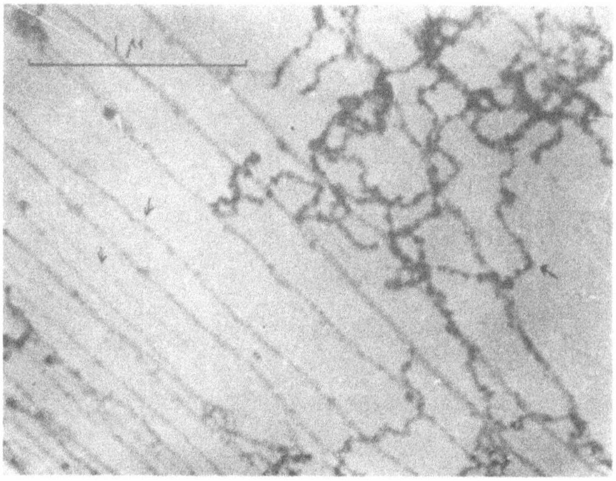

Abb. 4 Ib Das gleiche; Nucleoproteid-Fibrillen mit einem Durchmesser von 200—250 Å werden nach Entspiralisierung dünner und länger. Vergrößerung 45000.

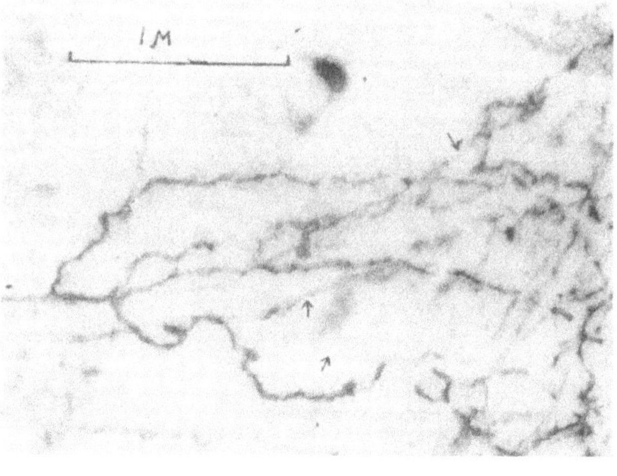

Abb. 4 Ic Das gleiche; zu erkennen ist, wie sich aus Nucleoproteid-Fribrillen mit einem Durchmesser von 200—250 Å feine Fäden mit einem Durchmesser von 50—70 Å bilden. Vergrößerung 45000.

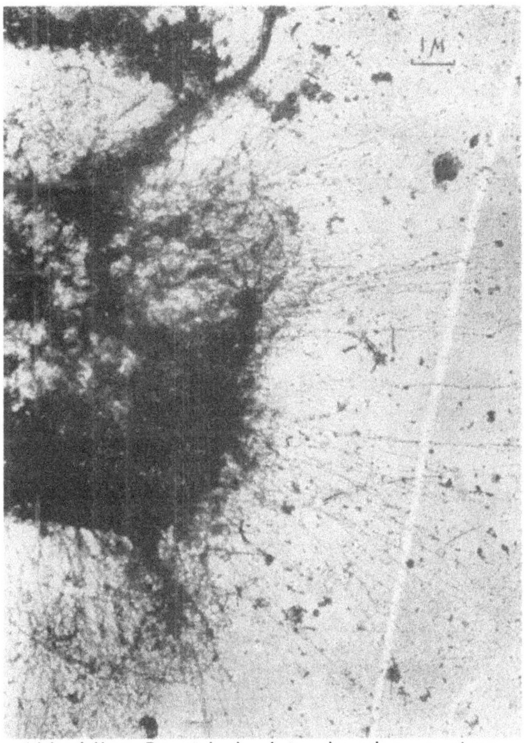

Abb. 4 II a Bereich des Interphasekernes einer Thymuszelle der Ratte. Sichtbar sind Nucleoproteid-Fibrillen. Vergrößerung 7500.

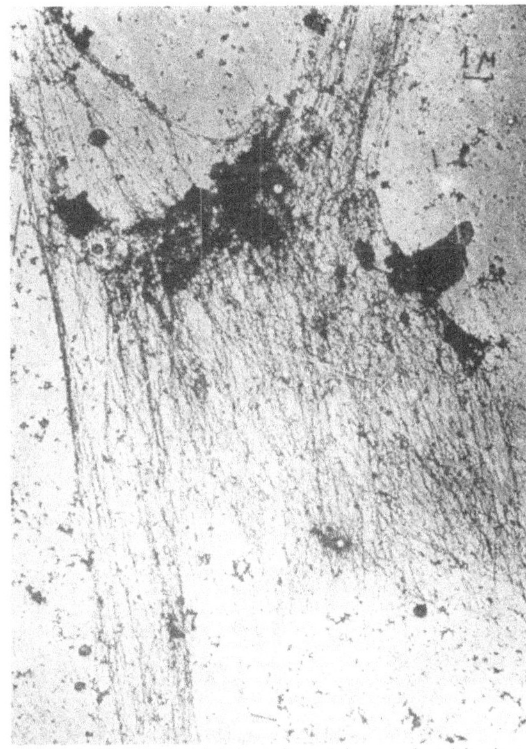

Abb. 4 II b Das gleiche; das sich aus den Nucleoproteid-Fibrillen nach ihrer Entspiralisierung bildende Netz. Vergrößerung 5000.

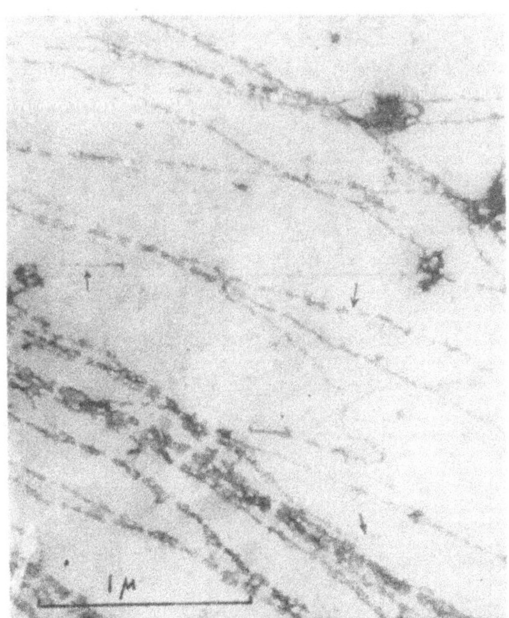

Abb. 4 II c Das gleiche. Es ist zu sehen, wie sich aus Nucleoproteid-Fibrillen mit einem Durchmesser von 150—200 Å Fäden mit einem Durchmesser von 50—60 Å bilden können, ähnlich den Fäden der SP-DNA (Abb. 3 b). Vergrößerung 45000.

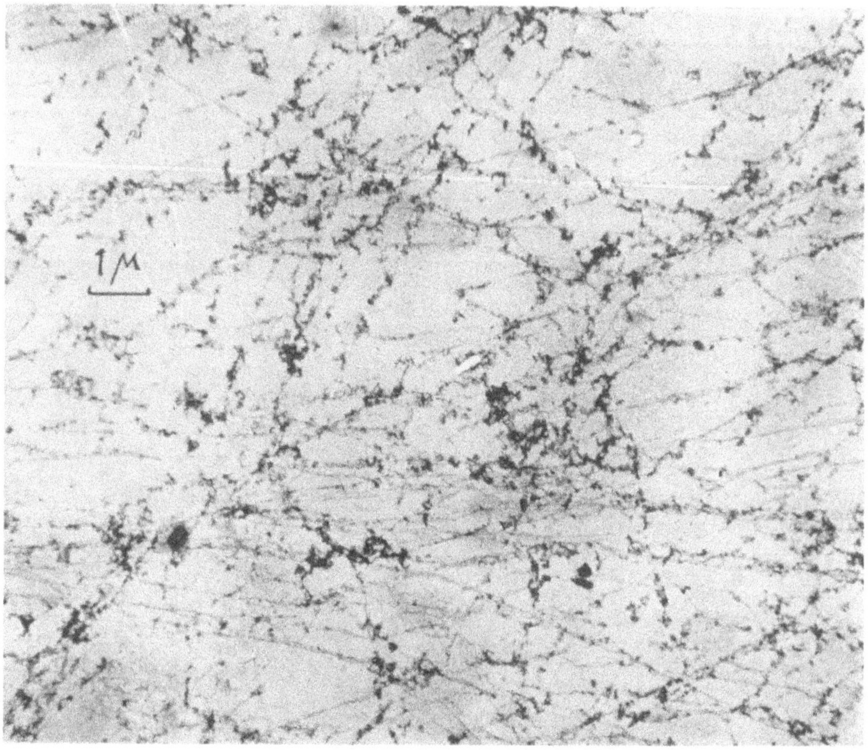

Abb. 5 SP-DNA aus Rattenthymus unmittelbar nach γ-Bestrahlung ihrer Lösung mit 15000 r (Kontrolle auf Abb. 3a), Vergrößerung 10500. Erkennbar sind bedeutende Konfigurationsveränderungen in der SP-DNA-Struktur. Bei dieser Bestrahlungsdosis erfolgt nach den elastoviskosimetr. Ergebnissen eine völlige Zerstörung der SP-DNA-Struktur. Mit Hilfe der elastoviskosimetr. Methode gelingt es, Strukturschäden der SP-DNA nach deren Bestrahlung in Lösungen bei Dosen von 100 festzustellen.

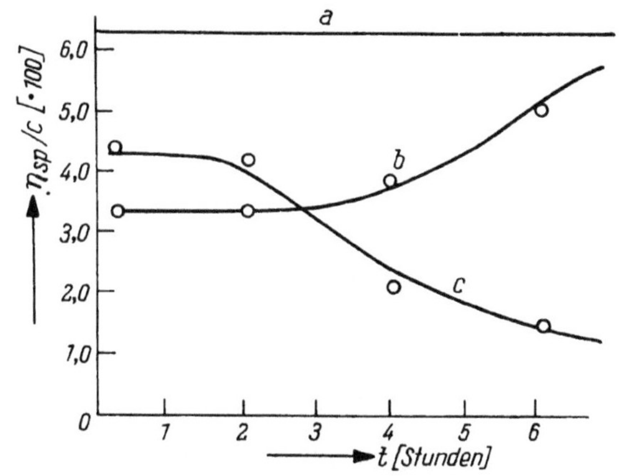

Abb. 6 Dynamik der Zustandsveränderungen übermolekularer DNA-Strukturen aus Rattenthymus. η_{spez}/c in Abhängigkeit von der Zeit t nach Injektion von 100 μg Adrenalin in die Bauchhöhle und nach γ-Bestrahlung der Tiere mit einer Dosis von 1000 r. *a* — Kontrolle; *b* — nach Einführen d. Adrenalins; zu erkennen ist eine reversible Veränderung d. übermolekularen Strukturen in der Zelle; *c* — Nach Bestrahlg. ist zu sehen, daß die Veränderung der übermolekularen DNA-Strukturen irreversibel wird.

D Nucleinsäuren

The Participation of Nucleotides in the Purine and Pyrimidine Isopliths in the DNA Molecule

V. HABERMANN

Institute for Medical Chemistry of the University, Plzen

SUMMARY

A method for the production of apyrimidinic DNA is described. Calf thymus DNA and E. coli C DNA are investigated; the relative frequency for the occurrence of pyrimidine (and of purine) clusters with one, two, three, four, five, and six and more nucleotides are given. There are significant differences between calculated (for random distribution) and actually found frequencies. For explaination a model is proposed.

We know several methods suitable for the determination of the pyrimidine nucleotides in the polynucleotide chain of DNA. So the pyrimidine nucleotide clusters can be prepared from DNA by mild acid hydrolysis [1)-2)] or more exactly by heating DNA with diphenylamine and formic acid [3)-4)]. Those degradation procedures represent, in fact, the two-step processes. (Fig. 1) In the first step an intermediary product - the apurinic DNA - is formed. This polymer is then cleaved into pyrimidine nucleotide clusters (isopliths).

In our laboratory we have developed a method for the isolation of purine nucleotide clusters from DNA. For this purpose the nucleic acid is freed of pyrimidines by hydrazinolysis and the apyrimidinic DNA is then degradated in alkali [5)-6)]. (Fig. 2)

The nucleotide clusters obtained by these methods could be readily separated according to the chain length by DEAE-cellulose chromatography in lithium chloride gradient [7)]. In experiments described here we separated the first five groups of isopliths only; the higher components were eluted together as one peak by 1 M lithium chloride.

The quantitative analysis of the participation of nucleotides in purine or pyrimidine isopliths of different length offers an opportunity to show some regularities in the structure of DNA, even though we cannot, in the meantime, differentiate the individual purine or pyrimidine nucleotides. In our considerations we take into account only purine and pyrimidine nucleotides without further differentiation.

First of all, it may be seen that the distribution of purine nucleotides into purine isopliths of different length agrees with the distribution of pyrimidine ones.

The distribution of purine nucleotides into purine isopliths of different length in calf thymus DNA [8)] is shown in table 1 as well as the distribution of pyrimidine nucleotides. It may be stated here that both these distributions are very similar,

Table 1
The Participation of Nucleotides in Purine and Pyrimidine Isopliths
The data in the table are given as % of total purine /or pyrimidine/ nucleotides

	Calf Thymus DNA		E. Coli C DNA		Our model system for:		R. D. model:
Isoplith:	Purine nucleotides	Pyrimidine nucleotides	Purine nucleotides	Pyrimidine nucleotides	n = 13	n = 6	§
p.N.p	17.3	17.9	27.3	28.5	20.6	28.8	25.0
$p./N.p/_2$	15.8	17.6	26.5	27.8	17.0	24.2	25.0
$p./N.p/_3$	15.4	15.5	19.8	18.5	14.4	18.2	18.8
$p./N.p/_4$	12.8	13.6	10.9	9.3	11.6	12.1	12.4
$p./N.p/_5$	10.9	9.9	6.3	5.9	9.6	7.6	7.6
Higher:	27.8	25.5	9.3	9.0	26.7	9.1	11.2

§ Random distributed model

which is in a good agreement with the WATSON and CRICK model of DNA. In this model consisting of two complementary chains united in the double helix the participation of purine and pyrimidine isopliths must be the same - opposite to each purine nucleotide sequence the analogous pyrimidine block would be observed.

Similar results were obtained in the case of E. coli C DNA; the distribution, however, differs markedly from the distribution of nucleotides in calf thymus DNA (table 1) (ref. [9]).

Beside this, I should draw your attention to the fact, that there are evident differences between these distributions and the values calculated for the statistical random distributed model. Especially markedly they are seen in the case of calf thymus DNA (table 1). DNA of E. coli C approaches to the random distribution rather closely; however, there are also some deviations (table 1).

It must be mentioned here that the random distributed model is not the most suitable one for DNA. First of all this model is calculated for the molecule consisting of an infinite number of nucleotides; the number of nucleotides in the DNA molecule is, in fact, limited. But this is not the main trouble. We can assume that many sequences included into the random distributed model are not realized in the polynucleotide chain. So, the statistical model takes into account even such extreme cases, when, e.g., one half of the chain contains purine nucleotides only, the other - pyrimidine ones only. We may believe that many similar extreme structures have been rejected during the biochemical evolution. Therefore the searching of a new model system, more closely related to the experimental data, may be very fruitful. It could help us to understand some principles which restrict the sequence variability of the DNA molecule.

Our experimental data offer an opportunity to make an effort in this respect. We have arranged a new model system which corresponds to the experimental data rather than the random distributed one. It is arranged in the following way: A certain number of nucleotides of the same type n is arranged in all possible ways so that no arrangement would be repeated, the sequence of the individual units being of no concern. Thus e.g. for n = 5 a total of 7 arrangements are possible:

aaaaa

aaaa + a

aaa + aa

aaa + a + a

aa + aa + a

aa + a + a + a

a + a + a + a + a

The whole set is composed of 20 sequences of 5 types:

aaaaa ..... 1 x

aaaa ...... 1 x

aaa .... 2 x

aa .... 4 x

a ....12 x

The probality of occurrence of the individual sequences is determined by the ratio of the participation of the sequence type studied to the total number of all sequence type studied to the total number of all sequences in the set investigated. The data on the participation of nucleotides in the individual sequences of different lenght could be calculated in a similar way.

It is evident that the results of such calculations depend considerably on the number of nucleotides n which are taken into consideration. We have arranged the model systems for different values of n. The distribution of purine and pyrimidine isopliths in calf thymus DNA corresponds most closely to a model system for n = 13 (table 1). It may be seen that our experimental model system agrees with experimental data better than the random distributed values.

The participation of nucleotides in E. coli C DNA is more related to random distribution, than in the calf thymus DNA, but also in this case our model system for n = 6 is slightly more suitable than the random distribution (table 1).

There may be several interpretations of the relation between our experimental model system and the structure of the DNA molecule.

First of all, it may be stated for calf thymus DNA, that the longest purine or pyrimidine isoplith in this molecule consists of 13 nucleotides.

A DNA molecule containing such a single purine and pyrimidine isoplith composed of 13 nucleotides each will also include 555 purine and 555 pyrimidine isopliths similarly as our experimental model system for n = 13. If now the purine and pyrimidine nucleotides in all these isopliths are added, the total number of nucleotides is 2626, which corresponds to a molecular weight of the polynucleotide chain of 810 000. The double stranded molecule should thus have a molecular weight in the vicinity of 1,6 x 10^6. This value agrees with some data about the minimal molecular weight of calf thymus DNA [10)-12)]. We suppose that the number n for which our experimental model system is derived may be thus a function of the molecular weight of a subunit of the DNA molecule.

Further, it may be supposed that the molecule of calf thymus DNA is built from basal coding clusters, each cluster containing a set of a small number of nucleotides, approximately 13.

L I T E R A T U R

1) SHAPIRO,H.S., E.CHARGAFF, Biochim. biophysica Acta 23, 451 (1957)
2) SPENCER,J.H., E.CHARGAFF, Biochim. biophysica Acta 51, 209 (1961)
3) BURTON,K., G.B.PETERSEN, Biochem. J. 75, 17 (1960)
4) BURTON,K., Biochem. J. 77, 547 (1960)
5) HABERMAN,V., Biochim. biophysica Acta 55, 999 (1962)
6) HABERMAN,V., Collect. czechoslov. chem. Commun. 28, 510 (1963)
7) TENER,G.M., P.T.GILHAM, W.E.RAZZELL, A.F.TURNER, H.G.KHORANA, Ann. N.Y. Acad. Sci. 81, 757 (1959)
8) HABERMAN,V., E.MAIDLOVÁ, Collect. czechoslov. chem. Commun., in the press
9) HABERMAN,V., S.HABERMANNOVÁ, M.CERHOVÁ, Biochim. biophysica Acta, in the press
10) KAHLER,H., J. physic. Colloid Chem. 52, 676 (1948)
11) CECIL,R., A.G.OGSTON, J. chem. Soc. 1948, 1382
12) CONWAY,B., L.GILBERT, J.A.V.BUTLER, J. chem. Soc. 1950, 3421.

Figure 1

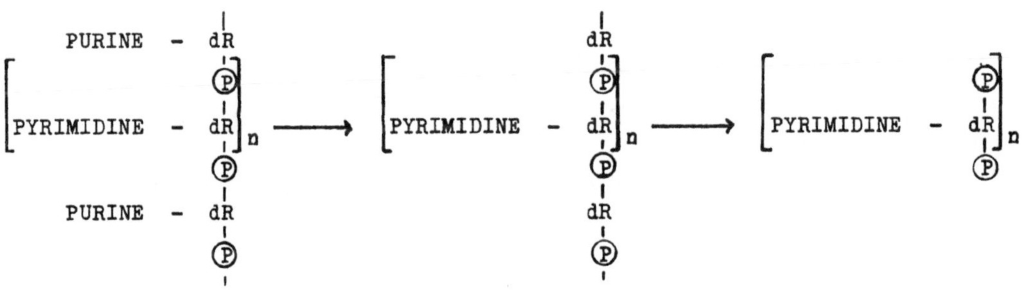

dR - deoxyribose, (P) - phosphate

Figure 2

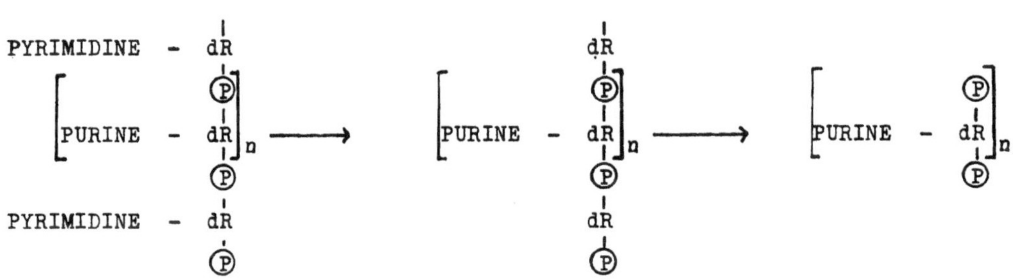

dR - deoxyribose, (P) - phosphate

D Nucleinsäuren

Eine neue Methode für die Trennung von Gemischen höherer Oligonucleotide

E. BUDOWSKY und V. DJEMUSCHKIN

Institut für Naturstoffchemie, Moskau

S U M M A R Y

A method for separation of higher oligonucleotides, involving ion-exchange chromatography consecutively with acid (p_H ca. 3) and alkaline (p_H 9) eluants is described. The method enables ready and precise resolution of practically any oligonucleotide mixtures, which can arise from nucleic acid after specific cleavage.

Eines der wichtigsten Probleme der Chemie und Biologie der Nucleinsäuren ist die Bestimmung der Nucleotidsequenz. In unserem Laboratorium lösen wir diese Probleme komplex, und zwar: 1) Isolierung der spezifischen löslichen Nucleinsäuren, 2) chemische und chemisch-enzymatische Methoden der spezifischen Spaltung der Nucleinsäure und 3) die Trennung der Gemische von Oligonucleotiden, die sich bei der spezifischen Spaltung der polymeren Kette bilden.

In diesem Artikel werden wir nur über die Trennung der Oligonucleotidgemische sprechen. Diese Frage ist jetzt besonders wichtig, weil auf den ersten zwei Gebieten schon große Erfolge erreicht sind. Es ist zu bemerken, daß eine bequeme Präzisionsmethode für die Fraktionierung der Oligonucleotide die Ansprüche zur Reinheit des Eingangspräparates und zur Spezifität der Spaltungsmethode herabsetzt.

Die Anwendung von Standardverfahren zur Lösung dieses Problems, und zwar die Anwendung von verschiedenen Ionenaustauschern und komplizierten Gradienten erlaubt nur die Fraktionierung von Gemischen niederer oder monotoner Oligonucleotide.

Im Falle von komplizierten Oligomergemischen (beispielsweise von RNA-Hydrolysaten, die durch Einwirkung von Pancreasribonuclease oder bei anderen spezifischen Spaltungen gewonnen werden) wird der Unterschied zwischen den Eigenschaften einzelner Komponenten so gering, daß die Teilung solcher Gemische mit bekannten Methoden unmöglich wird.

Jedoch die größte Information über die Primärstruktur der Polynucleotidkette geben eben die höheren Oligonucleotide.

Im Zusammenhang damit haben wir spezielle Forschungen unternommen, um eine Fraktionierungsmethode für die höheren Oligonucleotide auszuarbeiten.

Die Fraktionierung von Oligonucleotiden auf Ionenaustauschern ist ein komplizierter Prozeß, der im wesentlichen von drei Faktoren abhängt: 1) von der Ladung der Phosphatgruppen, 2) von der Ladung der ionisierten Nucleinbasen und 3) von der Adsorptionszusammenwirkung zwischen dem Oligomeren und der Ionenaustauschermatrix.

Im Falle der niederen Oligonucleotide sind die ersten zwei Faktoren die wichtigsten. Der erste Faktor (Ladung der Phosphatgruppen) verändert sich bei der Veränderung des p_H-Wertes für beliebige Oligonucleotide der Form X_nP_n oder X_nP_{n-1} monoton. Der zweite Faktor hängt vom p_H und der Nucleotidzusammensetzung des Oligonucleotids ab. Indem man den p_H-Wert des eluierenden Puffers im weiten Bereich ändert, kann man die Gesamtladung des Oligonucleotids ändern, wodurch die Festigkeit der Ionenbindung des Oligonucleotids mit dem Ionenaustauscher verändert wird.

Für die höheren Oligonucleotide wird der dritte Faktor zunehmend wichtiger, da die Differenzen der Gesamtladungen sich nivellieren im Zusammenhang damit, daß in höheren Oligomeren die Nucleotidzusammensetzung sich in weiten Bereichen verändern kann. Die gleichmäßige Veränderung des p_H-Wertes (bei beliebigen Gradienten) kann keine Verschiedenheiten in den Eigenschaften der Oligonucleotide zu Tage bringen. Deshalb ist die Fraktionierung von beliebigen Oligonucleotidgemischen mit Hilfe von einmaliger Chromatographie sehr fraglich. Sogar unter der Bedingung, daß man den Einfluß des dritten Faktors - und zwar die Adsorptionszusammenwirkung - durch Harnstoff (BELL, TOMLINSON und TENER [2]) beseitigt und die Versuche bei den günstigen p_H-Werten (p_H 7,8), wo der Ionisierungsgrad der Nucleinbasen am geringsten ist, durchführt, wirken die restlichen Ladungen der Nucleinbasen, indem sie das Bild verschleiern und die Teilung von Tetra- und höheren Oligonucleotiden unvollkommen machen.

Die Berücksichtigung der Adsorptionszusammenwirkung der Nucleinsäurekomponenten ist zur Zeit unmöglich. Man weiß nur, daß die Veränderungen der Tautomerie der Nucleinbasen einen Einfluß auf den Adsorptionsgrad ausüben, aber man kann nicht einmal die Richtung dieser Veränderungen voraussagen.

Die Adsorptionszusammenwirkung hängt mehr als die Summe der ersten zwei Faktoren von der Zusammensetzung und wahrscheinlich von der Sequenz der Mononucleotide in Oligomeren ab. Folglich können wir damit rechnen, unsere Aufgabe zu lösen und Gemische von Oligonucleotiden, die sich voneinander nicht nur in der Zusammensetzung, sondern auch in der Sequenz unterscheiden, zu teilen, indem wir die Adsorptionszusammenwirkung regulieren.

Bei der Fraktionierung von Oligonucleotidgemischen, die bei der spezifischen RNA-Spaltung entstehen, muß man, um klare quantitative Resultate zu bekommen, mit hohen Konzentrationen arbeiten, was unmöglich ist, wenn man im selben oder in einem ähnlichen System rechromatographiert.

Von diesen Gründen ausgehend haben wir für die Fraktionierung eines Modellgemisches von Oligonucleotiden (die bei der Spaltung von RNA durch Pancreasribonuclease entstehen) eine Methode der "Zweidimensionalen Ionenaustausch-Chromatographie" angewendet. Die erste Fraktionierung wurde im System von HURLBERT und POTTER in der Modifikation von BAJEV [1], d. h. auf Dowex-1 in der Formiatform durchgeführt. Der Ionenaustauscher wurde mittels der hydrodynamischen Methode von HAMILTON [3] fraktioniert. Wir benutzten die Fraktion 240-280 mesh (> 90 %). Die Fraktion (Abb. 1), die die höheren Oligonucleotide enthielt, rechromatographierten wir (nach Entsalzung) auf DEAE-Sephadex A-25 medium im Gradienten von Triäthylammoniumbikarbonat p_H 9,0. Aus Abb. 2 sehen wir, daß man mit dieser Methode praktisch alle Oligonucleotide aus diesem Gemisch lösen kann. Um eine gute Trennung zu bekommen, muß man mit hohen Säulen (Durchmesser /Höhe nicht weniger als 1:50) arbeiten. Wenn man kurze Säulen benutzt, dann gibt es nicht völlige Trennung (Abb. 2). Die Anwendung von Elutionssystemen mit kleineren p_H-Änderungen ergab entweder keine oder nur eine teilweise Teilung (Abb. 3).

Der Grad der Reinheit der isolierten Oligonucleotide wurde durch Papierchromatographie nach RUSCHIZKI und KNIGHT [4] oder Ionenaustauschchromatographie im Säuresystem (p_H 4,0) geprüft (Abb. 4 und 5). Diese Kontrolle bestätigte die Homogenität der isolierten Stoffe. Auf diese Weise hat sich die Methode der "Zweidimensionalen Ionenaustausch-Chromatographie" als eine bequeme Methode für die Trennung von komplizierten Oligonucleotidgemischen erwiesen. Wir können annehmen, daß nach dieser Methode, insbesondere in Verbindung mit der Methode von TOMLINSON und TENER, ein beliebiges Oligonucleotidgemisch, das bei der spezifischen Spaltung der RNA entsteht, aufgetrennt werden kann.

L I T E R A T U R

1) BAJEV,A., Report on the Symposium on Soluble Ribonucleic Acid, Moscow, 1963
2) BELL,D., R.TOMLINSON, G.TENER, Biochem. biophysic. Res. Commun. 10, 304 (1963)
3) HAMILTON,P., Analyt. Chem. 30, 914 (1958)
4) RUSCHIZKY,G., C. KNIGHT, Virology 11, 236 (1960)

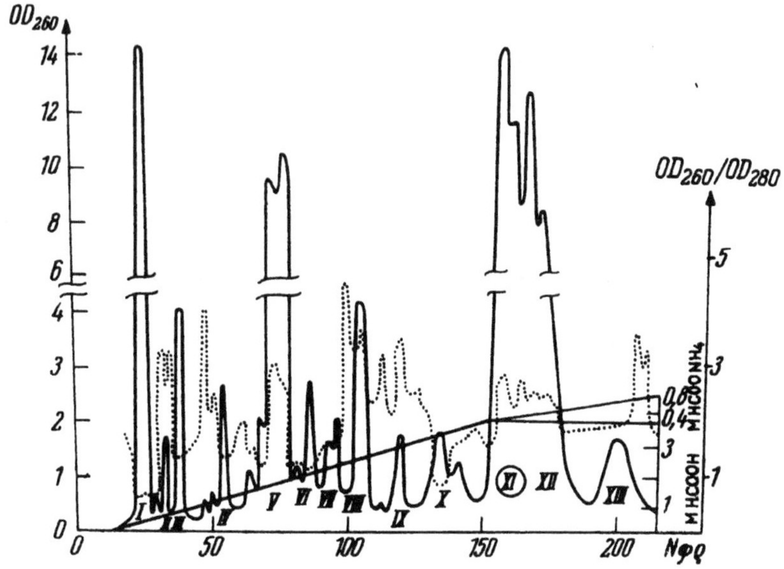

Abb.1 Die Trennung eines Pancreasribonuclease RNA-Hydrolysates auf Dowex-1x2 (240-280 mesch) in der Formiat-Form (2x34 cm). Fließgeschwindigkeit: 2ml/cm^2/min.

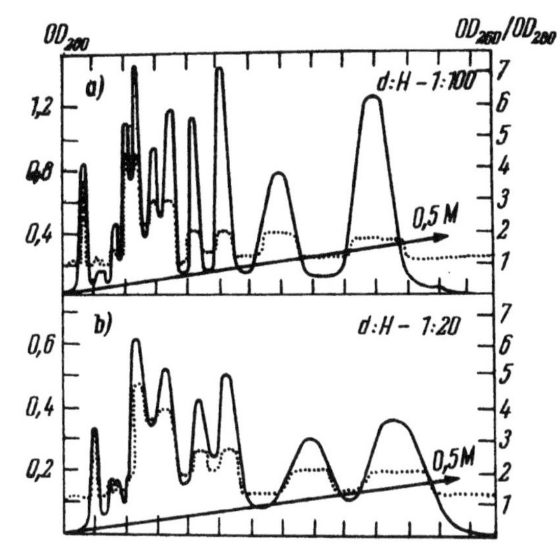

Abb.2 Einfluß der Größe der Säule auf die Trennung der Fraktion XI (s.Abb.1) auf DEAE-Sephadex A-25 medium in der Bikarbonat-Form, mit Triäthylammoniumbikarbonat-Puffer (pH 9 : linearer Gradient). Fließgesc windigkeit: 2ml/cm^2/min. a) Säule 0,8.80 cm b) Säule 0,8.15 cm.

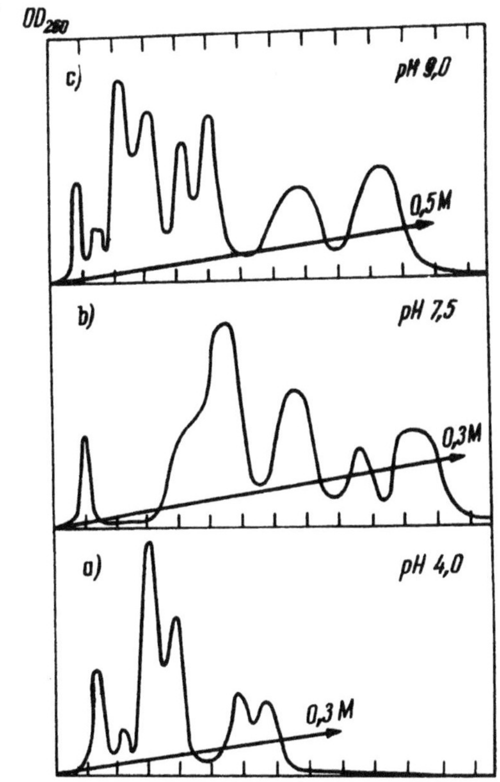

Abb.3 Chromatographieversuche zur Trennung der Fraktion XI (Abb.1) auf DEAE-Sephadex A-25 medium (0,8.15 cm), 2 ml/cm^2/min. Elution mit linearem Konzentrationsgradienten von: a) Triäthylammoniumacetat, pH 4, b)Triäthylammoniumbikarbonat, pH 7,5, c)Triäthylammoniumkarbonat pH 9.

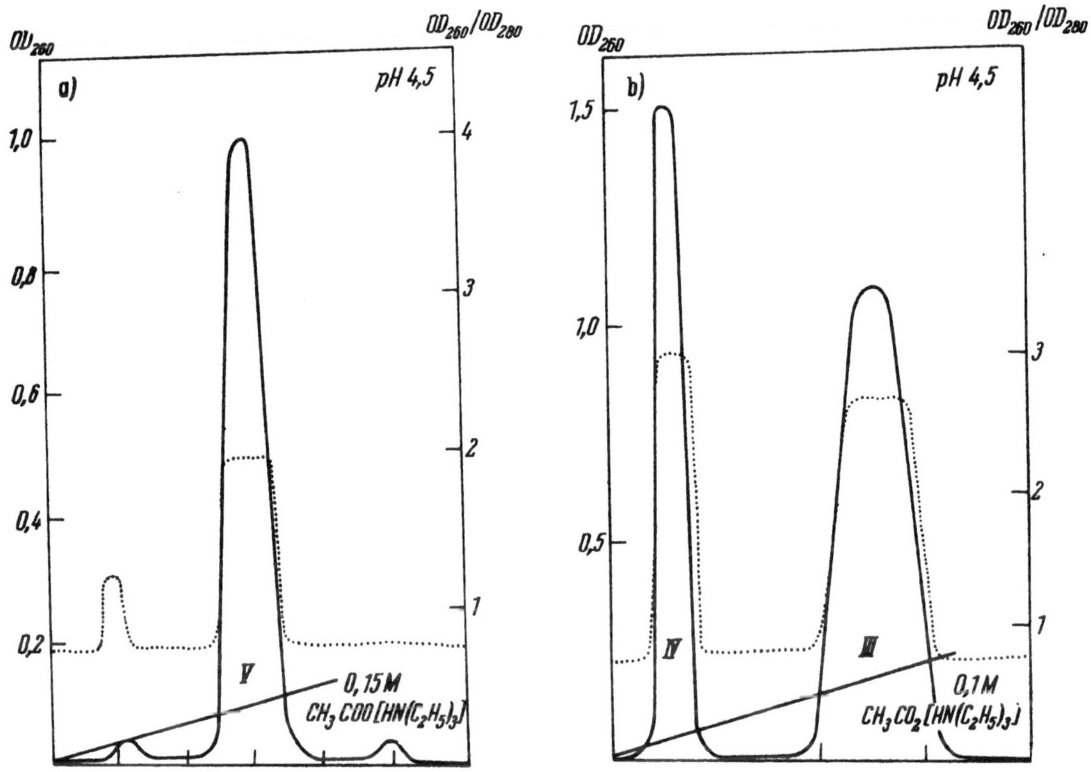

Abb.4 Die Rechromatographie der Fraktionen 5 (a) und 3-4 (b) (Abb.2a) auf DEAE-Sephadex A-25 medium in der Acetat-Form (Säule 0,8.15 cm). Elution mit linearen Konzentrationsgradienten von Triäthylammoniumacetat, pH 4,5.

D Nucleinsäuren

Die Anwendung der spektrophotometrischen Analyse bei der Untersuchung der Nucleotidsequenz der transfer-RNA

A. A. BAJEW, T. W. WENKSTERN und A. D. MIRSABEKOW

Institut für radiochemische und physikalisch-chemische Biologie der Akademie, Moskau

S U M M A R Y

The absorptionspectra of some oligonucleotides from the transfer RNA pancreatic RNase hydrolysate have been studied; among them the spectra of GpCp, ApCp, ApUp and (ApGp)Cp. The oligonucleotide spectra are additive in reference to the absorption spectra of the component nucleotides. Bromination is an effective tool for oligonucleotide identification. Some oligonucleotides, containing minor (odd) components (ΨUMP, 1-methylguanine, 2-methylamino-6-hydroxypurine, 2-dimethylamino-6-hydroxypurine) were also analysed. The spectra of $G^{M}p\,\Psi$ Up and $G^{M1}pG^{M2}pCp$ are quite characteristic. The authors suggest to use the direct spectrophotometric analysis for routine identification of the simpler oligonucleotides.

Im Mittelpunkt der Interessen des Laboratoriums für funktionelle Enzymologie des Instituts für radiochemische und physikalisch-chemische Biologie steht das Problem der Nucleotidsequenz der löslichen Ribonucleinsäuren (RNA). Die experimentelle Lösung dieser Aufgabe geschieht durch Spaltung der RNA-Moleküle in Fragmente mit Hilfe von Pyrimidyl- und Guanylribonucleasen +). Oligoribonucleotide, die bei der enzymatischen Hydrolyse entstehen, werden mittels Chromatographie auf Ionenaustauschern und Papier isoliert. Individuelle Oligonucleotide, die auf diese Art gewonnen werden, identifiziert man nach ihrer Nucleotidzusammensetzung und Nucleotidsequenz. Für diesen Zweck werden enzymatische, chemische und spektrophotometrische Methoden angewendet. Hier werden wir nur das Problem der spektrophotometrischen Analyse der Oligonucleotide erläutern.

Methodisches: die Spektren der homogenen Oligonucleotide, deren Reinheit durch Papierchromatographie geprüft wurde, nahmen wir auf dem selbstregistrierenden Spektrophotometer EpS-2 der Firma Hitachi (Japan) in 0,1 n HCl, 0,1 n KOH und nach dem Bromieren gegen die zugehörigen Kontrollen auf. Die Bromierung erfolgte nach SUZUKI und ITO [1], indem zur sauren Lösung (3 ml) gesättigtes, bzw. 20fach verdünntes Br_2-H_2O (0,1 ml) zugefügt wurde. Nach 10 Minuten wurde das überschüssige Brom durch 10 Minuten langes Belüften entfernt.

Für die Reinigung der Oligonucleotide, deren Reinheit eine hervorragende Rolle bei der spektrophotometrischen Analyse spielt, wurden 2 spezielle Methoden ausgearbeitet. Wir nutzten erstens die chromatographische Trennung auf Papier in Lösungssystemen, in denen das Nucleotid unbeweglich oder fast unbeweglich ist, aus. Diese Methode kann man

+) Polyribonucleotid-2-oligonucleotidtransferase, 2.7.7.16 und 3'-Guanylhydrolase der Ribonucleinsäure 3.1.4.8 (Report of the Commission on Enzymes, 1961)

als "chromatographisches Auswaschen" bezeichnen, da das Lösungssystem, das aus leichtflüchtigen Komponenten zusammengesetzt ist, verschiedene Beimengungen und Verunreinigungen auswäscht, während das Oligonucleotid am Startpunkt zurückbleibt. Die zweite Methode (11) besteht in der Sorbtion der Oligonucleotide auf Mikrosäulen mit Dowex-1 Cl'. Die Beimengungen und Verunreinigungen gehen durch den Ionenaustauscher, während das Oligonucleotid sorbiert und dann durch 0,1 n HCl eluiert wird.

Resultate. Die Spektrophotometrie der Stickstoffbasen und deren Derivate ist eine der wichtigsten Methoden, die zur Identifizierung dieser Verbindungen angewendet wird und die die vorhergehenden Etappen der chemischen und chromatographischen Analyse vollendet. In speziellen Arbeiten und Handbüchern kann man Absorptionsspektren der Purin- und Pyrimidinstickstoffbasen sowie von deren Nucleosiden und Nucleotiden finden. Die Absorptionsspektren der Oligonucleotide haben noch nicht das gehörige Interesse erweckt, obwohl dazu schon eine dringende Notwendigkeit besteht. Deshalb haben wir uns die Aufgabe gestellt, auf Grund der Kenntnisse über die Spektren der Oligonucleotide und ihren Veränderungen bei verschiedenen Bedingungen die Möglichkeiten der unmittelbaren spektrophotometrischen Identifizierung der Oligonucleotide herauszufinden. Dabei gingen wir davon aus, daß die Absorptionsspektren der Oligonucleotide in bezug auf die sie zusammensetzenden Mononucleotide annähernd additiv sind und der hypochrome Effekt im Gebiet des Maximums sich gleichmäßig auswirkt. In Abb. 1 ist das Spektrum des Trinucleotids (ApGp)Cp vor und nach der Hydrolyse angeführt; die Absorptionskurve des hydrolysierten Trinucleotids liegt etwas höher als die Kurve für das Ausgangsnucleotid; die Kurven gehen aber beinahe parallel; das zeugt von der annähernden Additivität des Spektrums. Es ist anzunehmen, daß die Additivität bei beliebigen Verhältnissen gültig ist und daß die Veränderung der Mononucleotide ihren Ausdruck in der Veränderung der Absorptionskurve des Oligonucleotids findet. Unsere Resultate zeigen, daß durch die Untersuchung der charakteristischen Spektrenveränderungen bei verschiedenen p_H und beim Bromieren die Nucleotidzusammensetzung der niederen Oligonucleotide auf Grund ihrer Spektren ermittelt werden kann [2]. In Abb. 2 sind die Absorptionskurven der vier Hauptribonucleotide und in Abb. 3 diejenigen der ΨUridylsäure in saurer und alkalischer Lösung und nach dem Bromieren wiedergegeben.

Besonders auffallend sind die hypsochrome Verschiebung und die starke Abnahme des Absorptionsmaximums beim Übergang vom sauren zum alkalischen Milieu für CMP; das Ausbleiben von merklichen Veränderungen für AMP; die starke Abnahme des Absorptionsmaximums ohne Verschiebung bei der Alkalisierung von UMP; die ausgeprägte Veränderung der Form der Absorptionskurve von GMP und endlich die scharfe bathochrome Verschiebung der Absorptionskurve für ΨUMP. Mit punktierten Linien sind die Kurven der bromierten Nucleotide gekennzeichnet. Die Bromierung wurde von vielen Forschern für die Identifizierung der Stickstoffbasen und ihrer Derivate angewendet [3)-9)]. Besonders nützlich ist diese Methode für die Differenzierung zwischen Adenin und Uracil und den entsprechenden Derivaten, da ihre Spektren sehr ähnlich sind, dagegen beim Bromieren ein unterschiedliches Verhalten zeigen: das Spektrum von AMP bleibt unverändert, während das Spektrum von UMP verschwindet. Das wurde von uns zur Identifizierung der Adenin- und Uracilderivate der säurelöslichen Nucleotide von E. coli ausgenutzt [10]. Zur Identifizierung können auch einige weniger auffällige Merkmale der Spektrenveränderungen herangezogen werden; so fallen z. B. der hohe Verlauf der Restkurve für GMP und das Auftreten eines neuen Maximums bei 290-300 mμ im Falle der unvollendeten Bromierung von CMP auf.

In Abb. 4 sind die Spektren von vier Oligonucleotiden im sauren und alkalischen Milieu und nach dem Bromieren angeführt.

Wir sehen die Spektrophotometrie der Oligonucleotide nach ihrer chromatographischen Trennung auf Ionenaustauschern und auf Papier als eine obligatorische Etappe ihrer Analyse an. Die Spektren, die in Abb. 3 angeführt sind, können als Illustration der spektrophotometrischen Identifizierung der Oligonucleotide dienen. Die Kurve 4 (I) zeigt das Spektrum des Dinucleotids GpCp. Schon die zweibuckelige Ausgangskurve bei p_H 2 ist sehr charakteristisch und zeugt von der Anwesenheit von GMP; die starke hypsochrome Verschiebung des rechten Zweiges bei der Alkalisierung und das Auftreten eines neuen Maximums bei 300 $m\mu$ nach dem Bromieren sprechen für die Anwesenheit von CMP.

Das Spektrum 4 (II) gehört dem Dinucleotid ApCp, das an folgenden Merkmalen zu erkennen ist: das breite Maximum, das zwischen 260 und 280 $m\mu$ gelegen ist, zeigt, daß es sich aus den Spektralkurven zweier Verbindungen zusammensetzt; eine davon muß ein Maximum bei 260, das andere ein Maximum bei 280 $m\mu$ haben. Die Anwesenheit von CMP wird weiter durch die starke Abnahme des Maximums, durch die hypsochrome Verschiebung des rechten Zweiges und durch das Auftreten eines neuen kleinen Maximums nach dem Bromieren bestätigt. Die Anwesenheit von AMP und nicht von UMP im Dinucleotid wird dadurch bewiesen, daß nach dem Bromieren das Spektrum nicht verschwindet, sondern sich in das Spektrum von AMP mit dem charakteristischen Maximum bei 257 $m\mu$ verwandelt.

Es ist schwer, im Spektrum 4 (III) (ApUp) die Komponenten zu erkennen, da die Maxima für AMP und UMP praktisch zusammenfallen. Aber die Verminderung der Absorption bei der Alkalisierung und eine noch größere Verminderung nach dem Bromieren sprechen dafür, daß das Oligonucleotid nicht nur aus AMP, sondern auch aus UMP zusammengesetzt ist. Im Falle der A-U-Polymeren kann die Größe der Absorptionsverringerung bei 260 $m\mu$ als Maß des relativen Gehaltes an AMP und UMP dienen.

Die vierte Kurve 4 (IV) gehört dem Trinucleotid (ApGp)Cp; da es aus 3 verschiedenen Komponenten besteht, äußert sich hier schon der Effekt der Summierung, indem das Spektrum sich dem Spektrum von RNA nähert. Nichtsdestoweniger zeugen die Veränderungen des Spektrums bei der Alkalisierung und beim Bromieren davon, daß im Oligonucleotid AMP und CMP vorhanden sind. Einige Trinucleotide mit sich wiederholenden Komponenten (z. B. GpGpCp) haben noch charakteristischere Spektren, die ihre Zusammensetzung auf den ersten Blick erkennen lassen.

Bei der Analyse der Hydrolysate, die durch Einwirkung von Pyrimidylribonuclease auf t-RNA gewonnen wurden, haben wir eine Reihe von Oligonucleotiden finden können, die die seltenen (minor) Komponenten enthalten [12]. Für diese Nucleotide sind natürlich die gleichen Gesetzmäßigkeiten charakteristisch, und bei ihrer Identifizierung haben wir dieselben Verfahren angewendet, d. h. Verfolgung der Spektren in saurem und alkalischem Milieu und nach der Bearbeitung mit Bromwasser. In diesem Falle sind ebenfalls die Absorptionskurven solcher Oligonucleotide besonders demonstrativ, die Komponenten mit charakteristischen Spektren oder Spektrenveränderungen enthalten. Zu den ersten gehören alle drei methylierten Derivate des Guanins, die sehr eigentümliche Spektren haben; zu den zweiten gehört ΨUMP, für das die ausgeprägte bathochrome Verschiebung des Absorptionsmaximums bei der Alkalisierung charakteristisch ist und seine Anwesenheit in einem beliebigen Oligonucleotid anzeigt.

Abb. 5 zeigt die Spektren zweier Oligonucleotide, welche ein großes Interesse vom Standpunkt der Primärstruktur von t-RNA und der Lokalisierung der seltenen Komponenten in diesem Molekül besitzen [12]. Eines der Oligonucleotide ist ein Trinucleotid, das CMP und zwei Methylderivate von Guanin (und zwar 2-Methylamino-6-hydroxypurin und 1-Methylguanin) enthält. Die Anwesenheit der methylierten Guanine äußert sich in der sonderbaren Form des Maximums im sauren und alkalischen Milieu und im hohen Verlauf der Restkurve nach dem Bromieren.

Das Spektrum des Dinucleotids $G^Mp\Psi Up$, wo G^M 2-Dimethylamino-6-hydroxypurin ist, hat ein sehr charakteristisches Ausgangsspektrum mit einer hohen Absorption im längeren Ultraviolett, was dem Spektrum von G^M zuzuordnen ist. Äußerst charakteristisch ist die ausgeprägte Veränderung des Spektrums beim Alkalisieren, die eine Folge der bathochromen Verschiebung von ΨUMP ist.

Aus dem Gesagten geht hervor, daß die Spektren der niederen Oligonucleotide und ihre Veränderungen unter verschiedenen Bedingungen mit Erfolg zur Identifizierung dieser Verbindungen herangezogen werden können.

Man darf die Möglichkeiten dieser Methode nicht unterschätzen, indem man die Identifizierung der Di- und Trinucleotide als eine Einzelfrage ansieht. Die Analyse der Nucleotidzusammensetzung und der Nucleotidsequenz der höheren Oligonucleotide kann man durch Gewinnung und Identifizierung von kürzeren Fragmenten vornehmen.

Außerdem ist zu bemerken, daß die spektrophotometrische Identifizierung bei verschiedenen p_H und nach dem Bromieren äußerst sparsam ist; alle 3 Spektren können mit einer Probe nacheinander aufgenommen werden, wozu 1,5 - 2 spektrophotometrische Einheiten notwendig sind. Das ist sehr wichtig, da die gewöhnlichen Oligonucleotide und besonders diejenigen Nucleotide, die die seltenen Komponenten enthalten, in geringen Mengen vorliegen. Eine besonders große Bedeutung wird dies bei der Analyse der homogenen t-RNA spielen. Die Hauptmethode für die Analyse der homogenen RNA wird die Methode der Fingerprints sein; die spektrophotometrische Identifizierung der Oligonucleotide, die aus solchen Chromatogrammen eluiert werden, wird von uns auch jetzt schon angewendet [11].

Abschließend kann man sagen, daß Di- und Trinucleotide mit Hilfe ihrer Spektren mit Gewißheit identifiziert werden können. Die Bedeutung dieser Behauptung erhellt daraus, daß ein beliebiges höheres Oligonucleotid durch enzymatische Hydrolyse in Fragmente gespalten werden kann, die aus zwei und drei Nucleotiden bestehen. Deshalb ist die Spektrophotometrie der niederen Oligonucleotide eine in bezug auf Zeit und Menge effektive und sparsame Methode zur Bestimmung der Nucleotidzusammensetzung und Nucleotidsequenz.

L I T E R A T U R

1) SUZUKI,T., E.ITO, J. Biochemistry 45,6,403 (1958)
2) WENKSTERN,T.W., A.D.MIRSABEKOW, W.I.GORSCHKOWA, A.A.BAJEW, Biochimia 28, 712 (1963)
3) WHEELER,H.L., T.B.JOHNSON, J. Biol. Chem. 3, 183 (1907)
4) CAPUTTO,R., L.F.LELOIR, C.E.CARDINI, A.C.PALADINI, J. Biol. Chem. 184, 333 (1950)
5) BERGKVIST,R., A.DEUTSCH, Acta Chem. Scand. 8, 1880 (1954)
6) HAMER,D., D.M.WALDRON, D.L.WOODHOUSE, Arch. Biochem. Biophys. 47, 272 (1953)
7) SOODAK,M., A.PIRCIO, L.R.CERECEDO, J. Biol. Chem. 181, 713 (1949)
8) DOWEDOWA,E.D., Biochimia 24, 414 (1959)
9) JONES,A.S., D.L.WOODHOUSE, Nature 183, 1603 (1959)
10) CHANINA,M.F., T.W.WENKSTERN, A.A.BAJEW, Biochimia, im Druck
11) KRUTILINA,A.I., T.W.WENKSTERN, A.A.BAJEW, Biochimia, im Druck
12) WENKSTERN, T.W., A.A.BAJEW, A.D.MIRSABEKOW, W.I.GORSCHKOWA, Dokladi Akademii Nauk 151, 220 (1963)

DISKUSSION-BEMERKUNGEN

SAUERBIER.- Ich glaube, die Schwierigkeiten bei der Basensequenzanalyse von transfer-RNA liegen weniger in der Trennung nach der Aminosäure-Spezifität oder in der Bestimmung der Basenzusammensetzung von Oligonucleotiden als in der "charakteristischen" Spaltung der RNA. Glauben Sie, daß Sie diese Spaltung schon praktisch durchführen können?

BAJEW.- Die spezifische Spaltung der s-RNA kann man mit Hilfe der RNase erzielen. Die gereinigte Pancreas-RNase spaltet die polymere Kette der s-RNA nur nach Pyrimidylresten. Wir konnten uns überzeugen, daß in der Mononucleotidfraktion der RNase als Hydrolysate der s-RNA nur UMP, CMP und ΨUMP anwesend sind. Das ist ein überzeugender Beweis der hohen Spezifität der enzymatischen Spaltung. Eine kleine Beimengung von AMP stammt von der partiellen und unspezifischen Spaltung des acceptierenden Trinucleotids der s-RNA, CpCpA. Die RNasen T_1 und T_2, die s-RNA nach Guanyl- bzw. nach Adenylresten spalten, sind hoch spezifisch. Es ist möglich, daß einige chemische Modifikationen der s-RNA die Spezifität der Pancreas-RNase erhöhen werden und daß die Spaltung der s-RNA nur nach einem Pyrimidylrest, z. B. nach UMP, vor sich gehen wird.

DELLWEG zu SAUERBIER.- Natürlich ist die Sequenzanalyse von transfer-RNA noch weit entfernt und sehr schwierig, das schließt aber doch nicht aus, daß wir uns einstweilen die später nötigen Grundlagen erarbeiten.

DELLWEG.- Ist es möglich, auch quantitative Bestimmungen zu machen, was ja zur Unterscheidung von homologen Nucleotiden wichtig wäre, als Beispiel AAC und AAAC?

BAJEW.- Das ist unmöglich. Die Methode ist nicht quantitativ.

BUDOWSKY.- Die Methode der oszillographischen Polarographie ist zu kompliziert und unzugänglich für gewöhnliche Laboratorien. Die spektrophotometrische Methode von BAJEW und Mitarbeitern ist einfach, bequem und aussichtsreich.

BUDOWSKY zu SAUERBIER.- Im Institut für Strahlen- und physikochemische Biochemie werden enzymatische, bei uns chemisch-enzymatische Methoden des spezifischen RNA-Abbaues ausgearbeitet. Es gibt auch bestimmte Ergebnisse bei der Isolierung spezifischer löslicher RNA.

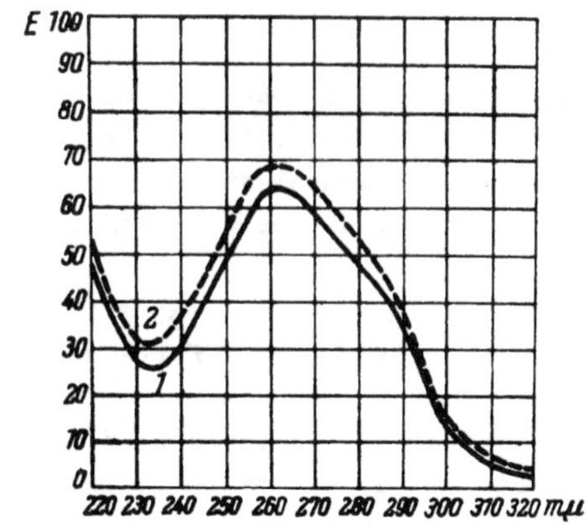

Abb.1 Absorptionskurve des Trinucleotides (ApGp)Cp
1 - vor der Hydrolyse 2 - nach der Hydrolyse

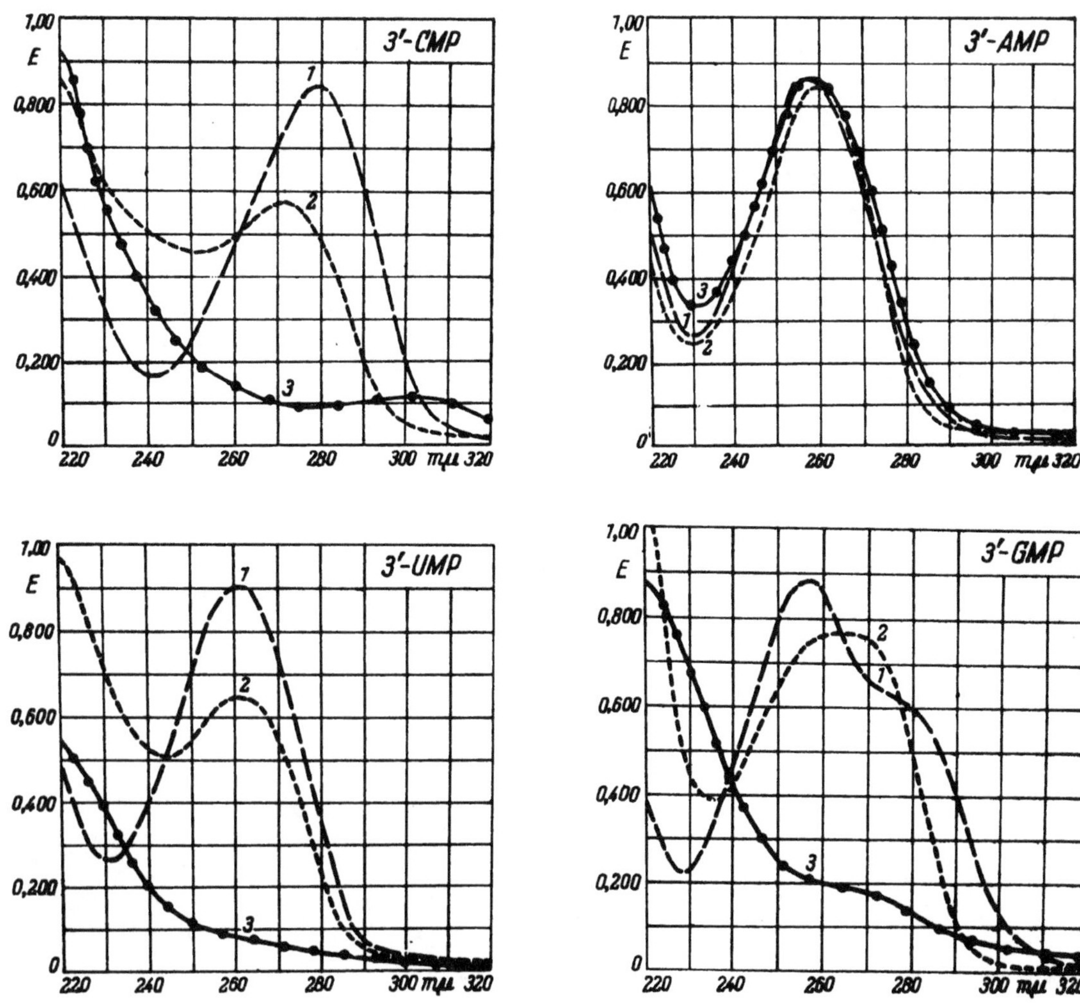

Abb.2 Absorptionskurven von CMP (I), AMP (II), UMP (III) und GMP (IV)
1 - in 0,1 N HCl 2 - in 0,1 N NaOH 3 - nach Bromieren.

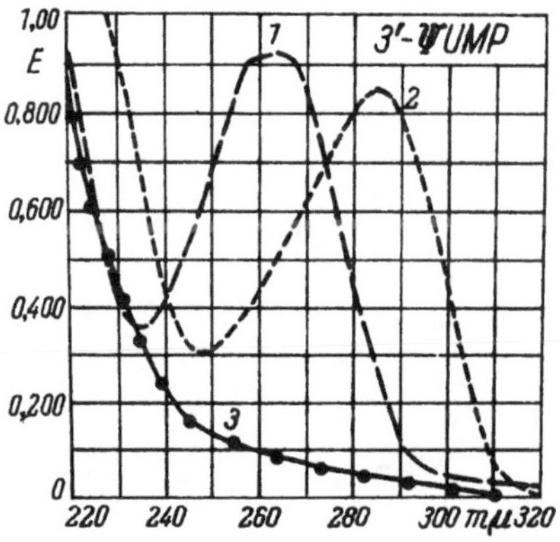

Abb.3 Absorptionskurve von ΨUMP
1 - in 0,1 N HCl   2 - in 0,1 N NaOH   3 - nach Bromieren.

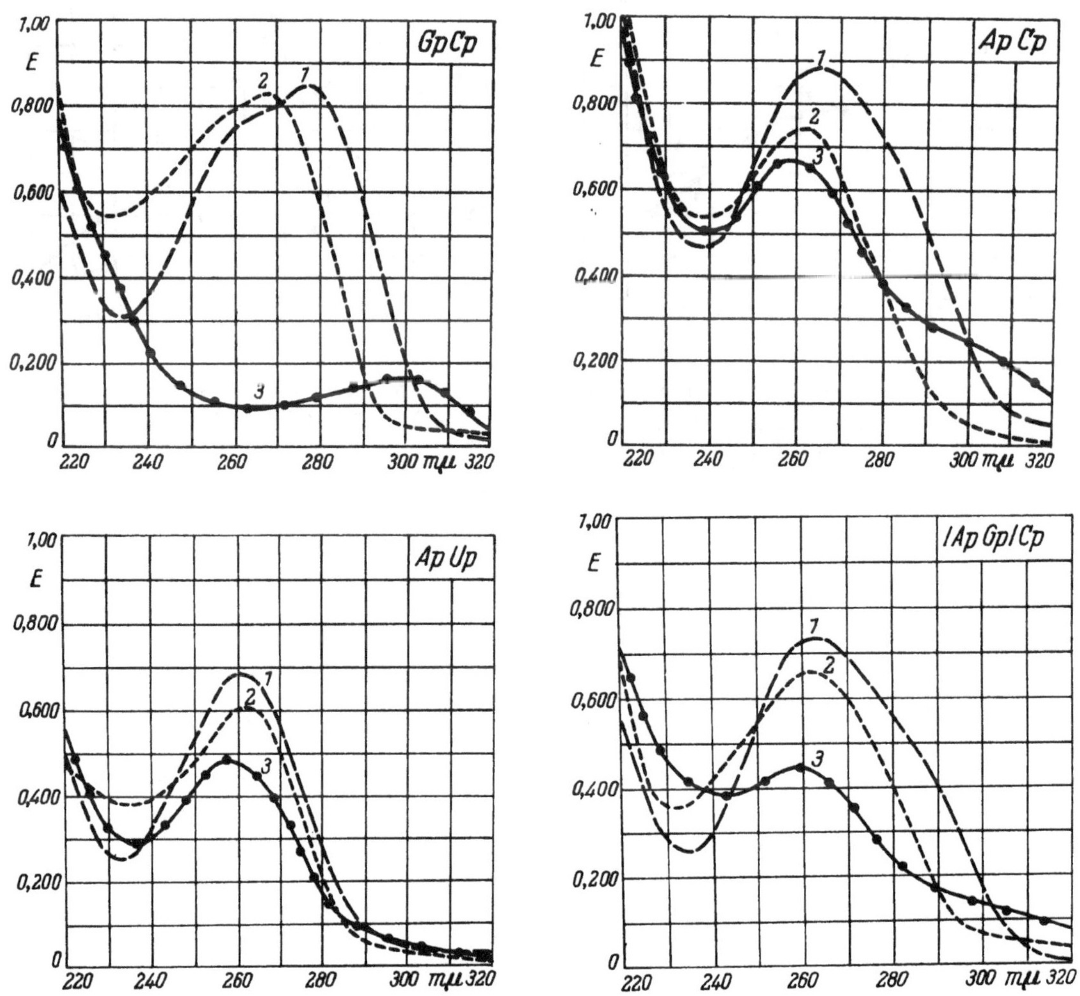

Abb.4 Absorptionskurven von GPCp(I), ApCp(II), ApUp(III) und (ApGp)Cp(IV)
1 - in 0,1 N HCl   2 - in 0,1 N NaOH   3 - nach Bromieren.

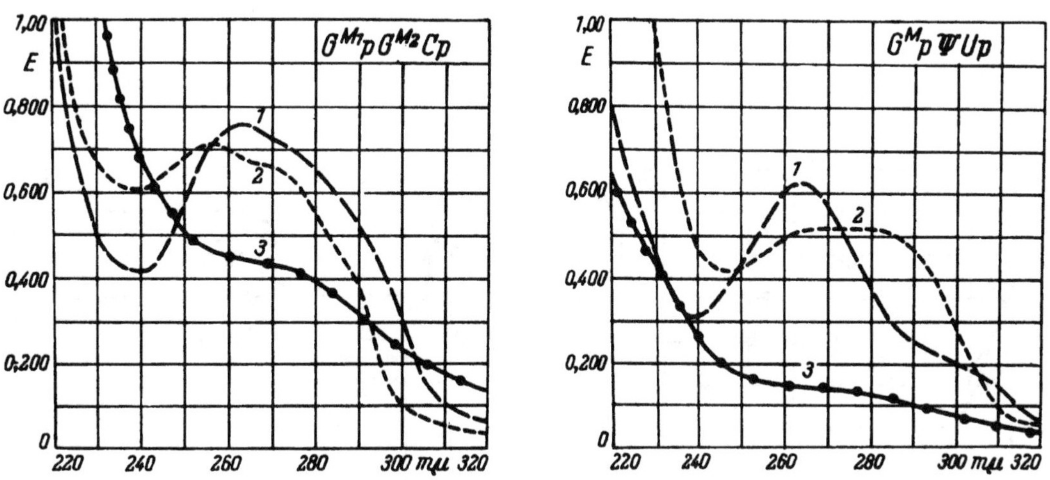

Abb.5 Absorptionskurven von G^M p G^{M2} p Cp(I) und G^MpΨ Up(II)
(Erklärung s.Text) 1 - in 0,1 N HCl 2 - in 0,1 N NaOH
3 - nach Bromieren.

Generaldiskussion III

Probleme physikochemischer Untersuchungen der Nucleinsäuren

Chairman: D. JORDAN (Adelaide)

D. JORDAN:

We have discussed during the course of the last day and a half many aspects of the chemistry of nucleic acids, the emphasis being on DNA rather than RNA. The thirty-one papers which have been presented serve to illustrate the variety and extent of the studies in progress on DNA structure and behaviour and are indicative of the progress being made. I think the results of our discussions can be summarised under six or so broad headings.

For many years, nucleic acid research was concentrated on the determination of the size and shape of the DNA molecule. With the realisation that much of this work was being carried out on degraded and heterogeneous samples such studies became of less significance. However, isolation techniques have improved and with the shift from cellular to bacterial and viral or phage DNA such studies have become of renewed interest and significance. It is evident from the results of Dr. ATEN's work that meaningful sedimentation and viscosity measurements can be made on relatively homogeneous T4 phage DNA. The molecular weight of this material is approximately ten times that believed to be common for DNA a few years ago (6×10^6), but is nevertheless much lower than some specimens of DNA isolated by special techniques. Estimates of the molecular weight of these specimens obtained by autoradiography are as high as 230×10^6. The isolation of these very high molecular weight samples of DNA raises new problems. The handling of this material without causing rupture by exposing it to shear gradients is most difficult and it is evident that if the chemistry of such preparations is to be studied new techniques must be developed. Other problems arise of a more theoretical nature since most formulae for determining molecular weights from, for example, sedimentation and viscosity measurements, have only been verified in the lower molecular weight range and their use at these very high molecular weights must be viewed with caution.

It has long been recognised that precise physico-chemical measurements have little value if made on heterogeneous specimens of DNA. One way round this problem is to work with relatively homogeneous DNA prepared from phage. However, the problem of the fractionation of the heterogeneous DNA from cells must be faced sooner or later. Fractionation on the albumen-kieselguhr column appears to be a great improvement over the Ecteola columns of BENDICH and meaningful relationships between GC content and Tm and molecular weight can apparently be made on the fractions. However, since the genetic information is associated with the base sequence which is contained within the molecule, it is somewhat surprising that any basic column can fractionate DNA on other than a molecular weight basis. What would be of more significance would be the fractionation of denatured, i.e. single strand, DNA on acid or perhaps purine-pyrimidine columns with subsequent renaturation of the fractions.

The hyperchromic effect is made use of in the characterisation of DNA by the determination of Tm. No complete theoretical interpretation of this effect has, however, yet been given. The studies of TINOCO, RHODES, NESBITT and BOLTON and WEISS generally lead to the belief that the hyperchromic effect can be explained in terms of dispersion forces only. The possibility of π-type interaction between the bases has also been investigated and it would appear that a plausible explanation of the spectroscopic behaviour of DNA can also be given on this basis. The ideas of LÖWDIN, which attempt to explain spontaneous mutation on the basis of double proton tunnelling must be viewed with caution since the presence of water in close association with the molecule appears to have been overlooked. Associated with this problem is that of the nature of the forces responsible for the stability of the double helix. Although the hydrogen bonds between the bases are responsible for the specificity of the base-pairs it is evident that dispersion forces and hydrophobic bonding play an important part in determining the stability of the helix.

Studies of the denaturation of DNA are of particular value. Not only is the mechanism of uncoiling and recoiling of the double helix of intrinsic interest, but also the process is of great biological importance as this process, or one closely related to it, must occur during DNA and RNA synthesis and also at cell division. Heat denaturation is now a reasonably well understood process in which it has been clearly demonstrated that separation of the double helix into single strands may occur if the conditions are favourable. Furthermore the reverse process, that of renaturation to form DNA of similar physical and biological properties to the original native DNA has been well described. The alternative way of denaturing DNA by solution in very low ionic strength solvents is less well understood and it would appear that at room temperature strand separation is not complete. Denaturation appears to occur more readily in heavy water as solvent, i.e. the double helix is less stable in this solvent than in water and protonation of the basic groups of the purines and pyrimidines also reduces the stability. Of particular interest is the influence of dye ions and of various cations on the stability, divalent ions as Ca^{2+} and Mg^{2+} generally cause an increase in stability but monobasic organic ions such as amino-acridine cause a marked decrease in the stability most probably due to the penetration of the double helix by the polycyclic nucleus. Metal ions, as copper ions, capable of complexing with the purine and pyrimidine bases, cause a marked decrease in the thermal stability of DNA. The effect of free purine and pyrimidine bases and of nucleosides has also been studied.

Further information on structure of DNA comes from the shape of the melting curve. A sharp profile indicates a homogeneity of the molecule whilst a broad profile indicates heterogeneity. Particular attention is being drawn both to the theory of the melting process and to the collection of more accurate data on the shapes of the melting profiles. Such studies should throw further light on the nature of the bonding forces in the helix.

Certain aspects of the changes in DNA structure produced by irradiation are now reasonably well understood. I was particularly interested in the work which was done in this institute by Dr. VENNER, in which he showed that UV-irradiation causes a decrease of Tm and a broadening of the melting curve. The specific formation of the dimer of thymine on irradiation of DNA gives information on the juxtaposition of thymine residues in the native DNA. The use of inhibitors also has yielded information on the part played by free radicals in the photodecomposition. One experimental procedure of particular interest is the use by Dr. BERG of polarographic techniques to follow the rapid reactions produced by irradiation, these studies appear to be yielding results of some significance.

Before throwing the meeting open to general discussion I would like to express my gratitude to Dr. BERG for inviting me to participate in this symposium, to congratulate him on the excellent way it has been organised and to express on your behalf our thanks for the great hospitality we have received.

H. BERG:

Ich möchte 3 Punkte herausstellen:

1. Gibt es eine generelle Abhängigkeit des Tm-Wertes vom Molekulargewicht der DNA?

2. Für die Erscheinung der Denaturierung sind viele Parameter von Bedeutung, insbesondere auch die Salzkonzentration. Erfolgt eine Renaturierung des DNA-Moleküls auch in reinem Wasser?

3. Wichtig erscheint die Frage nach der Empfindlichkeit der Tm-Methode. Professor TUMERMAN hat in Moskau eine Fluoreszenzmessung für die Verfolgung des thermischen Überganges der DNA angewandt. Die Methode beruht auf der Messung von Fluoreszenzunterschieden, die bei der Wechselwirkung von Monomerfarbstoffen mit der DNA-Doppelhelix und der Addition von Dimeren zu der denaturierten Form auftreten. Diese Methode soll um eine Zehnerpotenz empfindlicher als die übliche Tm-Technik sein.

D. CROTHERS:

By statistical mechanical calculation on the shape of the helix-coil transitions and in connection with work we found that the difficult prediction depends on the length of the molecule. We did experiments. One can plot the Tm and the breadth of the transition against the molecular weight of the DNA. We find that the melting temperature Tm remains constant over a molecular weight region from 8 to 6 in log units. At this point the melting temperature begins slowly to drop and drops in an accelerated manner down to the order as low as we went of about 5. The breadth of the transition is constant in the same regions and then it begins to increase in a similar way. Our molecular weight materials were obtained by shear degradation of T2-DNA. The reason for these observations seems to be that the dropping melting temperature is due to the strand separation becoming important statisticly; e.g. if you have long molecules then it only seems to be hold together in plane. If you have short molecular weights than the melting temperature drops, e.g. it's sensible. Look at the DNA molecule then the ends of the molecule are thermodynamically less stable than regions in the middle because of the stacking energy. There is a little unwinding at the ends. If one make enough ends than this makes at any given temperature more base pairs unbonded so that one finds that the breadth of the transition, that is the degree of unpairing at a temperature some below the midpoint of the transition, increases.

R. ROWND:

I think it's somewhat difficult to except the view for the following reason. It's increasingly clear by evidence from the past several years, that the denaturation of the DNA-molecule, that is the disruption in the helix occurs in localized regions of the molecule. During heating process as one goes into the region of the hypochromic change we have a molecule with structures in which localized regions are damaged whereas others are not, where the two chains are held together simply by nuclei containing the highest base composition in this particular molecule. For this reason when one does experiments on DNA in absorbance temperature measurements at the elevated temperature you measure a Tm value. If one measure the optical density at room temperature after cooling the solution, then the Tm-value is always greater by several degrees. The reason for this seems to be, that if measurements are made at the elevated temperature one measures a structure with localized regions of the molecule, whereas a species of the other type in which one has achieved essentially 100 % hyperchromicity is cooled down, this goes back almost completely to the native structure. So that, since these disordering is occurring over only localized regions and there is no precise estimate and appeared to be very small in comparison to 10^5 size in molecular weight.

D. CROTHERS:

That's well. I should have go in more details. We have made calculations based on the stacking energy of synthetic polynucleotides and we have made calculations of the estimated average length of such regions. There will be these regions existing all along a large molecule. We find that the value is of 200 to 400 base pairs of the average length. The region in which you really begin to see this rise in ΔT, when the molecular weight drops is to the order of 200-400 base pairs. So it is consistent that this in a way the regions in which an end effect becomes apparent is a measure of the stacking interaction in a DNA molecule; e.g. when you get to low molecular weights there is no melting in the middle then the end effect is important and that is just what one can see experimentally.

R. ROWND:

Would it not follow then from your argument that this temperature difference between the absorbance-Tm and say a viscosity-Tm measured at room temperature or a transforming activity-Tm, they should approach zero as you go down in molecular weight.

D. CROTHERS:

In fact it does.

R. ROWND:

In measurements which we have done a bacterial DNA of molecular weight of about a half a million there is still a difference in this.

D. CROTHERS:

We went down to molecular weights of the order of 200 000 and the difference of this point is only of the order of $0{,}2^{\circ}$ and this is beyond the limit of our accuracy. There is in fact a difference between these two transitions. In the order of high molecular weights there is a difference between the optical density transition and the strand-separation transition. The difference between these two temperatures decreases and seems to be going to zero as we go to molecular weights of the order 10^5.

H. BERG:

Kann jemand noch etwas zur Frage der Renaturierung in rein wäßriger Lösung, also ohne Salzzusatz berichten?

R. ROWND:

We have studied the ionic strength dependence of renaturation in SSC pH 7. The normal condition choosed for renaturation is in 1,9 times SSC. In earlier studies MARMUR had found that the rate of renaturation as a function of ionic strength reaches a saturation level. The rate of helix growth continuously increases as the ionic strength is raised up to 5 times and even 10 times SSC. In 0.5 SSC over three hours of annealing, which is sufficient amount of time for completely renatured bacterial DNA in 1.9 times SSC, there is essentially no interaction which occurs between the chains. In lower ionic strength then a corresponding not expected renaturation occurs and this is the point, which I compare to Professor JORDAN's studies, since he is working much lower than this.

H. BERG:

Wenn dieser Salzeinfluß von derartig entscheidender Bedeutung ist, dann wäre es an der Zeit, sich in dieser Hinsicht mehr mit der Theorie der Elektrolytlösungen und der Debye-Hückel-Theorie zu befassen. Bei der Klärung derart fundamentaler Fragen würde sie mithelfen können.

Ch. ZIMMER:

I think also the existence of the calculated number of 200 to 400 base pairs given by Dr. CROTHERS in a sequence of the DNA molecule in the transition region depends on ionic strength or not?

D. CROTHERS:

Apparently it does'nt. The reason for this seems to be because we take polynucleotides on which we base our computation. They are homogeneous. In this case the breath of the transition is almost independent of salt concentration.

Ch. ZIMMER:

The lower the concentration the greater is the broadening of the transition of a DNA molecule. If one goes to a very low counterion concentration the broadening is connected with the existence of a greater number of splitted regions as discussed by DOVE and DAVIDSON in one of their last papers. Therefore the helix length of intact regions should be shorter.

D. CROTHERS:

Yes, there may be so, but more a secondary effect. But the difficulty in DNA is, that those polymer is heterogeneous. So it's difficult to answer your question what is the length of these centers in DNA molecules. In the synthetic polynucleotides the stacking energy is very insensitive to ionic strength.

D. JORDAN:

Is this poly GC?

D. CROTHERS:

Yes, this is a polymer of that type, not of the AT type.

Ch. ZIMMER:

Another factor of the DNA environment is the pH. So the stacking energy of the DNA molecule should be pH dependent, because the energy depends on temperature and other conditions.

D. CROTHERS:

Yes, it's the matter of what are the interactions. To estimate the effect of pH or ionic strength at present is impossible. We can only do experiments.

Ch. ZIMMER:

We want to know more details about the sharpening of the transition produced by protonation of the DNA molecule, which was discussed by Professor JORDAN in his introduction.

J. SPONAR:

I'll make some remarks on the question of the dependence of the Tm value on molecular weight. It must not be necessarily the fact that the real molecular weight of the particle in the solution is equal to the unit in which the helix-coil transition occurs. There may be some heterogeneous bonds and so it may be somewhat complicated.

D. CROTHERS:

Yes, that's very true. Theoretically you have considered the whole molecule, and since the molecule is heterogeneous its a physical problem. In many ways the logic is simplified if one comes to a calculation of the average correlation length and one considers these to be more or less independent. But this is not a rigorous process. However, it may be a convenient way.

P. MUNK:

There is a question of all or none denaturation process. Are experiments with partially denatured DNA or not.

R. ROWND:

We have done a very sensitive study with partially denatured DNA. It's quite clear, that the rate of unwinding of the two strands of the DNA molecule is a very rapid process. E.g. in Bacillus subtilis DNA of a molecular weight 16–18 million with using biological hybrid DNA. Heating this material about 30 seconds in 0.1 SSC in which the Tm is about 70° C; boiling this material at 100° C the strands are completely separated within 30 seconds. So the unwinding of the two strands of the DNA molecule is a very rapid process. If one introduces a crosslink in a native molecule and boils this material one finds that this is essentially completely reversibly denatured. On the other hand in renaturation, as I showed you yesterday the rate of helix growth appears to require about 30 minutes at the optimal temperature for renaturation and below 50° C it's completely retarded. So, with crosslinked molecules helix growth is a very rapid process whereas in renatured DNA it appears that it is a slow process. The reason for this is, when this crosslinked molecule is heated to the elevated temperature one disrupts the secondary structure of the molecule. The strands prevented from separating and when this material is cooled this can reverse perhaps to the native structure or can collapse into a denatured structure depending on whether the formation of intra-chain base pairs is faster than the reformation of inter-chain base pairs. Now the evidence, I think, shows that the formation of inter chain base pairs is kinetically favoured. I showed yesterday in the plot of the biological activity which is related to the helical content versus temperature, that the point at which the activity is no longer restored is in this case 93° C. The Tm value in this solvent is 89° C. So we have a structure at lower temperature with two complementary strands, in which one has formed a nucleus over a small region. The remaining regions are involved in interstrand pairing and this prevents helical growth. Until 50° C density measurements and biological activity measurements indicate that there is no helical growth in species of this type. If one heats to 93° C one forms a structure in which the chains are held together not by an intra chain crosslink, but by the most resistant base sequences which occurs in the molecule. But even at 93° C one has a structure of this type you regain the maximum level of biological activity upon cooling this material. So I think, that where is originally this material within the type of metastable state and the intra chain secondary structure prevented helix growth. If these regions are disrupted then the growth of helix is a very rapid process and behaves just like a crosslinked DNA molecule. So, I think, we can conclude, that if there is a possibility of forming inter-chain or intra chain base pairs that the kinetically favoured process is the reformation of inter chain bonds. In the case of renaturation of DNA, when one takes the two strands completely apart, at the elevated temperature, and then anneales these at a lower temperature before these strands can interact, you have sufficient time for reforming the interstrand base pairs. So one gets slow helix growth. The other question is: Is DNA denaturation an all or none process. We have done experiments to see, that there would be very little of partially denatured species. We could produce in the process of denaturation metastable species which had both native and denatured regions within the same molecule. To do this we work in low ionic strength solvents (0.01 m NaCl), because the transition broadens somewhat under these conditions and becomes less cooperative. By heating this material somewhat above the Tm in this solvent and then banding in a density gradient we found, e.g. with T4 DNA that there is a mixture of native and denatured DNA. There is no overlap in the band profile. The material which was heated under these conditions had a band profile in which about 15 – 20 % of the material banded at an intermediate density. We then extended the studies to transforming DNA.

If one plots the results of transformance $\frac{c}{s}$ as a function of density across the band profile in the native regions one finds a value of 2.0, which is characteristic of the native DNA. In the fractions which are of increased bouyant densities due to increasing amount of metastable partially denatured regions this ratio of $\frac{c}{s}$ transformance increases to a value of about 7. If one reaches the denatured region the band profile falls back to the value characteristic for the denatured DNA. This would be on cooling a molecule, which are in a partially collapsed state at an elevated temperature. There is a fraction, which falls back to form metastable states but due to the formation of the inter strand base pairs and from the biological experiments it appears that there are specific regions within the individual molecules which are weaker and most susceptible to being denatured.

H. BERG:
Now, I come to another question. What is the highest molecular weight, which one can isolate from natural materials to day? And are there cells for physical measurements, e.g. diffusion and sedimentation cells, in which no shear degradation of DNA molecules occur.

D. CROTHERS:
To day one can show by special methods molecular weights of the order of 10^9. One can measure sedimentation and viscosity, and we get up to 10^8 in molecular weight of DNA of bacteriophages. We are now, as the work of Dr. ATEN shows, able to measure sedimentation coefficients with consistency in this regions, and we are also able to measure viscosity within this region with new technics in physical chemistry. This point is a problem of the material and a question of measurements. These natural polymers are quite sensitive to shear degradation. The difficulties in this range are the shear dependence of viscosity, the viscosity dependence on the rate of the flow, and the sedimentation coefficient depends on the rotor speed. One can do this measurements if one goes to low rates of flow.

J. ATEN:
We have now values of about 130×10^6 for T2-DNA. The values of the lower molecular weight region have been calibrated on light scattering measurements. Another possibility to do measurements is autoradiography.

P. MUNK:
I believe that the molecular weight wether in this region are not very reliable. In our measurements of streaming birefringence we have found that the DNA molecules of very high molecular weight interact. This interaction may influence the data of sedimentation, viscosity and other molecular weight determinations.

D. CROTHERS:
The values are all extrapolated to zero concentration and give linear extrapolations in this way.

P. MUNK:
It is not enough in this range.

P. JORDAN:
You mean, you can't get up by variation of concentration to remove these interactions?

P. MUNK:
We have measured the kinetic of streaming birefringence and learned that the concentration dependence are not good, and the kinetic effect was in one solution greater and in another solution smaller.

D. CROTHERS:

What is the sodium ion concentration?

P. MUNK:

It was SSC.

D. CROTHERS:

We are measuring at very much lower concentration and I know that your effects disappear again below.

P. MUNK:

We have two dependences, originally concentration dependence and the dependence of the hydrolysis kinetic effect. It was the only measure for the specificity in the reaction, the fall of birefringence with time.

Ch. ZIMMER:

It is known that low sodium ion concentration prevents somewhat aggregation. Did you measure the dependence on counterion concentration in your experiments?

P. MUNK:

No, we have not done this.

D. CROTHERS:

Well, unfortunately at very low sodium ion concentrations one gets considerable interaction. In hydrodynamic problems it is better to work in SSC.

H. VENNER:

Man könnte die Frage von Dr. BERG etwas modifizieren. Welche DNA kann der Biochemiker zur Verfügung stellen in bezug auf das höchst erreichbare mittlere Molekulargewicht. Dabei spielen noch andere Faktoren eine wesentliche Rolle. Meist sind die chemischen Daten von DNA-Proben unzureichend in den Mitteilungen verzeichnet. Insbesondere ist die Kenntnis von Protein- und Polysaccharidgehalten in DNA-Präparaten stets notwendig, wobei die Bestimmungsmethode von grundlegender Wichtigkeit ist. Es kommt nicht nur darauf an, eine DNA hohen Molekulargewichtes, sondern auch hochreines Material in der Hand zu haben. Deshalb wäre eine Diskussion hierüber sehr nützlich.

J. SPONAR:

We are quite unable to remove all proteins present in nucleic acid samples. This proteins may have also some structural effect, perhaps. The reproducibility of estimation methods is not enough.

J. ATEN:

In our phenol extraction method first used by MANDEL and HERSHEY the protein from T4 DNA is removed and gives a very reproducible result. The limit of viscosity measurements and sedimentation coefficients of T4 DNA samples from different T4 phages isolated by different peoples by different methods is always the same. Our chemists are not able to estimate protein contents exactly below 0.5 %. Our protein content of DNA samples is always below 0.5 %.

H. VENNER:

Die Reinigung der DNA aus Phagen ist offenbar weniger problematisch gegenüber der Isolierung nativer DNA aus der Mikroorganismenzelle. Bei der Isolierung von DNA aus Mikroorganismen gibt es im Prinzip zwei Möglichkeiten: mechanischer Zellwandabbau, dann ist Scherabbau der DNA zu erwarten, oder enzymatische Zellwandlyse, dann ist gleichlaufend eine Einwirkung zelleigener Enzyme auf die DNA möglich. Darin liegt die Schwierigkeit. Nach unseren Erfahrungen ist es am günstigsten, die Mikroorganismen vor der DNA-Isolierung in einen Protoplastenzustand zu überführen. Dadurch läßt sich hochmolekulare, möglichst schonend behandelte und von Proteinen bzw. anderen zelleigenen Substanzen freie DNA gewinnen.

H. BERG:

If we isolate DNA and store it in dry state the solubility can decrease after some months and the important question arises: what reactions are responsible for this effect?

J. ATEN:

I think, its better to store DNA in solution. In dry state the biological activity decreases in one day and is damaged.

D. JORDAN:

But the question is what is damaged!

H. BERG:

Der Physiker braucht ein stabiles System. Daher interessiert ihn, warum es außerhalb der Zelle instabil ist. Was ist wohl die Ursache für die Veränderung der DNA im festen Zustand?

R. ROWND:

If one have dried transforming DNA and redissolvs it in solution the biological activity is high. So the damage I don't think is very significant. But the biological activity is very sensitive to depurination and chain scission.

L. BLUMENFELD:

I think that there is a main difference between the denaturation process in the solid state and in the solution. We have different states of structure. It's important the water around the structure and this is quite different in these two states.

H. VENNER:

Wir haben auf Grund von Elementaranalysen an bis zur Gewichtskonstanz getrockneter DNA festgestellt, daß in jener Probe noch zu viel Wasserstoff und Sauerstoff enthalten ist, was auf eine festgebundene Wassermenge unter diesen Bedingungen schließen läßt.

J. ATEN:

I believe it's very difficult to say what happens with DNA. A DNA sample with molecular weight of 2×10^6, as I showed you yesterday, has the same biological activity as samples with higher molecular weight. Even if we go down to 700 000 we found the same biological activity. It's very difficult to answer the question of Dr. BERG.

R. ROWND:

It was shown by LITT, MARMUR, EPHRUSSI-TAYLOR and DOTY in 1958, that the biological activity of transforming DNA from 15×10^6 and down are very sensitive to changes in molecular weight. From an extrapolated value to zero activity in a plot of molecular weight versus biological activity it has been shown that transforming activity is very sensitive to molecular weight.

H. BERG:

Die Möglichkeiten der Untersuchung und Aussagen auf dem Gebiet der Nucleinsäureforschung sollen im folgenden kurz zusammengefaßt werden. Dabei sind Methoden gemeint, die große Aussagemöglichkeiten zulassen und deren Vertiefung deshalb wünschenswert ist.

DNA	Methode	Aussage über:
a) im festen Zustand	Elektronenstreuung	Struktur
	Elmi-Autoradiographie	Sequenz, Reduplikation
b) in Lösung	NMR-Spektroskopie	Hydratationszustand
	IR-Spektroskopie	und Wechselwirkung
c) im Zellverband	Autoradiographie	nativer Zustand
(d. h. in Verbindung mit anderen Zellsubstanzen)	Elektronenmikroskopie	(Nucleoproteide u. a.)

D. JORDAN:

"I think, we have heared from Dr. BERG what we have to do in the future. Perhaps in two years we come back here and discuss the results."

Abschluß und Ausblick

Wenn nun der Versuch unternommen werden soll, aus den Ergebnissen des Symposiums einige neuartige Entwicklungslinien aufzuzeigen, so kann das nur mit allem Vorbehalt geschehen, da gerade auf dem jüngeren Nucleinsäuregebiet ständig mit unerwarteten Phänomenen zu rechnen ist. Dennoch können folgende Feststellungen getroffen werden :

A ALLGEMEINE GRUNDLAGEN UND METHODEN

a) Erhebliche Steigerung der Meßgenauigkeit allgemein zur exakteren Fundierung der Aussagen über schonendst isolierte Präparate ist oberstes Gebot, wobei kein Aufwand gescheut werden darf (St.CLAESSON, Uppsala).
b) Als ein weites Feld für diffizile kinetische Messungen erweisen sich die Relaxationsphänomene (bei Änderung der Temperatur, des pH-Wertes und der Bestrahlung u.a.), wie es erstmals in den Beiträgen von K.E.REINERT (Jena) und H.Ch.SPATZ (Göttingen) angedeutet wurde.
c) Die Möglichkeiten der elektronenmikroskopischen Abbildung sind noch keineswegs erschöpft (J.RICHTER, Basel).
d) Zum Spurennachweis und für Lokalisierungen beginnt die Aktivierungsanalyse (J.ALTMANN, Wien) eine wesentliche Rolle zu spielen.
e) Die Analyse der inter- und intramolekularen Wechselwirkungen durch Fluoreszenz-, IR- und NMR-Methoden schreitet voran. Auch die Theorie befaßt sich verstärkt mit größeren Molekülen.

B POLYSACCHARIDE

Unter diesen befinden sich definierte unverzweigte Kettenmoleküle, die als Modelle zur Überprüfung von Theorien der Konformation und der Lösungseigenschaften dienen können (G.MEYERHOFF, Mainz, W.BURCHARD, Freiburg i.Br.).

C PROTEINE

Das besondere Interesse konzentriert sich auf den Zusammenhang zwischen Enzymaktivität und Konformation (R.LUMRY, Minneapolis). Die Photo- und Strahlenchemie kann infolge der Spezifität der -S-S-Brückenlyse eindeutigere Ergebnisse als bei anderen Makromolekülen erzielen.

D NUCLEINSÄUREN

Es standen sehr viele Fragen zur Debatte, die nur zu einem kleinen Teil beantwortet werden konnten :

a) Welche maximalen Molekulargewichte kann der Biochemiker aus Phagen oder aus Zellen isolieren und bis zu welcher Größe kann der Physiker diese in seinen Apparaturen ohne Denaturierung messen ?
b) Welchen Reaktionen unterliegt quasitrockene DNA im festen Zustand, die zur bekannten Löslichkeitsverminderung führen ?
c) Ist der T_M-Wert abhängig vom Molekulargewicht und kann er als Kriterium für die vollständige Renaturierung (nach thermischer Vorbehandlung) gelten ?
d) Welche Meßmethoden eignen sich für Relaxationsuntersuchungen ?
e) Welche Methoden versprechen die genauesten Aussagen über die Struktur der DNA als isoliertes trockenes Präparat, in der Lösung und in Zellorganellen auch im Hinblick auf eine physikalische Sequenzanalyse ?

Die steigende Kompliziertheit der Untersuchungsobjekte von den Proteinen bis zu den Nucleinsäuren warf zahlreiche Definitionsfragen auf, z.B. bei den Begriffen : Heterogenität, natives Molekül, Denaturierung u.a. und es wäre zu begrüßen, wenn z.B. die LUMRYschen Anregungen (vgl. Generaldiskussion II) aufgegriffen würden.

Gefördert durch die außerordentliche Aktualität der Thematik und das faszinierende Wesen der Nucleinsäuren,unterstützt durch die Abgeschiedenheit des Goethe-Schlosses in Dornburg nahm das Symposium den gewünschten Verlauf. Die Vertreter der wichtigsten Disziplinen kamen ins Gespräch mit Blickrichtung auf ein gemeinsames Zentralproblem. Die lückenlose Organisation schloß ein Zusatzprogramm (Besichtigungen, Ausflüge, Konzerte und Empfangsabend auf dem Fuchsturm) ein und bot den Gästen vielfältige Anregungen und die Möglichkeit zu vertrauensvollem Gedankenaustausch. Die sowjetische Delegation, die einen Querschnitt durch ihre molekularbiologischen Ergebnisse brachte, gab ihrerseits ein Abschluß-Bankett im Hotel "International", wo zwischen Ost und West Einigkeit darin herrschte, bald wieder zusammenzukommen. Die nächste derartige Gelegenheit in Jena wird im Mai (26.-29.) 1965 sein mit dem Thema: "Elektrochemische Methoden und Prinzipien in der Molekularbiologie".

Hermann BERG

Teilnehmer

A U S T R A L I E N

JORDAN,Prof. D., Department of Physical and Inorganic Chemistry,Univ., Adelaide

C S S R

FRIČ,I.Dipl.Ing.,C.Sc., Inst. für Organische Chemie u. Biochemie d.Akademie, Prag
MUNK, Dipl.Ing.P.(m.Frau), Inst.f.Makromolekulare Chemie d.Akad., Prag
ŠPONAR,J.Dipl.Ing.,C.Sc., Inst.f.Organische Chemie u. Biochemie d.Akademie, Prag
HABERMANN,Dr.V., Institut für Medizinische Chemie, Plzen
ZAHRADNIK,Dr.R., Institut für Physikalische Chemie d.Akademie, Prag

D E U T S C H E   B U N D E S R E P U B L I K

BODMANN,Dr.O., Institut für Physikalische Chemie der Universität, Mainz
BURCHARD,Dr.W., Institut für Makromolekulare Chemie d. Universität, Freiburg i.Br.
DELLWEG,Dr.H., Institut für Therapeutische Biochemie d. Universität, Frankfurt/M.
DOSE,Dr.K., Max-Planck-Institut für Biophysik, Frankfurt/M.
MEYERHOFF,Prof.G., Institut für Physikalische Chemie d. Universität, Mainz
SAUERBIER,Dr.W. (m.Frau), Institut für Genetik der Universität, Köln
SCHINKEL,Ing.O., Phywe AG, Göttingen
SPATZ,Dipl.Phys.H.Ch., Max-Planck-Institut f. Physikalische Chemie, Göttingen

D E U T S C H E   D E M O K R A T I S C H E   R E P U B L I K

AROLD,Dr.H., Institut für Organische Chemie u. Biochemie d. Universität, Jena
BERG,Dr.H.(m.Frau), Institut f.Mikrobiologie u.exp. Therapie d.Akad., Jena
BÖHME,Dr.H., Institut für Kulturpflanzenforschung, Gatersleben
BÖTTGER,Dr.I., Institut für Biochemie d. Pflanzen d. Akademie, Halle/S.
BRUNS,Prof.G., Institut f. Mikrobiologie u. exp. Therapie d. Akad., Jena
DAMASCHUN,Dipl.Phys., Physikalisches Institut d. Universität, Jena
DAMERAU, Dipl.Chem.W., Institut f. Biophysik d. Akademie, Berlin-Buch
DAUTZENBERG, Institut f. Faserstoff-Forschung d. Akademie, Teltow-Seehof
ECKART,Dipl.Phys.W., Physikalisch-Technisches Institut d. Akademie, Berlin
EINS,Dipl.Phys.S., Institut für Mikrobiologie u. exp. Therapie d. Akademie, Jena
FRITSCH,Dr.S., Institut für Mikrobiologie u. exp. Therapie d. Akademie, Jena
FRITZSCHE,Dr.H., Institut für Mikrobiologie u. exp. Therapie d. Akademie, Jena
GELLER,Dipl.Chem.K., Institut für Mikrobiologie u. exp. Therapie d.Akademie, Jena
GRALHEER,Dr.H., Institut für Tierseuchenforschung der Akademie, Insel Riems
GRIMM,Dr.H., Institut für Mikrobiologie u. exp. Therapie d. Akademie, Jena
GÜNTHER,Dipl.Phys., Institut für Mikrobiologie u. exp. Therapie d.Akad., Jena
HABERDITZL,Dr.W., Physikalisch-chemisches Institut d.Humboldt-Univ., Berlin
HANTSCHEL,Dr.H., Institut für Tierseuchenforschung d. Akademie, Insel Riems
HARTMANN,Dipl.Chem.M., Institut für Mikrobiologie u. exp. Therapie d.Akademie, Jena
HERRMANN,Dr.H., Institut für Mikrobiologie u. exp. Therapie d. Akademie, Jena
KAYSER,W., Institut für Mikrobiologie u. exp. Therapie d. Akademie, Jena
KRAMARCZYK,Dipl.Chem.K., Institut f. Mikrobiologie u. exp. Therapie d. Akad., Jena
KNÖLL,Prof.H., Institut für Mikrobiologie u. exp. Therapie d. Akademie, Jena
KRUMBIEGEL,Dipl.Phys.J., Institut für Biophysik d. Akademie, Berlin-Buch
LANGHAMMER,Dr.G., Institut für Physik und Physikalische Chem.d.Hochpolymeren der Akademie, Leipzig

LÖBER,Dipl.Phys.G., Institut für Mikrobiologie u. exp. Therapie d. Akademie, Jena
LÜCKE,Dipl.Phys.H., jetzt Institut für Tierseuchenforschung d. Akademie, Insel Riems
MÜLLER,Dr.R., Institut für Mikrobiologie u. exp. Therapie d. Akademie, Jena
MÜHLIG,Dipl.Phys.P., Institut für Mikrobiologie u. exp. Therapie d. Akademie, Jena
MUND,Ch., Institut für Mikrobiologie u. exp. Therapie d. Akademie, Jena
MÜNCHMEYER,Dipl.Chem.R., Institut f. Tierernährungslehre d. Humboldt-Univ., Berlin
NOACK,Dr.D., Institut für Mikrobiologie u. exp. Therapie d. Akademie, Jena
NÜSKE,Dr.R., Institut für Mikrobiologie u. exp. Therapie d. Akademie, Jena
PALM,Dr.K., Medizinische Akademie Erfurt, Medizinische Klinik, Erfurt
PARTHIER,Dr.B., Botanisches Institut der Universität, Halle/S.
REICH,Dr.G., Deutsches Lederinstitut, Freiberg/Sa.
REINERT,Dr.K.E., Institut für Mikrobiologie u. exp. Therapie d. Akademie, Jena
REINERT, Dipl.Chem.H., Institut für Mikrobiologie u. exp. Therapie d.Akad., Jena
ROST,Dipl.Biol.K., Institut für Mikrobiologie u. exp. Therapie d. Akad., Jena
SARFERT,E., Institut für Mikrobiologie u. exp. Therapie der Akademie, Jena
SCHÜTTE,Dr., Institut für Biochemie der Pflanzen d. Akademie, Halle/S.
SCHÜTZ,Dipl.Phys.H., Institut für Mikrobiologie u. exp. Therapie d. Akad., Jena
SCHICHT, Dipl.Biol.G., Institut für Mikrobiologie u. exp. Therapie d. Akad., Jena
SCHRÖDER,S., Institut für Mikrobiologie u. exp. Therapie d. Akademie, Jena
STAHL,Dr., Institut für Experimentelle Krebsforschung d. Akad., Berlin-Buch
TRESSELT,Dipl.Phys.D., Institut für Mikrobiologie u. exp. Therapie d. Akad., Jena
TRIEBEL,Dipl.Phys.H., Institut für Mikrobiologie u. exp. Therapie d. Akademie, Jena
TKOCZ,Dipl.Chem.H., Institut für Mikrobiologie u. exp. Therapie d. Akademie, Jena
UNANGST, Dr.D., Physikalisches Institut der Universität, Jena
VENNER,Dr.H., Institut für Mikrobiologie u. exp. Therapie der Akademie, Jena
WÄHNERT,Dipl.Chem.U., Institut für Mikrobiologie u. exp. Therapie der Akademie, Jena
WEBER,Dipl.Biol.H., Institut für Mikrobiologie u. exp. Therapie d. Akademie, Jena
WEILAND,Dipl.Chem.R., VEB Serumwerk Bernburg, Bernburg/S.
WEISS,DR.R., Institut für Mikrobiologie u. exp. Therapie d. Akademie, Jena
WOLLGIEHN,Dr.R., Institut f. Biochemie der Pflanzen d. Akademie, Halle/S.
WULF,Dr.K., Institut f. Faserstoff-Forschung d. Akademie, Teltow-Seehof
ZIMMER,Dr.Ch., Institut für Mikrobiologie u. exp. Therapie d. Akademie, Jena
ZIMMERMANN,Dr.G. (m.Frau), Forschungsinstitut f. Impfstoffe d.Akademie, Dessau

J U G O S L A W I E N

LAPANJE,Dr.S., Physikalisch-chemisches Laboratorium d. Universität, Ljubljana

N I E D E R L A N D E

ATEN,Dr.J., Medisch Biologisch Laboratorium RVO-TNO, Rijswijk

O E S T E R R E I C H

ALTMANN,Dr.J. (m.Frau), Inst.f.Biologie u. Landwirtschaft, Reaktorzentrum Seibersdorf, Wien

S C H W E D E N

CLAESSON,Prof.St., Physik.-chem. Institut der Universität, Uppsala
GRANATH,Dr.K., Pharmacia Ltd., Uppsala
LÖFROTH,Dr., Institut für Radiobiol., Royal Universität, Stockholm
NORBERG,P., Physikalisch-chemisches Institut der Universität, Uppsala
ÖHMAN,J., Physikalisch-chemisches Institut der Universität, Uppsala
STRÖMBERG,R., Physikalisch-chemisches Institut der Universität, Uppsala

S C H W E I Z

RICHTER,J., Physikalisch-chemisches Institut der Universität, Basel

U d S S R

BUTJAGIN,P., Institut für Chemische Physik der Akademie, Moskau
BAJEW,Dr.A., Inst.f.Strahlenbiol. u.Physik.-Chem.Biol. d. Akad., Moskau
BLUMENFELD,Prof.L., Inst.f.Chemische Physik d. Akademie, Moskau
BUDOWSKY,E., Institut für Naturstoffchemie, Moskau
GOTTICH,B., Inst.f.Strahlenbiol. u.Physik.-Chem.Biol. d. Akad., Moskau
GRISCHKEWITSCH,W., Inst.f.Strahlenbiol. u.Physik.-Chem.Biol. d. Akad., Moskau
LASCHENKO,Ju., Institut zur Erforschung neuer Antibiotika, Moskau
LASURKIN,Ju., Kurcatov's Institut für Atomenergie, Moskau
NICOLAJEWA, Institut für Chemische Physik, Moskau
SAMSONOW,Prof.G., Institut f. Hochmolekulare Verbindungen d. Akad., Leningrad
STRUČKOV,W., Institut für Biophysik der Akademie, Moskau
VARSHAVSKY,Prof.J., Inst.f.Strahlenbiol. u.Physik.-Chem.Biol. d. Akad., Moskau
ZWETKOW,W., Inst.f. Hochmolekulare Verbindungen d. Akad., Leningrad

U S A

CROTHERS,Dr.D.M. (m.Frau), z.Z.Max-Planck-Inst. f. Physikalische Chemie, Göttingen
HELMKAMP,Prof.G. (m.Frau), Department of Chemistry, Univ.of California, Riverside
LUMRY,Prof.R., School of Chem., University of Minnesota, Minneapolis
ROWND,Dr.R., Department of Chemistry, Harvard University, Cambridge, Mass.

www.ingramcontent.com/pod-product-compliance
Lightning Source LLC
LaVergne TN
LVHW081346011026
849161LV00013BB/3012